D0147986

SCHAUM'S OUTLINE OF

THEORY AND PROBLEMS

OF

STATE SPACE and LINEAR SYSTEMS

•

BY

DONALD M. WIBERG, Ph.D.

Associate Professor of Engineering
University of California, Los Angeles

SCHAUM'S OUTLINE SERIES

McGRAW-HILL BOOK COMPANY

New York, St. Louis, San Francisco, Düsseldorf, Johannesburg, Kuala Lumpur, London, Mexico, Montreal, New Delhi, Panama, Rio de Janeiro, Singapore, Sydney, and Toronto

QA
402
1971
012
gift
mab
11-11-21

Copyright © 1971 by McGraw-Hill, Inc. All Rights Reserved. Printed in the United States of America. No part of this publication may be reproduced, stored in a retrieval system, or transmitted, in any form or by any means, electronic, mechanical, photocopying, recording, or otherwise, without the prior written permission of the publisher.

07-070096-6

1 2 3 4 5 6 7 8 9 0 SHSH 7 5 4 3 2 1

Preface

The importance of state space analysis is recognized in fields where the time behavior of any physical process is of interest. The concept of state is comparatively recent, but the methods used have been known to mathematicians for many years. As engineering, physics, medicine, economics, and business become more cognizant of the insight that the state space approach offers, its popularity increases.

This book was written not only for upper division and graduate students, but for practicing professionals as well. It is an attempt to bridge the gap between theory and practical use of the state space approach to the analysis and design of dynamical systems. The book is meant to encourage the use of state space as a tool for analysis and design, in proper relation with other such tools. The state space approach is more general than the "classical" Laplace and Fourier transform theory. Consequently, state space theory is applicable to all systems that can be analyzed by integral transforms in time, and is applicable to many systems for which transform theory breaks down. Furthermore, state space theory gives a somewhat different insight into the time behavior of linear systems, and is worth studying for this aspect alone.

In particular, the state space approach is useful because: (1) linear systems with time-varying parameters can be analyzed in essentially the same manner as time-invariant linear systems, (2) problems formulated by state space methods can easily be programmed on a computer, (3) high-order linear systems can be analyzed, (4) multiple input–multiple output systems can be treated almost as easily as single input–single output linear systems, and (5) state space theory is the foundation for further studies in such areas as nonlinear systems, stochastic systems, and optimal control. These are five of the most important advantages obtained from the generalization and rigorousness that state space brings to the classical transform theory.

Because state space theory describes the time behavior of physical systems in a mathematical manner, the reader is assumed to have some knowledge of differential equations and of Laplace transform theory. Some classical control theory is needed for Chapter 8 only. No knowledge of matrices or complex variables is prerequisite.

The book may appear to contain too many theorems to be comprehensible and/or useful to the nonmathematician. But the theorems have been stated and proven in a manner suited to show the range of application of the ideas and their logical interdependence. Space that might otherwise have been devoted to solved problems has been used instead to present the physical motivation of the proofs. Consequently I give my strongest recommendation that the reader seek to understand the physical ideas underlying the proofs rather than to merely memorize the theorems. Since the emphasis is on applications, the book might not be rigorous enough for the pure mathematician, but I feel that enough information has been provided so that he can tidy up the statements and proofs himself.

The book has a number of novel features. Chapter 1 gives the fundamental ideas of state from an informal, physical viewpoint, and also gives a correct statement of linearity. Chapter 2 shows how to write transfer functions and ordinary differential equations in matrix notation, thus motivating the material on matrices to follow. Chapter 3 develops the important concepts of range space and null space in detail, for later application. Also exterior products (Grassmann algebra) are developed, which give insight into determinants, and which considerably shorten a number of later proofs. Chapter 4 shows how to actually solve for the Jordan form, rather than just proving its existence. Also a detailed treatment of pseudoinverses is given. Chapter 5 gives techniques for computation of transition matrices for high-order time-invariant systems, and contrasts this with a detailed development of transition matrices for time-varying systems. Chapter 6 starts with giving physical insight into controllability and observability of simple systems, and progresses to the point of giving algebraic criteria for time-varying systems. Chapter 7 shows how to reduce a system to its essential parameters. Chapter 8 is perhaps the most novel. Techniques from classical control theory are extended to time-varying, multiple input–multiple output linear systems using state space formulation. This gives practical methods for control system design, as well as analysis. Furthermore, the pole placement and observer theory developed can serve as an introduction to linear optimal control and to Kalman filtering. Chapter 9 considers asymptotic stability of linear systems, and the usual restriction of uniformity is dispensed with. Chapter 10 gives motivation for the quadratic optimal control problem, with special emphasis on the practical time-invariant problem and its associated computational techniques. Since Chapters 6, 8, and 9 precede, relations with controllability, pole placement, and stability properties can be explored.

The book has come from a set of notes developed for engineering course 122B at UCLA, originally dating from 1966. It was given to the publisher in June 1969. Unfortunately, the publication delay has dated some of the material. Fortunately, it also enabled a number of errors to be weeded out.

Now I would like to apologize because I have not included references, historical development, and credit to the originators of each idea. This was simply impossible to do because of the outline nature of the book.

I would like to express my appreciation to those who helped me write this book. Chapter 1 was written with a great deal of help from A. V. Balakrishnan. L. M. Silverman helped with Chapter 7 and P.K.C. Wang with Chapter 8. Interspersed throughout the book is material from a course given by R. E. Kalman during the spring of 1962 at Caltech. J. J. DiStefano, R. C. Erdmann, N. Levan, and K. Yao have used the notes as a text in UCLA course 122B and have given me suggestions. I have had discussions with R. E. Mortensen, M. M. Sholar, A. R. Stubberud, D. R. Vaughan, and many other colleagues. Improvements in the final draft were made through the help of the control group under the direction of J. Ackermann at the DFVLR in Oberpfaffenhofen, West Germany, especially by G. Grübel and R. Sharma. Also, I want to thank those UCLA students, too numerous to mention, that have served as guinea pigs and have caught many errors of mine. Ruthie Alperin was very efficient as usual while typing the text. David Beckwith, Henry Hayden, and Daniel Schaum helped publish the book in its present form. Finally, I want to express my appreciation of my wife Merideth and my children Erik and Kristin for their understanding during the long hours of involvement with the book.

DONALD M. WIBERG

University of California, Los Angeles
June 1971

CONTENTS

CONTENTS

Chapter 1

Meaning of State

1.1 INTRODUCTION TO STATE

To introduce the subject, let's take an informal, physical approach to the idea of state. (An exact mathematical approach is taken in more advanced texts.) First, we make a distinction between physical and abstract objects. A physical object is an object perceived by our senses whose time behavior we wish to describe, and its abstraction is the mathematical relationships that give some expression for its behavior. This distinction is made because, in making an abstraction, it is possible to lose some of the relationships that make the abstraction behave similar to the physical object. Also, not all mathematical relationships can be realized by a physical object.

The concept of state relates to those physical objects whose behavior can change with time, and to which a stimulus can be applied and the response observed. To predict the future behavior of the physical object under any input, a series of experiments could be performed by applying stimuli, or inputs, and observing the responses, or outputs. From these experiments we could obtain a listing of these inputs and their corresponding observed outputs, i.e. a list of input-output pairs. An input-output pair is an ordered pair of real time functions defined for all $t \geqq t_0$, where t_0 is the time the input is first applied. Of course segments of these input time functions must be consistent and we must agree upon what kind of functions to consider, but in this introduction we shall not go into these mathematical details.

Definition 1.1: The *state of a physical object* is any property of the object which relates input to output such that knowledge of the input time function for $t \geqq t_0$ and state at time $t = t_0$ completely determines a unique output for $t \geqq t_0$.

Example 1.1.

Consider a black box, Fig. 1-1, containing a switch to one of two voltage dividers. Intuitively, the state of the box is the position of the switch, which agrees with Definition 1.1. This can be ascertained by the experiment of applying a voltage V to the input terminal. Natural laws (Ohm's law) dictate that if the switch is in the lower position A, the output voltage is $V/2$, and if the switch is in the upper position B, the output voltage is $V/4$. Then the state A determines the input-output pair to be $(V, V/2)$, and the state B corresponds to $(V, V/4)$.

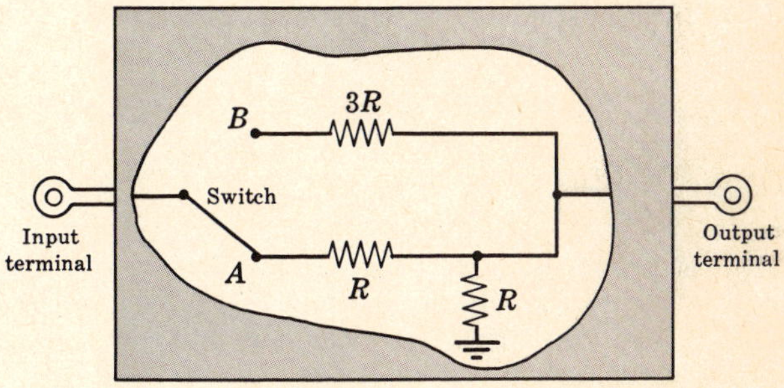

Fig. 1-1

1.2 STATE OF AN ABSTRACT OBJECT

The basic ideas contained in the above example can be extended to many physical objects and to the abstract relationships describing their time behavior. This will be done after abstracting the properties of physical objects such as the black box. For example, the color of the box has no effect on the experiment of applying a voltage. More subtly, the value of resistance R is immaterial if it is greater than zero. All that is needed is a listing of every input-output pair over all segments of time $t \geqq t_0$, and the corresponding states at time t_0.

Definition 1.2: An *abstract object* is the totality of input-output pairs that describe the behavior of a physical object.

Instead of a specific list of input time functions and their corresponding output time functions, the abstract object is usually characterized as a class of all time functions that obey a set of mathematical equations. This is in accord with the scientific method of hypothesizing an equation and then checking to see that the physical object behaves in a manner similar to that predicted by the equation. Hence we can often summarize the abstract object by using the mathematical equations representing physical laws.

The mathematical relations which summarize an abstract object must be *oriented*, in that m of the time functions that obey the relations must be designated inputs (denoted by the vector $\mathbf{u}$, having m elements u_i) and k of the time functions must be designated outputs (denoted by the vector $\mathbf{y}$, having k elements y_i). This need has nothing to do with causality, in that the outputs are not "caused" by the inputs.

Definition 1.3: The *state of an abstract object* is a collection of numbers which together with the input $\mathbf{u}(t)$ for all $t \geqq t_0$ uniquely determines the output $\mathbf{y}(t)$ for all $t \geqq t_0$.

In essence the state parametrizes the listing of input-output pairs. The state is the answer to the question "Given $\mathbf{u}(t)$ for $t \geqq t_0$ and the mathematical relationships of the abstract object, what additional information is needed to completely specify $\mathbf{y}(t)$ for $t \geqq t_0$?"

Example 1.2.

A physical object is the resistor-capacitor network shown in Fig. 1-2. An experiment is performed by applying a voltage $u(t)$, the input, and measuring a voltage $y(t)$, the output. Note that another experiment could be to apply $y(t)$ and measure $u(t)$, so that these choices are determined by the experiment.

The list of all input-output pairs for this example is the class of all functions $u(t), y(t)$ which satisfy the mathematical relationship

$$RC\,dy/dt + y = u \qquad (1.1)$$

This summarizes the abstract object. The solution of (1.1) is

$$y(t) = y(t_0)\,e^{(t_0-t)/RC} + \frac{1}{RC}\int_{t_0}^{t} e^{(\tau-t)/RC}u(\tau)\,d\tau \qquad (1.2)$$

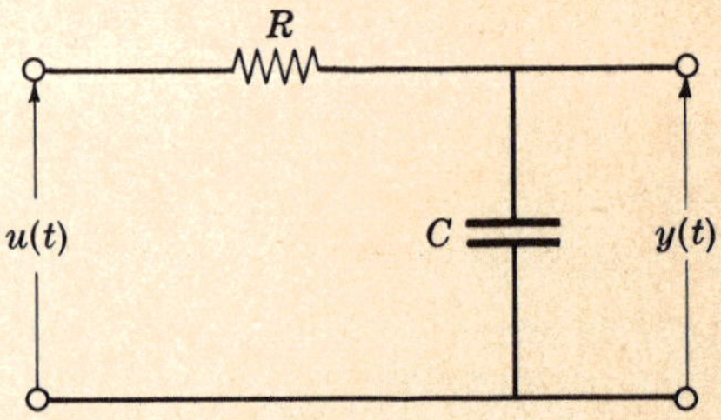

Fig. 1-2

This relationship explicitly exhibits the list of input-output pairs. For any input time function $u(\tau)$ for $\tau \geqq t_0$, the output time function $y(t)$ is uniquely determined by $y(t_0)$, a number at time t_0. Note the distinction between time functions and numbers. Thus the set of numbers $y(t_0)$ parametrizes all input-output pairs, and therefore is the state of the abstract object described by (1.1). Correspondingly, a choice of state of the RC network is the output voltage at time t_0.

Example 1.3.

The physical object shown in Fig. 1-3 is two RC networks in series. The pertinent equation is

$$R^2C^2\,d^2y/dt^2 + 2.5RC\,dy/dt + y = u \qquad (1.3)$$

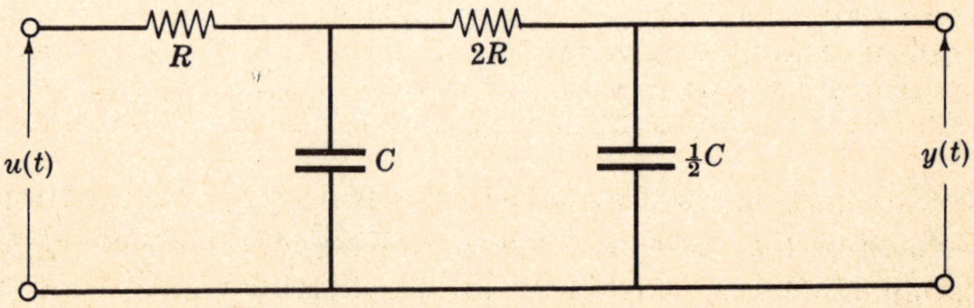

Fig. 1-3

with a solution

$$y(t) = \frac{y(t_0)}{3}[4e^{(t_0-t)/2RC} - e^{(t_0-t)2/RC}]$$
$$+ \frac{2}{3RC}\frac{dy}{dt}(t_0)[e^{(t_0-t)/2RC} - e^{(t_0-t)2/RC}]$$
$$+ \frac{2}{3RC}\int_{t_0}^{t}[e^{(\tau-t)/2RC} - e^{(\tau-t)2/RC}]u(\tau)\,d\tau \qquad (1.4)$$

Here the set of numbers $y(t_0)$ and $\frac{dy}{dt}(t_0)$ parametrizes the input-output pairs, and may be chosen as state. Physically, the voltage and its derivative across the smaller capacitor at time t_0 correspond to the state.

Definition 1.4: A *state variable*, denoted by the vector $\mathbf{x}(t)$, is the time function whose value at any specified time is the state of the abstract object at that time.

Note this difference in going from a set of numbers to a time function. The state can be a set consisting of an infinity of numbers (e.g. Problems 1.1 and 1.2), in which case the state variable is an infinite collection of time functions. However, in most cases considered in this book, the state is a set of n numbers and correspondingly $\mathbf{x}(t)$ is an n-vector function of time.

Definition 1.5: The *state space*, denoted by Σ, is the set of all $\mathbf{x}(t)$.

Example 1.4.

The state variable in Example 1.2 is $x(t) = y(t)$, whereas in Example 1.1 the state variable remains either A or B for all time.

Example 1.5.

The state variable in Example 1.3 is $\mathbf{x}(t) = \begin{pmatrix} y(t) \\ \frac{dy}{dt}(t) \end{pmatrix}$.

The state representation is not unique. There can be many different ways of expressing the relationship of input to output.

Example 1.6.

In Example 1.3, instead of the voltage and its derivative across the smaller capacitor, the state could be the voltage and its derivative across the larger capacitor, or the state could be the voltages across both capacitors.

There can exist inputs that do not influence the state, and, conversely, there can exist outputs that are not influenced by the state. These cases are called uncontrollable and unobservable, respectively, about which much more will be said in Chapter 6.

Example 1.7.

In Example 1.1, the physical object is state uncontrollable. No input can make the switch change positions. However, the switch position is observable. If the wire to the output were broken, it would be unobservable. A state that is both unobservable and uncontrollable makes no physical sense, since it cannot be detected by experiment. Examples 1.2 and 1.3 are both controllable and observable.

One more point to note is that we consider here only deterministic abstract objects. The problem of obtaining the state of an abstract object in which random processes are inputs, etc., is beyond the scope of this book. Consequently, all statements in the whole book are intended only for deterministic processes.

1.3 TRAJECTORIES IN STATE SPACE

The state variable $\mathbf{x}(t)$ is an explicit function of time, but also depends implicitly on the starting time t_0, the initial state $\mathbf{x}(t_0) = \mathbf{x}_0$, and the input $\mathbf{u}(\tau)$. This functional dependency can be written as $\mathbf{x}(t) = \boldsymbol{\phi}(t; t_0, \mathbf{x}_0, \mathbf{u}(\tau))$, called a trajectory. The trajectory can be plotted in n-dimensional state space as t increases from t_0, with t an implicit parameter. Often this plot can be made by eliminating t from the solutions to the state equation.

Example 1.8.

Given $x_1(t) = \sin t$ and $x_2(t) = \cos t$, squaring each equation and adding gives $x_1^2 + x_2^2 = 1$. This is a circle in the x_1x_2 plane with t an implicit parameter.

Example 1.9.

In Example 1.3, note that equation (*1.4*) depends on t, $u(\tau)$, $\mathbf{x}(t_0)$ and t_0, where $\mathbf{x}(t_0)$ is the vector with components $y(t_0)$ and $dy/dt(t_0)$. Therefore the trajectories $\boldsymbol{\phi}$ depend on these quantities.

Suppose now $u(t) = 0$ and $RC = 1$. Let $x_1 = y(t)$ and $x_2 = dy/dt$. Then $dx_1/dt = x_2$ and $d^2y/dt^2 = dx_2/dt$. Therefore $dt = dx_1/x_2$ and so $d^2y/dt^2 = x_2\,dx_2/dx_1$. Substituting these relationships into (*1.3*) gives

$$x_2\,dx_2/dx_1 + 2.5x_2 + x_1 = 0$$

which is independent of t. This has a solution

$$x_1 + 2x_2 = C(2x_1 + x_2)^4$$

where the constant $C = [x_1(t_0) + 2x_2(t_0)]/[2x_1(t_0) + x_2(t_0)]^4$. Typical trajectories in state space are shown in Fig. 1-4. The one passing through points $x_1(t_0) = 0$ and $x_2(t_0) = 1$ is drawn in bold. The arrows point in the direction of increasing time, and all trajectories eventually reach the origin for this particular stable system.

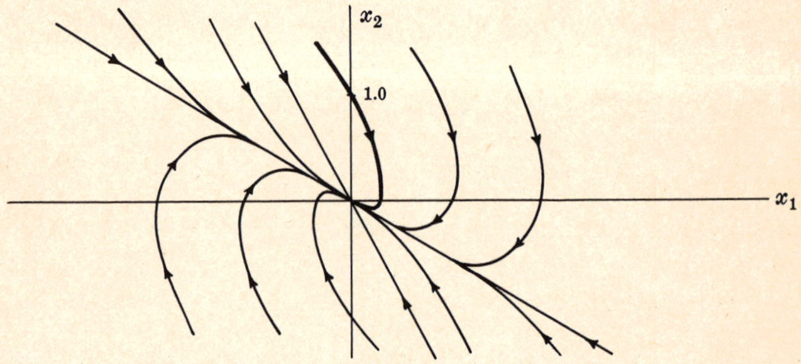

Fig. 1-4

1.4 DYNAMICAL SYSTEMS

In the foregoing we have assumed that an abstract object exists, and that sometimes we can find a set of oriented mathematical relationships that summarizes this listing of input and output pairs. Now suppose we are given a set of oriented mathematical relationships, do we have an abstract object? The answer to this question is not always affirmative, because there exist mathematical equations whose solutions do not result in abstract objects.

Example 1.10.

The oriented mathematical equation $y(t) = ju(t)$ cannot give an abstract object, because either the input or the output must be imaginary.

If a mathematical relationship always determines a real output $\mathbf{y}(t)$ existing for all $t \geqq t_0$ given any real input $\mathbf{u}(t)$ for all time t, then we can form an abstract object. Note that by supposing an input $u(t)$ for all past times as well as future times, we can form an abstract object from the equation for a delayor $y(t) = u(t - T)$. [See Problem 1.1.]

However, we can also form an abstract object from the equation for a predictor $y(t) = u(t+T)$. If we are to restrict ourselves to mathematical relations that can be mechanized, we must specifically rule out such relations whose present outputs depend on future values of the input.

Definition 1.6: A *dynamical system* is an oriented mathematical relationship in which:

(1) A real output $\mathbf{y}(t)$ exists for all $t \geqq t_0$ given a real input $\mathbf{u}(t)$ for all t.

(2) Outputs $\mathbf{y}(t)$ do not depend on inputs $\mathbf{u}(\tau)$ for $\tau > t$.

Given that we have a dynamical system relating $\mathbf{y}(t)$ to $\mathbf{u}(t)$, we would like to construct a set of mathematical relations defining a state $\mathbf{x}(t)$. We shall assume that a state space description can be found for the dynamical system of interest satisfying the following conditions (although such a construction may take considerable thought):

Condition 1: A real, unique output $\mathbf{y}(t) = \boldsymbol{\eta}(t, \boldsymbol{\phi}(t; t_0, \mathbf{x}_0, \mathbf{u}(\tau)), \mathbf{u}(t))$ exists for all $t > t_0$ given the state $\mathbf{x}_0$ at time t_0 and a real input $\mathbf{u}(\tau)$ for $\tau \geqq t_0$.

Condition 2: A unique trajectory $\boldsymbol{\phi}(t; t_0, \mathbf{x}_0, \mathbf{u}(\tau))$ exists for all $t > t_0$ given the state at time t_0 and a real input for all $t \geqq t_0$.

Condition 3: A unique trajectory starts from each state, i.e.

$$\lim_{t \to t_1} \boldsymbol{\phi}(t; t_1, \mathbf{x}(t_1), \mathbf{u}(\tau)) = \mathbf{x}(t_1) \quad \text{for all } t_1 \geqq t_0 \tag{1.5}$$

Condition 4: Trajectories satisfy the transition property

$$\boldsymbol{\phi}(t; t_0, \mathbf{x}(t_0), \mathbf{u}(\tau)) = \boldsymbol{\phi}(t; t_1, \mathbf{x}(t_1), \mathbf{u}(\tau)) \quad \text{for } t_0 < t_1 < t \tag{1.6}$$

where

$$\mathbf{x}(t_1) = \boldsymbol{\phi}(t_1; t_0, \mathbf{x}(t_0), \mathbf{u}(\tau)) \tag{1.7}$$

Condition 5: Trajectories $\boldsymbol{\phi}(t; t_0, \mathbf{x}_0, \mathbf{u}(\tau))$ do not depend on inputs $\mathbf{u}(\tau)$ for $\tau > t$.

Condition 1 gives the functional relationship $\mathbf{y}(t) = \boldsymbol{\eta}(t, \mathbf{x}(t), \mathbf{u}(t))$ between initial state and future input such that a unique output is determined. Therefore, with a proper state space description, it is not necessary to know inputs prior to t_0, but only the state at time t_0. The state at the initial time completely summarizes all the past history of the input.

Example 1.11.

In Example 1.2., it does not matter how the voltage across the capacitor was obtained in the past. All that is needed to determine the unique future output is the state and the future input.

Condition 2 insures that the state at a future time is uniquely determined. Therefore knowledge of the state at any time, not necessarily t_0, uniquely determines the output. For a given $\mathbf{u}(t)$, one and only one trajectory passes through each point in state space and exists for all finite $t \geqq t_0$. As can be verified in Fig. 1-4, one consequence of this is that the state trajectories do not cross one another. Also, notice that condition 2 does not require the state to be real, even though the input and output must be real.

Example 1.12.

The relation $dy/dt = u(t)$ is obviously a dynamical system. A state space description $dx/dt = ju(t)$ with output $y(t) = -jx(t)$ can be constructed satisfying conditions 1-5, yet the state is imaginary.

Condition 3 merely requires the state space description to be consistent, in that the starting point of the trajectory should correspond to the initial state. Condition 4 says that the input $\mathbf{u}(\tau)$ takes the system from a state $\mathbf{x}(t_0)$ to a state $\mathbf{x}(t)$, and if $\mathbf{x}(t_1)$ is on that trajectory, then the corresponding segment of the input will take the system from $\mathbf{x}(t_1)$ to $\mathbf{x}(t)$. Finally, condition 5 has been added to assure causality of the input-output relationship resulting from the state space description to correspond with the causality of the original dynamical system.

Example 1.13.

We can construct a state space description of equation (*1.1*) of Example 1.2 by defining a state $x(t) = y(t)$. Then condition 1 is satisfied as seen by examination of the solution, equation (*1.2*). Clearly the trajectory $\phi(t;\, t_0, x_0, u(\tau))$ exists and is unique given a specified t_0, x_0 and $u(\tau)$, so condition 2 is satisfied. Also, conditions 3 and 5 are satisfied. To check condition 4, given $x(t_0) = y(t_0)$ and $u(\tau)$ over $t_0 \leqq \tau \leqq t$, then

$$x(t) \;=\; x(t_1)e^{(t_1-t)/RC} + \frac{1}{RC}\int_{t_1}^{t} e^{(\tau_1-t)/RC}\, u(\tau_1)\, d\tau_1 \qquad (1.8)$$

where

$$x(t_1) \;=\; x(t_0)e^{(t_0-t_1)/RC} + \frac{1}{RC}\int_{t_0}^{t_1} e^{(\tau_0-t_1)/RC} u(\tau_0)\, d\tau_0 \qquad (1.9)$$

Substitution of (*1.9*) into (*1.8*) gives the previously obtained (*1.2*). Therefore the dynamical system (*1.1*) has a state space description satisfying conditions 1-5.

Henceforth, instead of "dynamical system with a state space description" we will simply say "system" and the rest will be understood.

1.5 LINEARITY AND TIME INVARIANCE

Definition 1.7: Given any two numbers α, β; two states $\mathbf{x}_1(t_0)$, $\mathbf{x}_2(t_0)$; two inputs $\mathbf{u}_1(\tau)$, $\mathbf{u}_2(\tau)$; and two corresponding outputs $\mathbf{y}_1(\tau)$, $\mathbf{y}_2(\tau)$ for $\tau \geqq t_0$. Then a system is *linear* if (1) the state $\mathbf{x}_3(t_0) = \alpha\mathbf{x}_1(t_0) + \beta\mathbf{x}_2(t_0)$, the output $\mathbf{y}_3(\tau) = \alpha\mathbf{y}_1(\tau) + \beta\mathbf{y}_2(\tau)$, and the input $\mathbf{u}_3(\tau) = \alpha\mathbf{u}_1(\tau) + \beta\mathbf{u}_2(\tau)$ can appear in the oriented abstract object and (2) both $\mathbf{y}_3(\tau)$ and $\mathbf{x}_3(\tau)$ correspond to the state $\mathbf{x}_3(t_0)$ and input $\mathbf{u}_3(\tau)$.

The operators $\boldsymbol{\phi}(t;\, t_0, \mathbf{x}_0, \mathbf{u}(\tau)) = \mathbf{x}(t)$ and $\eta(t; \boldsymbol{\phi}(t; t_0, \mathbf{x}_0, \mathbf{u}(\tau))) = \mathbf{y}(t)$ are linear on $\{\mathbf{u}(\tau)\} \oplus \{\mathbf{x}(t_0)\}$ is an equivalent statement.

Example 1.14.

In Example 1.2,

$$y_1(t) \;=\; x_1(t) \;=\; x_1(t_0)e^{(t_0-t)/RC} + \frac{1}{RC}\int_{t_0}^{t} e^{(\tau-t)/RC} u_1(\tau)\, d\tau$$

$$y_2(t) \;=\; x_2(t) \;=\; x_2(t_0)e^{(t_0-t)/RC} + \frac{1}{RC}\int_{t_0}^{t} e^{(\tau-t)/RC} u_2(\tau)\, d\tau$$

are the corresponding outputs $y_1(t)$ and $y_2(t)$ to the states $x_1(t)$ and $x_2(t)$ with inputs $u_1(\tau)$ and $u_2(\tau)$. Since any magnitude of voltage is permitted in this idealized system, any state $x_3(t) = \alpha x_1(t) + \beta x_2(t)$, any input $u_3(t) = \alpha u_1(t) + \beta u_2(t)$, and any output $y_3(t) = \alpha y_1(t) + \beta y_2(t)$ will appear in the list of input-output pairs that form the abstract object. Therefore part (1) of Definition 1.7 is satisfied. Furthermore, let's look at the response generated by $x_3(t_0)$ and $u_3(\tau)$.

$$\begin{aligned} y(t) &= x_3(t_0)e^{(t_0-t)/RC} + \frac{1}{RC}\int_{t_0}^{t} e^{(\tau-t)/RC} u_3(\tau)\, d\tau \\ &= [\alpha x_1(t_0) + \beta x_2(t_0)]e^{(t_0-t)/RC} + \frac{1}{RC}\int_{t_0}^{t} e^{(\tau-t)/RC}\,[\alpha u_1(\tau) + \beta u_2(\tau)]\, d\tau \\ &= \alpha y_1(t) + \beta y_2(t) \;=\; y_3(t) \end{aligned}$$

Since $y_3(t) = x_3(t)$, both the future output and state correspond to $x_3(t_0)$ and $u_3(t)$ and the system is linear.

Example 1.15.

Consider the system of Example 1.1. For some α and β there is no state equal to $\alpha A + \beta B$, where A and B are the switch positions. Consequently the system violates condition (1) of Definition 1.7 and is not linear.

Example 1.16.

Given the system $dx/dt = 0$, $y = u \cos x$. Then $y_1(t) = u_1(t) \cos x_1(t_0)$ and $y_2(t) = u_2(t) \cos x_2(t_0)$. The state $x_3(t) = x_3(t_0) = \alpha x_1(t_0) + \beta x_2(t_0)$ and is linear, but the output

$$y(t) = [\alpha u_1(t) + \beta u_2(t)] \cos [\alpha x_1(t_0) + \beta x_2(t_0)] \neq \alpha y_1(t) + \beta y_2(t)$$

except in special cases like $x_1(t_0) = x_2(t_0) = 0$, so the system is not linear.

If a system is linear, then superposition holds for nonzero $\mathbf{u}(t)$ with $\mathbf{x}(t_0) = \mathbf{0}$ and also for nonzero $\mathbf{x}(t_0)$ with $\mathbf{u}(t) = \mathbf{0}$ but not both together. In Example 1.14, with zero initial voltage on the capacitor, the response to a biased a-c voltage input (constant + sin ωt) could be calculated as the response to a constant voltage input plus the response to an unbiased a-c voltage input. Also, note from Example 1.16 that even if superposition does hold for nonzero $\mathbf{u}(t)$ with $\mathbf{x}(t_0) = \mathbf{0}$ and for nonzero $\mathbf{x}(t_0)$ with $\mathbf{u}(t) = \mathbf{0}$, the system may still not be linear.

Definition 1.8: A system is *time-invariant* if the time axis can be translated and an equivalent system results.

One test for time-invariance is to compare the original output with the shifted output. First, shift the input time function by T seconds. Starting from the same initial state $\mathbf{x}_0$ at time $t_0 + T$, does $\mathbf{y}(t+T)$ of the shifted system equal $\mathbf{y}(t)$ of the original system?

Example 1.17.

Given the nonlinear differential equation

$$\frac{dx}{dt} = x^2 + u^2$$

with $x(6) = a$. Let $\tau = t - 6$ so that $d\tau = dt$ and

$$\frac{dx}{d\tau} = x^2 + u^2$$

where $x(\tau = 0) = a$, resulting in the same system.

If the nonlinear equation for the state x were changed to

$$\frac{dx}{dt} = tx^2 + u^2$$

with the substitution $\tau = t - 6$, then

$$\frac{dx}{d\tau} = \tau x^2 + u^2 + 6x^2$$

and the appearance of the last term on the right gives a different system. Therefore this is a time-varying nonlinear system. Equations with explicit functions of t as coefficients multiplying the state will usually be time-varying.

1.6 SYSTEMS CONSIDERED

This book will consider only time-invariant and time-varying linear dynamical systems described by sets of differential or difference equations of finite order. We shall see in the next chapter that in this case the state variable $\mathbf{x}(t)$ is an n-vector and the system is linear.

Example 1.18.

A time-varying linear differential system of order n with one input and one output is described by the equation

$$\frac{d^n y}{dt^n} + \alpha_1(t)\frac{d^{n-1}y}{dt^{n-1}} + \cdots + \alpha_n(t)y = \beta_0(t)\frac{d^n u}{dt^n} + \cdots + \beta_n(t)u \tag{1.10}$$

Example 1.19.

A time-varying linear difference system of order n with one input and one output is described by the equation

$$y(k+n) + \alpha_1(k)\,y(k+n-1) + \cdots + \alpha_n(k)\,y(k) = \beta_0(k)\,u(k+n) + \cdots + \beta_n(k)\,u(k) \qquad (1.11)$$

The values of $\alpha(k)$ depend on the step (k) of the process, in a way analogous to which the $\alpha(t)$ depend on t in the previous example.

1.7 LINEARIZATION OF NONLINEAR SYSTEMS

State space techniques are especially applicable to time-varying linear systems. In this section we shall find out why time-varying linear systems are of such practical importance. Comparatively little design of systems is performed from the time-varying point of view at present, but state space methods offer great promise for the future.

Consider a set of n nonlinear differential equations of first order:

$$\begin{aligned} dy_1/dt &= f_1(y_1, y_2, \ldots, y_n, u, t) \\ dy_2/dt &= f_2(y_1, y_2, \ldots, y_n, u, t) \\ &\cdots\cdots\cdots\cdots \\ dy_n/dt &= f_n(y_1, y_2, \ldots, y_n, u, t) \end{aligned}$$

A nonlinear equation of nth order $d^n y/dt^n = g(y, dy/dt, \ldots, d^{n-1}y/dt^{n-1}, u, t)$ can always be written in this form by defining $y_1 = y$, $dy/dt = y_2, \ldots, d^{n-1}y/dt^{n-1} = y_n$. Then a set of n first order nonlinear differential equations can be obtained as

$$\begin{aligned} dy_1/dt &= y_2 \\ dy_2/dt &= y_3 \\ &\cdots\cdots\cdots\cdots \qquad (1.12) \\ dy_{n-1}/dt &= y_n \\ dy_n/dt &= g(y_1, y_2, \ldots, y_n, u, t) \end{aligned}$$

Example 1.20.

To reduce the second order nonlinear differential equation $d^2y/dt^2 - 2y^3 + u\,dy/dt = 0$ to two first order nonlinear differential equations, define $y = y_1$ and $dy/dt = y_2$. Then

$$\begin{aligned} dy_1/dt &= y_2 \\ dy_2/dt &= 2y_1^3 - uy_2 \end{aligned}$$

Suppose a solution can be found (perhaps by computer) to equations (1.12) for some initial conditions $y_1(t_0), y_2(t_0), \ldots, y_n(t_0)$ and some input $w(t)$. Denote this solution as the trajectory $\boldsymbol{\phi}(t; w(t), y_1(t_0), \ldots, y_n(t_0), t_0)$. Suppose now that the initial conditions are changed:

$$y(t_0) = y_1(t_0) + x_1(t_0), \quad \frac{dy}{dt}(t_0) = y_2(t_0) + x_2(t_0), \quad \ldots, \quad \frac{d^{n-1}y}{dt^{n-1}}(t_0) = y_n(t_0) + x_n(t_0)$$

where $x_1(t_0)$, $x_2(t_0)$ and $x_n(t_0)$ are small. Furthermore, suppose the input is changed slightly to $u(t) = w(t) + v(t)$ where $v(t)$ is small. To satisfy the differential equations,

$$\begin{aligned} d(\phi_1 + x_1)/dt &= f_1(\phi_1 + x_1, \phi_2 + x_2, \ldots, \phi_n + x_n, w + v, t) \\ d(\phi_2 + x_2)/dt &= f_2(\phi_1 + x_1, \phi_2 + x_2, \ldots, \phi_n + x_n, w + v, t) \\ &\cdots\cdots\cdots\cdots \\ d(\phi_n + x_n)/dt &= f_n(\phi_1 + x_1, \phi_2 + x_2, \ldots, \phi_n + x_n, w + v, t) \end{aligned}$$

If $f_1, f_2, \ldots, f_n$ can be expanded about $\phi_1, \phi_2, \ldots, \phi_n$ and w using Taylor's theorem for several variables, then neglecting higher order terms we obtain

$$\frac{d\phi_1}{dt} + \frac{dx_1}{dt} = f_1(\phi_1, \phi_2, \ldots, \phi_n, w, t) + \frac{\partial f_1}{\partial y_1}x_1 + \frac{\partial f_1}{\partial y_2}x_2 + \cdots + \frac{\partial f_1}{\partial y_n}x_n + \frac{\partial f_1}{\partial u}v$$

$$\frac{d\phi_2}{dt} + \frac{dx_2}{dt} = f_2(\phi_1, \phi_2, \ldots, \phi_n, w, t) + \frac{\partial f_2}{\partial y_1}x_1 + \frac{\partial f_2}{\partial y_2}x_2 + \cdots + \frac{\partial f_2}{\partial y_n}x_n + \frac{\partial f_2}{\partial u}v$$

$$\cdots\cdots\cdots\cdots\cdots\cdots\cdots\cdots\cdots\cdots\cdots\cdots$$

$$\frac{d\phi_n}{dt} + \frac{dx_n}{dt} = f_n(\phi_1, \phi_2, \ldots, \phi_n, w, t) + \frac{\partial f_n}{\partial y_1}x_1 + \frac{\partial f_n}{\partial y_2}x_2 + \cdots + \frac{\partial f_n}{\partial y_n}x_n + \frac{\partial f_n}{\partial u}v$$

where each $\partial f_i/\partial y_j$ is the partial derivative of $f_i(y_1, y_2, \ldots, y_n, u, t)$ with respect to y_j, evaluated at $y_1 = \phi_1$, $y_2 = \phi_2$, $\ldots$, $y_n = \phi_n$ and $u = w$. Now, since each ϕ_i satisfies the original equation, then $d\phi_i/dt = f_i$ can be canceled to leave

$$\frac{d}{dt}\begin{pmatrix} x_1 \\ x_2 \\ \vdots \\ x_n \end{pmatrix} = \begin{pmatrix} \partial f_1/\partial y_1 & \partial f_1/\partial y_2 & \ldots & \partial f_1/\partial y_n \\ \partial f_2/\partial y_1 & \partial f_2/\partial y_2 & \ldots & \partial f_2/\partial y_n \\ \cdots & \cdots & \cdots & \cdots \\ \partial f_n/\partial y_1 & \partial f_n/\partial y_2 & \ldots & \partial f_n/\partial y_n \end{pmatrix} \begin{pmatrix} x_1 \\ x_2 \\ \vdots \\ x_n \end{pmatrix} + \begin{pmatrix} \partial f_1/\partial u \\ \partial f_2/\partial u \\ \vdots \\ \partial f_n/\partial u \end{pmatrix} v$$

which is, in general, a time-varying linear differential equation, so that the nonlinear equation has been linearized. Note this procedure is valid only for $x_1, x_2, \ldots, x_n$ and v small enough so that the higher order terms in the Taylor's series can be neglected. The matrix of $\partial f_i/\partial y_j$ evaluated at $y_j = \phi_j$ is called the Jacobian matrix of the vector $\mathbf{f}(\mathbf{y}, \mathbf{u}, t)$.

Example 1.21.

Consider the system of Example 1.20 with initial conditions $y(t_0) = 1$ and $\dot{y}(t_0) = -1$ at $t_0 = 1$. If the particular input $w(t) = 0$, we obtain the trajectories $\phi_1(t) = t^{-1}$ and $\phi_2(t) = -t^{-2}$. Since $f_1 = y_2$, then $\partial f_1/\partial y_1 = 0$, $\partial f_1/\partial y_2 = 1$ and $\partial f_1/\partial u = 0$. Since $f_2 = 2y_1^3 - uy_2$, then $\partial f_2/\partial y_1 = 6y_1^2$, $\partial f_2/\partial y_2 = -u$ and $\partial f_2/\partial u = -y_2$. Hence for initial conditions $y(t_0) = 1 + x_1(t_0)$, $\dot{y}(t_0) = -1 + x_2(t_0)$ and inputs $u = v(t)$, we obtain

$$\frac{d}{dt}\begin{pmatrix} x_1 \\ x_2 \end{pmatrix} = \begin{pmatrix} 0 & 1 \\ 6t^{-2} & 0 \end{pmatrix}\begin{pmatrix} x_1 \\ x_2 \end{pmatrix} + \begin{pmatrix} 0 \\ t^{-2} \end{pmatrix} v$$

This linear equation gives the solution $y(t) = x_1(t) + t^{-1}$ and $dy/dt = x_2 - t^{-2}$ for the original nonlinear equation, and is valid as long as x_1, x_2 and v are small.

Example 1.22.

Given the system $dy/dt = ky - y^2 + u$. Taking $u(t) = 0$, we can find two constant solutions $\phi(t) = 0$ and $\psi(t) = k$. The equation for small motions $x(t)$ about $\phi(t) = 0$ is $dx/dt = kx + u$ so that $y(t) \approx x(t)$, and the equation for small motions $x(t)$ about $\psi(t) = k$ is $dx/dt = -kx + u$ so that $y(t) \approx k + x(t)$.

Solved Problems

1.1. Given a delay line whose output is a voltage input delayed T seconds. What is the physical object, the abstract object, the state variable and the state space? Also, is it controllable, observable and a dynamical system with a state space description?

The physical object is the delay line for an input $u(t)$ and an output $y(t) = u(t-T)$. This equation is the abstract object. Given an input time function $u(t)$ for all t, the output $y(t)$ is defined for $t \geq t_0$, so it is a dynamical system. To completely specify the output given only $u(t)$ for $t \geq t_0$, the voltages already inside the delay line must be known. Therefore, the state at time t_0 is $x(t_0) = u_{[t_0-T,\, t_0)}$, where the notation $u_{[t_2,\, t_1)}$ means the time function $u(\tau)$ for τ in the interval $t_2 \leq \tau < t_1$. For $\epsilon > 0$ as small as we please, $u_{[t_0-T,\, t_0)}$ can be considered as the uncountably infinite set of numbers

$$\{u(t_0-T),\ u(t_0-T+\epsilon),\ \ldots,\ u(t_0-\epsilon)\} \quad = \quad u_{[t_0-T,\, t_0)} \quad = \quad x(t_0)$$

In this sense we can consider the state as consisting of an infinity of numbers. Then the state variable is the infinite set of time functions

$$x(t) \quad = \quad u_{[t-T,\, t)} \quad = \quad \{u(t-T),\ u(t-T+\epsilon),\ \ldots,\ u(t-\epsilon)\}$$

The state space is the space of all time functions T seconds long, perhaps limited by the breakdown voltage of the delay line.

An input $u(t)$ for $t_0 \leq t < t_0 + T$ will completely determine the state T seconds later, and any state will be observed in the output after T seconds, so the system is both observable and controllable. Finally, $x(t)$ is uniquely made up of $x(t-\tau)$ shifted τ seconds plus the input over τ seconds, so that the mathematical relation $y(t) = u(t-T)$ gives a system with a state space description.

1.2. Given the uncontrollable partial differential equation (diffusion equation)

$$\frac{\partial y(r,t)}{\partial t} \quad = \quad \frac{\partial^2 y(r,t)}{\partial r^2} + 0u(r,t)$$

with boundary conditions $y(0,t) = y(1,t) = 0$. What is the state variable?

The solution to this zero input equation for $t \geq t_0$ is

$$y(r,t) \quad = \quad \sum_{n=1}^{\infty} c_n e^{-n^2\pi^2(t-t_0)} \sin n\pi r$$

where $c_n = \int_0^1 y(r,t_0) \sin n\pi r \, dr$. All that is needed to determine the output is $y(r,t_0)$, so that $y(r,t)$ is a choice for the state at any time t. Since $y(r,t)$ must be known for almost all r in the interval $0 \leq r \leq 1$, the state can be considered as an infinity of numbers similar to the case of Problem 1.1.

1.3. Given the mathematical equation $(dy/dt)^2 = y^2 + 0u$. Is this a dynamical system with a state space description?

A real output exists for all $t \geq t_0$ for any $u(t)$, so it is a dynamical system. The equation can be written as $dy/dt = s(t)y$, where $s(t)$ is a member of a set of time functions that take on the value $+1$ or -1 at any fixed time t. Hence knowledge of $y(t_0)$ and $s(t)$ for $t \geq t_0$ uniquely specify $y(t)$ and they are the state.

1.4. Plot the trajectories of

$$\frac{d^2y}{dt^2} \quad = \quad \begin{Bmatrix} 0 & \text{if } (dy/dt - y \geq 1) \\ -y & \text{if } (dy/dt - y < 1) \end{Bmatrix} + 0u$$

Changing variables to $y = x_1$ and $dy/dt = x_2$, the mathematical relation becomes

$$x_2\, dx_2/dx_1 \; = \; 0 \qquad \text{or} \qquad x_2\, dx_2/dx_1 \; = \; -x_1$$

The former equation can be integrated immediately to $x_2(t) = x_2(t_0)$, a straight line in the phase plane. The latter equation is solved by multiplying by $dx_1/2$ to obtain $x_2\,dx_2/2 + x_1\,dx_1/2 = 0$. This can be integrated to obtain $x_2^2(t) + x_1^2(t) = x_2^2(t_0) + x_1^2(t_0)$. The result is an equation of circles in the phase plane. The straight lines lie to the left of the line $x_2 - x_1 = 1$ and the circles to the right, as plotted in Fig. 1-5.

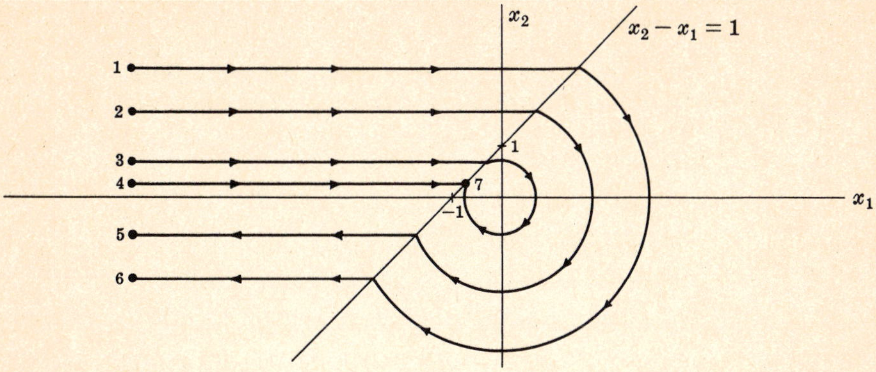

Fig. 1-5

Note x_1 increases for $x_2 > 0$ (positive velocity) and decreases for $x_2 < 0$, giving the motion of the system in the direction of the arrows as t increases. For instance, starting at the initial conditions $x_1(t_0)$ and $x_2(t_0)$ corresponding to the point numbered 1, the system moves along the outer trajectory to the point 6. Similarly point 2 moves to point 5. However, starting at either point 3 or point 4, the system goes to point 7 where the system motion in the next instant is not determined. At point 7 the output $y(t)$ does not exist for future times, so that this is not a dynamical system.

1.5. Given the electronic device diagrammed in Fig. 1-6, with a voltage input $u(t)$ and a voltage output $y(t)$. The resistors R have constant values. For $t_0 \leq t < t_1$, the switch S is open; and for $t \geq t_1$, S is closed. Is this system linear?

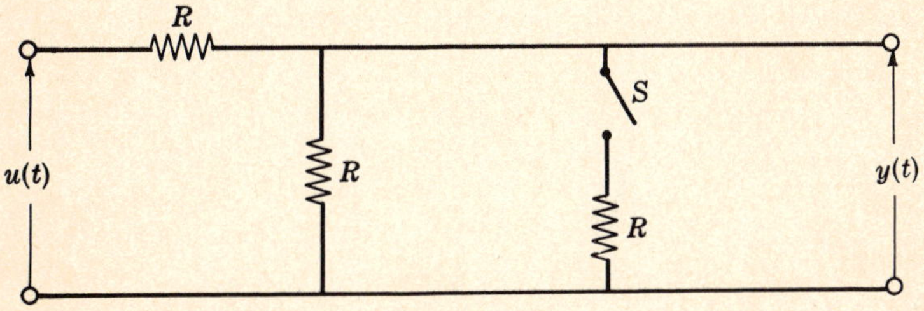

Fig. 1-6

Referring to Definition 1.7, it becomes apparent the first thing to do is find the state. No other information is necessary to determine the output given the input, so there is no state, i.e. the dimension of the state space is zero. This problem is somewhat analogous to Example 1.1, except that the position of the switch is specified at each instant of time.

To see if the system is linear, since $x(t) = 0$ for all time, we only need assume two inputs $u_1(t)$ and $u_2(t)$. Then $y_1(t) = u_1(t)/2$ for $t_0 \leq t < t_1$ and $y_1(t) = u_1(t)/3$ for $t \geq t_1$. Similarly $y_2(t) = u_2(t)/2$ or $u_2(t)/3$. Now assume an input $\alpha u_1(t) + \beta u_2(t)$. The output is $[\alpha u_1(t) + \beta u_2(t)]/2$ for $t_0 \leq t < t_1$ and $[\alpha u_1(t) + \beta u_2(t)]/3$ for $t > t_1$. Substituting $y_1(t)$ and $y_2(t)$, the output is $\alpha y_1(t) + \beta y_2(t)$, showing that superposition does hold and that the system is linear. The switch S can be considered a time-varying resistor whose resistance is infinite for $t < t_1$ and zero for $t \geq t_1$. Therefore Fig. 1-6 depicts a time-varying linear device.

1.6. Given the electronic device of Problem 1.5 (Fig. 1-6), with a voltage input $u(t)$ and a voltage output $y(t)$. The resistors R have constant values. However, now the position of the switch S depends on $y(t)$. Whenever $y(t)$ is positive, the switch S is open; and whenever $y(t)$ is negative, the switch S is closed. Is this system linear?

Again there is no state, and only superposition for zero state but nonzero input need be investigated. The input-output relationship is now

$$y(t) \quad = \quad [5u(t) + u(t) \operatorname{sgn} u(t)]/12$$

where $\operatorname{sgn} u = +1$ if u is positive and -1 if u is negative. Given two inputs u_1 and u_2 with resultant outputs y_1 and y_2 respectively, an output y with an input $u_3 = \alpha u_1 + \beta u_2$ is expressible as

$$y \quad = \quad [5(\alpha u_1 + \beta u_2) + (\alpha u_1 + \beta u_2) \operatorname{sgn}(\alpha u_1 + \beta u_2)]/12$$

To be linear, $\alpha y + \beta y_2$ must be equal y which would be true only if

$$\alpha u_1 \operatorname{sgn} u_1 + \beta u_2 \operatorname{sgn} u_2 \quad = \quad (\alpha u_1 + \beta u_2) \operatorname{sgn}(\alpha u_1 + \beta u_2)$$

This equality holds only in special cases, such as $\operatorname{sgn} u_1 = \operatorname{sgn} u_2 = \operatorname{sgn}(\alpha u_1 + \beta u_2)$, so that the system is not linear.

1.7. Given the abstract object characterized by

$$y(t) \quad = \quad x_0 e^{t_0 - t} \; + \; \int_{t_0}^{t} e^{\tau - t} u(\tau)\, d\tau$$

Is this time-varying?

This abstract object is that of Example 1.2, with $RC = 1$ in equation (*1.1*). By the same procedure used in Example 1.16, it can also be shown time-invariant. However, it can also be shown time-invariant by the test given after Definition 1.8. The input time function $u(t)$ is shifted by T, to become $\hat{u}(t)$. Then as can be seen in Fig. 1-7,

$$\hat{u}(t) \quad = \quad u(t - T)$$

Starting from the same initial state at time $t_0 + T$,

$$\hat{y}(\sigma) \quad = \quad x_0 e^{t_0 + T - \sigma} \; + \; \int_{t_0 + T}^{\sigma} e^{t - \sigma} u(\tau - T)\, d\tau$$

Let $\xi = \tau - T$:

$$\hat{y}(\sigma) \quad = \quad x_0 e^{t_0 + T - \sigma} \; + \; \int_{t_0}^{\sigma - T} e^{\xi + T - \sigma} u(\xi)\, d\xi$$

Evaluating $\hat{y}$ at $\sigma = t + T$ gives

$$\hat{y}(t + T) \quad = \quad x_0 e^{t_0 - t} \; + \; \int_{t_0}^{t} e^{\xi - t} u(\xi)\, d\xi$$

which is identical with the output $y(t)$.

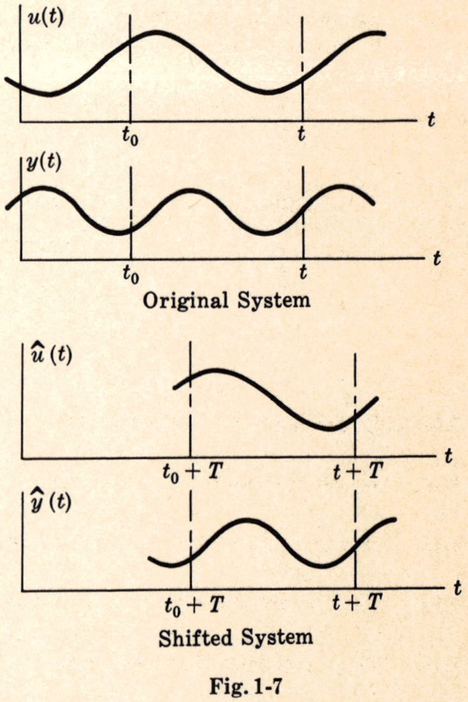

Fig. 1-7

Supplementary Problems

1.8. Given the spring-mass system shown in Fig. 1-8. What is the physical object, the abstract object, and the state variable?

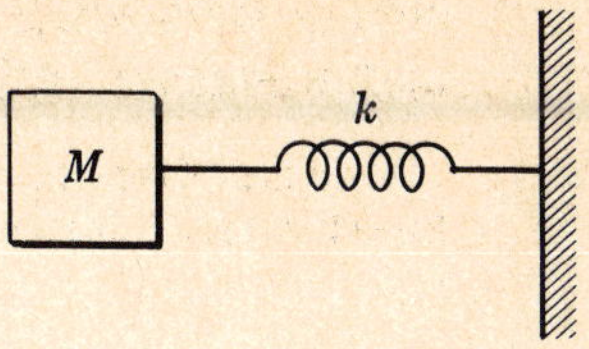

Fig. 1-8

1.9. Given the hereditary system $y(t) = \int_{-\infty}^{t} K(t,\tau)\, u(\tau)\, d\tau$ where $K(t,\tau)$ is some single-valued continuously differentiable function of t and τ. What is the state variable? Is the system linear? Is the system time-varying?

1.10. Given the discrete time system $x(n+1) = x(n) + u(n)$, the series of inputs $u(0), u(1), \ldots, u(k), \ldots,$ and the state at step 3, $x(3)$. Find the state variable $x(m)$ at any step $m \geqq 0$.

1.11. An abstract object is characterized by $y(t) = u(t)$ for $t_0 \leqq t < t_1$, and by $dy/dt = du/dt$ for $t \geqq t_1$. It is given that this abstract object will permit discontinuities in $y(t)$ at t_1. What is the dimension of the state space for $t_0 \leqq t < t_1$ and for $t \geqq t_1$?

1.12. Verify the solution *(1.2)* of equation *(1.1)*, and then verify the solution *(1.3)* of equation *(1.4)*. Finally, verify the solution $x_1 + 2x_2 = C(2x_1 + x_2)^4$ of Example 1.9.

1.13. Draw the trajectories in two dimensional state space of the system $\ddot{y} + y = 0$.

1.14. Given the circuit diagram of Fig. 1-9(*a*), where the nonlinear device NL has the voltage-current relation shown in Fig. 1-9(*b*). A mathematical equation is formed using $i = C\dot{v}$ and $v = f(i) = f(C\dot{v})$:

$$\dot{v} = (1/C)f^{-1}(v)$$

where f^{-1} is the inverse function. Also, $v(t_0)$ is taken to be the initial voltage on the capacitors. Is this mathematical relation a dynamical system with a state space description?

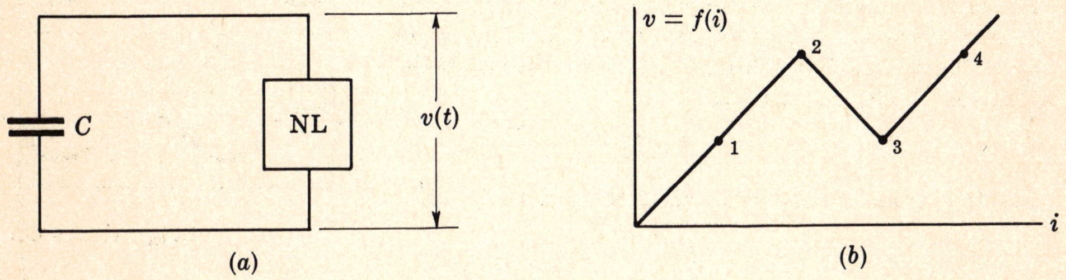

Fig. 1-9

1.15. Is the mathematical equation $y^2 + 1 = u$ a dynamical system?

1.16. Is the system $dy/dt = t^2 y$ time-varying? Is it linear?

1.17. Is the system $dy/dt = 1/y$ time-varying? Is it linear?

1.18. Verify that the system of Example 1.16 is nonlinear.

1.19. Show equation *(1.10)* is linear.

1.20. Show equation *(1.10)* is time-invariant if the coefficients α_i and β_j for $i = 1, 2, \ldots, n$ and $j = 0, 1, \ldots, n$, are not functions of time.

1.21. Given $dx_1/dt = x_1 + 2$, $dx_2/dt = x_2 + u$, $y = x_1 + x_2$.

(*a*) Does this system have a state space description?

(*b*) What is the input-output relation?

(*c*) Is the system linear?

1.22. What is the state space description for the anticipatory system $y(t) = u(t+T)$ in which only condition 5 for dynamical systems is violated?

1.23. Is the system $dx/dt = e^t u$, $y = e^{-t}x$ time-varying?

1.24. What is the state space description for the differentiator $y = du/dt$?

1.25. Is the equation $y = f(t)u$ a dynamical system given values of $f(t)$ for $t_0 \leqq t \leqq t_1$ only?

Answers to Supplementary Problems

1.8. The physical object is the spring-mass system, the abstract object is all x obeying $M\ddot{x} + kx = 0$, and the state variable is the vector having elements $x(t)$ and dx/dt. This system has a zero input.

1.9. It is *not* possible to represent

$$y(t) \quad = \quad y(t_0) + \int_{t_0}^{t} K(t,\tau)\,u(\tau)\,d\tau, \qquad \text{unless } K(t,\tau) = K(t_0,\tau)$$

(This is true for $t_0 \geqq T$ for the delay line.) For a general $K(t,\tau)$, the state at time t must be taken as $u(\tau)$ for $-\infty < \tau < t$. The system is linear and time varying, unless $K(t,\tau) = K(t-\tau)$, in which case it is time-invariant.

1.10. $x(k) = x(3) + \sum_{i=3}^{k-1} u(i)$ for $k = 4, 5, \ldots$ and $x(k) = x(3) - \sum_{i=1}^{3-k} u(3-i)$ for $k = 1, 2, 3$. Note we need not "tie" ourselves to an "initial" condition because any one of the values $x(i)$ will be the state for $i = 0, 1, \ldots n$.

1.11. The dimension of the state space is zero for $t_0 \leqq t < t_1$, and one-dimensional for $t \geqq t_1$. Because the state space is time-dependent in general, it must be a family of sets for each time t. Usually it is possible to consider a single set of input-output pairs over all t, i.e. the state space is time-invariant. Abstract objects possessing this property are called uniform abstract objects. This problem illustrates a nonuniform abstract object.

1.12. Plugging the solutions into the equations will verify them.

1.13. The trajectories are circles. It is a dynamical system.

1.14. It is a dynamical system, because $v(t)$ is real and defined for all $t \geqq t_0$. However, care must be taken in giving a state space description, because $f^{-1}(v)$ is not single valued. The state space description must include a means of determining which of the lines 1-2, 2-3 or 3-4 a particular voltage corresponds to.

1.15. No, because the input $u < 1$ results in an imaginary output.

1.16. It is linear and time-varying.

1.17. It is nonlinear and time-invariant.

1.21. (*a*) Yes

(*b*) $y(t) \quad = \quad e^{t-t_0}[x_1(t_0) + x_2(t_0)] + \int_{t_0}^{t} e^{t-\tau}[u(\tau) + 2]\,d\tau$

(*c*) No

1.22. No additional knowledge is needed other than the input, so that state is zero dimensional and the state space description is $y(t) = u(t+T)$. It is not a dynamical system because it is not realizable physically if $u(t)$ is unknown in advance for all $t \geq t_0$. However, its state space description violates only condition 5, so that other equations besides dynamical systems can have a state space description if the requirement of causality is waived.

1.23. Yes. $y(t) = e^{-t}x_0 + \int_{t_0}^{t} e^{(\tau - t)}\, u(\tau)\, d\tau$, and the contribution of the initial condition x_0 depends on when the system is started. If $x_0 = 0$, the system is equivalent to $dx/dt = -x + u$ where $y = x$, which is time-invariant.

1.24. If we define

$$du/dt = \lim_{\epsilon \to 0} [u(t+\epsilon) - u(t)]/\epsilon$$

so that $y(t_0)$ is defined, then the state space is zero dimensional and knowledge of $u(t)$ determines $y(t)$ for all $t \geq t_0$. Other definitions of du/dt may require knowledge of $u(t)$ for $t_0 - \epsilon \leq t \leq t_0$, which would be the state in that case.

1.25. Obviously $y(t)$ is not defined for $t > t_1$, so that as stated the equation is not a dynamical system. However, if the behavior of engineering interest lies between t_0 and t_1, merely append $y = 0u$ for $t \geq t_1$ to the equation and a dynamical system results.

Chapter 2

Methods for Obtaining the State Equations

2.1 FLOW DIAGRAMS

Flow diagrams are a simple diagrammatical means of obtaining the state equations. Because only linear differential or difference equations are considered here, only four basic objects are needed. The utility of flow diagrams results from the fact that no differentiating devices are permitted.

Definition 2.1: A *summer* is a diagrammatical abstract object having n inputs $u_1(t), u_2(t), \ldots, u_n(t)$ and one output $y(t)$ that obey the relationship

$$y(t) = \pm u_1(t) \pm u_2(t) \pm \cdots \pm u_n(t)$$

where the sign is positive or negative as indicated in Fig. 2-1, for example.

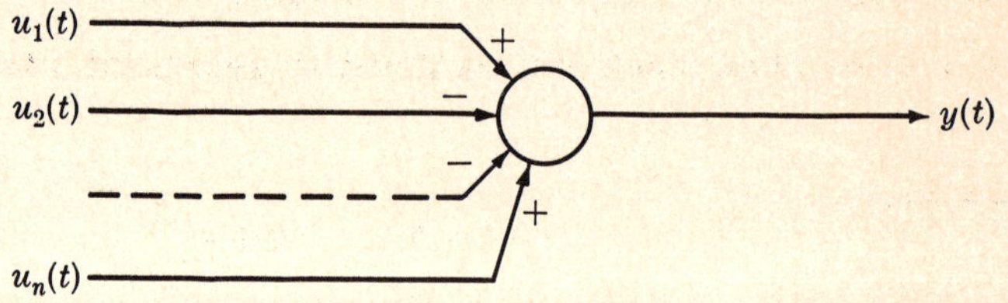

Fig. 2-1. Summer

Definition 2.2: A *scalor* is a diagrammatical abstract object having one input $u(t)$ and one output $y(t)$ such that the input is scaled up or down by the time function $\alpha(t)$ as indicated in Fig. 2-2. The output obeys the relationship $y(t) = \alpha(t)\,u(t)$.

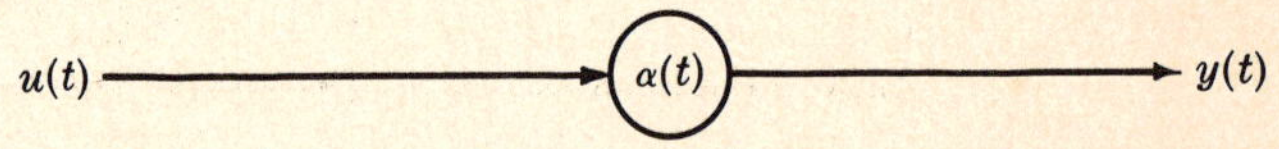

Fig. 2-2. Scalor

Definition 2.3: An *integrator* is a diagrammatical abstract object having one input $u(t)$, one output $y(t)$, and perhaps an initial condition $y(t_0)$ which may be shown or not, as in Fig. 2-3. The output obeys the relationship

$$y(t) = y(t_0) + \int_{t_0}^{t} u(\tau)\,d\tau$$

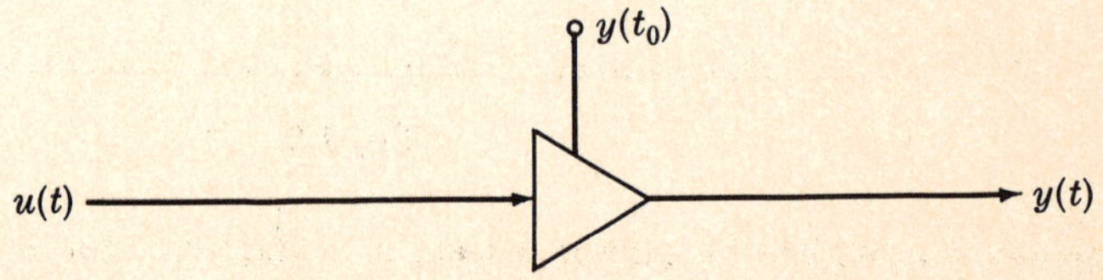

Fig. 2-3. Integrator at Time t

Definition 2.4: A *delayor* is a diagrammatical abstract object having one input $u(k)$, one output $y(k)$, and perhaps an initial condition $y(l)$ which may be shown or not, as in Fig. 2-4. The output obeys the relationship

$$y(j+l+1) = u(j+l) \quad \text{for } j = 0, 1, 2, \ldots$$

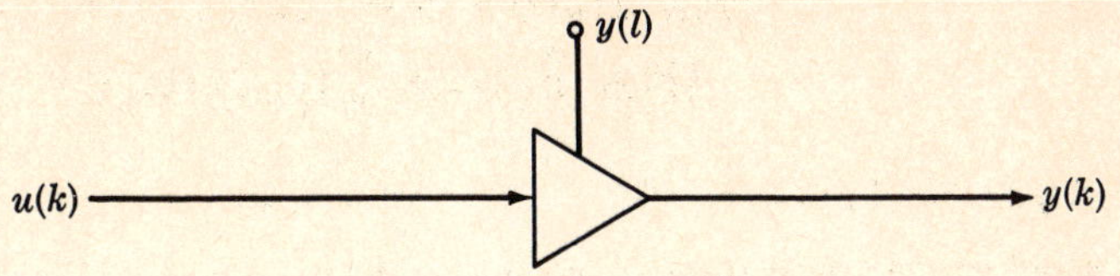

Fig. 2-4. Delayor at Time k

2.2 PROPERTIES OF FLOW DIAGRAMS

Any set of time-varying or time-invariant linear differential or difference equations of the form *(1.10)* or *(1.11)* can be represented diagrammatically by an interconnection of the foregoing elements. Also, any given transfer function can also be represented merely by rewriting it in terms of *(1.10)* or *(1.11)*. Furthermore, multiple input and/or multiple output systems can be represented in an analogous manner.

Equivalent interconnections can be made to represent the same system.

Example 2.1.

Given the system

$$dy/dt = \alpha y + \alpha u \tag{2.1}$$

with initial condition $y(t_0)$. An interconnection for this system is shown in Fig. 2-5.

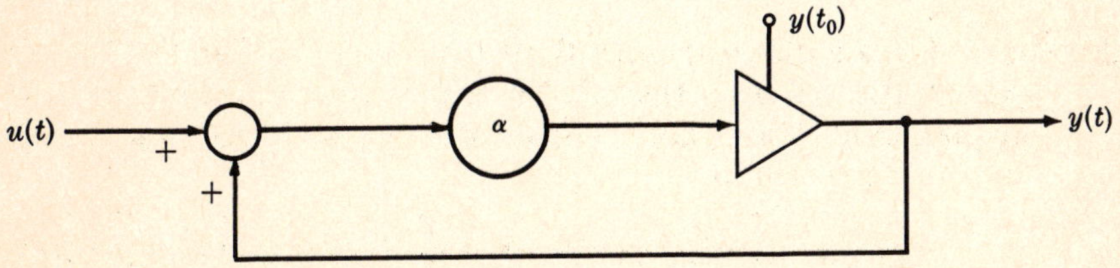

Fig. 2-5

Since α is a constant function of time, the integrator and scalor can be interchanged if the initial condition is adjusted accordingly, as shown in Fig. 2-6.

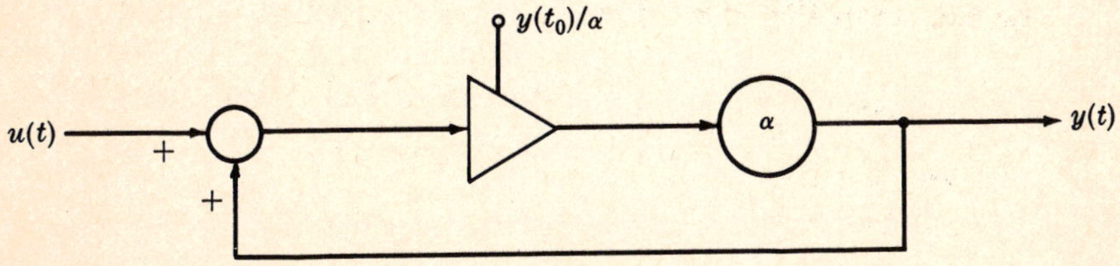

Fig. 2-6

This interchange could not be done if α were a general function of time. In certain special cases it is possible to use integration by parts to accomplish this interchange. If $\alpha(t) = t$, then *(2.1)* can be integrated to

$$y(t) = y(t_0) + \int_{t_0}^{t} \tau[y(\tau) + u(\tau)]\, d\tau$$

Using integration by parts,

$$y(t) \quad = \quad y(t_0) - \int_{t_0}^{t}\int_{t_0}^{\tau} [y(\xi)+u(\xi)]\,d\xi\,d\tau \;+\; t\int_{t_0}^{t} [y(\tau)+u(\tau)]\,d\tau$$

which gives the alternate flow diagram shown in Fig. 2-7.

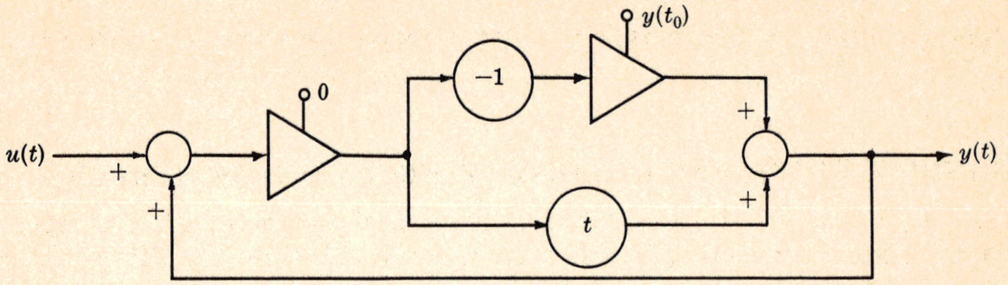

Fig. 2-7

Integrators are used in continuous time systems, delayors in discrete time (sampled data) systems. Discrete time diagrams can be drawn by considering the analogous continuous time system, and vice versa. For time-invariant systems, the diagrams are almost identical, but the situation is not so easy for time-varying systems.

Example 2.2.

Given the discrete time system

$$y(k+l+1) \quad = \quad \alpha y(k+l) + \alpha u(k+l) \tag{2.2}$$

with initial condition $y(l)$. The analogous continuous time systems is equation (*2.1*), where d/dt takes the place of a unit advance in time. This is more evident by taking the Laplace transform of (*2.1*),

$$sY(s) - y(t_0) \quad = \quad \alpha Y(s) + \alpha U(s)$$

and the $\mathcal{Z}$ transform of (*2.2*),

$$zY(z) - zy(l) \quad = \quad \alpha Y(z) + \alpha U(z)$$

Hence from Fig. 2-5 the diagram for (*2.2*) can be drawn immediately as in Fig. 2-8.

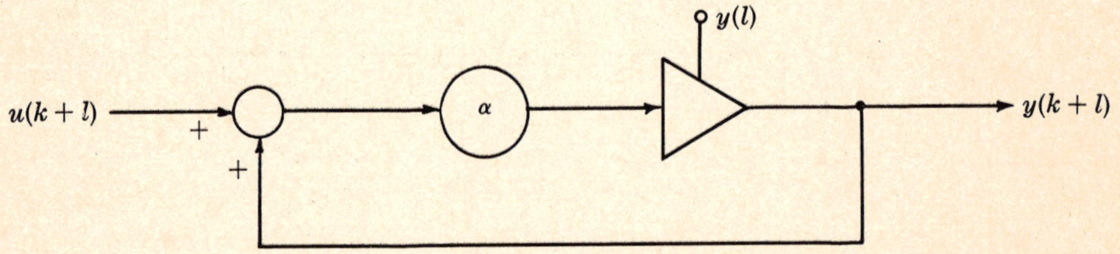

Fig. 2-8

If the initial condition of the integrator or delayor is arbitrary, the output of that integrator or delayor can be taken to be a state variable.

Example 2.3.

The state variable for (*2.1*) is $y(t)$, the output of the integrator. To verify this, the solution to equation (*2.1*) is

$$y(t) \quad = \quad y(t_0)\,e^{\alpha(t-t_0)} + \alpha\int_{t_0}^{t} e^{\alpha(t-\tau)}\,u(\tau)\,d\tau$$

Note $y(t_0)$ is the state at t_0, so the state variable is $y(t)$.

Example 2.4.

The state variable for equation (*2.2*) is $y(k+l)$, the output of the delayor. This can be verified in a manner similar to the previous example.

Example 2.5.

From Fig. 2-7, the state is the output of the second integrator only, because the initial condition of the first integrator is specified to be zero. This is true because Fig. 2-7 and Fig. 2-5 are equivalent systems.

Example 2.6.

A summer or a scalor has no state associated with it, because the output is completely determined by the input.

2.3 CANONICAL FLOW DIAGRAMS FOR TIME-INVARIANT SYSTEMS

Consider a general time-invariant linear differential equation with one input and one output, with the letter p denoting the time derivative d/dt. Only the differential equations need be considered, because by Section 2.2 discrete time systems follow analogously.

$$p^n y + \alpha_1 p^{n-1} y + \cdots + \alpha_{n-1} p y + \alpha_n y \;=\; \beta_0 p^n u + \beta_1 p^{n-1} u + \cdots + \beta_{n-1} p u + \beta_n u \qquad (2.3)$$

This can be rewritten as

$$p^n(y - \beta_0 u) + p^{n-1}(\alpha_1 y - \beta_1 u) + \cdots + p(\alpha_{n-1} y - \beta_{n-1} u) + \alpha_n y - \beta_n u \;=\; 0$$

because $\alpha_i p^{n-i} y = p^{n-i} \alpha_i y$, which is not true if α_i depends on time. Dividing through by p^n and rearranging gives

$$y \;=\; \beta_0 u + \frac{1}{p}(\beta_1 u - \alpha_1 y) + \cdots + \frac{1}{p^{n-1}}(\beta_{n-1} u - \alpha_{n-1} y) + \frac{1}{p^n}(\beta_n u - \alpha_n y) \qquad (2.4)$$

from which the flow diagram shown in Fig. 2-9 can be drawn starting with the output y at the right and working to the left.

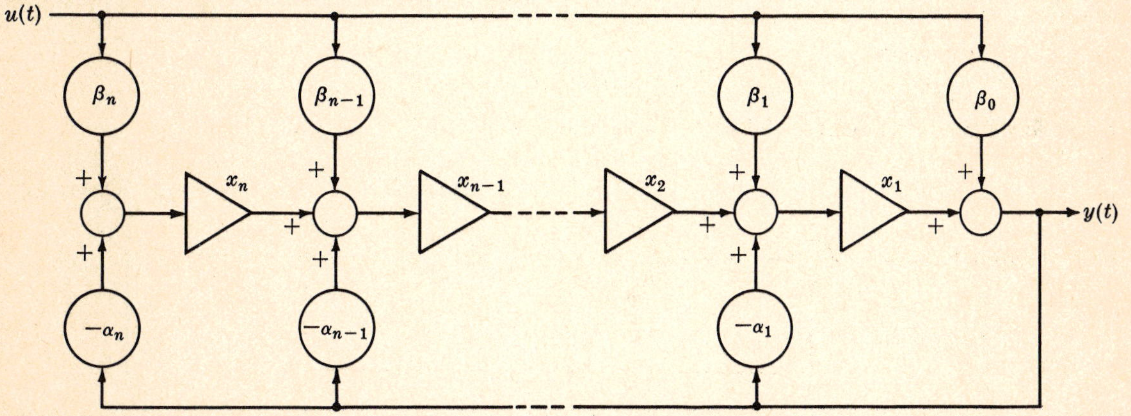

Fig. 2-9. Flow Diagram of the First Canonical Form

The output of each integrator is labeled as a state variable.

The summer equations for the state variables have the form

$$\begin{aligned}
y &= x_1 + \beta_0 u \\
\dot{x}_1 &= -\alpha_1 y + x_2 + \beta_1 u \\
\dot{x}_2 &= -\alpha_2 y + x_3 + \beta_2 u \\
&\cdots\cdots\cdots\cdots\cdots\cdots \\
\dot{x}_{n-1} &= -\alpha_{n-1} y + x_n + \beta_{n-1} u \\
\dot{x}_n &= -\alpha_n y + \beta_n u
\end{aligned} \qquad (2.5)$$

Using the first equation in (*2.5*) to eliminate y, the differential equations for the state variables can be written in the canonical matrix form

$$\frac{d}{dt}\begin{pmatrix} x_1 \\ x_2 \\ \vdots \\ x_{n-1} \\ x_n \end{pmatrix} = \begin{pmatrix} -\alpha_1 & 1 & 0 & \dots & 0 \\ -\alpha_2 & 0 & 1 & \dots & 0 \\ \dots & \dots & \dots & \dots & \dots \\ -\alpha_{n-1} & 0 & 0 & \dots & 1 \\ -\alpha_n & 0 & 0 & \dots & 0 \end{pmatrix}\begin{pmatrix} x_1 \\ x_2 \\ \vdots \\ x_{n-1} \\ x_n \end{pmatrix} + \begin{pmatrix} \beta_1 - \alpha_1\beta_0 \\ \beta_2 - \alpha_2\beta_0 \\ \vdots \\ \beta_{n-1} - \alpha_{n-1}\beta_0 \\ \beta_n - \alpha_n\beta_0 \end{pmatrix} u \qquad (2.6)$$

We will call this the first canonical form. Note the 1s above the diagonal and the α's down the first column of the $n \times n$ matrix. Also, the output can be written in terms of the state vector

$$y = (1\ 0\ \dots\ 0\ 0)\begin{pmatrix} x_1 \\ x_2 \\ \vdots \\ x_{n-1} \\ x_n \end{pmatrix} + \beta_0 u \qquad (2.7)$$

Note this form can be written down directly from the original equation (*2.3*).

Another useful form can be obtained by turning the first canonical flow diagram "backwards." This change is accomplished by reversing all arrows and integrators, interchanging summers and connection points, and interchanging input and output. This is a heuristic method of deriving a specific form that will be developed further in Chapter 7.

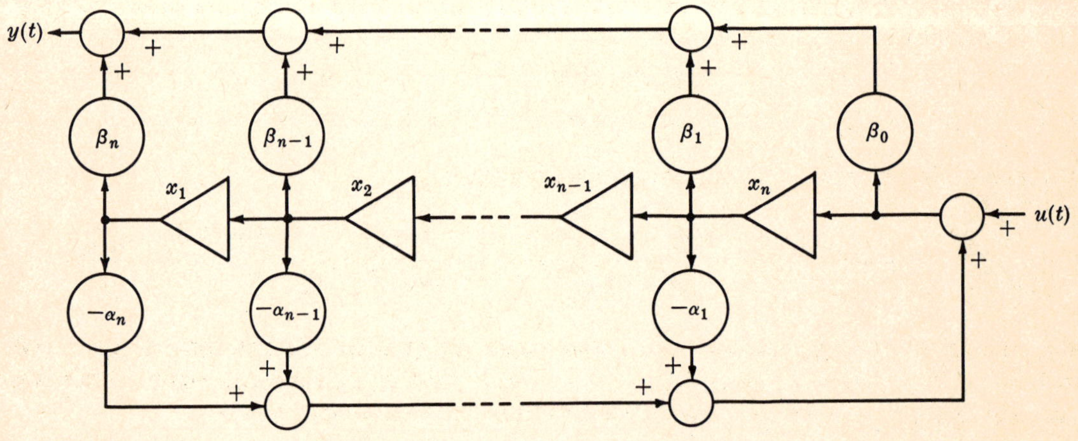

Fig. 2-10. Flow Diagram of the Second Canonical (Phase-variable) Form

Here the output of each integrator has been relabeled. The equations for the state variables are now

$$\begin{aligned} \dot{x}_1 &= x_2 \\ \dot{x}_2 &= x_3 \\ &\cdots\cdots\cdots\cdots \\ \dot{x}_{n-1} &= x_n \\ \dot{x}_n &= -\alpha_1 x_n - \alpha_2 x_{n-1} - \cdots - \alpha_{n-1}x_2 - \alpha_n x_1 + u \\ y &= \beta_n x_1 + \beta_{n-1}x_2 + \cdots + \beta_1 x_n + \beta_0[u - \alpha_1 x_n - \cdots - \alpha_{n-1}x_2 - \alpha_n x_1] \end{aligned} \qquad (2.8)$$

In matrix form, (2.8) may be written as

$$\frac{d}{dt}\begin{pmatrix} x_1 \\ x_2 \\ \vdots \\ x_{n-1} \\ x_n \end{pmatrix} = \begin{pmatrix} 0 & 1 & 0 & \cdots & 0 \\ 0 & 0 & 1 & \cdots & 0 \\ \cdots & \cdots & \cdots & \cdots & \cdots \\ 0 & 0 & 0 & \cdots & 1 \\ -\alpha_n & -\alpha_{n-1} & -\alpha_{n-2} & \cdots & -\alpha_1 \end{pmatrix}\begin{pmatrix} x_1 \\ x_2 \\ \vdots \\ x_{n-1} \\ x_n \end{pmatrix} + \begin{pmatrix} 0 \\ 0 \\ \vdots \\ 0 \\ 1 \end{pmatrix} u \tag{2.9}$$

and

$$y = (\beta_n - \alpha_n\beta_0 \quad \beta_{n-1} - \alpha_{n-1}\beta_0 \quad \cdots \quad \beta_1 - \alpha_1\beta_0)\begin{pmatrix} x_1 \\ x_2 \\ \vdots \\ x_{n-1} \\ x_n \end{pmatrix} + \beta_0 u \tag{2.10}$$

This will be called the second canonical form, or phase-variable canonical form. Here the 1s are above the diagonal but the α's go across the bottom row of the $n \times n$ matrix. By eliminating the state variables $\mathbf{x}$, the general input-output relation (2.3) can be verified. The phase-variable canonical form can also be written down upon inspection of the original differential equation (2.3).

2.4 JORDAN FLOW DIAGRAM

The general time-invariant linear differential equation (2.3) for one input and one output can be written as

$$y = \frac{\beta_0 p^n + \beta_1 p^{n-1} + \cdots + \beta_{n-1}p + \beta_n}{p^n + \alpha_1 p^{n-1} + \cdots + \alpha_{n-1}p + \alpha_n} u \tag{2.11}$$

By dividing once by the denominator, this becomes

$$y = \beta_0 u + \frac{(\beta_1 - \alpha_1\beta_0)p^{n-1} + (\beta_2 - \alpha_2\beta_0)p^{n-2} + \cdots + (\beta_{n-1} - \alpha_{n-1}\beta_0)p + \beta_n - \alpha_n\beta_0}{p^n + \alpha_1 p^{n-1} + \cdots + \alpha_{n-1}p + \alpha_n} u \tag{2.12}$$

Consider first the case where the denominator polynomial factors into distinct poles λ_i, $i = 1, 2, \ldots, n$. Distinct means $\lambda_i \neq \lambda_j$ for $i \neq j$, that is, no repeated roots. Because most practical systems are stable, the λ_i usually have negative real parts.

$$p^n + \alpha_1 p^{n-1} + \cdots + \alpha_{n-1}p + \alpha_n = (p - \lambda_1)(p - \lambda_2)\cdots(p - \lambda_n) \tag{2.13}$$

A partial fraction expansion can now be made having the form

$$y = \beta_0 u + \frac{\rho_1}{p - \lambda_1}u + \frac{\rho_2}{p - \lambda_2}u + \cdots + \frac{\rho_n}{p - \lambda_n}u \tag{2.14}$$

Here the residue ρ_i can be calculated as

$$\rho_i = \frac{(\beta_1 - \alpha_1\beta_0)\lambda_i^{n-1} + (\beta_2 - \alpha_2\beta_0)\lambda_i^{n-2} + \cdots + (\beta_{n-1} - \alpha_{n-1}\beta_0)\lambda_i + (\beta_n - \alpha_n\beta_0)}{(\lambda_i - \lambda_1)(\lambda_i - \lambda_2)\cdots(\lambda_i - \lambda_{i-1})(\lambda_i - \lambda_{i+1})\cdots(\lambda_i - \lambda_n)} \tag{2.15}$$

The partial fraction expansion (2.14) gives a very simple flow diagram, shown in Fig. 2-11 following.

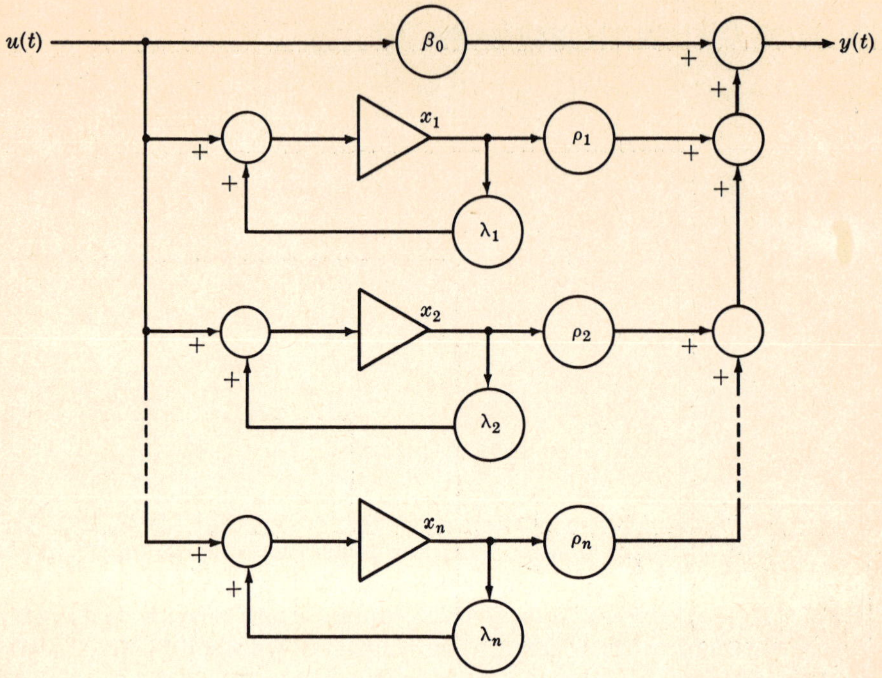

Fig. 2-11. Jordan Flow Diagram for Distinct Roots

Note that because ρ_i and λ_i can be complex numbers, the states x_i are complex-valued functions of time. The state equations assume the simple form

$$\begin{aligned} \dot{x}_1 &= \lambda_1 x_1 + u \\ \dot{x}_2 &= \lambda_2 x_2 + u \\ &\cdots\cdots\cdots\cdots \\ \dot{x}_n &= \lambda_n x_n + u \\ y &= \beta_0 u + \rho_1 x_1 + \rho_2 x_2 + \cdots + \rho_n x_n \end{aligned} \tag{2.16}$$

Consider now the general case. For simplicity, only one multiple root (actually one Jordan block, see Section 4.4, page 73) will be considered, because the results are easily extended to the general case. Then the denominator in (2.12) factors to

$$p^n + \alpha_1 p^{n-1} + \cdots + \alpha_{n-1} p + \alpha_n = (p-\lambda_1)^\nu (p-\lambda_{\nu+1}) \cdots (p-\lambda_n) \tag{2.17}$$

instead of (2.13). Here there are ν identical roots. Performing the partial fraction expansion for this case gives

$$y = \beta_0 u + \frac{\rho_1 u}{(p-\lambda_1)^\nu} + \frac{\rho_2 u}{(p-\lambda_1)^{\nu-1}} + \cdots + \frac{\rho_\nu u}{p-\lambda_1} + \frac{\rho_{\nu+1} u}{p-\lambda_{\nu+1}} + \cdots + \frac{\rho_n u}{p-\lambda_n} \tag{2.18}$$

The residues at the multiple roots can be evaluated as

$$\rho_k = \frac{1}{(k-1)!}\left[\frac{d^{k-1}}{dp^{k-1}}(p-\lambda_1)^\nu f(p)\right]_{p=\lambda_1} \qquad k = 1, 2, \ldots, \nu \tag{2.19}$$

where $f(p)$ is the polynomial fraction in p from (2.12). This gives the flow diagram shown in Fig. 2-12 following.

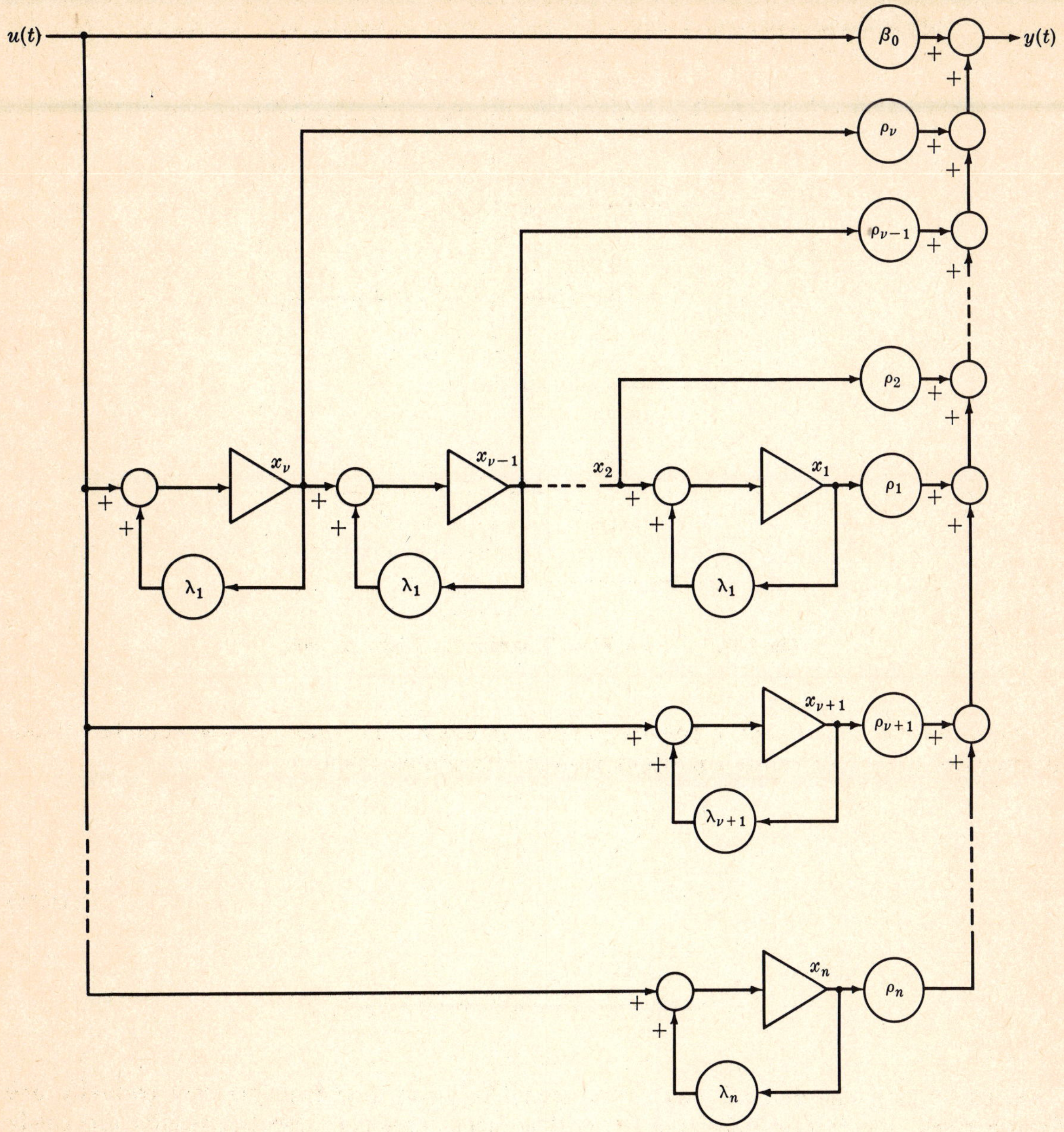

Fig. 2-12. Jordan Flow Diagram with One Multiple Root

The state equations are then

$$
\begin{aligned}
\dot{x}_1 &= \lambda_1 x_1 + x_2 \\
\dot{x}_2 &= \lambda_1 x_2 + x_3 \\
&\cdots\cdots\cdots\cdots\cdots\cdots \\
\dot{x}_{\nu-1} &= \lambda_1 x_{\nu-1} + x_\nu \\
\dot{x}_\nu &= \lambda_1 x_\nu + u \\
\dot{x}_{\nu+1} &= \lambda_{\nu+1} x_{\nu+1} + u \\
&\cdots\cdots\cdots\cdots\cdots\cdots \\
\dot{x}_n &= \lambda_n x_n + u \\
y &= \beta_0 u + \rho_1 x_1 + \rho_2 x_2 + \cdots + \rho_n x_n
\end{aligned}
\tag{2.20}
$$

The matrix differential equations associated with this Jordan form are

$$\frac{d}{dt}\begin{pmatrix} x_1 \\ x_2 \\ \vdots \\ x_{\nu-1} \\ x_\nu \\ x_{\nu+1} \\ \vdots \\ x_n \end{pmatrix} = \begin{pmatrix} \lambda_1 & 1 & 0 & \ldots & 0 & 0 & 0 & \ldots & 0 \\ 0 & \lambda_1 & 1 & \ldots & 0 & 0 & 0 & \ldots & 0 \\ \cdots & & & & & & & & \cdots \\ 0 & 0 & 0 & \ldots & \lambda_1 & 1 & 0 & \ldots & 0 \\ 0 & 0 & 0 & \ldots & 0 & \lambda_1 & 0 & \ldots & 0 \\ 0 & 0 & 0 & \ldots & 0 & 0 & \lambda_{\nu+1} & \ldots & 0 \\ \cdots & & & & & & & & \cdots \\ 0 & 0 & 0 & \ldots & 0 & 0 & 0 & \ldots & \lambda_n \end{pmatrix}\begin{pmatrix} x_1 \\ x_2 \\ \vdots \\ x_{\nu-1} \\ x_\nu \\ x_{\nu+1} \\ \vdots \\ x_n \end{pmatrix} + \begin{pmatrix} 0 \\ 0 \\ \vdots \\ 0 \\ 1 \\ 1 \\ \vdots \\ 1 \end{pmatrix} u \qquad (2.21)$$

and

$$y = (\rho_1 \; \rho_2 \; \cdots \; \rho_n)\begin{pmatrix} x_1 \\ x_2 \\ \vdots \\ x_n \end{pmatrix} + \beta_0 u \qquad (2.22)$$

In the $n \times n$ matrix, there is a diagonal row of ones above each λ_1 on the diagonal, and then the other λ's follow on the diagonal.

Example 2.7.

Derive the Jordan form of the differential system

$$\ddot{y} + 2\dot{y} + y = \dot{u} + u \qquad (2.23)$$

Equation (2.23) can be written as $y = \frac{(p+1)}{(p+1)^2}u$ whose partial fraction expansion gives

$$y = \frac{0}{(p+1)^2}u + \frac{1}{p+1}u$$

Figure 2-13 is then the flow diagram.

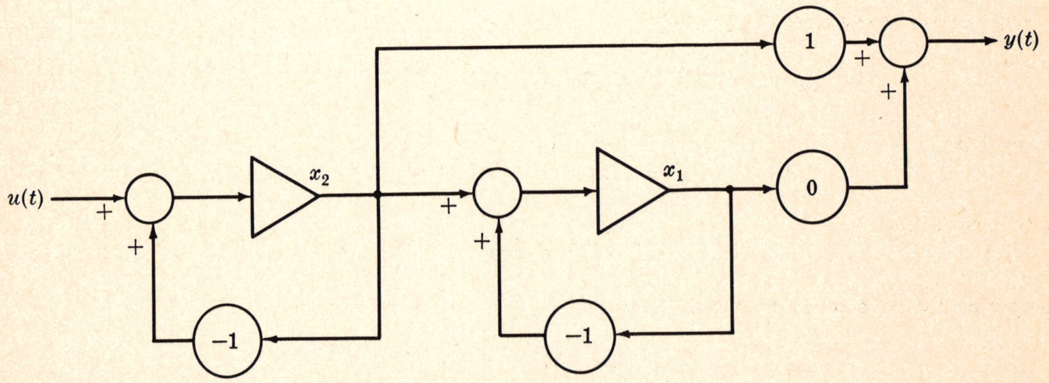

Fig. 2-13

Because the scalor following x_1 is zero, this state is unobservable. The matrix state equations in Jordan form are

$$\frac{d}{dt}\begin{pmatrix} x_1 \\ x_2 \end{pmatrix} = \begin{pmatrix} -1 & 1 \\ 0 & -1 \end{pmatrix}\begin{pmatrix} x_1 \\ x_2 \end{pmatrix} + \begin{pmatrix} 0 \\ 1 \end{pmatrix} u$$

$$y = (0 \;\; 1)\begin{pmatrix} x_1 \\ x_2 \end{pmatrix}$$

2.5 TIME-VARYING SYSTEMS

Finding the state equations of a time-varying system is not as easy as for time-invariant systems. However, the procedure is somewhat analogous, and so only one method will be given here.

The general time-varying differential equation of order n with one input and one output is shown again for convenience.

$$\frac{d^n y}{dt^n} + \alpha_1(t)\frac{d^{n-1}y}{dt^{n-1}} + \cdots + \alpha_n(t)y = \beta_0(t)\frac{d^n u}{dt^n} + \cdots + \beta_n(t)u \tag{1.10}$$

Differentiability of the coefficients a suitable number of times is assumed. Proceeding in a manner somewhat similar to the second canonical form, we shall try defining states as in (*2.8*). However, an amount $\gamma_i(t)$ [to be determined] of the input $u(t)$ enters into all the states.

$$\begin{aligned}
\dot{x}_1 &= x_2 + \gamma_1(t)u \\
\dot{x}_2 &= x_3 + \gamma_2(t)u \\
&\cdots\cdots\cdots\cdots\cdots\cdots\cdots\cdots\cdots \\
\dot{x}_{n-1} &= x_n + \gamma_{n-1}(t)u \\
\dot{x}_n &= -\alpha_1(t)x_n - \alpha_2(t)x_{n-1} - \cdots - \alpha_n(t)x_1 + \gamma_n(t)u \\
y &= x_1 + \gamma_0(t)u
\end{aligned} \tag{2.24}$$

By differentiating y n times and using the relations for each state, each of the unknown γ_i can be found. In the general case,

$$\begin{aligned}
\gamma_0(t) &= \beta_0(t) \\
\gamma_i(t) &= \beta_i(t) - \sum_{k=0}^{i-1}\sum_{j=0}^{i-k}\frac{(n+j-i)!}{j!\,(n-i)!}\alpha_{i-k-j}(t)\frac{d^j\gamma_k(t)}{dt^j}
\end{aligned} \tag{2.25}$$

Example 2.8.

Consider the second order equation

$$\frac{d^2y}{dt^2} + \alpha_1(t)\frac{dy}{dt} + \alpha_2(t)y = \beta_0(t)\frac{d^2u}{dt^2} + \beta_1(t)\frac{du}{dt} + \beta_2(t)u \tag{2.26}$$

Then by (*2.24*)

$$y = x_1 + \gamma_0(t)u \tag{2.27}$$

and differentiating,

$$\dot{y} = \dot{x}_1 + \dot{\gamma}_0(t)u + \gamma_0(t)\dot{u} \tag{2.28}$$

Substituting the first relation of (*2.24*) into (*2.28*) gives

$$\dot{y} = x_2 + [\gamma_1(t) + \dot{\gamma}_0(t)]u + \gamma_0(t)\dot{u} \tag{2.29}$$

Differentiating again,

$$\ddot{y} = \dot{x}_2 + [\dot{\gamma}_1(t) + \ddot{\gamma}_0(t)]u + [\gamma_1(t) + 2\dot{\gamma}_0(t)]\dot{u} + \gamma_0(t)\ddot{u} \tag{2.30}$$

From (*2.24*) we have

$$\dot{x}_2 = -\alpha_1(t)x_2 - \alpha_2(t)x_1 + \gamma_2(t)u \tag{2.31}$$

Now substituting (*2.27*), (*2.29*) and (*2.30*) into (*2.31*) yields

$$\begin{aligned}
&\ddot{y} - [\dot{\gamma}_1(t) + \ddot{\gamma}_0(t)]u - [\gamma_1(t) + 2\dot{\gamma}_0(t)]\dot{u} - \gamma_0(t)\ddot{u} \\
&\quad = -\alpha_1(t)\{\dot{y} - [\gamma_1(t) + \dot{\gamma}_0(t)]u - \gamma_0(t)\dot{u}\} - \alpha_2(t)[y - \gamma_0(t)u] + \gamma_2(t)u
\end{aligned} \tag{2.32}$$

Equating coefficients in (*2.26*) and (*2.32*),

$$\gamma_2 + \dot{\gamma}_1 + \ddot{\gamma}_0 + (\gamma_1 + \dot{\gamma}_0)\alpha_1 + \gamma_0\alpha_2 = \beta_2 \tag{2.33}$$

$$\gamma_1 + 2\dot{\gamma}_0 + \alpha_1\gamma_0 = \beta_1 \tag{2.34}$$

$$\gamma_0 = \beta_0 \tag{2.35}$$

Substituting (*2.35*) into (*2.34*),

$$\gamma_1 = \beta_1 - \alpha_1\beta_0 - 2\dot{\beta}_0 \tag{2.36}$$

and putting (*2.35*) and (*2.36*) into (*2.33*),

$$\gamma_2 = \ddot{\beta}_0 - \dot{\beta}_1 + \beta_2 + (\alpha_1\beta_0 + 2\dot{\beta}_0 - \beta_1)\alpha_1 + \beta_0\dot{\alpha}_1 - \beta_0\alpha_2 \tag{2.37}$$

Using equation (*2.24*), the matrix state equations become

$$\frac{d}{dt}\begin{pmatrix} x_1 \\ x_2 \\ \vdots \\ x_{n-1} \\ x_n \end{pmatrix} = \begin{pmatrix} 0 & 1 & 0 & \cdots & 0 \\ 0 & 0 & 1 & \cdots & 0 \\ \cdots & \cdots & \cdots & \cdots & \cdots \\ 0 & 0 & 0 & \cdots & 1 \\ -\alpha_n(t) & -\alpha_{n-1}(t) & -\alpha_{n-2}(t) & \cdots & -\alpha_1(t) \end{pmatrix}\begin{pmatrix} x_1 \\ x_2 \\ \vdots \\ x_{n-1} \\ x_n \end{pmatrix} + \begin{pmatrix} \gamma_1(t) \\ \gamma_2(t) \\ \vdots \\ \gamma_{n-1}(t) \\ \gamma_n(t) \end{pmatrix} u \tag{2.38}$$

$$y = (1 \; 0 \; \ldots \; 0)\begin{pmatrix} x_1 \\ x_2 \\ \vdots \\ x_{n-1} \\ x_n \end{pmatrix} + \gamma_0(t)u$$

2.6 GENERAL STATE EQUATIONS

Multiple input–multiple output systems can be put in the same canonical forms as single input–single output systems. Due to complexity of notation, they will not be considered here. The input becomes a vector $\mathbf{u}(t)$ and the output a vector $\mathbf{y}(t)$. The components are the inputs and outputs, respectively. Inspection of matrix equations (*2.6*), (*2.9*), (*2.21*) and (*2.38*) indicates a similarity of form. Accordingly a general form for the state equations of a linear differential system of order n with m inputs and k outputs is

$$\begin{aligned} d\mathbf{x}/dt &= \mathbf{A}(t)\mathbf{x} + \mathbf{B}(t)\mathbf{u} \\ \mathbf{y} &= \mathbf{C}(t)\mathbf{x} + \mathbf{D}(t)\mathbf{u} \end{aligned} \tag{2.39}$$

where $\mathbf{x}(t)$ is an n-vector,
$\mathbf{u}(t)$ is an m-vector,
$\mathbf{y}(t)$ is a k-vector,
$\mathbf{A}(t)$ is an $n \times n$ matrix,
$\mathbf{B}(t)$ is an $n \times m$ matrix,
$\mathbf{C}(t)$ is a $k \times n$ matrix,
$\mathbf{D}(t)$ is a $k \times m$ matrix.

In a similar manner a general form for discrete time systems is

$$\begin{aligned} \mathbf{x}(n+1) &= \mathbf{A}(n)\,\mathbf{x}(n) + \mathbf{B}(n)\,\mathbf{u}(n) \\ \mathbf{y}(n) &= \mathbf{C}(n)\,\mathbf{x}(n) + \mathbf{D}(n)\,\mathbf{u}(n) \end{aligned} \tag{2.40}$$

where the dimensions are also similar to the continuous time case.

Specifically, if the system has only one input u and one output y, the differential equations for the system are

$$d\mathbf{x}/dt = \mathbf{A}(t)\mathbf{x} + \mathbf{b}(t)u$$

$$y = \mathbf{c}^\dagger(t)\mathbf{x} + d(t)u$$

and similarly for discrete time systems. Here $\mathbf{c}(t)$ is taken to be a column vector, and $\mathbf{c}^\dagger(t)$ denotes the complex conjugate transpose of the column vector. Hence $\mathbf{c}^\dagger(t)$ is a row vector, and $\mathbf{c}^\dagger(t)\mathbf{x}$ is a scalar. Also, since u, y and $d(t)$ are not boldface, they are scalars.

Since these state equations are matrix equations, to analyze their properties a knowledge of matrix analysis is needed before progressing further.

Solved Problems

2.1. Find the matrix state equations in the first canonical form for the linear time-invariant differential equation

$$\ddot{y} + 5\dot{y} + 6y = \dot{u} + u \tag{2.41}$$

with initial conditions $y(0) = y_0$, $\dot{y}(0) = \dot{y}_0$. Also find the initial conditions on the state variables.

Using $p = d/dt$, equation (2.41) can be written as $p^2y + 5py + 6y = pu + u$. Dividing by p^2 and rearranging,

$$y = \frac{1}{p}(u - 5y) + \frac{1}{p^2}(u - 6y)$$

The flow diagram of Fig. 2-14 can be drawn starting from the output at the right.

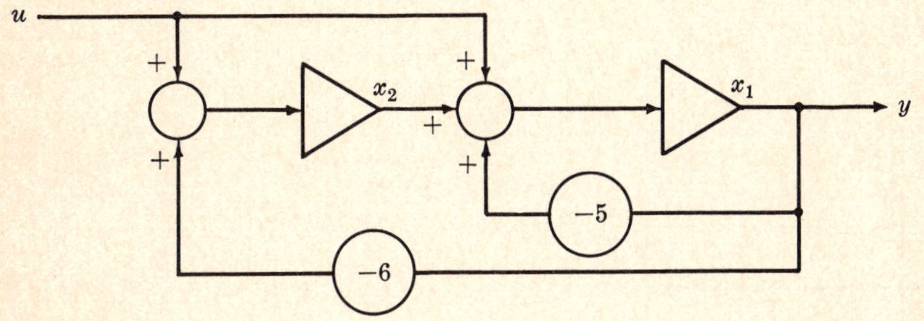

Fig. 2-14

Next, the outputs of the integrators are labeled the state variables x_1 and x_2 as shown. Now an equation can be formed using the summer on the left:

$$\dot{x}_2 = -6y + u$$

Similarly, an equation can be formed using the summer on the right:

$$\dot{x}_1 = x_2 - 5y + u$$

Also, the output equation is $y = x_1$. Substitution of this back into the previous equations gives

$$\dot{x}_1 = -5x_1 + x_2 + u$$

$$\dot{x}_2 = -6x_1 + u \tag{2.42}$$

The state equations can then be written in matrix notation as

$$\frac{d}{dt}\begin{pmatrix} x_1 \\ x_2 \end{pmatrix} = \begin{pmatrix} -5 & 1 \\ -6 & 0 \end{pmatrix}\begin{pmatrix} x_1 \\ x_2 \end{pmatrix} + \begin{pmatrix} 1 \\ 1 \end{pmatrix} u$$

with the output equation

$$y = (1 \quad 0)\begin{pmatrix} x_1 \\ x_2 \end{pmatrix}$$

The initial conditions on the state variables must be related to y_0 and $\dot{y}_0$, the given output initial conditions. The output equation is $x_1(t) = y(t)$, so that $x_1(0) = y(0) = y_0$. Also, substituting $y(t) = x_1(t)$ into (2.42) and setting $t = 0$ gives

$$\dot{y}(0) = -5y(0) + x_2(0) + u(0)$$

Use of the given initial conditions determines

$$x_2(0) = \dot{y}_0 + 5y_0 - u(0)$$

These relationships for the initial conditions can also be obtained by referring to the flow diagram at time $t = 0$.

2.2. **Find the matrix state equations in the second canonical form for the equation (2.41) of Problem 2.1, and the initial conditions on the state variables.**

The flow diagram (Fig. 2-14) of the previous problem is turned "backwards" to get the flow diagram of Fig. 2-15.

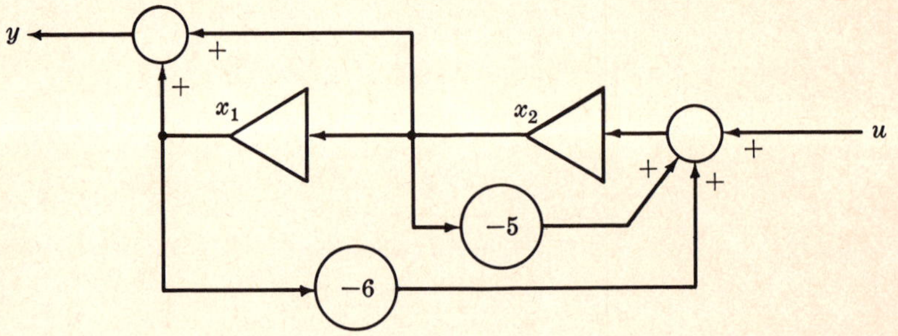

Fig. 2-15

The outputs of the integrators are labeled x_1 and x_2 as shown. These state variables are different from those in Problem 2.1, but are also denoted x_1 and x_2 to keep the state vector $\mathbf{x}(t)$ notation, as is conventional. Then looking at the summers gives the equations

$$y = x_1 + x_2 \tag{2.43}$$

$$\dot{x}_2 = -6x_1 - 5x_2 + u \tag{2.44}$$

Furthermore, the input to the left integrator is

$$\dot{x}_1 = x_2 \tag{2.45}$$

This gives the state equations

$$\frac{d}{dt}\begin{pmatrix} x_1 \\ x_2 \end{pmatrix} = \begin{pmatrix} 0 & 1 \\ -6 & -5 \end{pmatrix}\begin{pmatrix} x_1 \\ x_2 \end{pmatrix} + \begin{pmatrix} 0 \\ 1 \end{pmatrix} u$$

and

$$y = (1 \quad 1)\begin{pmatrix} x_1 \\ x_2 \end{pmatrix}$$

The initial conditions are found using (2.43),

$$y_0 = x_1(0) + x_2(0) \tag{2.46}$$

and its derivative

$$\dot{y}_0 = \dot{x}_1(0) + \dot{x}_2(0)$$

Use of (2.44) and (2.45) then gives

$$\dot{y}_0 = x_2(0) - 6x_1(0) - 5x_2(0) + u(0) \tag{2.47}$$

Equations (2.46) and (2.47) can be solved for the initial conditions

$$x_1(0) = -2y_0 - \tfrac{1}{2}\dot{y}_0 + \tfrac{1}{2}u(0)$$

$$x_2(0) = 3y_0 + \tfrac{1}{2}\dot{y}_0 - \tfrac{1}{2}u(0)$$

2.3. Find the matrix state equations in Jordan canonical form for equation (2.41) of Problem 2.1, and the initial conditions on the state variables.

The transfer function is

$$y = \frac{p+1}{p^2+5p+6}u = \frac{p+1}{(p+2)(p+3)}u$$

A partial fraction expansion gives

$$y = \frac{-1}{p+2}u + \frac{2}{p+3}u$$

From this the flow diagram can be drawn:

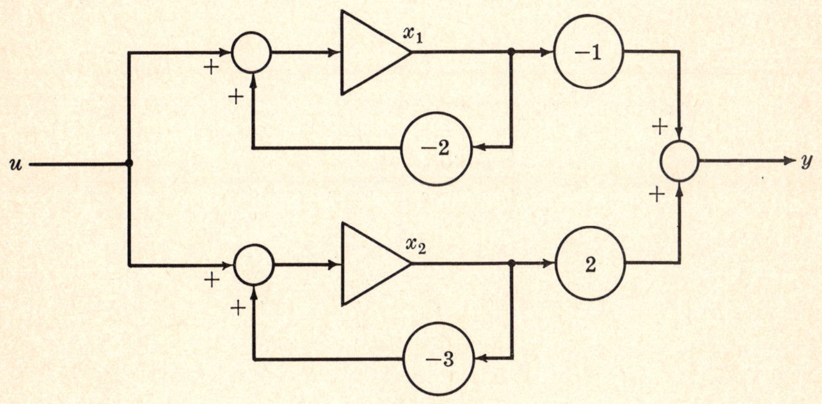

Fig. 2-16

The state equations can then be written from the equalities at each summer:

$$\frac{d}{dt}\begin{pmatrix} x_1 \\ x_2 \end{pmatrix} = \begin{pmatrix} -2 & 0 \\ 0 & -3 \end{pmatrix}\begin{pmatrix} x_1 \\ x_2 \end{pmatrix} + \begin{pmatrix} 1 \\ 1 \end{pmatrix} u$$

$$y = (-1 \quad 2)\begin{pmatrix} x_1 \\ x_2 \end{pmatrix}$$

From the output equation and its derivative at time t_0,

$$y_0 = 2x_2(0) - x_1(0)$$

$$\dot{y}_0 = 2\dot{x}_2(0) - \dot{x}_1(0)$$

The state equation is used to eliminate $\dot{x}_1(0)$ and $\dot{x}_2(0)$:

$$\dot{y}_0 = 2x_1(0) - 6x_2(0) + u(0)$$

Solving these equations for $x_1(0)$ and $x_2(0)$ gives

$$x_1(0) = u(0) - 3y_0 - \dot{y}_0$$

$$x_2(0) = \tfrac{1}{2}[u(0) - 2y_0 - \dot{y}_0]$$

2.4. Given the state equations

$$\frac{d}{dt}\begin{pmatrix} x_1 \\ x_2 \end{pmatrix} = \begin{pmatrix} 0 & 1 \\ -6 & -5 \end{pmatrix}\begin{pmatrix} x_1 \\ x_2 \end{pmatrix} + \begin{pmatrix} 0 \\ 1 \end{pmatrix} u$$

$$y = (1\ \ 1)\begin{pmatrix} x_1 \\ x_2 \end{pmatrix}$$

Find the differential equation relating the input to the output.

In operator notation, the state equations are

$$px_1 = x_2$$

$$px_2 = -6x_1 - 5x_2 + u$$

$$y = x_1 + x_2$$

Eliminating x_1 and x_2 then gives

$$p^2y + 5py + 6y = pu + u$$

This is equation (*2.41*) of Example 2.1 and the given state equations were derived from this equation in Example 2.2. Therefore this is a way to check the results.

2.5. Given the feedback system of Fig. 2-17 find a state space representation of this closed loop system.

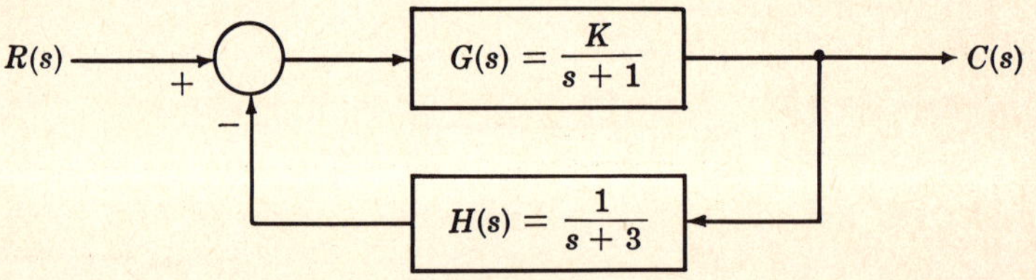

Fig. 2-17

The transfer function diagram is almost in flow diagram form already. Using the Jordan canonical form for the plant $G(s)$ and the feedback $H(s)$ separately gives the flow diagram of Fig. 2-18.

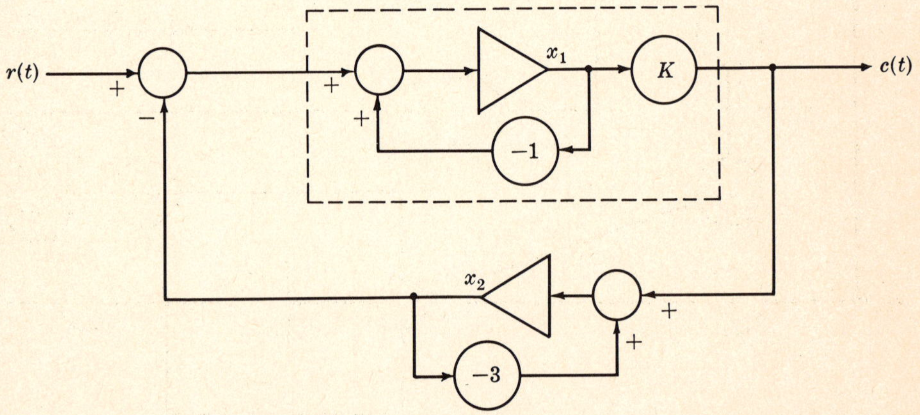

Fig. 2-18

Note $G(s)$ in Jordan form is enclosed by the dashed lines. Similarly the part for $H(s)$ was drawn, and then the transfer function diagram is used to connect the parts. From the flow diagram,

$$\frac{d}{dt}\begin{pmatrix} x_1 \\ x_2 \end{pmatrix} = \begin{pmatrix} -1 & -1 \\ K & -3 \end{pmatrix}\begin{pmatrix} x_1 \\ x_2 \end{pmatrix} + \begin{pmatrix} 1 \\ 0 \end{pmatrix} r(t), \qquad c(t) = (K\ \ 0)\begin{pmatrix} x_1 \\ x_2 \end{pmatrix}$$

2.6. Given the linear, time-invariant, multiple input–multiple output, discrete-time system

$$y_1(n+2) + \alpha_1 y_1(n+1) + \alpha_2 y_1(n) + \gamma_2 y_2(n+1) + \gamma_3 y_2(n) = \beta_1 u_1(n) + \delta_1 u_2(n)$$
$$y_2(n+1) + \gamma_1 y_2(n) + \alpha_3 y_1(n+1) + \alpha_4 y_1(n) = \beta_2 u_1(n) + \delta_2 u_2(n)$$

Put this in the form

$$\mathbf{x}(n+1) = \mathbf{A}\mathbf{x}(n) + \mathbf{B}\mathbf{u}(n)$$
$$\mathbf{y}(n) = \mathbf{C}\mathbf{x}(n) + \mathbf{D}\mathbf{u}(n)$$

where $\mathbf{y}(n) = \begin{pmatrix} y_1(n) \\ y_2(n) \end{pmatrix}$, $\mathbf{u}(n) = \begin{pmatrix} u_1(n) \\ u_2(n) \end{pmatrix}$.

The first canonical form will be used. Putting the given input-output equations into z operations (analogous to p operations of continuous time system),

$$z^2 y_1 + \alpha_1 z y_1 + \alpha_2 y_1 + \gamma_2 z y_2 + \gamma_3 y_2 = \beta_1 u_1 + \delta_1 u_2$$
$$z y_2 + \gamma_1 y_2 + \alpha_3 z y_1 + \alpha_4 y_1 = \beta_2 u_1 + \delta_2 u_2$$

Dividing by z^2 and z respectively and solving for y_1 and y_2,

$$y_1 = \frac{1}{z}(-\alpha_1 y_1 - \gamma_2 y_2) + \frac{1}{z^2}(\beta_1 u_1 + \delta_1 u_2 - \alpha_2 y_1 - \gamma_3 y_2)$$

$$y_2 = -\alpha_3 y_1 + \frac{1}{z}(\beta_2 u_1 + \delta_2 u_2 - \gamma_1 y_2 - \alpha_4 y_1)$$

Starting from the right, the flow diagram can be drawn as shown in Fig. 2-19.

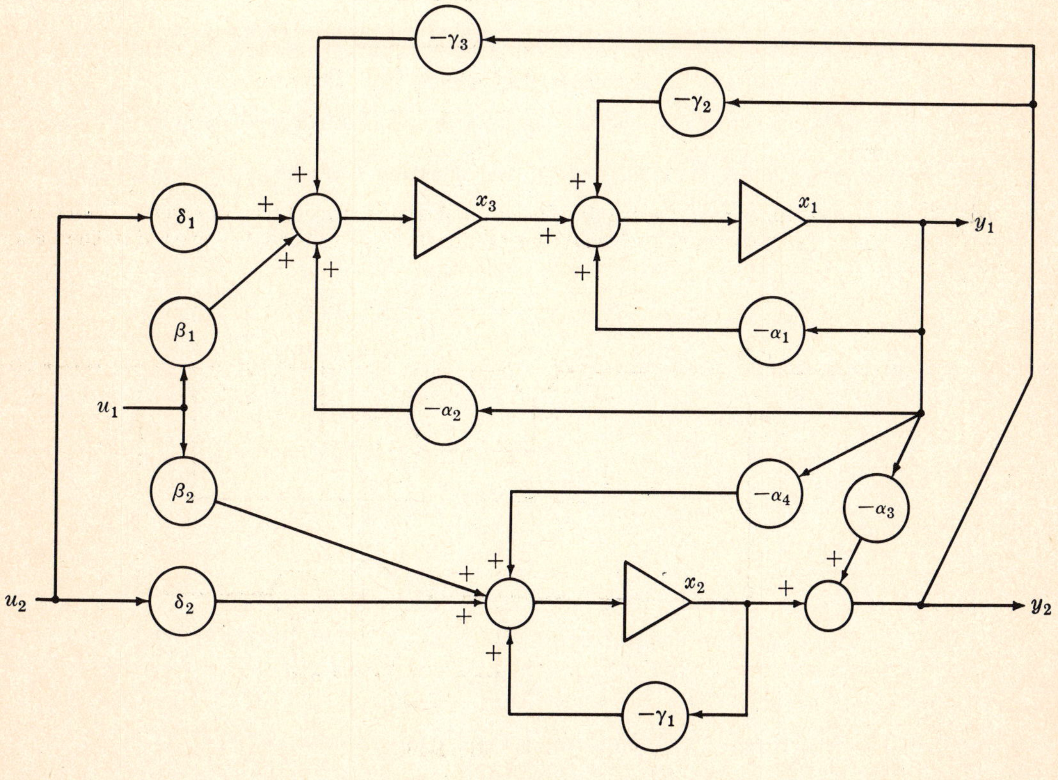

Fig. 2-19

Any more than three delayors with arbitrary initial conditions are not needed because a fourth such delayor would result in an unobservable or uncontrollable state. From this diagram the state equations are found to be

$$\begin{pmatrix} x_1(n+1) \\ x_2(n+1) \\ x_3(n+1) \end{pmatrix} = \begin{pmatrix} -\alpha_1 + \gamma_2\alpha_3 & -\gamma_2 & 1 \\ -\alpha_4 & -\gamma_1 & 0 \\ -\alpha_2 + \gamma_3\alpha_3 & -\gamma_3 & 0 \end{pmatrix} \begin{pmatrix} x_1(n) \\ x_2(n) \\ x_3(n) \end{pmatrix} + \begin{pmatrix} 0 & 0 \\ \beta_2 & \delta_2 \\ \beta_1 & \delta_1 \end{pmatrix} \begin{pmatrix} u_1(n) \\ u_2(n) \end{pmatrix}$$

$$\begin{pmatrix} y_1(n) \\ y_2(n) \end{pmatrix} = \begin{pmatrix} 1 & 0 & 0 \\ -\alpha_3 & 1 & 0 \end{pmatrix} \begin{pmatrix} x_1(n) \\ x_2(n) \\ x_3(n) \end{pmatrix}$$

2.7. Write the matrix state equations for a general time-varying second order discrete time equation, i.e. find matrices $\mathbf{A}(n)$, $\mathbf{B}(n)$, $\mathbf{C}(n)$, $\mathbf{D}(n)$ such that

$$\mathbf{x}(n+1) = \mathbf{A}(n)\,\mathbf{x}(n) + \mathbf{B}(n)\,\mathbf{u}(n)$$

$$\mathbf{y}(n) = \mathbf{C}(n)\,\mathbf{x}(n) + \mathbf{D}(n)\,\mathbf{u}(n) \qquad (2.48)$$

given the discrete time equation

$$y(n+2) + \alpha_1(n)\,y(n+1) + \alpha_2(n)\,y(n) = \beta_0(n)\,u(n+2) + \beta_1(n)\,u(n+1) + \beta_2(n)\,u(n) \qquad (2.49)$$

Analogously with the continuous time equations, try

$$x_1(n+1) = x_2(n) + \gamma_1(n)\,u(n) \qquad (2.50)$$

$$x_2(n+1) = -\alpha_1(n)\,x_2(n) - \alpha_2(n)\,x_1(n) + \gamma_2(n)\,u(n) \qquad (2.51)$$

$$y(n) = x_1(n) + \gamma_0(n)\,u(n) \qquad (2.52)$$

Stepping (2.52) up one and substituting (2.50) gives

$$y(n+1) = x_2(n) + \gamma_1(n)\,u(n) + \gamma_0(n+1)\,u(n+1) \qquad (2.53)$$

Stepping (2.53) up one and substituting (2.51) yields

$$y(n+2) = -\alpha_1(n)\,x_2(n) - \alpha_2(n)\,x_1(n) + \gamma_2(n)\,u(n) + \gamma_1(n+1)\,u(n+1) + \gamma_0(n+2)\,u(n+2) \qquad (2.54)$$

Substituting (2.52), (2.53), (2.54) into (2.49) and equating coefficients of $u(n)$, $u(n+1)$ and $u(n+2)$ gives

$$\gamma_0(n) = \beta_0(n-2)$$

$$\gamma_1(n) = \beta_1(n-1) - \alpha_1(n)\,\beta_0(n-2)$$

$$\gamma_2(n) = \beta_2(n) - \alpha_1(n)\,\beta_1(n-1) + [\alpha_1(n)\alpha_1(n-1) - \alpha_2(n)]\beta_0(n-2)$$

In matrix form this is

$$\begin{pmatrix} x_1(n+1) \\ x_2(n+1) \end{pmatrix} = \begin{pmatrix} 0 & 1 \\ -\alpha_2(n) & -\alpha_1(n) \end{pmatrix} \begin{pmatrix} x_1(n) \\ x_2(n) \end{pmatrix} + \begin{pmatrix} \gamma_1(n) \\ \gamma_2(n) \end{pmatrix} u(n)$$

$$y(n) = (1 \quad 0) \begin{pmatrix} x_1(n) \\ x_2(n) \end{pmatrix} + \gamma_0(n)\,u(n)$$

2.8. Given the time-varying second order continuous time equation with zero input,

$$\ddot{y} + \alpha_1(t)\dot{y} + \alpha_2(t)y = 0 \qquad (2.55)$$

write this in a form corresponding to the first canonical form

$$\begin{pmatrix} \dot{x}_1 \\ \dot{x}_2 \end{pmatrix} = \begin{pmatrix} -\alpha_1(t) & 1 \\ -\alpha_2(t) & 0 \end{pmatrix} \begin{pmatrix} x_1 \\ x_2 \end{pmatrix} \qquad (2.56)$$

$$y = (\gamma_1(t) \quad \gamma_2(t)) \begin{pmatrix} x_1 \\ x_2 \end{pmatrix} \qquad (2.57)$$

To do this, differentiate (*2.57*) and substitute for $\dot{x}_1$ and $\dot{x}_2$ from (*2.56*),

$$\dot{y} = (\dot{\gamma}_1 - \gamma_1\alpha_1 - \gamma_2\alpha_2)x_1 + (\gamma_1 + \dot{\gamma}_2)x_2 \qquad (2.58)$$

Differentiating again and substituting for x_1 and x_2 as before,

$$\ddot{y} = (\ddot{\gamma}_1 - 2\alpha_1\dot{\gamma}_1 - \dot{\alpha}_1\gamma_1 - \dot{\alpha}_2\gamma_2 - 2\alpha_2\dot{\gamma}_2 - \alpha_2\gamma_1 + \alpha_1\alpha_2\gamma_2 + \alpha_1^2\gamma_1)x_1 + (2\dot{\gamma}_1 - \alpha_1\gamma_1 - \alpha_2\gamma_2 + \ddot{\gamma}_2)x_2 \qquad (2.59)$$

Substituting (*2.57*), (*2.58*) and (*2.59*) into (*2.55*) and equating coefficients of x_1 and x_2 gives the equations

$$\ddot{\gamma}_1 - \alpha_1\dot{\gamma}_1 + (\alpha_2 - \dot{\alpha}_1)\gamma_1 = 0$$

$$\ddot{\gamma}_2 + (\alpha_1 - 2\alpha_2)\dot{\gamma}_2 + (\alpha_2^2 - \alpha_2\alpha_1 + \alpha_2 - \dot{\alpha}_2)\gamma_2 + 2\dot{\gamma}_1 - \alpha_2\gamma_1 = 0$$

In this case, $\gamma_1(t)$ may be taken to be zero, and any non-trivial $\gamma_2(t)$ satisfying

$$\ddot{\gamma}_2 + (\alpha_1 - 2\alpha_2)\dot{\gamma}_2 + (\alpha_2^2 - \alpha_2\alpha_1 + \alpha_2 - \dot{\alpha}_2)\gamma_2 = 0 \qquad (2.60)$$

will give the desired canonical form.

This problem illustrates the utility of the given time-varying form (*2.38*). It may always be found by differentiating known functions of time. Other forms usually involve the solution of equations such as (*2.60*), which may be quite difficult, or require differentiation of the $\alpha_i(t)$. Addition of an input contributes even more difficulty. However, in a later chapter forms analogous to the first canonical form will be given.

Supplementary Problems

2.9. Given the discrete time equation $y(n+2) + 3y(n+1) + 2y(n) = u(n+1) + 3u(n)$, find the matrix state equations in (i) the first canonical form, (ii) the second canonical form, (iii) the Jordan canonical form.

2.10. Find a matrix state equation for the multiple input–multiple output system

$$\ddot{y}_1 + \alpha_1\dot{y}_1 + \alpha_2 y_1 + \gamma_3\dot{y}_2 + \gamma_4 y_2 = \beta_1 u_1 + \delta_1 u_2$$

$$\ddot{y}_2 + \gamma_1\dot{y}_2 + \gamma_2 y_2 + \alpha_3\dot{y}_1 + \alpha_4 y_1 = \beta_2 u_1 + \delta_2 u_2$$

2.11. Write the matrix state equations for the system of Fig. 2-20, using Fig. 2-20 directly.

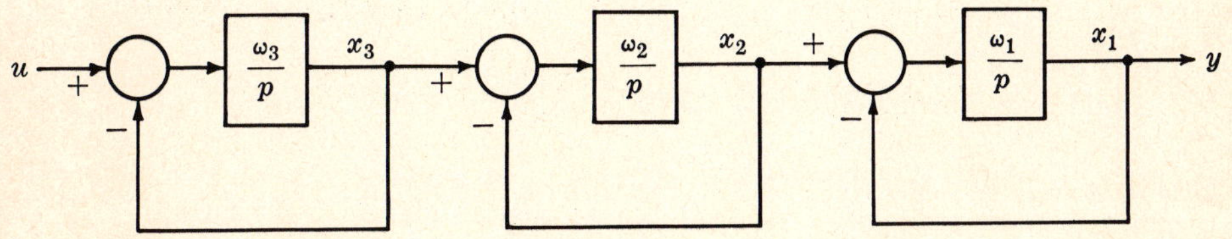

Fig. 2-20

2.12. Consider the plant dynamics and feedback compensation illustrated in Fig. 2-21. Assuming the initial conditions are specified in the form $v(0)$, $\dot{v}(0)$, $\ddot{v}(0)$, $\dddot{v}(0)$, $w(0)$ and $\dot{w}(0)$, write a state space equation for the plant plus compensation in the form $\dot{\mathbf{x}} = \mathbf{A}\mathbf{x} + \mathbf{B}u$ and show the relation between the specified initial conditions and the components of $\mathbf{x}(0)$.

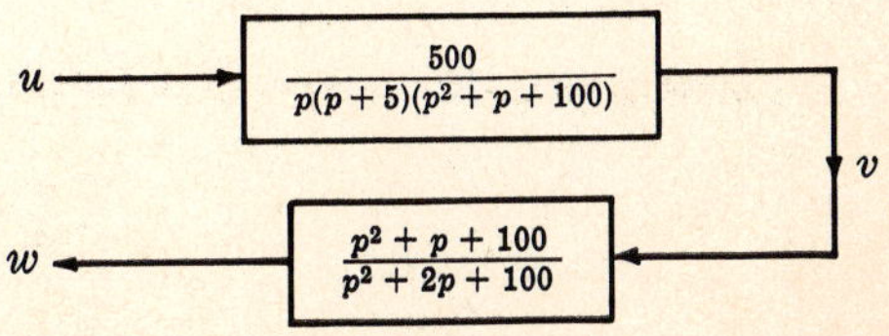

Fig. 2-21

2.13. The procedure of turning "backwards" the flow diagram of the first canonical form to obtain the second canonical form was never justified. Do this by verifying that equations (*2.8*) satisfy the original input-output relationship (*2.3*). Why can't the time-varying flow diagram corresponding to equations (*2.24*) be turned backwards to get another canonical form?

2.14. Given the linear time-varying differential equation

$$\ddot{y} + \alpha_1(t)\dot{y} + \alpha_2(t)y = \beta_0(t)\ddot{u} + \beta_1(t)\dot{u} + \beta_2(t)u$$

with the initial conditions $y(0) = y_0$, $\dot{y}(0) = \dot{y}_0$.

(i) Draw the flow diagram and indicate the state variables.

(ii) Indicate on the flow diagram the values of the scalors at all times.

(iii) Write down the initial conditions of the state variables.

2.15. Verify equation (*2.37*) using equation (*2.25*).

2.16. The simplified equations of a d-c motor are

$$\text{Motor Armature:} \quad Ri + L\frac{di}{dt} = V - K_f\frac{d\theta}{dt}$$

$$\text{Inertial Load:} \quad K_t i = J\frac{d^2\theta}{dt^2} + B\frac{d\theta}{dt}$$

Obtain a matrix state equation relating the input voltage V to the output shaft angle θ using a state vector

$$\mathbf{x} = \begin{pmatrix} i \\ \theta \\ d\theta/dt \end{pmatrix}$$

2.17. The equations describing the time behavior of the neutrons in a nuclear reactor are

$$\text{Prompt Neutrons:} \quad l\dot{n} = (\rho(t) - \beta)n + \sum_{i=1}^{6} \lambda_i C_i$$

$$\text{Delayed Neutrons:} \quad \dot{C}_i = \beta_i n - \lambda_i C_i$$

where $\beta = \sum_{i=1}^{6} \beta_1$ and $\rho(t)$ is the time-varying reactivity, perhaps induced by control rod motion. Write the matrix state equations.

2.18. Assume the simplified equations of motion for a missile are given by

$$\text{Lateral Translation:} \quad \ddot{z} + K_1\phi + K_2\alpha + K_3\beta = 0$$

$$\text{Rotation:} \quad \ddot{\phi} + K_4\alpha + K_5\beta = 0$$

$$\text{Angle of Attack:} \quad \alpha = \phi - K_6\dot{z}$$

$$\text{Rigid Body Bending Moment:} \quad M(l) = K_\alpha(l)\alpha + K_B(l)\beta$$

where z = lateral translation, ϕ = attitude, α = angle of attack, β = engine deflection, $M(l)$ = bending moment at station l. Obtain a state space equation in the form

$$\frac{d\mathbf{x}}{dt} = \mathbf{A}\mathbf{x} + \mathbf{B}u, \qquad \mathbf{y} = \mathbf{C}\mathbf{x} + \mathbf{D}u$$

where $u = \beta$ and $\mathbf{y} = \begin{pmatrix} \alpha \\ M(l) \end{pmatrix}$, taking as the state vector $\mathbf{x}$

$$\text{(i)} \quad \mathbf{x} = \begin{pmatrix} z \\ \dot{z} \\ \phi \\ \dot{\phi} \end{pmatrix} \qquad \text{(ii)} \quad \mathbf{x} = \begin{pmatrix} \dot{z} \\ \phi \\ \dot{\phi} \end{pmatrix} \qquad \text{(iii)} \quad \mathbf{x} = \begin{pmatrix} z \\ \alpha \\ \phi \\ \dot{\phi} \end{pmatrix}$$

2.19. Consider the time-varying electrical network of Fig. 2-22. The voltages across the inductors and the current in the capacitors can be expressed by the relations

$$e_a - e_1 = \frac{d}{dt}(L_1 i_1) = L_1 \frac{di_1}{dt} + i_1 \frac{dL_1}{dt}$$

$$i_1 - i_2 = \frac{d}{dt}(C_1 e_1) = C_1 \frac{de_1}{dt} + e_1 \frac{dC_1}{dt}$$

$$e_1 - e_b = \frac{d}{dt}(L_2 i_2) = L_2 \frac{di_2}{dt} + i_2 \frac{dL_2}{dt}$$

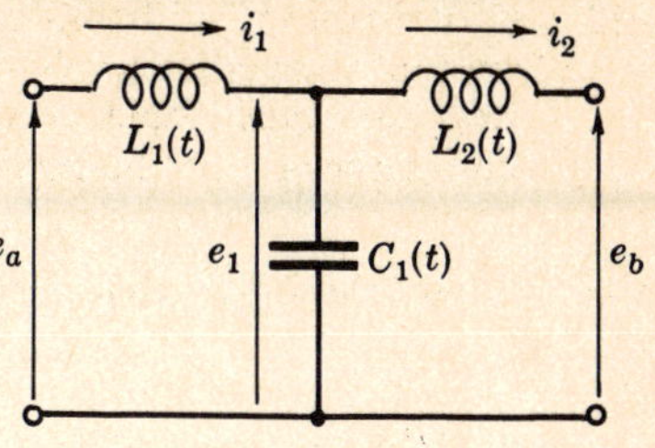

Fig. 2-22

It is more convenient to take as states the inductor fluxes

$$p_1 = \int_{t_0}^{t} (e_a - e_1)\,dt + p_1(t_0)$$

$$p_2 = \int_{t_0}^{t} (e_1 - e_b)\,dt + p_2(t_0)$$

and the capacitor charge

$$q_1 = \int_{t_0}^{t} (i_1 - i_2)\,dt + q_1(t_0)$$

Obtain a state space equation for the network in the form

$$\frac{d\mathbf{x}}{dt} = \mathbf{A}(t)\mathbf{x} + \mathbf{B}(t)\mathbf{u}$$

$$\mathbf{y}(t) = \mathbf{C}(t)\mathbf{x} + \mathbf{D}(t)\mathbf{u}$$

where the state vector $\mathbf{x}$, input vector $\mathbf{u}$, and output vector $\mathbf{y}$ are

(i) $$\mathbf{x} = \begin{pmatrix} p_1 \\ q_1 \\ p_2 \end{pmatrix} \qquad \mathbf{u} = \begin{pmatrix} e_a \\ e_b \end{pmatrix} \qquad \mathbf{y} = \begin{pmatrix} i_1 \\ e_1 \\ i_2 \end{pmatrix}$$

(ii) $$\mathbf{x} = \begin{pmatrix} i_1 \\ e_1 \\ i_2 \end{pmatrix} \qquad \mathbf{u} = \begin{pmatrix} e_a \\ e_b \end{pmatrix} \qquad \mathbf{y} = \begin{pmatrix} i_1 \\ e_1 \\ i_2 \end{pmatrix}$$

2.20. Given the quadratic Hamiltonian $H = \frac{1}{2}\mathbf{q}^T\mathbf{V}\mathbf{q} + \frac{1}{2}\mathbf{p}^T\mathbf{T}\mathbf{p}$ where $\mathbf{q}$ is a vector of n generalized coordinates, $\mathbf{p}$ is a vector of corresponding conjugate momenta, $\mathbf{V}$ and $\mathbf{T}$ are $n \times n$ matrices corresponding to the kinetic and potential energy, and the superscript ^{T} on a vector denotes transpose. Write a set of matrix state equations to describe the state.

Answers to Supplementary Problems

2.9. (i) $$\mathbf{x}(n+1) = \begin{pmatrix} -3 & 1 \\ -2 & 0 \end{pmatrix}\mathbf{x}(n) + \begin{pmatrix} 1 \\ 3 \end{pmatrix} u(n); \qquad y(n) = (1 \;\; 0)\,\mathbf{x}(n)$$

(ii) $$\mathbf{x}(n+1) = \begin{pmatrix} 0 & 1 \\ -2 & -3 \end{pmatrix}\mathbf{x}(n) + \begin{pmatrix} 0 \\ 1 \end{pmatrix} u(n); \qquad y(n) = (3 \;\; 1)\,\mathbf{x}(n)$$

(iii) $$\mathbf{x}(n+1) = \begin{pmatrix} -2 & 0 \\ 0 & -1 \end{pmatrix}\mathbf{x}(n) + \begin{pmatrix} 1 \\ 1 \end{pmatrix} u(n); \qquad y(n) = (-1 \;\; 2)\,\mathbf{x}(n)$$

2.10. Similar to first canonical form

$$\dot{\mathbf{x}} = \begin{pmatrix} -\alpha_1 & 1 & -\gamma_3 & 0 \\ -\alpha_2 & 0 & -\gamma_4 & 0 \\ -\alpha_3 & 0 & -\gamma_1 & 1 \\ -\alpha_4 & 0 & -\gamma_2 & 0 \end{pmatrix}\mathbf{x} + \begin{pmatrix} 0 & 0 \\ \beta_1 & \delta_1 \\ 0 & 0 \\ \beta_2 & \delta_2 \end{pmatrix}\mathbf{u}$$

$$\mathbf{y} = \begin{pmatrix} 1 & 0 & 0 & 0 \\ 0 & 0 & 1 & 0 \end{pmatrix}\mathbf{x}$$

or similar to second canonical form

$$\dot{\mathbf{x}} = \begin{pmatrix} 0 & 1 & 0 & 0 \\ -\alpha_2 & -\alpha_1 & -\alpha_4 & -\alpha_3 \\ 0 & 0 & 0 & 0 \\ -\gamma_4 & -\gamma_3 & -\gamma_2 & -\gamma_1 \end{pmatrix}\mathbf{x} + \begin{pmatrix} 0 & 0 \\ \beta_1 & \delta_1 \\ 0 & 0 \\ \beta_2 & \delta_2 \end{pmatrix}\mathbf{u}$$

$$\mathbf{y} = \begin{pmatrix} 1 & 0 & 0 & 0 \\ 0 & 0 & 1 & 0 \end{pmatrix}\mathbf{x}$$

2.11.

$$\frac{d}{dt}\begin{pmatrix} x_1 \\ x_2 \\ x_3 \end{pmatrix} = \begin{pmatrix} -\omega_1 & +\omega_1 & 0 \\ 0 & -\omega_2 & +\omega_2 \\ 0 & 0 & -\omega_3 \end{pmatrix}\begin{pmatrix} x_1 \\ x_2 \\ x_3 \end{pmatrix} + \begin{pmatrix} 0 \\ 0 \\ \omega_3 \end{pmatrix} u$$

$$y = (1 \;\; 0 \;\; 0)\begin{pmatrix} x_1 \\ x_2 \\ x_3 \end{pmatrix}$$

2.12. For $x_1 = v$, $x_2 = \dot{v}$, $x_3 = \ddot{v}$, $x_4 = \dddot{v}$, $x_5 = w$, $x_6 = \dot{w}$ the initial conditions result immediately and

$$\mathbf{x} = \begin{pmatrix} 0 & 1 & 0 & 0 & 0 & 0 \\ 0 & 0 & 1 & 0 & 0 & 0 \\ 0 & 0 & 0 & 1 & 0 & 0 \\ 0 & -500 & -105 & -6 & 0 & 0 \\ 0 & 0 & 0 & 0 & 0 & 1 \\ 100 & 1 & 1 & 0 & -100 & -2 \end{pmatrix}\mathbf{x} + \begin{pmatrix} 0 \\ 0 \\ 0 \\ 500 \\ 0 \\ 0 \end{pmatrix} u$$

2.13. The time-varying flow diagram cannot be turned backwards to obtain another canonical form because the order of multiplication by a time-varying coefficient and an integration cannot be interchanged.

2.14. (i)

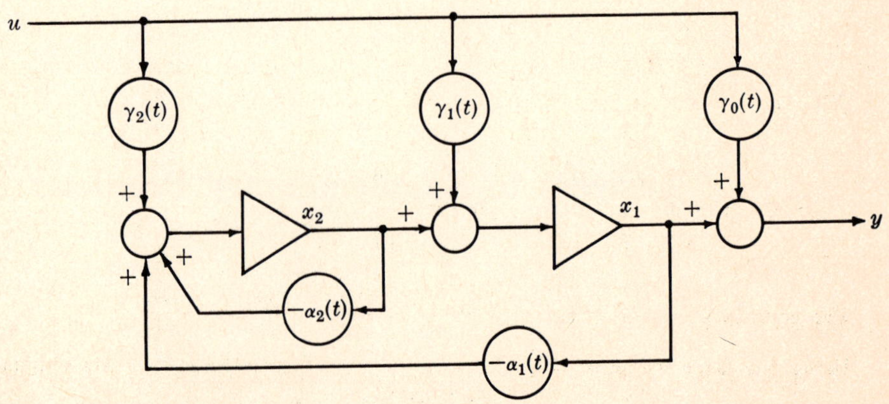

Fig. 2-23

(ii) The values of γ_0, γ_1 and γ_2 are given by equations (*2.33*), (*2.34*) and (*2.35*).

(iii) $x_1(0) = y_0 - \gamma_0(0)\,u(0) \qquad x_2(0) = \dot{y}_0 - (\dot{\gamma}_0(0) + \gamma_1(0))\,u(0) - \gamma_0(0)\,\dot{u}(0)$

2.16. $$\frac{d}{dt}\begin{pmatrix} i \\ \theta \\ d\theta/dt \end{pmatrix} = \begin{pmatrix} -RL^{-1} & 0 & -K_L L^{-1} \\ 0 & 1 & 0 \\ K_t J^{-1} & 0 & -BJ^{-1} \end{pmatrix}\begin{pmatrix} i \\ \theta \\ d\theta/dt \end{pmatrix} + \begin{pmatrix} L^{-1} \\ 0 \\ 0 \end{pmatrix} V \qquad \theta = (0\ 1\ 0)\begin{pmatrix} i \\ \theta \\ d\theta/dt \end{pmatrix}$$

2.17. $$\frac{d}{dt}\begin{pmatrix} n \\ C_1 \\ \vdots \\ C_6 \end{pmatrix} = \begin{pmatrix} (\rho(t)-\beta)l^{-1} & \lambda_1 & \cdots & \lambda_6 \\ \beta_1 & -\lambda_1 & \cdots & 0 \\ \cdots & \cdots & \cdots & \cdots \\ \beta_6 & 0 & \cdots & -\lambda_6 \end{pmatrix}\begin{pmatrix} n \\ C_1 \\ \vdots \\ C_6 \end{pmatrix}$$

2.18. (i) $$\mathbf{A} = \begin{pmatrix} 0 & 1 & 0 & 0 \\ 0 & K_2K_6 & -(K_1+K_2) & 0 \\ 0 & 0 & 0 & 1 \\ 0 & K_4K_6 & -K_4 & 0 \end{pmatrix}$$

$$\mathbf{B} = \begin{pmatrix} 0 \\ -K_3 \\ 0 \\ -K_5 \end{pmatrix} \qquad \mathbf{C} = \begin{pmatrix} 0 & -K_6 & 1 & 0 \\ 0 & -K_6K_\alpha & K_\alpha & 0 \end{pmatrix} \qquad \mathbf{D} = \begin{pmatrix} 0 \\ K_\beta \end{pmatrix}$$

(ii) $$\mathbf{A} = \begin{pmatrix} K_2K_6 & -(K_1+K_2) & 0 \\ 0 & 0 & 1 \\ K_4K_6 & -K_4 & 0 \end{pmatrix} \qquad \mathbf{B} = \begin{pmatrix} -K_3 \\ 0 \\ -K_5 \end{pmatrix}$$

$$\mathbf{C} = \begin{pmatrix} -K_6 & 1 & 0 \\ -K_4K_6 & K_\alpha & 0 \end{pmatrix} \qquad \mathbf{D} = \begin{pmatrix} 0 \\ K_\beta \end{pmatrix}$$

(iii) $$\mathbf{A} = \begin{pmatrix} 0 & -K_6^{-1} & K_6^{-1} & 0 \\ 0 & K_2K_6 & K_1K_6 & 1 \\ 0 & 0 & 0 & 1 \\ 0 & -K_4 & 0 & 0 \end{pmatrix} \qquad \mathbf{B} = \begin{pmatrix} 0 \\ K_3K_6 \\ 0 \\ -K_5 \end{pmatrix}$$

$$\mathbf{C} = \begin{pmatrix} 0 & 1 & 0 & 0 \\ 0 & K_\alpha & 0 & 0 \end{pmatrix} \qquad \mathbf{D} = \begin{pmatrix} 0 \\ K_\beta \end{pmatrix}$$

2.19. (i) $$\mathbf{A} = \begin{pmatrix} 0 & -C_1^{-1} & 0 \\ L_1^{-1} & 0 & -L_2^{-1} \\ 0 & C_1^{-1} & 0 \end{pmatrix} \qquad \mathbf{B} = \begin{pmatrix} 1 & 0 \\ 0 & 0 \\ 0 & -1 \end{pmatrix}$$

$$\mathbf{C} = \begin{pmatrix} L_1^{-1} & 0 & 0 \\ 0 & C_1^{-1} & 0 \\ 0 & 0 & L_2^{-1} \end{pmatrix} \qquad \mathbf{D} = \begin{pmatrix} 0 & 0 \\ 0 & 0 \\ 0 & 0 \end{pmatrix}$$

(ii) $$\mathbf{A} = \begin{pmatrix} -\dot{L}_1L_1^{-1} & -L_1^{-1} & 0 \\ C_1^{-1} & -\dot{C}_1C_1^{-1} & -C_1^{-1} \\ 0 & L_2^{-1} & -\dot{L}_2L_2^{-1} \end{pmatrix} \qquad \mathbf{B} = \begin{pmatrix} L_1^{-1} & 0 \\ 0 & 0 \\ 0 & -L_2^{-1} \end{pmatrix}$$

$$\mathbf{C} = \begin{pmatrix} 1 & 0 & 0 \\ 0 & 1 & 0 \\ 0 & 0 & 1 \end{pmatrix} \qquad \mathbf{D} = \begin{pmatrix} 0 & 0 \\ 0 & 0 \\ 0 & 0 \end{pmatrix}$$

2.20. The equations of motion are $$\frac{dq_i}{dt} = \frac{\partial H}{\partial p_i} \qquad \frac{dp_i}{dt} = -\frac{\partial H}{\partial q_i}$$

Using the state vector $\mathbf{x} = (q_1 \ \ldots \ q_n \ p_1 \ \ldots \ p_n)^T$, the matrix state equations are

$$\dot{\mathbf{x}} = \begin{pmatrix} \mathbf{0} & \mathbf{T} \\ -\mathbf{V} & \mathbf{0} \end{pmatrix}\mathbf{x}$$

Chapter 3

Elementary Matrix Theory

3.1 INTRODUCTION

This and the following chapter present the minimum amount of matrix theory needed to comprehend the material in the rest of the book. It is recommended that even those well versed in matrix theory glance over the text of these next two chapters, if for no other reason than to become familiar with the notation and philosophy. For those not so well versed, the presentation is terse and oriented towards later use. For a more comprehensive presentation of matrix theory, we suggest study of textbooks solely concerned with the subject.

3.2 BASIC DEFINITIONS

Definition 3.1: A *matrix,* denoted by a capital boldfaced letter, such as $\mathbf{A}$ or $\mathbf{\Phi}$, or by the notation $\{\mathbf{a}_i\}$ or $\{a_{ij}\}$, is a rectangular array of elements that are members of a field or ring, such as real numbers, complex numbers, polynomials, functions, etc. The kind of element will usually be clear from the context.

Example 3.1.

An example of a matrix is

$$\mathbf{A} = \begin{pmatrix} 0 & 2 & j \\ \pi & x^2 & \sin t \end{pmatrix} = \begin{pmatrix} a_{11} & a_{12} & a_{13} \\ a_{21} & a_{22} & a_{23} \end{pmatrix} = \{a_{ij}\}$$

Definition 3.2: A *row* of a matrix is the set of all elements through which one horizontal line can be drawn.

Definition 3.3: A *column* of a matrix is the set of all elements through which one vertical line can be drawn.

Example 3.2.

The rows of the matrix of Example 3.1 are $(0 \;\; 2 \;\; j)$ and $(\pi \;\; x^2 \;\; \sin t)$. The columns of this matrix are $\begin{pmatrix} 0 \\ \pi \end{pmatrix}$, $\begin{pmatrix} 2 \\ x^2 \end{pmatrix}$ and $\begin{pmatrix} j \\ \sin t \end{pmatrix}$.

Definition 3.4: A *square matrix* has the same number of rows and columns.

Definition 3.5: The *order* of a matrix is $m \times n$, read m by n, if it has m rows and n columns.

Definition 3.6: A *scalar,* denoted by a letter that is not in boldface type, is a 1×1 matrix. In other words, it is one element. When part of a matrix $\mathbf{A}$, the notation a_{ij} means the particular element in the ith row and jth column.

Definition 3.7: A *vector,* denoted by a lower case boldfaced letter, such as $\mathbf{a}$, or with its contents displayed in braces, such as $\{a_i\}$, is a matrix with only one row or only one column. Usually $\mathbf{a}$ denotes a column vector, and $\mathbf{a}^T$ a row vector.

Definition 3.8: The *diagonal* of a square matrix is the set of all elements a_{ij} of the matrix in which $i = j$. In other words, it is the set of all elements of a square matrix through which can pass a diagonal line drawn from the upper left hand corner to the lower right hand corner.

Example 3.3.

Given the matrix

$$\mathbf{B} = \{b_{ij}\} = \begin{pmatrix} b_{11} & b_{12} & b_{13} \\ b_{21} & b_{22} & b_{23} \\ b_{31} & b_{32} & b_{33} \end{pmatrix}$$

The diagonal is the set of elements through which the solid line is drawn, b_{11}, b_{22}, b_{33}, and not those sets determined by the dashed lines.

Definition 3.9: The *trace* of a square matrix $\mathbf{A}$, denoted $\operatorname{tr}\mathbf{A}$, is the sum of all elements on the diagonal of $\mathbf{A}$.

$$\operatorname{tr}\mathbf{A} = \sum_{i=1}^{n} a_{ii}$$

3.3 BASIC OPERATIONS

Definition 3.10: Two matrices are *equal* if their corresponding elements are equal. $\mathbf{A} = \mathbf{B}$ means $a_{ij} = b_{ij}$ for all i and j. The matrices must be of the same order.

Definition 3.11: *Matrix addition* is performed by adding corresponding elements. $\mathbf{A} + \mathbf{B} = \mathbf{C}$ means $a_{ij} + b_{ij} = c_{ij}$, for all i and j. *Matrix subtraction* is analogously defined. The matrices must be of the same order.

Definition 3.12: *Matrix differentiation* or *integration* means differentiate or integrate each element, since differentiation and integration are merely limiting operations on sums, which have been defined by Definition 3.11.

Addition of matrices is associative and commutative, i.e. $\mathbf{A} + (\mathbf{B} + \mathbf{C}) = (\mathbf{A} + \mathbf{B}) + \mathbf{C}$ and $\mathbf{A} + \mathbf{B} = \mathbf{B} + \mathbf{A}$. In this and all the foregoing respects, the operations on matrices have been natural extensions of the corresponding operations on scalars and vectors. However, recall there is no multiplication of two vectors, and the closest approximations are the dot (scalar) product and the cross product. Matrix multiplication is an extension of the dot product $\mathbf{a} \cdot \mathbf{b} = a_1 b_1 + a_2 b_2 + \cdots + a_n b_n$, a scalar.

Definition 3.13: To perform *matrix multiplication,* the element c_{ij} of the product matrix $\mathbf{C}$ is found by taking the dot product of the ith row of the left matrix $\mathbf{A}$ and the jth column of the right matrix $\mathbf{B}$, where $\mathbf{C} = \mathbf{AB}$, so that

$$c_{ij} = \sum_{k=1}^{n} a_{ik} b_{kj}.$$

Note that this definition requires the left matrix ($\mathbf{A}$) to have the same number of columns as the right matrix ($\mathbf{B}$) has rows. In this case the matrices $\mathbf{A}$ and $\mathbf{B}$ are said to be compatible. It is undefined in other cases, excepting when one of the matrices is 1×1, i.e. a scalar. In this case each of the elements is multiplied by the scalar, e.g. $\alpha\mathbf{B} = \{\alpha b_{ij}\}$ for all i and j.

Example 3.4.

The vector equation $\mathbf{y} = \mathbf{Ax}$, when $\mathbf{y}$ and $\mathbf{x}$ are 2×1 matrices, i.e. column vectors, is

$$\begin{pmatrix} y_1 \\ y_2 \end{pmatrix} = \begin{pmatrix} a_{11} & a_{12} \\ a_{21} & a_{22} \end{pmatrix} \begin{pmatrix} x_1 \\ x_2 \end{pmatrix}$$

where $y_i = \sum_{k=1}^{2} a_{ik}x_k$ for $i = 1$ and 2. But suppose $\mathbf{x} = \mathbf{Bz}$, so that $x_k = \sum_{j=1}^{2} b_{kj}z_j$. Then

$$y_i = \sum_{k=1}^{2} a_{ik}\left(\sum_{j=1}^{2} b_{kj}z_j\right) = \sum_{j=1}^{2} c_{ij}z_j$$

so that $\mathbf{y} = \mathbf{A}(\mathbf{Bz}) = \mathbf{Cz}$, where $\mathbf{AB} = \mathbf{C}$.

Example 3.4 can be extended to show $(\mathbf{AB})\mathbf{C} = \mathbf{A}(\mathbf{BC})$, i.e. matrix multiplication is associative. But, in general, matrix multiplication is *not* commutative, $\mathbf{AB} \neq \mathbf{BA}$. Also, there is *no* matrix division.

Example 3.5.

To show $\mathbf{AB} \neq \mathbf{BA}$, consider

$$\mathbf{AB} = \begin{pmatrix} 1 & 2 \\ 3 & 0 \end{pmatrix}\begin{pmatrix} 1 & 1 \\ 2 & 2 \end{pmatrix} = \begin{pmatrix} 5 & 5 \\ 3 & 3 \end{pmatrix} = \mathbf{D}$$

$$\mathbf{BA} = \begin{pmatrix} 1 & 1 \\ 2 & 2 \end{pmatrix}\begin{pmatrix} 1 & 2 \\ 3 & 0 \end{pmatrix} = \begin{pmatrix} 4 & 2 \\ 8 & 4 \end{pmatrix} = \mathbf{C}$$

and note $\mathbf{D} \neq \mathbf{C}$. Furthermore, to show there is no division, consider

$$\mathbf{BF} = \begin{pmatrix} 1 & 1 \\ 2 & 2 \end{pmatrix}\begin{pmatrix} 0 & 0 \\ 4 & 2 \end{pmatrix} = \begin{pmatrix} 4 & 2 \\ 8 & 4 \end{pmatrix} = \mathbf{C}$$

Since $\mathbf{BA} = \mathbf{BF} = \mathbf{C}$, "division" by $\mathbf{B}$ would imply $\mathbf{A} = \mathbf{F}$, which is certainly not true.

Suppose we have the *vector* equation $\mathbf{Ax} = \mathbf{Bx}$, where $\mathbf{A}$ and $\mathbf{B}$ are $n \times n$ matrices. It can be concluded that $\mathbf{A} = \mathbf{B}$ only if $\mathbf{x}$ is an *arbitrary* n-vector. For, if $\mathbf{x}$ is arbitrary, we may choose $\mathbf{x}$ successively as $\mathbf{e}_1, \mathbf{e}_2, \ldots, \mathbf{e}_n$ and find that the column vectors $\mathbf{a}_1 = \mathbf{b}_1$, $\mathbf{a}_2 = \mathbf{b}_2, \ldots, \mathbf{a}_n = \mathbf{b}_n$. Here $\mathbf{e}_i$ are unit vectors, defined after Definition 3.17, page 41.

Definition 3.14: To *partition* a matrix, draw a vertical and/or horizontal line between two rows or columns and consider the subset of elements formed as individual matrices, called *submatrices*.

As long as the submatrices are compatible, i.e. have the correct order so that addition and multiplication are possible, the submatrices can be treated as elements in the basic operations.

Example 3.6.

A 3×3 matrix $\mathbf{A}$ can be partitioned into a 2×2 matrix $\mathbf{A}_{11}$, a 1×2 matrix $\mathbf{A}_{21}$, a 2×1 matrix $\mathbf{A}_{12}$, and a 1×1 matrix $\mathbf{A}_{22}$.

$$\mathbf{A} = \left(\begin{array}{cc|c} a_{11} & a_{12} & a_{13} \\ a_{21} & a_{22} & a_{23} \\ \hline a_{31} & a_{32} & a_{33} \end{array}\right) = \left(\begin{array}{c|c} \mathbf{A}_{11} & \mathbf{A}_{12} \\ \hline \mathbf{A}_{21} & \mathbf{A}_{22} \end{array}\right)$$

A similarly partitioned 3×3 matrix $\mathbf{B}$ adds as

$$\mathbf{A} + \mathbf{B} = \left(\begin{array}{c|c} \mathbf{A}_{11} + \mathbf{B}_{11} & \mathbf{A}_{12} + \mathbf{B}_{12} \\ \hline \mathbf{A}_{21} + \mathbf{B}_{21} & \mathbf{A}_{22} + \mathbf{B}_{22} \end{array}\right)$$

and multiplies as

$$\mathbf{AB} = \left(\begin{array}{c|c} \mathbf{A}_{11}\mathbf{B}_{11} + \mathbf{A}_{12}\mathbf{B}_{21} & \mathbf{A}_{11}\mathbf{B}_{12} + \mathbf{A}_{12}\mathbf{B}_{22} \\ \hline \mathbf{A}_{21}\mathbf{B}_{11} + \mathbf{A}_{22}\mathbf{B}_{21} & \mathbf{A}_{21}\mathbf{B}_{12} + \mathbf{A}_{22}\mathbf{B}_{22} \end{array}\right)$$

Facility with partitioned matrix operations will often save time and give insight.

3.4 SPECIAL MATRICES

Definition 3.15: The *zero matrix,* denoted **0**, is a matrix whose elements are all zeros.

Definition 3.16: A *diagonal matrix* is a square matrix whose off-diagonal elements are all zeros.

Definition 3.17: The *unit matrix,* denoted **I**, is a diagonal matrix whose diagonal elements are all ones.

Sometimes the order of **I** is indicated by a subscript, e.g. $\mathbf{I}_n$ is an $n \times n$ matrix. *Unit vectors,* denoted $\mathbf{e}_i$, have a one as the ith element and all other elements zero, so that $\mathbf{I}_m = (\mathbf{e}_1 \mid \mathbf{e}_2 \mid \ldots \mid \mathbf{e}_m)$.

Note $\mathbf{IA} = \mathbf{AI} = \mathbf{A}$, where **A** is any compatible matrix.

Definition 3.18: An *upper triangular* matrix has all zeros below the diagonal, and a *lower triangular* matrix has all zeros above the diagonal. The diagonal elements need not be zero.

An upper triangular matrix added to or multiplied by an upper triangular matrix results in an upper triangular matrix, and similarly for lower triangular matrices.

Definition 3.19: A *transpose matrix,* denoted $\mathbf{A}^T$, is the matrix resulting from an interchange of rows and columns of a given matrix **A**. If $\mathbf{A} = \{a_{ij}\}$, then $\mathbf{A}^T = \{a_{ji}\}$, so that the element in the ith row and jth column of **A** becomes the element in the jth row and ith column of $\mathbf{A}^T$.

Definition 3.20: The *complex conjugate transpose* matrix, denoted $\mathbf{A}^\dagger$, is the matrix whose elements are complex conjugates of the elements of $\mathbf{A}^T$.

Note $(\mathbf{AB})^T = \mathbf{B}^T\mathbf{A}^T$ and $(\mathbf{AB})^\dagger = \mathbf{B}^\dagger\mathbf{A}^\dagger$.

Definition 3.21: A matrix **A** is *symmetric* if $\mathbf{A} = \mathbf{A}^T$.

Definition 3.22: A matrix **A** is *Hermitian* if $\mathbf{A} = \mathbf{A}^\dagger$.

Definition 3.23: A matrix **A** is *normal* if $\mathbf{A}^\dagger\mathbf{A} = \mathbf{A}\mathbf{A}^\dagger$.

Definition 3.24: A matrix **A** is *skew-symmetric* if $\mathbf{A} = -\mathbf{A}^T$.

Note that for the different cases: Hermitian $\mathbf{A} = \mathbf{A}^\dagger$, skew Hermitian $\mathbf{A} = -\mathbf{A}^\dagger$, real symmetric $\mathbf{A}^T = \mathbf{A}$, real skew-symmetric $\mathbf{A} = -\mathbf{A}^T$, unitary $\mathbf{A}^\dagger\mathbf{A} = \mathbf{I}$, diagonal **D**, or orthogonal $\mathbf{A}^\dagger\mathbf{A} = \mathbf{D}$, the matrix is normal.

3.5 DETERMINANTS AND INVERSE MATRICES

Definition 3.25: The *determinant* of an $n \times n$ (square) matrix $\{a_{ij}\}$ is the sum of the signed products of all possible combinations of n elements, where each element is taken from a different row and column.

$$\det \mathbf{A} \quad = \quad \sum^{n!} (-1)^p a_{1p_1} a_{2p_2} \cdots a_{np_n} \qquad (3.1)$$

Here $p_1, p_2, \ldots, p_n$ is a permutation of $1, 2, \ldots, n$, and the sum is taken over all possible permutations. A permutation is a rearrangement of $1, 2, \ldots, n$ into some other order, such as $2, n, \ldots, 1$, that is obtained by successive transpositions. A transposition is the interchange of places of two (and only two) numbers in the list $1, 2, \ldots, n$. The exponent p of -1 in (*3.1*) is the number of transpositions it takes to go from the natural order of $1, 2, \ldots, n$ to $p_1, p_2, \ldots, p_n$. There are $n!$ possible permutations of n numbers, so each determinant is the sum of $n!$ products.

Example 3.7.

To find the determinant of a 3×3 matrix, all possible permutations of $1, 2, 3$ must be found. Performing one transposition at a time, the following table can be formed.

p	p_1, p_2, p_3
0	1, 2, 3
1	3, 2, 1
2	2, 3, 1
3	2, 1, 3
4	3, 1, 2
5	1, 3, 2

This table is not unique in that for $p = 1$, possible entries are also $1, 3, 2$ and $2, 1, 3$. However, these entries can result only from an odd p, so that the sign of each product in a determinant is unique. Since there are $3! = 6$ terms, all possible permutations are given in the table. Notice at each step only two numbers are interchanged. Using the table and (*3.1*) gives

$$\det \mathbf{A} = (-1)^0 a_{11}a_{22}a_{33} + (-1)^1 a_{12}a_{21}a_{33} + (-1)^2 a_{12}a_{23}a_{31} + (-1)^3 a_{13}a_{22}a_{31} + (-1)^4 a_{13}a_{21}a_{32} + (-1)^5 a_{11}a_{23}a_{32}$$

***Theorem 3.1*:** Let $\mathbf{A}$ be an $n \times n$ matrix. Then $\det(\mathbf{A}^T) = \det(\mathbf{A})$.

Proof is given in Problem 3.3, page 59.

***Theorem 3.2*:** Given two $n \times n$ matrices $\mathbf{A}$ and $\mathbf{B}$. Then $\det(\mathbf{AB}) = (\det \mathbf{A})(\det \mathbf{B})$.

Proof of this theorem is most easily given using exterior products, defined in Section 3.13, page 56. The proof is presented in Problem 3.15, page 65.

Definition 3.26: *Elementary row* (*or column*) *operations* are:

(i) Interchange of two rows (or columns).

(ii) Multiplication of a row (or column) by a scalar.

(iii) Adding a scalar times a row (or column) to another row (column).

To perform an elementary row (column) operation on an $n \times m$ matrix $\mathbf{A}$, calculate the product $\mathbf{EA}$ ($\mathbf{AE}$ for columns), where $\mathbf{E}$ is an $n \times n (m \times m)$ matrix found by performing the elementary operation on the $n \times n (m \times m)$ unit matrix $\mathbf{I}$. The matrix $\mathbf{E}$ is called an *elementary matrix*.

Example 3.8.

Consider the 2×2 matrix $\mathbf{A} = \{a_{ij}\}$. To interchange two rows, interchange the two rows in $\mathbf{I}$ to obtain $\mathbf{E}$.

$$\mathbf{EA} = \begin{pmatrix} 0 & 1 \\ 1 & 0 \end{pmatrix} \begin{pmatrix} a_{11} & a_{12} \\ a_{21} & a_{22} \end{pmatrix} = \begin{pmatrix} a_{21} & a_{22} \\ a_{11} & a_{12} \end{pmatrix}$$

To add 6 times the second column to the first, multiply

$$\mathbf{AE} = \begin{pmatrix} a_{11} & a_{12} \\ a_{21} & a_{22} \end{pmatrix} \begin{pmatrix} 1 & 0 \\ 6 & 1 \end{pmatrix} = \begin{pmatrix} a_{11} + 6a_{12} & a_{12} \\ a_{21} + 6a_{22} & a_{22} \end{pmatrix}$$

Using Theorem 3.2 on the product $\mathbf{AE}$ or $\mathbf{EA}$, it can be found that (i) interchange of two rows or columns changes the sign of a determinant, i.e. $\det(\mathbf{AE}) = -\det \mathbf{A}$, (ii) multiplication of a row or column by a scalar α multiplies the determinant by α, i.e. $\det(\mathbf{AE}) = \alpha \det \mathbf{A}$, and (iii) adding a scalar times a row to another row does not change the value of the determinant, i.e. $\det(\mathbf{AE}) = \det \mathbf{A}$.

Taking the value of α in (ii) to be zero, it can be seen that a matrix containing a row or column of zeros has a zero determinant. Furthermore, if two rows or columns are identical or multiples of one another, then use of (iii) will give a zero row or column, so that the determinant is zero.

Each elementary matrix $\mathbf{E}$ always has an inverse $\mathbf{E}^{-1}$, found by undoing the row or column operation of $\mathbf{I}$. Of course an exception is $\alpha = 0$ in (ii).

Example 3.9.

The inverse of $\mathbf{E} = \begin{pmatrix} 0 & 1 \\ 1 & 0 \end{pmatrix}$ is $\mathbf{E}^{-1} = \begin{pmatrix} 0 & 1 \\ 1 & 0 \end{pmatrix}$. The inverse of $\mathbf{E} = \begin{pmatrix} 1 & 0 \\ 6 & 1 \end{pmatrix}$ is $\mathbf{E}^{-1} = \begin{pmatrix} 1 & 0 \\ -6 & 1 \end{pmatrix}$.

Definition 3.27: The determinant of the matrix formed by deleting the ith row and the jth column of the matrix $\mathbf{A}$ is the *minor* of the element a_{ij}, denoted $\det \mathbf{M}_{ij}$. The *cofactor* $c_{ij} = (-1)^{i+j} \det \mathbf{M}_{ij}$.

Example 3.10.

The minor of a_{22} of a 3×3 matrix $\mathbf{A}$ is $\det \mathbf{M}_{22} = a_{11}a_{33} - a_{13}a_{31}$. The cofactor $c_{22} = (-1)^4 \det \mathbf{M}_{22} = \det \mathbf{M}_{22}$.

Theorem 3.3: The Laplace expansion of the determinant of an $n \times n$ matrix $\mathbf{A}$ is $\det \mathbf{A} = \sum_{i=1}^{n} a_{ij}c_{ij}$ for any column j or $\det \mathbf{A} = \sum_{j=1}^{n} a_{ij}c_{ij}$ for any row i.

Proof of this theorem is presented as part of the proof of Theorem 3.21, page 57.

Example 3.11.

The Laplace expansion of a 3×3 matrix $\mathbf{A}$ about the second column is

$$\begin{aligned} \det \mathbf{A} &= a_{12}c_{12} + a_{22}c_{22} + a_{32}c_{32} \\ &= -a_{12}(a_{21}a_{33} - a_{23}a_{31}) + a_{22}(a_{11}a_{33} - a_{13}a_{31}) - a_{32}(a_{11}a_{23} - a_{13}a_{21}) \end{aligned}$$

Corollary 3.4: The determinant of a triangular $n \times n$ matrix equals the product of the diagonal elements.

Proof is by induction. The corollary is obviously true for $n = 1$. For arbitrary n, the Laplace expansion about the nth row (or column) of an $n \times n$ upper (or lower) triangular matrix gives $\det \mathbf{A} = a_{nn}c_{nn}$. By assumption, $c_{nn} = a_{11}a_{22} \cdots a_{n-1,n-1}$, proving the corollary.

Explanation of the induction method: First the hypothesis is shown true for $n = n_0$, where n_0 is a fixed number. ($n_0 = 1$ in the foregoing proof.) Then assume the hypothesis is true for an arbitrary n and show it is true for $n+1$. Let $n = n_0$, for which it is known true, so that it is true for $n_0 + 1$. Then let $n = n_0 + 1$, so that it is true for $n_0 + 2$, etc. In this manner the hypothesis is shown true for all $n \geqq n_0$.

Corollary 3.5: The determinant of a diagonal matrix equals the product of the diagonal elements.

This is true because a diagonal matrix is also a triangular matrix.

The Laplace expansion for a matrix whose kth row equals the ith row is $\det \mathbf{A} = \sum_{j=1}^{n} a_{kj}c_{ij}$ for $k \neq i$, and for a matrix whose kth column equals the jth column the Laplace expansion is $\det \mathbf{A} = \sum_{i=1}^{n} a_{ik}c_{ij}$. But these determinants are zero since $\mathbf{A}$ has two identical rows or columns.

Definition 3.28: The *Kronecker delta* $\delta_{ij} = 1$ if $i = j$ and $\delta_{ij} = 0$ if $i \neq j$.

Using the Kronecker notation,

$$\sum_{i=1}^{n} a_{ik}c_{ij} \quad = \quad \sum_{l=1}^{n} a_{kl}c_{jl} \quad = \quad \delta_{kj} \det \mathbf{A} \tag{3.2}$$

Definition 3.29: The *adjugate matrix* of $\mathbf{A}$ is $\operatorname{adj} \mathbf{A} = \{c_{ij}\}^T$, the transpose of the matrix of cofactors of $\mathbf{A}$.

The adjugate is sometimes called "adjoint", but this term is saved for Definition 5.2. Then (*3.2*) can be written in matrix notation as

$$\mathbf{A} \operatorname{adj} \mathbf{A} \;=\; \mathbf{I} \det \mathbf{A} \tag{3.3}$$

Definition 3.30: An $m \times n$ matrix $\mathbf{B}$ is a *left inverse* of the $n \times m$ matrix $\mathbf{A}$ if $\mathbf{BA} = \mathbf{I}_m$, and an $m \times n$ matrix $\mathbf{C}$ is a *right inverse* if $\mathbf{AC} = \mathbf{I}_n$. $\mathbf{A}$ is said to be *nonsingular* if it has both a left and a right inverse.

If $\mathbf{A}$ has both a left inverse $\mathbf{B}$ and a right inverse $\mathbf{C}$, $\mathbf{C} = \mathbf{IC} = \mathbf{BAC} = \mathbf{BI} = \mathbf{B}$. Since $\mathbf{BA} = \mathbf{I}_m$ and $\mathbf{AC} = \mathbf{I}_n$, and if $\mathbf{C} = \mathbf{B}$, then $\mathbf{A}$ must be square. Furthermore suppose a nonsingular matrix $\mathbf{A}$ has two inverses $\mathbf{G}$ and $\mathbf{H}$. Then $\mathbf{G} = \mathbf{GI} = \mathbf{GAH} = \mathbf{IH} = \mathbf{H}$ so that a nonsingular matrix $\mathbf{A}$ must be square and have a unique inverse denoted $\mathbf{A}^{-1}$ such that $\mathbf{A}^{-1}\mathbf{A} = \mathbf{AA}^{-1} = \mathbf{I}$. Finally, use of Theorem 3.2 gives $\det \mathbf{A} \det \mathbf{A}^{-1} = \det \mathbf{I} = 1$, so that if $\det \mathbf{A} = 0$, $\mathbf{A}$ can have no inverse.

Theorem 3.6: **Cramer's rule.** Given an $n \times n$ (square) matrix $\mathbf{A}$ such that $\det \mathbf{A} \neq 0$. Then

$$\mathbf{A}^{-1} \quad = \quad \frac{1}{\det \mathbf{A}} \operatorname{adj} \mathbf{A}$$

The proof follows immediately from equation (*3.3*).

Example 3.12.

The inverse of a 2×2 matrix $\mathbf{A}$ is

$$\mathbf{A}^{-1} \quad = \quad \frac{1}{a_{11}a_{22} - a_{12}a_{21}} \begin{pmatrix} a_{22} & -a_{12} \\ -a_{21} & a_{11} \end{pmatrix}$$

Another and usually faster means of obtaining the inverse of nonsingular matrix $\mathbf{A}$ is to use elementary row operations to reduce $\mathbf{A}$ in the partitioned matrix $\mathbf{A} | \mathbf{I}$ to the unit matrix. To reduce $\mathbf{A}$ to a unit matrix, interchange rows until $a_{11} \neq 0$. Denote the interchange by $\mathbf{E}_1$. Divide the first row by a_{11}, denoting this row operation by $\mathbf{E}_2$. Then $\mathbf{E}_2\mathbf{E}_1\mathbf{A}$ has a one in the upper left hand corner. Multiply the first row by a_{i1} and subtract it from the ith row for $i = 2, 3, \ldots, n$, denoting this operation by $\mathbf{E}_3$. The first column of $\mathbf{E}_3\mathbf{E}_2\mathbf{E}_1\mathbf{A}$ is then the unit vector $\mathbf{e}_1$. Next, interchange rows $\mathbf{E}_3\mathbf{E}_2\mathbf{E}_1\mathbf{A}$ until the element in the second row and column is nonzero. Then divide the second row by this element and subtract from all other rows until the unit vector $\mathbf{e}_2$ is obtained. Continue in this manner until $\mathbf{E}_m \cdots \mathbf{E}_1\mathbf{A} = \mathbf{I}$. Then $\mathbf{E}_m \cdots \mathbf{E}_1 = \mathbf{A}^{-1}$, and operating on $\mathbf{I}$ by the same row operations will produce $\mathbf{A}^{-1}$. Furthermore, $\det \mathbf{A}^{-1} = \det \mathbf{E}_1 \det \mathbf{E}_2 \cdots \det \mathbf{E}_m$ from Theorem 3.2.

Example 3.13.

To find the inverse of $\begin{pmatrix} 0 & 1 \\ -1 & 1 \end{pmatrix}$, adjoin the unit matrix to obtain

$$\left(\begin{array}{cc|cc} 0 & 1 & 1 & 0 \\ -1 & 1 & 0 & 1 \end{array}\right)$$

Interchange rows ($\det \mathbf{E}_1 = -1$):

$$\left(\begin{array}{cc|cc} -1 & 1 & 0 & 1 \\ 0 & 1 & 1 & 0 \end{array}\right)$$

Divide the first row by -1 ($\det \mathbf{E}_2 = -1$):

$$\left(\begin{array}{cc|cc} 1 & -1 & 0 & -1 \\ 0 & 1 & 1 & 0 \end{array}\right)$$

It turns out the first column is already $\mathbf{e}_1$, and all that is necessary to reduce this matrix to $\mathbf{I}$ is to add the second row to the first ($\det \mathbf{E}_3 = 1$).

$$\left(\begin{array}{cc|cc} 1 & 0 & 1 & -1 \\ 0 & 1 & 1 & 0 \end{array}\right)$$

The matrix to the right of the partition line is the inverse of the original matrix, which has a determinant equal to $[(-1)(-1)(1)]^{-1} = 1$.

3.6 VECTOR SPACES

Not all matrix equations have solutions, and some have more than one.

Example 3.14.

(*a*) The matrix equation

$$\begin{pmatrix} 1 \\ 2 \end{pmatrix} (\xi) = \begin{pmatrix} 0 \\ 1 \end{pmatrix}$$

has no solution because no ξ exists that satisfies the two equations written out as

$$\begin{aligned} \xi &= 0 \\ 2\xi &= 1 \end{aligned}$$

(*b*) The matrix equation

$$\begin{pmatrix} 0 \\ 0 \end{pmatrix} (\xi) = \begin{pmatrix} 0 \\ 0 \end{pmatrix}$$

is satisfied for any ξ.

To find the necessary and sufficient conditions for existence and uniqueness of solutions of matrix equations, it is necessary to extend some geometrical ideas. These ideas are apparent for vectors of 2 and 3 dimensions, and can be extended to include vectors having an arbitrary number of elements.

Consider the vector $(2 \;\; 3)$. Since the elements are real, they can be represented as points in a plane. Let $(\xi_1 \;\; \xi_2) = (2 \;\; 3)$. Then this vector can be represented as the point in the ξ_1, ξ_2 plane shown in Fig. 3-1.

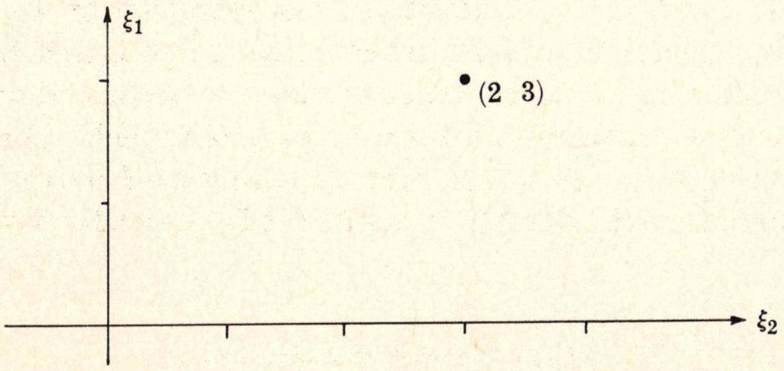

Fig. 3-1. Representation of (2 3) in the Plane

If the real vector were (1 2 3), it could be represented as a point in $(\xi_1\ \xi_2\ \xi_3)$ space by drawing the ξ_3 axis out of the page. Higher dimensions, such as (1 2 3 4), are harder to draw but can be imagined. In fact some vectors have an infinite number of elements. This can be included in the discussion, as can the case where the elements of the vector are other than real numbers.

Definition 3.31: Let $\mathcal{V}_n$ be the *set of all vectors with n components.* Let $\mathbf{a}_1$ and $\mathbf{a}_2$ be vectors having n components, i.e. $\mathbf{a}_1$ and $\mathbf{a}_2$ are in $\mathcal{V}_n$. This is denoted $\mathbf{a}_1 \in \mathcal{V}_n$, $\mathbf{a}_2 \in \mathcal{V}_n$. Given arbitrary scalars β_1 and β_2, it is seen that $(\beta_1\mathbf{a}_1 + \beta_2\mathbf{a}_2) \in \mathcal{V}_n$, i.e. an arbitrary linear combination of $\mathbf{a}_1$ and $\mathbf{a}_2$ is in $\mathcal{V}_n$.

$\mathcal{V}_1$ is an infinite line and $\mathcal{V}_2$ is an infinite plane. To represent diagrammatically these and, in general, $\mathcal{V}_n$ for any n, one uses the area enclosed by a closed curve. Let $\mathcal{U}$ be a set of vectors in $\mathcal{V}_n$. This can be represented as shown in Fig. 3-2.

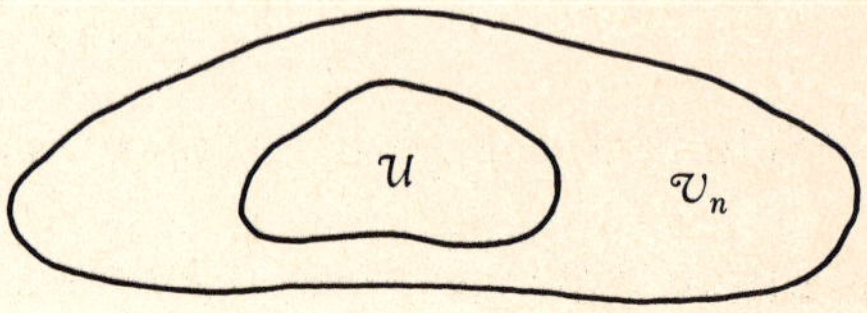

Fig. 3-2. A Set of Vectors $\mathcal{U}$ in $\mathcal{V}_n$

Definition 3.32: A set of vectors $\mathcal{U}$ in $\mathcal{V}_n$ is *closed under addition* if, given any $\mathbf{a}_1 \in \mathcal{U}$ and any $\mathbf{a}_2 \in \mathcal{U}$, then $(\mathbf{a}_1 + \mathbf{a}_2) \in \mathcal{U}$.

Example 3.15.

(*a*) Given $\mathcal{U}$ is the set of all 3-vectors whose elements are integers. This subset of $\mathcal{V}_3$ is closed under addition because the sum of any two 3-vectors whose elements are integers is also a 3-vector whose elements are integers.

(*b*) Given $\mathcal{U}$ is the set of all 2-vectors whose first element is unity. This set is not closed under addition because the sum of two vectors in $\mathcal{U}$ must give a vector whose first element is two.

Definition 3.33: A set of vectors $\mathcal{U}$ in $\mathcal{V}_n$ is *closed under scalar multiplication* if, given any vector $\mathbf{a} \in \mathcal{U}$ and any arbitrary scalar β, then $\beta\mathbf{a} \in \mathcal{U}$. The scalar β can be a real or complex number.

Example 3.16.

Given $\mathcal{U}$ is the set of all 3-vectors whose second and third elements are zero. Any scalar times any vector in $\mathcal{U}$ gives another vector in $\mathcal{U}$, so $\mathcal{U}$ is closed under scalar multiplication.

Definition 3.34: A set of n-vectors $\mathcal{U}$ in $\mathcal{V}_n$ that contains at least one vector is called a *vector space* if it is (1) closed under addition and (2) closed under scalar multiplication.

If $\mathbf{a} \in \mathcal{U}$, where $\mathcal{U}$ is a vector space, then $0\mathbf{a} = \mathbf{0} \in \mathcal{U}$ because $\mathcal{U}$ is closed under scalar multiplication. Hence the zero vector is in every vector space.

Given $\mathbf{a}_1, \mathbf{a}_2, \ldots, \mathbf{a}_n$, then the set of all linear combinations of the $\mathbf{a}_i$ is a vector space (*linear manifold*).

$$\mathcal{U} = \left\{\sum_{i=1}^{n} \beta_i \mathbf{a}_i\right\} \quad \text{for all } \beta_i$$

3.7 BASES

Definition 3.35: A vector space $\mathcal{U}$ in $\mathcal{V}_n$ is *spanned* by the vectors $\mathbf{a}_1, \mathbf{a}_2, \ldots, \mathbf{a}_k$ (k need not equal n) if (1) $\mathbf{a}_1 \in \mathcal{U}, \mathbf{a}_2 \in \mathcal{U}, \ldots, \mathbf{a}_k \in \mathcal{U}$ and (2) every vector in $\mathcal{U}$ is a linear combination of the $\mathbf{a}_1, \mathbf{a}_2, \ldots, \mathbf{a}_k$.

Example 3.17.

Given a vector space $\mathcal{U}$ in $\mathcal{V}_3$ to be the set of all 3-vectors whose third element is zero. Then (1 2 0), (1 1 0) and (0 1 0) span $\mathcal{U}$ because any vector in $\mathcal{U}$ can be represented as $(\alpha\ \beta\ 0)$, and

$$(\alpha\ \beta\ 0) \;=\; (\beta-\alpha)(1\ 2\ 0) + (2\alpha-\beta)(1\ 1\ 0) + 0(0\ 1\ 0)$$

Definition 3.36: Vectors $\mathbf{a}_1, \mathbf{a}_2, \ldots, \mathbf{a}_k \in \mathcal{V}_n$ are *linearly dependent* if there exist scalars $\beta_1, \beta_2, \ldots, \beta_k$ not all zero such that $\beta_1\mathbf{a}_1 + \beta_2\mathbf{a}_2 + \cdots + \beta_k\mathbf{a}_k = \mathbf{0}$.

Example 3.18.

The three vectors of Example 3.17 are linearly dependent because

$$+1(1\ 2\ 0) - 1(1\ 1\ 0) - 1(0\ 1\ 0) \;=\; (0\ 0\ 0)$$

Note that any set of vectors that contains the zero vector is a linearly dependent set.

Definition 3.37: A set of vectors are *linearly independent* if they are not linearly dependent.

***Theorem 3.7*:** If and only if the column vectors $\mathbf{a}_1, \mathbf{a}_2, \ldots, \mathbf{a}_n$ of a square matrix $\mathbf{A}$ are linearly dependent, then $\det \mathbf{A} = 0$.

Proof: If the column vectors of $\mathbf{A}$ are linearly dependent, from Definition 3.36 for some $\beta_1, \beta_2, \ldots, \beta_n$ not all zero we get $\mathbf{0} = \beta_1\mathbf{a}_1 + \beta_2\mathbf{a}_2 + \cdots + \beta_n\mathbf{a}_n$. Denote one nonzero β as β_i. Then

$$\mathbf{AE} \;=\; (\mathbf{a}_1 \mid \mathbf{a}_2 \mid \ldots \mid \mathbf{a}_i \mid \ldots \mid \mathbf{a}_n)\begin{pmatrix} 1 & 0 & \ldots & \beta_1 & \ldots & 0 \\ 0 & 1 & \ldots & \beta_2 & \ldots & 0 \\ \cdots & \cdots & \cdots & \cdots & \cdots & \cdots \\ 0 & 0 & \ldots & \beta_i & \ldots & 0 \\ \cdots & \cdots & \cdots & \cdots & \cdots & \cdots \\ 0 & 0 & \ldots & \beta_n & \ldots & 1 \end{pmatrix} \;=\; (\mathbf{a}_1 \mid \mathbf{a}_2 \mid \ldots \mid \mathbf{0} \mid \ldots \mid \mathbf{a}_n)$$

Since a matrix with a zero column has a zero determinant, use of the product rule of Theorem 3.2 gives $\det \mathbf{A} \det \mathbf{E} = 0$. Because $\det \mathbf{E} = \beta_i \neq 0$, then $\det \mathbf{A} = 0$.

Next, suppose the column vectors of $\mathbf{A}$ are linearly independent. Then so are the column vectors of $\mathbf{EA}$, for any elementary row operation $\mathbf{E}$. Proceeding stepwise as on page 44, we find $\mathbf{E}_1, \ldots, \mathbf{E}_m$ such that $\mathbf{E}_m \cdots \mathbf{E}_1\mathbf{A} = \mathbf{I}$. (Each step can be carried out since the column under consideration is not a linear combination of the preceding columns.) Hence, $(\det \mathbf{E}_m)\cdots(\det \mathbf{E}_1)(\det \mathbf{A}) = 1$, so that $\det \mathbf{A} \neq 0$.

Using this theorem it is possible to determine if $\mathbf{a}_1, \mathbf{a}_2, \ldots, \mathbf{a}_k$, $k \leq n$, are linearly dependent. Calculate all the $k \times k$ determinants formed by deleting all combinations of $n-k$ rows. If and only if all determinants are zero, the set of vectors is linearly dependent.

Example 3.19.

Consider $\begin{pmatrix} 1 \\ 2 \\ 3 \end{pmatrix}$ and $\begin{pmatrix} 2 \\ 4 \\ 0 \end{pmatrix}$. Deleting the bottom row gives $\det\begin{pmatrix} 1 & 2 \\ 2 & 4 \end{pmatrix} = 0$. Deleting the top row gives $\det\begin{pmatrix} 2 & 4 \\ 3 & 0 \end{pmatrix} = -12$. Hence the vectors are linearly independent. There is no need to check the determinant formed by deleting the middle row.

Definition 3.38: A set of n-vectors $\mathbf{a}_1, \mathbf{a}_2, \ldots, \mathbf{a}_k$ form a *basis* for $\mathcal{U}$ if (1) they span $\mathcal{U}$ and (2) they are linearly independent.

Example 3.20.

Any two of the three vectors given in Examples 3.17 and 3.18 form a basis of the given vector space, since (1) they span $\mathcal{U}$ as shown and (2) any two are linearly independent. To verify this for (1 2 0) and (1 1 0), set

$$\beta_1(1\ \ 2\ \ 0) + \beta_2(1\ \ 1\ \ 0) \;=\; (0\ \ 0\ \ 0)$$

This gives the equations $\beta_1 + \beta_2 = 0$, $2\beta_1 + \beta_2 = 0$. The only solution is that β_1 and β_2 are both zero, which violates the conditions of the definition of linear dependence.

Example 3.21.

Any three noncoplanar vectors in three-dimensional Euclidean space form a basis of $\mathcal{V}_3$ (not necessarily the orthogonal vectors). However, note that this definition has been abstracted to include vector spaces that can be subspaces of Euclidean space. Since conditions on the solution of algebraic equations are the goal of this section, it is best to avoid strictly geometric concepts and remain with the more abstract ideas represented by the definitions.

Consider $\mathcal{U}$ to be any infinite plane in three-dimensional Euclidean space $\mathcal{V}_3$. Any two noncolinear vectors in this plane form a basis for $\mathcal{U}$.

Theorem 3.8: If $\mathbf{a}_1, \mathbf{a}_2, \ldots, \mathbf{a}_k$ are a basis of $\mathcal{U}$, then every vector in $\mathcal{U}$ is expressible *uniquely* as a linear combination of $\mathbf{a}_1, \mathbf{a}_2, \ldots, \mathbf{a}_k$.

The key word here is uniquely. The proof is given in Problem 3.6.

To express any vector in $\mathcal{U}$ uniquely, a basis is needed. Suppose we are given a set of vectors that span $\mathcal{U}$. The next theorem is used in constructing a basis from this set.

Theorem 3.9: Given nonzero vectors $\mathbf{a}_1, \mathbf{a}_2, \ldots, \mathbf{a}_m \in \mathcal{V}_n$. The set is linearly dependent if and only if some $\mathbf{a}_k$, for $1 < k \leqq m$, is a linear combination of $\mathbf{a}_1, \mathbf{a}_2, \ldots, \mathbf{a}_{k-1}$.

Proof of this is given in Problem 3.7. This theorem states the given vectors need only be considered in order for the determination of linear dependency. We need not check and see that each $\mathbf{a}_k$ is linearly independent from the remaining vectors.

Example 3.22.

Given the vectors (1 −1 0), (−2 2 0) and (1 0 0). They are linearly dependent because (−2 2 0) = −2(1 −1 0). We need not check whether (1 0 0) can be formed as a linear combination of the first two vectors.

To construct a basis from a given set of vectors $\mathbf{a}_1, \mathbf{a}_2, \ldots, \mathbf{a}_m$ that span a vector space $\mathcal{U}$, test to see if $\mathbf{a}_2$ is a linear combination of $\mathbf{a}_1$. If it is, delete it from the set. Then test if $\mathbf{a}_3$ is a linear combination of $\mathbf{a}_1$ and $\mathbf{a}_2$, or only $\mathbf{a}_1$ if $\mathbf{a}_2$ has been deleted. Next test $\mathbf{a}_4$, etc., and in this manner delete all linearly dependent vectors from the set in order. The remaining vectors in the set form a basis of $\mathcal{U}$.

Theorem 3.10: Given a vector space $\mathcal{U}$ with a basis $\mathbf{a}_1, \mathbf{a}_2, \ldots, \mathbf{a}_m$ and with another basis $\mathbf{b}_1, \mathbf{b}_2, \ldots, \mathbf{b}_l$. Then $m = l$.

Proof: Note $\mathbf{a}_1, \mathbf{b}_1, \mathbf{b}_2, \ldots, \mathbf{b}_l$ are a linearly dependent set of vectors. Using Theorem 3.9 delete the vector $\mathbf{b}_k$ that is linearly dependent on $\mathbf{a}_1, \mathbf{b}_1, \ldots, \mathbf{b}_{k-1}$. Then $\mathbf{a}_1, \mathbf{b}_1, \ldots, \mathbf{b}_{k-1}, \mathbf{b}_{k+1}, \ldots, \mathbf{b}_l$ still span $\mathcal{U}$. Next note $\mathbf{a}_2, \mathbf{a}_1, \mathbf{b}_1, \ldots, \mathbf{b}_{k-1}, \mathbf{b}_{k+1}, \ldots, \mathbf{b}_l$ are a linearly dependent set. Delete another $\mathbf{b}$-vector such that the set still spans $\mathcal{U}$. Continuing in this manner gives $\mathbf{a}_l, \ldots, \mathbf{a}_2, \mathbf{a}_1$ span $\mathcal{U}$. If $l < m$, there is an $\mathbf{a}_{l+1}$ that is a linear combination of $\mathbf{a}_l, \ldots, \mathbf{a}_2, \mathbf{a}_1$. But the hypothesis states the $\mathbf{a}$-vectors are a basis, so they all must be linearly independent, hence $l \geqq m$. Interchanging the $\mathbf{b}$- and $\mathbf{a}$-vectors in the argument gives $m \geqq l$, proving the theorem.

Since all bases in a vector space $\mathcal{U}$ contain the same number of vectors, we can give the following definition.

Definition 3.39: A vector space $\mathcal{U}$ has *dimension* n if and only if a basis of $\mathcal{U}$ consists of n vectors.

Note that this extends the intuitive definition of dimension to a *sub*space of $\mathcal{V}_m$.

3.8 SOLUTION OF SETS OF LINEAR ALGEBRAIC EQUATIONS

Now the means are at hand to examine the solution of matrix equations. Consider the matrix equation

$$\mathbf{0} \quad = \quad \mathbf{Ax} \quad = \quad (\mathbf{a}_1 \mid \mathbf{a}_2 \mid \ldots \mid \mathbf{a}_n)\begin{pmatrix} \xi_1 \\ \vdots \\ \xi_n \end{pmatrix} \quad = \quad \sum_{i=1}^{n} \xi_i \mathbf{a}_i$$

If all ξ_i are zero, $\mathbf{x} = \mathbf{0}$ is the trivial solution, which can be obtained in all cases. To obtain a nontrivial solution, some of the ξ_i must be nonzero, which means the $\mathbf{a}_i$ must be linearly dependent, by definition. Consider the set of all solutions of $\mathbf{Ax} = \mathbf{0}$. Is this a vector space?

(1) Does the set contain at least one element?

Yes, because $\mathbf{x} = \mathbf{0}$ is always one solution.

(2) Are solutions closed under addition?

Yes, because if $\mathbf{Az} = \mathbf{0}$ and $\mathbf{Ay} = \mathbf{0}$, then the sum $\mathbf{x} = \mathbf{z} + \mathbf{y}$ is a solution of $\mathbf{Ax} = \mathbf{0}$.

(3) Are solutions closed under scalar multiplication?

Yes, because if $\mathbf{Ax} = \mathbf{0}$, then $\beta\mathbf{x}$ is a solution of $\mathbf{A}(\beta\mathbf{x}) = \mathbf{0}$.

So the set of all solutions of $\mathbf{Ax} = \mathbf{0}$ is a vector space.

Definition 3.40: The vector space of all solutions of $\mathbf{Ax} = \mathbf{0}$ is called the *null space* of $\mathbf{A}$.

***Theorem 3.11*:** If an $m \times n$ matrix $\mathbf{A}$ has n columns with r linearly independent columns, then the null space of $\mathbf{A}$ has dimension $n - r$.

The proof is given in Problem 3.8.

Definition 3.41: The dimension of the null space of $\mathbf{A}$ is called the *nullity* of $\mathbf{A}$.

Corollary 3.12: If $\mathbf{A}$ is an $n \times n$ matrix with n linearly independent columns, the null space has dimension zero. Hence the solution $\mathbf{x} = \mathbf{0}$ is unique.

***Theorem 3.13*:** The dimension of the vector space spanned by the row vectors of a matrix is equal to the dimension of the vector space spanned by the column vectors.

See Problem 3.9 for the proof.

Example 3.23.
Given the matrix $\begin{pmatrix} 1 & 2 & 3 \\ 2 & 4 & 6 \end{pmatrix}$. It has one independent row vector and therefore must have only one independent column vector.

Definition 3.42: The vector space of all $\mathbf{y}$ such that $\mathbf{Ax} = \mathbf{y}$ for some $\mathbf{x}$ is called the *range space* of $\mathbf{A}$.

It is left to the reader to verify that the range space is a vector space.

Example 3.24.

The range space and the null space may have other vectors in common in addition to the zero vector. Consider

$$\mathbf{A} = \begin{pmatrix} 0 & 1 \\ 0 & 0 \end{pmatrix} \qquad \mathbf{b} = \begin{pmatrix} \beta \\ 0 \end{pmatrix} \qquad \mathbf{c} = \begin{pmatrix} \gamma \\ \beta \end{pmatrix}$$

Then $\mathbf{Ab} = \mathbf{0}$, so $\mathbf{b}$ is in the null space; and $\mathbf{Ac} = \mathbf{b}$, so $\mathbf{b}$ is also in the range space.

Definition 3.43: The *rank* of the $m \times n$ matrix $\mathbf{A}$ is the dimension of the range space of $\mathbf{A}$.

***Theorem 3.14*:** The rank of $\mathbf{A}$ equals the maximum number of linearly independent column vectors of $\mathbf{A}$, i.e. the range space has dimension r.

The proof is given in Problem 3.10. Note the dimension of the range space plus the dimension of the null space $= n$ for an $m \times n$ matrix. This provides a means of determining the rank of $\mathbf{A}$. Determinants can be used to check the linear dependency of the row or column vectors.

***Theorem 3.15*:** Given an $m \times n$ matrix $\mathbf{A}$. Then $\operatorname{rank} \mathbf{A} = \operatorname{rank} \mathbf{A}^T = \operatorname{rank} \mathbf{A}^T\mathbf{A} = \operatorname{rank} \mathbf{A}\mathbf{A}^T$.

See Problem 3.12 for the proof.

***Theorem 3.16*:** Given an $m \times n$ matrix $\mathbf{A}$ and an $n \times k$ matrix $\mathbf{B}$, then $\operatorname{rank} \mathbf{AB} \leq \operatorname{rank} \mathbf{A}$, $\operatorname{rank} \mathbf{AB} \leq \operatorname{rank} \mathbf{B}$. Also, if $\mathbf{B}$ is nonsingular, $\operatorname{rank} \mathbf{AB} = \operatorname{rank} \mathbf{A}$; and if $\mathbf{A}$ is nonsingular, $\operatorname{rank} \mathbf{AB} = \operatorname{rank} \mathbf{B}$.

See Problem 3.13 for the proof.

3.9 GENERALIZATION OF A VECTOR

From Definition 3.7, a vector is defined as a matrix with only one row or column. Here we generalize the concept of a vector space and of a vector.

Definition 3.44: A set $\mathcal{U}$ of objects $\mathbf{x}, \mathbf{y}, \mathbf{z}, \ldots$ is called a *linear vector space* and the objects $\mathbf{x}, \mathbf{y}, \mathbf{z}, \ldots$ are called *vectors* if and only if for all complex numbers α and β and all objects $\mathbf{x}$, $\mathbf{y}$ and $\mathbf{z}$ in $\mathcal{U}$:

(1) $\mathbf{x} + \mathbf{y}$ is in $\mathcal{U}$,

(2) $\mathbf{x} + \mathbf{y} = \mathbf{y} + \mathbf{x}$,

(3) $(\mathbf{x} + \mathbf{y}) + \mathbf{z} = \mathbf{x} + (\mathbf{y} + \mathbf{z})$,

(4) for each $\mathbf{x}$ and $\mathbf{y}$ in $\mathcal{U}$ there is a unique $\mathbf{z}$ in $\mathcal{U}$ such that $\mathbf{x} + \mathbf{z} = \mathbf{y}$,

(5) $\alpha\mathbf{x}$ is in $\mathcal{U}$,

(6) $\alpha(\beta\mathbf{x}) = (\alpha\beta)\mathbf{x}$,

(7) $1\mathbf{x} = \mathbf{x}$,

(8) $\alpha(\mathbf{x} + \mathbf{y}) = \alpha\mathbf{x} + \alpha\mathbf{y}$,

(9) $(\alpha + \beta)\mathbf{x} = \alpha\mathbf{x} + \beta\mathbf{x}$.

The vectors of Definition 3.7 (n-vectors) and the vector space of Definition 3.34 satisfy this definition. Sometimes α and β are restricted to be real (linear vector spaces over the field of real numbers) but for generality they are taken to be complex here.

Example 3.25.

The set $\mathcal{U}$ of time functions that are linear combinations of $\sin t$, $\sin 2t$, $\sin 3t, \ldots$ is a linear vector space.

Example 3.26.

The set of all solutions to $d\mathbf{x}/dt = \mathbf{A}(t)\mathbf{x}$ is a linear vector space, but the set of all solutions to $d\mathbf{x}/dt = \mathbf{A}(t)\mathbf{x} + \mathbf{B}(t)\mathbf{u}$ for fixed $\mathbf{u}(t)$ does not satisfy (1) or (5) of Definition 3.44, and is not a linear vector space.

Example 3.27.

The set of all complex valued discrete time functions $x(nT)$ for $n = 0, 1, \ldots$ is a linear vector space, as is the set of all complex valued continuous time functions $x(t)$.

All the concepts of linear independence, basis, null space, range space, etc., extend immediately to a general linear vector space.

Example 3.28.

The functions $\sin t$, $\sin 2t$, $\sin 3t, \ldots$ form a basis for the linear vector space $\mathcal{U}$ of Example 3.25, and so the dimension of $\mathcal{U}$ is countably infinite.

3.10 DISTANCE IN A VECTOR SPACE

The concept of distance can be extended to a general vector space. To compare the distance from one point to another, no notion of an origin need be introduced. Furthermore, the ideas of distance involve two points (vectors), and the "yardstick" measuring distance may very well change in length as it is moved from point to point. To abstract this concept of distance, we have

Definition 3.45: A *metric*, denoted $\rho(\mathbf{a}, \mathbf{b})$, is any scalar function of two vectors $\mathbf{a} \in \mathcal{U}$ and $\mathbf{b} \in \mathcal{U}$ with the properties

(1) $\rho(\mathbf{a}, \mathbf{b}) \geqq 0$ (distance is always positive),

(2) $\rho(\mathbf{a}, \mathbf{b}) = 0$ if and only if $\mathbf{a} = \mathbf{b}$ (zero distance if and only if the points coincide),

(3) $\rho(\mathbf{a}, \mathbf{b}) = \rho(\mathbf{b}, \mathbf{a})$ (distance from $\mathbf{a}$ to $\mathbf{b}$ is the same as distance from $\mathbf{b}$ to $\mathbf{a}$),

(4) $\rho(\mathbf{a}, \mathbf{b}) + \rho(\mathbf{b}, \mathbf{c}) \geqq \rho(\mathbf{a}, \mathbf{c})$ (triangle inequality).

Example 3.29.

(*a*) An example of a metric for n-vectors $\mathbf{a}$ and $\mathbf{b}$ is

$$\rho(\mathbf{a}, \mathbf{b}) = \left[\frac{(\mathbf{a}-\mathbf{b})^\dagger(\mathbf{a}-\mathbf{b})}{1 + (\mathbf{a}-\mathbf{b})^\dagger(\mathbf{a}-\mathbf{b})}\right]^{1/2}$$

(*b*) For two real continuous scalar time functions $x(t)$ and $y(t)$ for $t_0 \leqq t \leqq t_1$, one metric is

$$\rho(x, y) = \left[\int_{t_0}^{t_1} [x(t) - y(t)]^2\, dt\right]^{1/2}$$

Further requirements are sometimes needed. By introducing the idea of an origin and by making a metric have constant length, we have the following definition.

Definition 3.46: The *norm*, denoted $\|\mathbf{a}\|$, of a vector $\mathbf{a}$ is a metric from the origin to the vector $\mathbf{a} \in \mathcal{U}$, with the additional property that the "yardstick" is not a function of $\mathbf{a}$. In other words a norm satisfies requirements (1)-(4) of a metric (Definition 3.45), with $\mathbf{b}$ understood to be zero, and has the additional requirement

(5) $\|\alpha\mathbf{a}\| = |\alpha|\,\|\mathbf{a}\|$.

The other four properties of a metric read, for a norm,

(1) $\|\mathbf{a}\| \geqq 0$,

(2) $\|\mathbf{a}\| = 0$ if and only if $\mathbf{a} = \mathbf{0}$,

(3) $\|\mathbf{a}\| = \|\mathbf{a}\|$ (trivial),

(4) $\|\mathbf{a}\| + \|\mathbf{c}\| \geqq \|\mathbf{a}+\mathbf{c}\|$.

Example 3.30.

A norm in $\mathcal{V}_2$ is $\|(a_1 \;\; a_2)\|_2 = \|\mathbf{a}\|_2 = \sqrt{|a_1|^2 + |a_2|^2}$.

Example 3.31.

A norm in $\mathcal{V}_2$ is $\|(a_1 \;\; a_2)\|_1 = \|\mathbf{a}\|_1 = |a_1| + |a_2|$.

Example 3.32.

A norm in $\mathcal{V}_n$ is the Euclidean norm $\|\mathbf{a}\|_2 = \sqrt{\mathbf{a}^\dagger\mathbf{a}} = \sqrt{\sum_{i=1}^{n} |a_i|^2}$.

In the above examples the subscripts 1 and 2 distinguish different norms.

Any positive monotone increasing concave function of a norm is a metric from $\mathbf{0}$ to $\mathbf{a}$.

Definition 3.47: The *inner product*, denoted $(\mathbf{x}, \mathbf{y})$, of any two vectors $\mathbf{a}$ and $\mathbf{b}$ in $\mathcal{U}$ is a *complex* scalar function of $\mathbf{a}$ and $\mathbf{b}$ such that given any complex numbers α and β,

(1) $(\mathbf{a}, \mathbf{a}) \geqq 0$,

(2) $(\mathbf{a}, \mathbf{a}) = 0$ if and only if $\mathbf{a} = \mathbf{0}$,

(3) $(\mathbf{a}, \mathbf{b})^* = (\mathbf{b}, \mathbf{a})$,

(4) $(\alpha\mathbf{a} + \beta\mathbf{b}, \mathbf{c}) = \alpha^*(\mathbf{a}, \mathbf{c}) + \beta^*(\mathbf{b}, \mathbf{c})$.

An inner product is sometimes known as a scalar or dot product.

Note $(\mathbf{a}, \mathbf{a})$ is real, $(\mathbf{a}, \alpha\mathbf{b}) = \alpha(\mathbf{a}, \mathbf{b})$ and $(\mathbf{a}, \mathbf{0}) = 0$.

Example 3.33.

An inner product in $\mathcal{V}_n$ is $(\mathbf{a}, \mathbf{b}) = \mathbf{a}^\dagger\mathbf{b} = \sum_{i=1}^{n} a_i^* b_i$.

Example 3.34.

An inner product in the vector space $\mathcal{U}$ of time functions of Example 3.25 is

$$(x, y) = \int_0^{\pi} x^*(t)\, y(t)\, dt$$

Definition 3.48: Two vectors $\mathbf{a}$ and $\mathbf{b}$ are *orthogonal* if $(\mathbf{a}, \mathbf{b}) = 0$.

Example 3.35.

In three-dimensional space, two vectors are perpendicular if they are orthogonal for the inner product of Example 3.33.

Theorem 3.17: For any inner product $(\mathbf{a}, \mathbf{b})$ the Schwarz inequality $|(\mathbf{a}, \mathbf{b})|^2 \leqq (\mathbf{a}, \mathbf{a})(\mathbf{b}, \mathbf{b})$ holds, and furthermore the equality holds if and only if $\mathbf{a} = \alpha\mathbf{b}$ or $\mathbf{a}$ or $\mathbf{b}$ is the zero vector.

Proof: If $\mathbf{a}$ or $\mathbf{b}$ is the zero vector, the equality holds, so take $\mathbf{b} \neq \mathbf{0}$. Then for any scalar β,

$$0 \leqq (\mathbf{a} + \beta\mathbf{b}, \mathbf{a} + \beta\mathbf{b}) = (\mathbf{a}, \mathbf{a}) + \beta^*(\mathbf{b}, \mathbf{a}) + \beta(\mathbf{a}, \mathbf{b}) + |\beta|^2(\mathbf{b}, \mathbf{b})$$

where the equality holds if and only if $\mathbf{a} + \beta\mathbf{b} = \mathbf{0}$. Setting $\beta = -(\mathbf{b}, \mathbf{a})/(\mathbf{b}, \mathbf{b})$ and rearranging gives the Schwarz inequality.

Example 3.36.

Using the inner product of Example 3.34,

$$\left|\int_0^\pi x^*(t)\, y(t)\, dt\right|^2 \;\le\; \left(\int_0^\pi |x^*(t)|^2\, dt\right)\left(\int_0^\pi |y(t)|^2\, dt\right)$$

Theorem 3.18: For any inner product, $\sqrt{(\mathbf{a}+\mathbf{b}, \mathbf{a}+\mathbf{b})} \le \sqrt{(\mathbf{a}, \mathbf{a})} + \sqrt{(\mathbf{b}, \mathbf{b})}$.

Proof:
$$(\mathbf{a}+\mathbf{b}, \mathbf{a}+\mathbf{b}) = |(\mathbf{a}+\mathbf{b}, \mathbf{a}+\mathbf{b})|$$
$$= |(\mathbf{a}, \mathbf{a}) + (\mathbf{b}, \mathbf{a}) + (\mathbf{a}, \mathbf{b}) + (\mathbf{b}, \mathbf{b})| \le |(\mathbf{a}, \mathbf{a})| + |(\mathbf{b}, \mathbf{a})| + |(\mathbf{a}, \mathbf{b})| + |(\mathbf{b}, \mathbf{b})|$$

Use of the Schwarz inequality then gives

$$(\mathbf{a}+\mathbf{b}, \mathbf{a}+\mathbf{b}) \le (\mathbf{a}, \mathbf{a}) + 2\sqrt{(\mathbf{a}, \mathbf{a})(\mathbf{b}, \mathbf{b})} + (\mathbf{b}, \mathbf{b})$$

and taking the square root of both sides proves the theorem.

This theorem shows that the square root of the inner product $(\mathbf{a}, \mathbf{a})$ satisfies the triangle inequality. Together with the requirements of the Definition 3.47 of an inner product, it can be seen that $\sqrt{(\mathbf{a}, \mathbf{a})}$ satisfies the Definition 3.46 of a norm.

Definition 3.49: The *natural norm*, denoted $\|\mathbf{a}\|_2$, of a vector $\mathbf{a}$ is $\|\mathbf{a}\|_2 = \sqrt{(\mathbf{a}, \mathbf{a})}$.

Definition 3.50: An *orthonormal* set of vectors $\mathbf{a}_1, \mathbf{a}_2, \ldots, \mathbf{a}_k$ is a set for which the inner product

$$(\mathbf{a}_i, \mathbf{a}_j) = \begin{Bmatrix} 0 & \text{if } i \ne j \\ 1 & \text{if } i = j \end{Bmatrix} = \delta_{ij}$$

where δ_{ij} is the Kronecker delta.

Example 3.37.

An orthonormal set of basis vectors in $\mathcal{V}_n$ is the set of unit vectors $\mathbf{e}_1, \mathbf{e}_2, \ldots, \mathbf{e}_n$, where $\mathbf{e}_i$ is defined as

$$\mathbf{e}_i = \begin{pmatrix} 0 \\ \vdots \\ 0 \\ 1 \\ 0 \\ \vdots \\ 0 \end{pmatrix} \leftarrow i\text{th position}$$

Given any set of k vectors, how can an orthonormal set of basis vectors be formed from them? To illustrate the procedure, suppose there are only two vectors, $\mathbf{a}_1$ and $\mathbf{a}_2$. First choose either one and make its length unity:

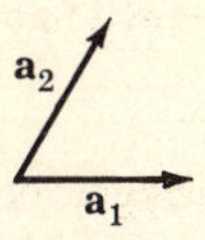

becomes

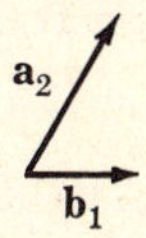

where $\mathbf{b}_1 = \gamma_1 \mathbf{a}_1$

Because $\mathbf{b}_1$ must have length unity, $\gamma_1 = \|\mathbf{a}_1\|_2^{-1}$. Now the second vector must be broken up into its components:

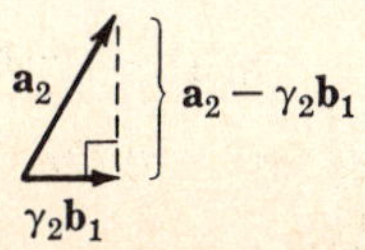

Here $\gamma_2 \mathbf{b}_1$ is the component of $\mathbf{a}_2$ along $\mathbf{b}_1$, and $\mathbf{a}_2 - \gamma_2 \mathbf{b}_1$ is the component perpendicular to $\mathbf{b}_1$. To find γ_2, note that this is the dot, or inner, product $\gamma_2 = \mathbf{b}_1^\dagger \cdot \mathbf{a}_2$. Finally, the second orthonormal vector is constructed by letting $\mathbf{a}_2 - \gamma_2 \mathbf{b}_1$ have unit length:

$$\mathbf{b}_2 = \frac{\mathbf{a}_2 - \gamma_2 \mathbf{b}_1}{\|\mathbf{a}_2 - \gamma_2 \mathbf{b}_1\|_2}$$

This process is called the *Gram-Schmit orthonormalization procedure.* The orthonormal $\mathbf{b}_j$ is constructed from $\mathbf{a}_j$ and the preceding $\mathbf{b}_i$ for $i = 1, \ldots, j-1$ according to

$$\mathbf{b}_j = \frac{\mathbf{a}_j - \sum_{i=1}^{j-1} (\mathbf{b}_i, \mathbf{a}_j)\mathbf{b}_i}{\|\mathbf{a}_j - \sum_{i=1}^{j-1} (\mathbf{b}_i, \mathbf{a}_j)\mathbf{b}_i\|_2}$$

Example 3.38.

Given the vectors $\mathbf{a}_1^T = (1 \;\; 1 \;\; 0)$, $\mathbf{a}_2^T = (1 \;\; -2 \;\; 1)/\sqrt{2}$. Then $\|\mathbf{a}_1\|_2 = \sqrt{2}$, so $\mathbf{b}_1^T = (1 \;\; 1 \;\; 0)/\sqrt{2}$. By the formula, the numerator is

$$\mathbf{a}_2 - (\mathbf{b}_1, \mathbf{a}_2)\mathbf{b}_1 = \frac{1}{\sqrt{2}}\begin{pmatrix} 1 \\ -2 \\ 1 \end{pmatrix} - \frac{1}{\sqrt{2}}(1 \;\; 1 \;\; 0)\frac{1}{\sqrt{2}}\begin{pmatrix} 1 \\ -2 \\ 1 \end{pmatrix}\frac{1}{\sqrt{2}}\begin{pmatrix} 1 \\ 1 \\ 0 \end{pmatrix} = \frac{1}{2\sqrt{2}}\begin{pmatrix} 3 \\ -3 \\ 2 \end{pmatrix}$$

Making this have length unity gives $\mathbf{b}_2^T = (3 \;\; -3 \;\; 2)/\sqrt{22}$.

***Theorem 3.19*:** Any finite dimensional linear vector space in which an inner product exists has an orthonormal basis.

Proof: Use the Gram-Schmit orthonormalization procedure on the set of all vectors in $\mathcal{U}$, or any set that spans $\mathcal{U}$, to construct the orthonormal basis.

3.11 RECIPROCAL BASIS

Given a set of basis vectors $\mathbf{b}_1, \mathbf{b}_2, \ldots, \mathbf{b}_n$ for the vector space $\mathcal{V}_n$, an arbitrary vector $\mathbf{x} \in \mathcal{V}_n$ can be expressed uniquely as

$$\mathbf{x} = \beta_1 \mathbf{b}_1 + \beta_2 \mathbf{b}_2 + \cdots + \beta_n \mathbf{b}_n \tag{3.4}$$

Definition 3.51: Given a set of basis vectors $\mathbf{b}_1, \mathbf{b}_2, \ldots, \mathbf{b}_n$ for $\mathcal{V}_n$, a *reciprocal basis* is a set of vectors $\mathbf{r}_1, \mathbf{r}_2, \ldots, \mathbf{r}_n$ such that the inner product

$$(\mathbf{r}_i, \mathbf{b}_j) = \delta_{ij} \qquad i, j = 1, 2, \ldots, n \tag{3.5}$$

Because of this relationship, defining $\mathbf{R}$ as the matrix made up of the $\mathbf{r}_i$ as column vectors and $\mathbf{B}$ as the matrix with $\mathbf{b}_i$ as column vectors, we have

$$\mathbf{R}^\dagger \mathbf{B} = \mathbf{I}$$

which is obtained by partitioned multiplication and use of *(3.5)*.

Since $\mathbf{B}$ has n linearly independent column vectors, $\mathbf{B}$ is nonsingular so that $\mathbf{R}$ is uniquely determined as $\mathbf{R} = (\mathbf{B}^{-1})^\dagger$. This demonstrates the set of $\mathbf{r}_i$ exists, and the set is a basis for $\mathcal{V}_n$ because $\mathbf{R}^{-1} = \mathbf{B}^\dagger$ exists so that all n of the $\mathbf{r}_i$ are linearly independent.

Having a reciprocal basis, the coefficients β in *(3.4)* can conveniently be expressed. Taking the inner product with an arbitrary $\mathbf{r}_i$ on both sides of *(3.4)* gives

$$(\mathbf{r}_i, \mathbf{x}) = \beta_1(\mathbf{r}_i, \mathbf{b}_1) + \beta_2(\mathbf{r}_i, \mathbf{b}_2) + \cdots + \beta_n(\mathbf{r}_i, \mathbf{b}_n)$$

Use of the property *(3.5)* then gives $\beta_i = (\mathbf{r}_i, \mathbf{x})$.

Note that an orthonormal basis is its own reciprocal basis. This is what makes "breaking a vector into its components" easy for an orthonormal basis, and indicates how to go about it for a non-orthonormal basis in $\mathcal{V}_n$.

3.12 MATRIX REPRESENTATION OF A LINEAR OPERATOR

Definition 3.52: A *linear operator* L is a function whose domain is the whole of a vector space $\mathcal{U}_1$ and whose range is contained in a vector space $\mathcal{U}_2$, such that for $\mathbf{a}$ and $\mathbf{b}$ in $\mathcal{U}_1$ and scalars α and β,

$$L(\alpha\mathbf{a}+\beta\mathbf{b}) = \alpha L(\mathbf{a}) + \beta L(\mathbf{b})$$

Example 3.39.

An $m \times n$ matrix $\mathbf{A}$ is a linear operator whose domain is the whole of $\mathcal{V}_n$ and whose range is in $\mathcal{V}_m$, i.e. it transforms n-vectors into m-vectors.

Example 3.40.

Consider the set $\mathcal{U}$ of time functions that are linear combinations of $\sin nt$, for $n = 1, 2, \ldots$. Then any $x(t)$ in $\mathcal{U}$ can be expressed as $x(t) = \sum_{n=1}^{\infty} \xi_n \sin nt$. The operation of integrating with respect to time is a linear operator, because

$$\int [\alpha x_1(t) + \beta x_2(t)]\, dt = \alpha \int x_1(t)\, dt + \beta \int x_2(t)\, dt$$

Example 3.41.

The operation of rotating a vector in $\mathcal{V}_2$ by an angle ϕ is a linear operator of $\mathcal{V}_2$ onto $\mathcal{V}_2$. Rotation of a vector $\alpha\mathbf{a}+\beta\mathbf{b}$ is the same as rotation of both $\mathbf{a}$ and $\mathbf{b}$ first and then adding α times the rotated $\mathbf{a}$ plus β times the rotated $\mathbf{b}$.

Theorem 3.20: Given a vector space $\mathcal{U}_1$ with a basis $\mathbf{b}_1, \mathbf{b}_2, \ldots, \mathbf{b}_n, \ldots$, and a vector space $\mathcal{U}_2$ with a basis $\mathbf{c}_1, \mathbf{c}_2, \ldots, \mathbf{c}_m, \ldots$. Then the linear operator L whose domain is $\mathcal{U}_1$ and whose range is in $\mathcal{U}_2$ can be represented by a matrix $\{\gamma_{ji}\}$ whose ith column consists of the components of $L(\mathbf{b}_i)$ relative to the basis $\mathbf{c}_1, \mathbf{c}_2, \ldots, \mathbf{c}_m, \ldots$.

Proof will be given for n dimensional $\mathcal{U}_1$ and m dimensional $\mathcal{U}_2$. Consider any $\mathbf{x}$ in $\mathcal{U}_1$. Then $\mathbf{x} = \sum_{i=1}^{n} \beta_i \mathbf{b}_i$. Furthermore since $L(\mathbf{b}_i)$ is a vector in $\mathcal{U}_2$, determine γ_{ji} such that $L(\mathbf{b}_i) = \sum_{j=1}^{m} \gamma_{ji}\mathbf{c}_j$. But

$$L(\mathbf{x}) = L\left(\sum_{i=1}^{n} \beta_i \mathbf{b}_i\right) = \sum_{i=1}^{n} \beta_i L(\mathbf{b}_i) = \sum_{i=1}^{n} \beta_i \sum_{j=1}^{m} \gamma_{ji}\mathbf{c}_j$$

Hence the jth component of $L(\mathbf{x})$ relative to the basis $\{\mathbf{c}_j\}$ equals $\sum_{i=1}^{n} \gamma_{ji}\beta_i$, i.e. the matrix $\{\gamma_{ji}\}$ times a vector whose components are the β_i.

Example 3.42.

The matrix representation of the rotation by an angle ϕ of Example 3.41 can be found as follows. Consider the basis $\mathbf{e}_1 = (1 \;\; 0)$ and $\mathbf{e}_2 = (0 \;\; 1)$ of $\mathcal{V}_2$. Then any $\mathbf{x} = (\beta_1 \;\; \beta_2) = \beta_1\mathbf{e}_1 + \beta_2\mathbf{e}_2$. Rotating $\mathbf{e}_1$ clockwise by an angle ϕ gives $L(\mathbf{e}_1) = (\cos\phi)\mathbf{e}_1 - (\sin\phi)\mathbf{e}_2$, and similarly $L(\mathbf{e}_2) = (\cos\phi)\mathbf{e}_2 + (\sin\phi)\mathbf{e}_1$, so that $\gamma_{11} = \cos\phi$, $\gamma_{21} = -\sin\phi$, $\gamma_{12} = \sin\phi$, $\gamma_{22} = \cos\phi$. Therefore rotation by an angle ϕ can be represented by

$$\mathbf{L} = \begin{pmatrix} \cos\phi & \sin\phi \\ -\sin\phi & \cos\phi \end{pmatrix}$$

Example 3.43.

An elementary row or column operation on a matrix $\mathbf{A}$ is represented by the elementary matrix $\mathbf{E}$ as given in the remarks following Definition 3.26.

The null space, range space, rank, etc., of linear transformation are obvious extensions of the definitions for its matrix representation.

3.13 EXTERIOR PRODUCTS

The advantages of exterior products have become widely recognized only recently, but the concept was introduced in 1844 by H. G. Grassmann. In essence it is a generalization of the cross product. We shall only consider its application to determinants here, but it is part of the general framework of tensor analysis, in which there are many applications.

First realize that a function of an m-vector can itself be a vector. For example, $\mathbf{z} = \mathbf{Ax} + \mathbf{By}$ illustrates a vector function $\mathbf{z}$ of the vectors $\mathbf{x}$ and $\mathbf{y}$. An exterior product $\boldsymbol{\phi}^p$ is a vector function of p m-vectors, $\mathbf{a}_1, \mathbf{a}_2, \ldots, \mathbf{a}_p$. However, $\boldsymbol{\phi}^p$ is an element in a *generalized* vector space, in that it satisfies Definition 3.44, but has no components that can be arranged as in Definition 3.7 except in special cases.

The functional dependency of $\boldsymbol{\phi}^p$ upon $\mathbf{a}_1, \mathbf{a}_2, \ldots, \mathbf{a}_p$ is written as $\boldsymbol{\phi}^p = \mathbf{a}_i \wedge \mathbf{a}_j \wedge \cdots \wedge \mathbf{a}_k$, where it is understood there are p of the $\mathbf{a}$'s. Each $\mathbf{a}$ is separated by a $\wedge$, read "wedge", which is Grassmann notation.

Definition 3.53: Given m-vectors $\mathbf{a}_i$ in $\mathcal{U}$, where $\mathcal{U}$ has dimension n. An *exterior product* $\boldsymbol{\phi}^p = \mathbf{a}_i \wedge \mathbf{a}_j \wedge \cdots \wedge \mathbf{a}_k$ for $p = 0, 1, 2, \ldots, n$ is a vector in an abstract vector space, denoted $\wedge^p\mathcal{U}$, such that for any complex numbers α and β,

$$(\alpha\mathbf{a}_i + \beta\mathbf{a}_j) \wedge \mathbf{a}_k \wedge \cdots \wedge \mathbf{a}_l = \alpha(\mathbf{a}_i \wedge \mathbf{a}_k \wedge \cdots \wedge \mathbf{a}_l) + \beta(\mathbf{a}_j \wedge \mathbf{a}_k \wedge \cdots \wedge \mathbf{a}_l) \qquad (3.6)$$

$$\mathbf{a}_i \wedge \cdots \wedge \mathbf{a}_j \wedge \cdots \wedge \mathbf{a}_k \wedge \cdots \wedge \mathbf{a}_l = -\mathbf{a}_i \wedge \cdots \wedge \mathbf{a}_k \wedge \cdots \wedge \mathbf{a}_j \wedge \cdots \wedge \mathbf{a}_l \qquad (3.7)$$

$$\mathbf{a}_i \wedge \mathbf{a}_j \wedge \cdots \wedge \mathbf{a}_k \neq \mathbf{0} \text{ if } \mathbf{a}_i, \mathbf{a}_j, \ldots, \mathbf{a}_k \text{ are linearly independent.}$$

Equation (*3.6*) says $\boldsymbol{\phi}^p$ is *multilinear*, and (*3.7*) says $\boldsymbol{\phi}^p$ is *alternating*.

Example 3.44.

The case $p = 0$ and $p = 1$ are degenerate, in that $\wedge^0\mathcal{U}$ is the space of all complex numbers and $\wedge^1\mathcal{U} = \mathcal{U}$, the original vector space of m-vectors having dimension n. The first nontrivial example is then $\wedge^2\mathcal{U}$. Then equations (*3.6*) and (*3.7*) become the bilinearity property

$$(\alpha\mathbf{a}_i + \beta\mathbf{a}_j) \wedge \mathbf{a}_k = \alpha(\mathbf{a}_i \wedge \mathbf{a}_k) + \beta(\mathbf{a}_j \wedge \mathbf{a}_k) \qquad (3.8)$$

and also

$$\mathbf{a}_i \wedge \mathbf{a}_j = -\mathbf{a}_j \wedge \mathbf{a}_i \qquad (3.9)$$

Interchanging the vectors in (*3.8*) according to (*3.9*) and multiplying by -1 gives

$$\mathbf{a}_k \wedge (\alpha\mathbf{a}_i + \beta\mathbf{a}_j) = \alpha(\mathbf{a}_k \wedge \mathbf{a}_i) + \beta(\mathbf{a}_k \wedge \mathbf{a}_j) \qquad (3.10)$$

By (*3.10*) we see that $\mathbf{a}_i \wedge \mathbf{a}_j$ is linear in either $\mathbf{a}_i$ or $\mathbf{a}_j$ (bilinear) but not both because in general $(\alpha\mathbf{a}_i) \wedge (\alpha\mathbf{a}_j) \neq \alpha(\mathbf{a}_i \wedge \mathbf{a}_j)$.

Note that setting $\mathbf{a}_j = \mathbf{a}_i$ in (*3.9*) gives $\mathbf{a}_i \wedge \mathbf{a}_i = \mathbf{0}$. Furthermore if and only if $\mathbf{a}_j$ is a linear combination of $\mathbf{a}_i$, $\mathbf{a}_i \wedge \mathbf{a}_j = \mathbf{0}$.

Let $\mathbf{b}_1, \mathbf{b}_2, \ldots, \mathbf{b}_n$ be a basis of $\mathcal{U}$. Then $\mathbf{a}_i = \sum_{k=1}^{n} \alpha_k\mathbf{b}_k$ and $\mathbf{a}_j = \sum_{l=1}^{n} \gamma_l\mathbf{b}_l$, so that

$$\mathbf{a}_i \wedge \mathbf{a}_j = \Big(\sum_{k=1}^{n} \alpha_k\mathbf{b}_k\Big) \wedge \Big(\sum_{l=1}^{n} \gamma_l\mathbf{b}_l\Big) = \sum_{k=1}^{n}\sum_{l=1}^{n} \alpha_k\gamma_l(\mathbf{b}_k \wedge \mathbf{b}_l)$$

Since $\mathbf{b}_k \wedge \mathbf{b}_k = \mathbf{0}$ and $\mathbf{b}_k \wedge \mathbf{b}_l = -\mathbf{b}_l \wedge \mathbf{b}_k$ for $k > l$, this sum can be rearranged to

$$\mathbf{a}_i \wedge \mathbf{a}_j = \sum_{l=1}^{n} \sum_{k=1}^{l-1} (\alpha_k \gamma_l - \alpha_l \gamma_k) \mathbf{b}_k \wedge \mathbf{b}_l \tag{3.11}$$

Because $\mathbf{a}_i \wedge \mathbf{a}_j$ is an arbitrary vector in $\wedge^2 \mathcal{U}$, and (3.11) is a linear combination of $\mathbf{b}_k \wedge \mathbf{b}_l$, then the vectors $\mathbf{b}_k \wedge \mathbf{b}_l$ for $1 \leq k < l \leq n$ form a basis for $\wedge^2 \mathcal{U}$. Summing over all possible k and l satisfying this relation shows that the dimension of $\wedge^2 \mathcal{U}$ is $n(n-1)/2 = \binom{n}{2}$.

Similar to the case $\wedge^2 \mathcal{U}$, if any $\mathbf{a}_i$ is a linear combination of the other **a**'s, the exterior product is zero and otherwise not. Furthermore the exterior products $\mathbf{b}_i \wedge \mathbf{b}_j \wedge \cdots \wedge \mathbf{b}_k$ for $1 \leq i < j < \cdots < k \leq n$ form a basis for $\wedge^p \mathcal{U}$, so that the dimension of $\wedge^p \mathcal{U}$ is

$$\frac{n!}{(n-p)!\,p!} = \binom{n}{p} \tag{3.12}$$

In particular the only basis vector for $\wedge^n \mathcal{U}$ is $\mathbf{b}_1 \wedge \mathbf{b}_2 \wedge \cdots \wedge \mathbf{b}_n$, so $\wedge^n \mathcal{U}$ is one-dimensional. Thus the exterior product containing n vectors of an n-dimensional space $\mathcal{U}$ is a scalar.

Definition 3.54: The *determinant of a linear operator* L whose domain and range is the vector space $\mathcal{U}$ with a basis $\mathbf{b}_1, \mathbf{b}_2, \ldots, \mathbf{b}_n$ is defined as

$$\det L = \frac{L(\mathbf{b}_1) \wedge L(\mathbf{b}_2) \wedge \cdots \wedge L(\mathbf{b}_n)}{\mathbf{b}_1 \wedge \mathbf{b}_2 \wedge \cdots \wedge \mathbf{b}_n} \tag{3.13}$$

The definiiton is completely independent of the matrix representation of L. If L is an $n \times n$ matrix $\mathbf{A}$ with column vectors $\mathbf{a}_1, \mathbf{a}_2, \ldots, \mathbf{a}_n$, and since $\mathbf{e}_1, \mathbf{e}_2, \ldots, \mathbf{e}_n$ are a basis for $\mathcal{V}_n$,

$$\det \mathbf{A} = \frac{\mathbf{A}\mathbf{e}_1 \wedge \mathbf{A}\mathbf{e}_2 \wedge \cdots \wedge \mathbf{A}\mathbf{e}_n}{\mathbf{e}_1 \wedge \mathbf{e}_2 \wedge \cdots \wedge \mathbf{e}_n} = \frac{\mathbf{a}_1 \wedge \mathbf{a}_2 \wedge \cdots \wedge \mathbf{a}_n}{\mathbf{e}_1 \wedge \mathbf{e}_2 \wedge \cdots \wedge \mathbf{e}_n}$$

Without loss of generality, we may define $\mathbf{e}_1 \wedge \mathbf{e}_2 \wedge \cdots \wedge \mathbf{e}_n = 1$, so that

$$\det \mathbf{A} = \mathbf{a}_1 \wedge \mathbf{a}_2 \wedge \cdots \wedge \mathbf{a}_n \tag{3.14}$$

and $\det \mathbf{I} = \mathbf{e}_1 \wedge \mathbf{e}_2 \wedge \cdots \wedge \mathbf{e}_n = 1$.

Now note that the Grassmann notation has a built-in multiplication process.

Definition 3.55: For an exterior product $\boldsymbol{\phi}^p = \mathbf{a}_1 \wedge \cdots \wedge \mathbf{a}_p$ in $\wedge^p \mathcal{U}$ and an exterior product $\boldsymbol{\psi}^q = \mathbf{c}_1 \wedge \cdots \wedge \mathbf{c}_q$ in $\wedge^q \mathcal{U}$, define *exterior multiplication* as

$$\boldsymbol{\phi}^p \wedge \boldsymbol{\psi}^q = \mathbf{a}_1 \wedge \cdots \wedge \mathbf{a}_p \wedge \mathbf{c}_1 \wedge \cdots \wedge \mathbf{c}_q \tag{3.15}$$

This definition is consistent with the Definition 3.53 for exterior products, and so $\boldsymbol{\phi}^p \wedge \boldsymbol{\psi}^q$ is itself an exterior product.

Also, if $m = n$ then $\mathbf{a}_1 \wedge \cdots \wedge \mathbf{a}_{n-2} \wedge \mathbf{a}_{n-1}$ is an n-vector since $\boldsymbol{\phi}^{n-1}$ has dimension n from (3.12), and must equal some vector in $\mathcal{U}$ since $\mathcal{U}$ and $\wedge^{n-1}\mathcal{U}$ must coincide.

Theorem 3.21. **(Laplace expansion).** Given an $n \times n$ matrix $\mathbf{A}$ with column vectors $\mathbf{a}_i$. Then $\det \mathbf{A} = \mathbf{a}_1 \wedge \mathbf{a}_2 \wedge \cdots \wedge \mathbf{a}_n = \mathbf{a}_1^T(\mathbf{a}_2 \wedge \cdots \wedge \mathbf{a}_n)$.

Proof: Let $\mathbf{e}_i$ be the ith unit vector in $\mathcal{V}_n$ and $\boldsymbol{\epsilon}_j$ be the jth unit vector in $\mathcal{V}_{n-1}$, i.e. $\mathbf{e}_i$ has n components and $\boldsymbol{\epsilon}_j$ has $n-1$ components. Then $\mathbf{a}_i = a_{1i}\mathbf{e}_1 + a_{2i}\mathbf{e}_2 + \cdots + a_{ni}\mathbf{e}_n$ so that

$$\mathbf{a}_1 \wedge \mathbf{a}_2 \wedge \cdots \wedge \mathbf{a}_n = a_{11}\mathbf{e}_1 \wedge \mathbf{a}_2 \wedge \cdots \wedge \mathbf{a}_n + a_{21}\mathbf{e}_2 \wedge \mathbf{a}_2 \wedge \cdots \wedge \mathbf{a}_n + \cdots + a_{n1}\mathbf{e}_n \wedge \mathbf{a}_2 \wedge \cdots \wedge \mathbf{a}_n$$

The first exterior product in the sum on the right can be written

$$\mathbf{e}_1 \wedge \mathbf{a}_2 \wedge \cdots \wedge \mathbf{a}_n = \mathbf{e}_1 \wedge (a_{12}\mathbf{e}_1 + a_{22}\mathbf{e}_2 + \cdots + a_{n2}\mathbf{e}_n) \wedge \cdots \wedge (a_{1n}\mathbf{e}_1 + \cdots + a_{nn}\mathbf{e}_n)$$

Using the multilinearity property gives

$$\begin{aligned}\mathbf{e}_1 \wedge \mathbf{a}_2 \wedge \cdots \wedge \mathbf{a}_n &\\ &= \mathbf{e}_1 \wedge \begin{pmatrix} 0 \\ a_{22} \\ \vdots \\ a_{n2} \end{pmatrix} \wedge \cdots \wedge \begin{pmatrix} 0 \\ a_{2n} \\ \vdots \\ a_{nn} \end{pmatrix} \\ &= \mathbf{e}_1 \wedge \left[a_{22}\begin{pmatrix} 0 \\ \boldsymbol{\epsilon}_1 \end{pmatrix} + \cdots + a_{n2}\begin{pmatrix} 0 \\ \boldsymbol{\epsilon}_{n-1} \end{pmatrix}\right] \wedge \cdots \wedge \left[a_{2n}\begin{pmatrix} 0 \\ \boldsymbol{\epsilon}_1 \end{pmatrix} + \cdots + a_{nn}\begin{pmatrix} 0 \\ \boldsymbol{\epsilon}_{n-1} \end{pmatrix}\right]\end{aligned}$$

Since $\det \mathbf{I}_n = 1 = \det \mathbf{I}_{n-1}$, then

$$\mathbf{e}_1 \wedge \begin{pmatrix} 0 \\ \boldsymbol{\epsilon}_1 \end{pmatrix} \wedge \cdots \wedge \begin{pmatrix} 0 \\ \boldsymbol{\epsilon}_{n-1} \end{pmatrix} = \boldsymbol{\epsilon}_1 \wedge \cdots \wedge \boldsymbol{\epsilon}_{n-1}$$

Performing exactly the same multilinear operations on the left hand side as on the right hand side gives

$$\begin{aligned}\mathbf{e}_1 \wedge \left[a_{22}\begin{pmatrix} 0 \\ \boldsymbol{\epsilon}_1 \end{pmatrix} + \cdots + a_{n2}\begin{pmatrix} 0 \\ \boldsymbol{\epsilon}_{n-1} \end{pmatrix}\right] \wedge \cdots \wedge \left[a_{2n}\begin{pmatrix} 0 \\ \boldsymbol{\epsilon}_1 \end{pmatrix} + \cdots + a_{nn}\begin{pmatrix} 0 \\ \boldsymbol{\epsilon}_n \end{pmatrix}\right] &\\ = [a_{22}\boldsymbol{\epsilon}_1 + \cdots + a_{n2}\boldsymbol{\epsilon}_{n-1}] \wedge \cdots \wedge [a_{2n}\boldsymbol{\epsilon}_1 + \cdots + a_{nn}\boldsymbol{\epsilon}_n] &\\ = \begin{pmatrix} a_{22} \\ \vdots \\ a_{n2} \end{pmatrix} \wedge \cdots \wedge \begin{pmatrix} a_{2n} \\ \vdots \\ a_{nn} \end{pmatrix} = \det M_{11} = c_{11}\end{aligned}$$

where c_{11} is the cofactor of a_{11}. Similarly,

$$\mathbf{e}_j \wedge \mathbf{a}_2 \wedge \cdots \wedge \mathbf{a}_n = c_{j1} \quad \text{and} \quad \mathbf{a}_1 \wedge \mathbf{a}_2 \wedge \cdots \wedge \mathbf{a}_n = \sum_{j=1}^{n} a_{j1}c_{j1} = (c_{11}\ c_{21}\ \ldots\ c_{n1})\mathbf{a}_1$$

so that $\mathbf{a}_2 \wedge \cdots \wedge \mathbf{a}_n = (c_{11}\ c_{21}\ \ldots\ c_{n1})^T$ and Theorem 3.21 is nothing more than a statement of the Laplace expansion of the determinant about the first column. The use of column interchanges generalizes the proof of the Laplace expansion about any column, and use of $\det \mathbf{A} = \det \mathbf{A}^T$ provides the proof of expansion about any row.

Solved Problems

3.1. Multiply the following matrices, where $\mathbf{a}_1, \mathbf{a}_2, \mathbf{b}_1$ and $\mathbf{b}_2$ are column vectors with n elements.

(i) $(1\ \ 2)\begin{pmatrix} 3 \\ 4 \end{pmatrix}$ (iii) $\begin{pmatrix} \mathbf{a}_1^T \\ \hline \mathbf{a}_2^T \end{pmatrix}(\mathbf{b}_1 \mid \mathbf{b}_2)$ (v) $\begin{pmatrix} \lambda_1 & 0 \\ 0 & \lambda_2 \end{pmatrix}\begin{pmatrix} \mathbf{a}_1^T \\ \hline \mathbf{a}_2^T \end{pmatrix}$

(ii) $\begin{pmatrix} 1 \\ 2 \end{pmatrix}(3\ \ 4)$ (iv) $(\mathbf{a}_1 \mid \mathbf{a}_2)\begin{pmatrix} \mathbf{b}_1^T \\ \hline \mathbf{b}_2^T \end{pmatrix}$ (vi) $(\mathbf{a}_1 \mid \mathbf{a}_2)\begin{pmatrix} \lambda_1 & 0 \\ 0 & \lambda_2 \end{pmatrix}$

Using the rule for multiplication from Definition 3.13, and realizing that multiplication of a $k \times n$ matrix times an $n \times m$ matrix results in a $k \times m$ matrix, we have

(i) $(1 \times 3 + 2 \times 4) = (11)$ (iii) $\begin{pmatrix} \mathbf{a}_1^T\mathbf{b}_1 & \mathbf{a}_1^T\mathbf{b}_2 \\ \hline \mathbf{a}_2^T\mathbf{b}_1 & \mathbf{a}_2^T\mathbf{b}_2 \end{pmatrix}$ (v) $\begin{pmatrix} \lambda_1\mathbf{a}_1^T \\ \hline \lambda_2\mathbf{a}_2^T \end{pmatrix}$

(ii) $\begin{pmatrix} (1 \times 3) & (1 \times 4) \\ (2 \times 3) & (2 \times 4) \end{pmatrix} = \begin{pmatrix} 3 & 4 \\ 6 & 8 \end{pmatrix}$ (iv) $(\mathbf{a}_1\mathbf{b}_1^T) + (\mathbf{a}_2\mathbf{b}_2^T)$ (vi) $(\lambda_1\mathbf{a}_1 \mid \lambda_2\mathbf{a}_2)$

3.2. Find the determinant of $\mathbf{A}$ (*a*) by direct computation, (*b*) by using elementary row and column operations, and (*c*) by Laplace's expansion, where

$$\mathbf{A} = \begin{pmatrix} 1 & 0 & 0 & 2 \\ 1 & 2 & 0 & 6 \\ 1 & 3 & 1 & 8 \\ 0 & 0 & 0 & 2 \end{pmatrix}$$

(*a*) To facilitate direct computation, form the table

p	p_1, p_2, p_3, p_4				p	p_1, p_2, p_3, p_4			
0	1	2	3	4	12	3	1	2	4
1	1	2	4	3	13	3	1	4	2
2	1	3	4	2	14	3	2	4	1
3	1	3	2	4	15	3	2	1	4
4	1	4	2	3	16	3	4	1	2
5	1	4	3	2	17	3	4	2	1
6	2	4	3	1	18	4	3	2	1
7	2	4	1	3	19	4	3	1	2
8	2	3	1	4	20	4	2	1	3
9	2	3	4	1	21	4	2	3	1
10	2	1	4	3	22	4	1	3	2
11	2	1	3	4	23	4	1	2	3

There are $4! = 24$ terms in this table. Then

$$\begin{aligned} \det \mathbf{A} = {} & +1\cdot2\cdot1\cdot2 - 1\cdot2\cdot8\cdot0 + 1\cdot0\cdot8\cdot0 - 1\cdot0\cdot3\cdot2 \\ & + 1\cdot6\cdot3\cdot0 - 1\cdot6\cdot1\cdot0 + 0\cdot6\cdot1\cdot0 - 0\cdot6\cdot1\cdot0 + 0\cdot0\cdot1\cdot2 \\ & - 0\cdot0\cdot8\cdot0 + 0\cdot1\cdot8\cdot0 - 0\cdot1\cdot1\cdot2 + 0\cdot1\cdot3\cdot2 - 0\cdot1\cdot8\cdot0 \\ & + 0\cdot2\cdot8\cdot0 - 0\cdot2\cdot1\cdot2 + 0\cdot6\cdot1\cdot0 - 0\cdot6\cdot3\cdot0 + 2\cdot0\cdot3\cdot0 \\ & - 2\cdot0\cdot1\cdot0 + 2\cdot2\cdot1\cdot0 - 2\cdot2\cdot1\cdot0 + 2\cdot1\cdot1\cdot0 - 2\cdot1\cdot3\cdot0 \\ = {} & 4 \end{aligned}$$

(*b*) Using elementary row and column operations, subtract 1, 3 and 4 times the bottom row from the first, second and third rows respectively. This reduces $\mathbf{A}$ to a triangular matrix, whose determinant is the product of its diagonal elements $1\cdot2\cdot1\cdot2 = 4$.

(*c*) Using the Laplace expansion about the bottom row,

$$\det \mathbf{A} = 2 \det \begin{pmatrix} 1 & 0 & 0 \\ 1 & 2 & 0 \\ 1 & 3 & 1 \end{pmatrix} = 2\cdot2 = 4$$

3.3. Prove that $\det(\mathbf{A}^T) = \det \mathbf{A}$, where $\mathbf{A}$ is an $n \times n$ matrix.

If $\mathbf{A} = \{a_{ij}\}$, then $\mathbf{A}^T = \{a_{ji}\}$ so that $\det(\mathbf{A}^T) = \sum^{n!} (-1)^p a_{p_1 1} a_{p_2 2} \cdots a_{p_n n}$.

Since a determinant is all possible combinations of products of elements where only one element is taken from each row and column, the individual terms in the sum are the same. Therefore the only question is the sign of each product. Consider a typical term from a 3×3 matrix: $a_{31}a_{12}a_{23}$, i.e. $p_1 = 3$, $p_2 = 1$, $p_3 = 2$. Permute the elements through $a_{12}a_{31}a_{23}$ to $a_{12}a_{23}a_{31}$, so that the row numbers are in natural 1, 2, 3 order instead of the column numbers. From this, it can be concluded in general that it takes exactly the same number of permutations to undo $p_1, p_2, \ldots, p_n$ to $1, 2, \ldots, n$ as it does to permute $1, 2, \ldots, n$ to obtain $p_1, p_2, \ldots, p_n$. Therefore p must be the same for each product term in the series, and so the determinants are equal.

3.4. A Vandermonde matrix **V** has the form

$$\mathbf{V} = \begin{pmatrix} 1 & 1 & \cdots & 1 \\ \theta_1 & \theta_2 & \cdots & \theta_n \\ \cdots & \cdots & \cdots & \cdots \\ \theta_1^{n-1} & \theta_2^{n-1} & \cdots & \theta_n^{n-1} \end{pmatrix}$$

Prove **V** is nonsingular if and only if $\theta_i \neq \theta_j$ for $i \neq j$.

This will be proven by showing

$$\det \mathbf{V} = (\theta_2 - \theta_1)(\theta_3 - \theta_2)(\theta_3 - \theta_1)\cdots(\theta_n - \theta_{n-1})(\theta_n - \theta_{n-2})\cdots(\theta_n - \theta_1) = \prod_{1 \leq i < j \leq n} (\theta_j - \theta_i)$$

For $n = 2$, $\det \mathbf{V} = \theta_2 - \theta_1$, which agrees with the hypothesis. By induction if the hypothesis can be shown true for n given it is true for $n-1$, then the hypothesis holds for $n \geq 2$. Note each term of $\det \mathbf{V}$ will contain one and only one element from the nth column, so that

$$\det \mathbf{V} = \gamma_0 + \gamma_1\theta_n + \cdots + \gamma_{n-1}\theta_n^{n-1}$$

If $\theta_n = \theta_i$ for $i = 1, 2, \ldots, n-1$, then $\det \mathbf{V} = 0$ because two columns are equal. Therefore $\theta_1, \theta_2, \ldots, \theta_{n-1}$ are the roots of the polynomial, and

$$\gamma_0 + \gamma_1\theta_n + \cdots + \gamma_{n-1}\theta_n^{n-1} = \gamma_{n-1}(\theta_n - \theta_1)(\theta_n - \theta_2)\cdots(\theta_n - \theta_{n-1})$$

But γ_{n-1} is the cofactor of θ_n^{n-1} by the Laplace expansion; hence

$$\gamma_{n-1} = \det \begin{pmatrix} 1 & 1 & \cdots & 1 \\ \theta_1 & \theta_2 & \cdots & \theta_{n-1} \\ \cdots & \cdots & \cdots & \cdots \\ \theta_1^{n-2} & \theta_2^{n-2} & \cdots & \theta_{n-1}^{n-1} \end{pmatrix}$$

By assumption that the hypothesis holds for $n-1$,

$$\gamma_{n-1} = \prod_{1 \leq i < j \leq n-1} (\theta_j - \theta_i)$$

Combining these relations gives

$$\det \mathbf{V} = (\theta_n - \theta_1)(\theta_n - \theta_2)\cdots(\theta_n - \theta_{n-1}) \prod_{1 \leq i < j \leq n-1} (\theta_j - \theta_i)$$

3.5. Show $\det \begin{pmatrix} \mathbf{A} & \mathbf{B} \\ \mathbf{0} & \mathbf{C} \end{pmatrix} = \det \mathbf{A} \det \mathbf{C}$, where **A** and **C** are $n \times n$ and $m \times m$ matrices respectively.

Either $\det \mathbf{A} = 0$ or $\det \mathbf{A} \neq 0$. If $\det \mathbf{A} = 0$, then the column vectors of **A** are linearly dependent. Hence the column vectors of $\begin{pmatrix} \mathbf{A} \\ \mathbf{0} \end{pmatrix}$ are linearly dependent, so that

$$\det \begin{pmatrix} \mathbf{A} & \mathbf{B} \\ \mathbf{0} & \mathbf{C} \end{pmatrix} = 0$$

and the hypothesis holds.

If $\det \mathbf{A} \neq 0$, then $\mathbf{A}^{-1}$ exists and

$$\begin{pmatrix} \mathbf{A} & \mathbf{B} \\ \mathbf{0} & \mathbf{C} \end{pmatrix} = \begin{pmatrix} \mathbf{A} & \mathbf{0} \\ \mathbf{0} & \mathbf{I} \end{pmatrix} \begin{pmatrix} \mathbf{I} & \mathbf{0} \\ \mathbf{0} & \mathbf{C} \end{pmatrix} \begin{pmatrix} \mathbf{I} & \mathbf{A}^{-1}\mathbf{B} \\ \mathbf{0} & \mathbf{I} \end{pmatrix}$$

The rightmost matrix is an upper triangular matrix, so its determinant is the product of the diagonal elements which is unity. Furthermore, repeated use of the Laplace expansion about the diagonal elements of **I** gives

$$\det \begin{pmatrix} \mathbf{A} & \mathbf{0} \\ \mathbf{0} & \mathbf{I} \end{pmatrix} = \det \mathbf{A} \qquad \det \begin{pmatrix} \mathbf{I} & \mathbf{0} \\ \mathbf{0} & \mathbf{C} \end{pmatrix} = \det \mathbf{C}$$

Use of the product rule of Theorem 3.2 then gives the proof.

3.6. Show that if $\mathbf{a}_1, \mathbf{a}_2, \ldots, \mathbf{a}_k$ are a basis of $\mathcal{U}$, then every vector in $\mathcal{U}$ is expressible uniquely as a linear combination of $\mathbf{a}_1, \mathbf{a}_2, \ldots, \mathbf{a}_k$.

Let $\mathbf{x}$ be an arbitrary vector in $\mathcal{U}$. Because $\mathbf{x}$ is in $\mathcal{U}$, and $\mathcal{U}$ is spanned by the basis vectors $\mathbf{a}_1, \mathbf{a}_2, \ldots, \mathbf{a}_k$ by definition, the question is one of uniqueness. If there are two or more linear combinations of $\mathbf{a}_1, \mathbf{a}_2, \ldots, \mathbf{a}_k$ that represent $\mathbf{x}$, they can be written as

$$\mathbf{x} = \sum_{i=1}^{k} \beta_i \mathbf{a}_i$$

and

$$\mathbf{x} = \sum_{i=1}^{k} \alpha_i \mathbf{a}_i$$

Subtracting one expression from the other gives

$$\mathbf{0} = (\beta_1 - \alpha_1)\mathbf{a}_1 + (\beta_2 - \alpha_2)\mathbf{a}_2 + \cdots + (\beta_k - \alpha_k)\mathbf{a}_k$$

Because the basis consists of linearly independent vectors, the only way this can be an equality is for $\beta_i = \alpha_i$. Therefore all representations are the same, and the theorem is proved.

Note that both properties of a basis were used here. If a set of vectors did not span $\mathcal{U}$, all vectors would not be expressible as a linear combination of the set. If the set did span $\mathcal{U}$ but were linearly dependent, a representation of other vectors would not be unique.

3.7. Given the set of nonzero vectors $\mathbf{a}_1, \mathbf{a}_2, \ldots, \mathbf{a}_m$ in $\mathcal{V}_n$. Show that the set is linearly dependent if and only if some $\mathbf{a}_k$, for $1 < k \le m$, is a linear combination of $\mathbf{a}_1, \mathbf{a}_2, \ldots, \mathbf{a}_{k-1}$.

If part: If $\mathbf{a}_k$ is a linear combination of $\mathbf{a}_1, \mathbf{a}_2, \ldots, \mathbf{a}_{k-1}$, then

$$\mathbf{a}_k = \sum_{i=1}^{k-1} \beta_i \mathbf{a}_i$$

where the β_i are not all zero since $\mathbf{a}_k$ is nonzero. Then

$$\mathbf{0} = \beta_1\mathbf{a}_1 + \beta_2\mathbf{a}_2 + \cdots + \beta_{k-1}\mathbf{a}_{k-1} + (-1)\mathbf{a}_k + 0\mathbf{a}_{k+1} + \cdots + 0\mathbf{a}_m$$

which satisfies the definition of linear dependence.

Only if part: If the set is linearly dependent, then

$$\mathbf{0} = \beta_1\mathbf{a}_1 + \beta_2\mathbf{a}_2 + \cdots + \beta_k\mathbf{a}_k + \cdots + \beta_m\mathbf{a}_m$$

where not all the β_i are zero. Find that nonzero β_k such that all $\beta_i = 0$ for $i > k$. Then the linear combination is

$$\mathbf{a}_k = \beta_k^{-1} \sum_{i=1}^{k-1} \beta_i \mathbf{a}_i$$

3.8. Show that if an $m \times n$ matrix $\mathbf{A}$ has n columns with at most r linearly independent columns, then the null space of $\mathbf{A}$ has dimension $n - r$.

Because $\mathbf{A}$ is $m \times n$, then $\mathbf{a}_i \in \mathcal{V}_m$, $\mathbf{x} \in \mathcal{V}_n$. Renumber the $\mathbf{a}_i$ so that the first r are the independent ones. The rest of the column vectors can be written as

$$\begin{aligned} \mathbf{a}_{r+1} &= \beta_{11}\mathbf{a}_1 + \beta_{12}\mathbf{a}_2 + \cdots + \beta_{1r}\mathbf{a}_r \\ &\cdots\cdots\cdots\cdots\cdots \\ \mathbf{a}_n &= \beta_{n-r,1}\mathbf{a}_1 + \beta_{n-r,2}\mathbf{a}_2 + \cdots + \beta_{n-r,r}\mathbf{a}_r \end{aligned} \qquad (3.16)$$

because $\mathbf{a}_{r+1}, \ldots, \mathbf{a}_n$ are linearly dependent and can therefore be expressed as a linear combination of the linearly independent column vectors. Construct the $n - r$ vectors $\mathbf{x}_1, \mathbf{x}_2, \ldots, \mathbf{x}_{n-r}$ such that

$$\mathbf{x}_1 = \begin{pmatrix} \beta_{11} \\ \beta_{12} \\ \vdots \\ \beta_{1r} \\ -1 \\ 0 \\ \vdots \\ 0 \end{pmatrix}, \qquad \mathbf{x}_2 = \begin{pmatrix} \beta_{21} \\ \beta_{22} \\ \vdots \\ \beta_{2r} \\ 0 \\ -1 \\ \vdots \\ 0 \end{pmatrix}, \qquad \ldots, \qquad \mathbf{x}_{n-r} = \begin{pmatrix} \beta_{n-r,1} \\ \beta_{n-r,2} \\ \vdots \\ \beta_{n-r,r} \\ 0 \\ 0 \\ \vdots \\ -1 \end{pmatrix}$$

Note that $\mathbf{Ax}_1 = \mathbf{0}$ by the first equation of *(3.16)*, $\mathbf{Ax}_2 = \mathbf{0}$ by the second equation of *(3.16)*, etc., so these are $n-r$ solutions.

Now it will be shown that these vectors are a basis for the null space of $\mathbf{A}$. First, they must be linearly independent because of the different positions of the -1 in the bottom part of the vectors. To show they are a basis, then, it must be shown that all solutions can be expressed as a linear combination of the $\mathbf{x}_i$, i.e. it must be shown the $\mathbf{x}_i$ span the null space of $\mathbf{A}$. Consider an arbitrary solution $\mathbf{x}$ of $\mathbf{Ax} = \mathbf{0}$. Then

$$\mathbf{x} = \begin{pmatrix} \xi_1 \\ \vdots \\ \xi_r \\ \xi_{r+1} \\ \xi_{r+2} \\ \vdots \\ \xi_n \end{pmatrix} = -\xi_{r+1}\begin{pmatrix} \beta_{11} \\ \vdots \\ \beta_{1r} \\ -1 \\ 0 \\ \vdots \\ 0 \end{pmatrix} - \xi_{r+2}\begin{pmatrix} \beta_{21} \\ \vdots \\ \beta_{2r} \\ 0 \\ -1 \\ \vdots \\ 0 \end{pmatrix} - \cdots - \xi_n\begin{pmatrix} \beta_{n-r,1} \\ \vdots \\ \beta_{n-r,r} \\ 0 \\ 0 \\ \vdots \\ -1 \end{pmatrix} + \begin{pmatrix} \sigma_1 \\ \vdots \\ \sigma_r \\ 0 \\ 0 \\ \vdots \\ 0 \end{pmatrix}$$

Or, in vector notation, $$\mathbf{x} = \sum_{i=1}^{n-r} -\xi_{r+i}\mathbf{x}_i + \mathbf{s}$$

where $\mathbf{s}$ is a remainder if the $\mathbf{x}_i$ do not span the null space of $\mathbf{A}$. If $\mathbf{s} = \mathbf{0}$, then the $\mathbf{x}_i$ do span the null space of $\mathbf{A}$. Check that the last $n-r$ elements of $\mathbf{s}$ are zero by writing the vector equality as a set of scalar equations.

Multiply both sides of the equation by $\mathbf{A}$. $$\mathbf{Ax} = \sum_{i=1}^{n-r} -\xi_{r+i}\mathbf{Ax}_i + \mathbf{As}$$

Since $\mathbf{Ax} = 0$ and $\mathbf{Ax}_i = \mathbf{0}$, $\mathbf{As} = \mathbf{0}$. Writing this out in terms of the column vectors of $\mathbf{A}$ gives $\sum_{i=1}^{r} \sigma_i \mathbf{a}_i = \mathbf{0}$. But these column vectors are linearly independent, so $\sigma_i = 0$. Hence the $n-r$ $\mathbf{x}_i$ are a basis, so the null space of $\mathbf{A}$ has dimension $n-r$.

3.9. Show that the dimension of the vector space spanned by the row vectors of a matrix is equal to the dimension of the vector space spanned by the column vectors.

Without loss of generality let the first r column vectors $\mathbf{a}_i$ be linearly independent and let s of the row vectors $\mathbf{a}_i$ be linearly independent. Partition $\mathbf{A}$ as follows:

$$\mathbf{A} = \left(\begin{array}{cccc|ccc} a_{11} & a_{12} & \cdots & a_{1r} & a_{1,r+1} & \cdots & a_{1n} \\ \cdots & \cdots & \cdots & \cdots & \cdots & \cdots & \cdots \\ a_{r1} & a_{r2} & \cdots & a_{rr} & a_{r,r+1} & \cdots & a_{rn} \\ a_{r+1,1} & a_{r+1,2} & \cdots & a_{r+1,r} & a_{r+1,r+1} & \cdots & a_{r+1,n} \\ \hline \cdots & \cdots & \cdots & \cdots & \cdots & \cdots & \cdots \\ a_{m1} & a_{m2} & \cdots & a_{mr} & a_{m,r+1} & \cdots & a_{mn} \end{array}\right)$$

$$= \left(\begin{array}{c|ccc} \mathbf{x}_1 & a_{1,r+1} & \cdots & a_{1n} \\ \cdots & \cdots & \cdots & \cdots \\ \mathbf{x}_r & a_{r,r+1} & \cdots & a_{rn} \\ \mathbf{x}_{r+1} & a_{r+1,r+1} & \cdots & a_{r+1,n} \\ \cdots & \cdots & \cdots & \cdots \\ \mathbf{x}_m & a_{m,r+1} & \cdots & a_{mn} \end{array}\right)$$

$$= \begin{pmatrix} \mathbf{y}_1 & \cdots & \mathbf{y}_r & \mathbf{y}_{r+1} & \cdots & \mathbf{y}_n \\ \hline \cdots & \cdots & \cdots & \cdots & \cdots & \cdots \\ a_{m1} & \cdots & a_{mr} & a_{m,r+1} & \cdots & a_{mn} \end{pmatrix}$$

so that $\mathbf{x}_i = (a_{i1}\ a_{i2}\ \ldots\ a_{ir})$ and $\mathbf{y}_j^T = (a_{1j}\ a_{2j}\ \ldots\ a_{r+1,j})$. Since the $\mathbf{x}_i$ are r-vectors, $\sum_{i=1}^{r+1} b_i\mathbf{x}_i = \mathbf{0}$ for some nonzero b_i. Let the vector $\mathbf{b}^T = (b_1\ b_2\ \ldots\ b_{r+1})$ so that

$$\mathbf{0} = \sum_{i=1}^{r+1} b_i\mathbf{x}_i = \left(\sum_{i=1}^{r+1} b_i a_{i1}\ \sum_{i=1}^{r+1} b_i a_{i2}\ \ldots\ \sum_{i=1}^{r+1} b_i a_{ir}\right)$$

$$= (\mathbf{b}^T\mathbf{y}_1\ \mathbf{b}^T\mathbf{y}_2\ \ldots\ \mathbf{b}^T\mathbf{y}_r)$$

Therefore $\mathbf{b}^T\mathbf{y}_j = 0$ for $j = 1, 2, \ldots, r$.

Since the last $n - r$ column vectors $\mathbf{a}_i$ are linearly dependent, $\mathbf{a}_i = \sum_{j=1}^{r} \alpha_{ij}\mathbf{a}_j$ for $i = r+1, \ldots, n$. Then $\mathbf{y}_i = \sum_{j=1}^{r} \alpha_{ij}\mathbf{y}_j$ so that $\mathbf{b}^T\mathbf{y}_i = \sum_{j=1}^{r} \alpha_{ij}\mathbf{b}^T\mathbf{y}_j = 0$ for $i = r+1, \ldots, n$. Hence

$$\mathbf{0} = (\mathbf{b}^T\mathbf{y}_1\ \mathbf{b}^T\mathbf{y}_2\ \ldots\ \mathbf{b}^T\mathbf{y}_r\ \mathbf{b}^T\mathbf{y}_{r+1}\ \ldots\ \mathbf{b}^T\mathbf{y}_n) = b_1\mathbf{a}_1 + b_2\mathbf{a}_2 + \cdots + b_{r+1}\mathbf{a}_{r+1}$$

Therefore $r+1$ of the row vectors $\mathbf{a}_i$ are linearly dependent, so that $s \leq r$. Now consider $\mathbf{A}^T$. The same argument leads to $r \leq s$, so that $r = s$.

3.10. Show that the rank of the $m \times n$ matrix $\mathbf{A}$ equals the maximum number of linearly independent column vectors of $\mathbf{A}$.

Let there be r linearly independent column vectors, and without loss of generality let them be $\mathbf{a}_1, \mathbf{a}_2, \ldots, \mathbf{a}_r$. The $\mathbf{a}_i = \sum_{j=1}^{r} \alpha_{ij}\mathbf{a}_j$ for $i = r+1, \ldots, n$. Any $\mathbf{y}$ in the range space of $\mathbf{A}$ can be expressed in terms of an arbitrary $\mathbf{x}$ as

$$\mathbf{y} = \mathbf{A}\mathbf{x} = \sum_{i=1}^{n} \mathbf{a}_i x_i = \sum_{i=1}^{r} \mathbf{a}_i x_i + \sum_{i=r+1}^{n} \left(\sum_{j=1}^{r} \alpha_{ij}\mathbf{a}_j\right) x_i = \sum_{i=1}^{r} \left(x_i + \sum_{k=r+1}^{n} \alpha_{ki}x_k\right) \mathbf{a}_i$$

This shows the $\mathbf{a}_i$ for $i = 1, \ldots, r$ span the range space of $\mathbf{A}$, and since they are linearly independent they are a basis, so the rank of $\mathbf{A} = r$.

3.11. For an $m \times n$ matrix $\mathbf{A}$, give necessary and sufficient conditions for the existence and uniqueness of the solutions of the matrix equations $\mathbf{A}\mathbf{x} = \mathbf{0}$, $\mathbf{A}\mathbf{x} = \mathbf{b}$ and $\mathbf{A}\mathbf{X} = \mathbf{B}$ in terms of the column vectors of $\mathbf{A}$.

For $\mathbf{A}\mathbf{x} = \mathbf{0}$, the solution $\mathbf{x} = \mathbf{0}$ always exists. A necessary and sufficient condition for uniqueness of the solution $\mathbf{x} = \mathbf{0}$ is that the column vectors $\mathbf{a}_1, \ldots, \mathbf{a}_n$ are linearly independent. To show the necessity: If $\mathbf{a}_i$ are dependent, by definition of linearly dependent some nonzero ξ_i exist such that $\sum_{i=1}^{n} \mathbf{a}_i\xi_i = \mathbf{0}$. Then there exists another solution $\mathbf{x} = (\xi_1\ \ldots\ \xi_n) \neq \mathbf{0}$. To show sufficiency: If the $\mathbf{a}_i$ are independent, only zero ξ_i exist such that $\sum_{i=1}^{n} \mathbf{a}_i\xi_i = \mathbf{A}\mathbf{x} = \mathbf{0}$.

For $\mathbf{A}\mathbf{x} = \mathbf{b}$, rewrite as $\mathbf{b} = \sum_{i=1}^{n} \mathbf{a}_i x_i$. Then from Problem 3.10 a necessary and sufficient condition for existence of solution is that $\mathbf{b}$ lie in the range space of $\mathbf{A}$, i.e. the space spanned by the column vectors. To find conditions on the uniqueness of solutions, write one solution as $(\eta_1\ \eta_2\ \ldots\ \eta_n)$ and another as $(\xi_1\ \xi_2\ \ldots\ \xi_n)$. Then $\mathbf{b} = \sum_{i=1}^{n} \mathbf{a}_i\eta_i = \sum_{i=1}^{n} \mathbf{a}_i\xi_i$ so that $\mathbf{0} = \sum_{i=1}^{n} (\eta_1 - \xi_i)\mathbf{a}_i$. The solution is unique if and only if $\mathbf{a}_1, \ldots, \mathbf{a}_n$ are linearly independent.

Whether or not $\mathbf{b} = \mathbf{0}$, necessary and sufficient conditions for existence and uniqueness of solution to $\mathbf{Ax} = \mathbf{b}$ are that $\mathbf{b}$ lie in the vector space spanned by the column vectors of $\mathbf{A}$ and that the column vectors are linearly independent.

Since $\mathbf{AX} = \mathbf{B}$ can be written as $\mathbf{Ax}_j = \mathbf{b}_j$ for each column vector $\mathbf{x}_j$ of $\mathbf{X}$ and $\mathbf{b}_j$ of $\mathbf{B}$, by the preceding it is required that each $\mathbf{b}_j$ lie in the range space of $\mathbf{A}$ and that all the column vectors form a basis for the range space of $\mathbf{A}$.

3.12. Given an $m \times n$ matrix $\mathbf{A}$. Show $\operatorname{rank} \mathbf{A} = \operatorname{rank} \mathbf{A}^T = \operatorname{rank} \mathbf{A}^T\mathbf{A} = \operatorname{rank} \mathbf{AA}^T$.

By Theorem 3.14, the rank of $\mathbf{A}$ equals the number r of linearly independent column vectors of $\mathbf{A}$. Hence the dimension of the vector space spanned by the column vectors equals r. By Theorem 3.13, then the dimension of the vector space spanned by the row vectors equals r. But the row vectors of $\mathbf{A}$ are the column vectors of $\mathbf{A}^T$, so $\mathbf{A}^T$ has rank r.

To show $\operatorname{rank} \mathbf{A} = \operatorname{rank} \mathbf{A}^T\mathbf{A}$, note both $\mathbf{A}$ and $\mathbf{A}^T\mathbf{A}$ have n columns. Then consider any vector $\mathbf{y}$ in the null space of $\mathbf{A}$, i.e. $\mathbf{Ay} = \mathbf{0}$. Then $\mathbf{A}^T\mathbf{Ay} = \mathbf{0}$, so that $\mathbf{y}$ is also in the null space of $\mathbf{A}^T\mathbf{A}$. Now consider any vector $\mathbf{z}$ in the null space of $\mathbf{A}^T\mathbf{A}$, i.e. $\mathbf{A}^T\mathbf{Az} = \mathbf{0}$. Then $\mathbf{z}^T\mathbf{A}^T\mathbf{Az} = \|\mathbf{Az}\|_2^2 = 0$, so that $\mathbf{Az} = \mathbf{0}$, i.e. $\mathbf{z}$ is also in the null space of $\mathbf{A}$. Therefore the null space of $\mathbf{A}$ is equal to the null space of $\mathbf{A}^T\mathbf{A}$, and has some dimension k. Use of Theorem 3.11 gives $\operatorname{rank} \mathbf{A} = n - k = \operatorname{rank} \mathbf{A}^T\mathbf{A}$. Substitution of $\mathbf{A}^T$ for $\mathbf{A}$ in this expression gives $\operatorname{rank} \mathbf{A}^T = \operatorname{rank} \mathbf{AA}^T$.

3.13. Given an $m \times n$ matrix $\mathbf{A}$ and an $n \times k$ matrix $\mathbf{B}$, show that $\operatorname{rank} \mathbf{AB} \leq \operatorname{rank} \mathbf{A}$, $\operatorname{rank} \mathbf{AB} \leq \operatorname{rank} \mathbf{B}$. Also show that if $\mathbf{B}$ is nonsingular, $\operatorname{rank} \mathbf{AB} = \operatorname{rank} \mathbf{A}$, and that if $\mathbf{A}$ is nonsingular, $\operatorname{rank} \mathbf{AB} = \operatorname{rank} \mathbf{B}$.

Let $\operatorname{rank} \mathbf{A} = r$, so that $\mathbf{A}$ has r linearly indepedendent column vectors $\mathbf{a}_1, \ldots, \mathbf{a}_r$. Then $\mathbf{a}_i = \sum_{j=1}^{r} \alpha_{ij}\mathbf{a}_j$ for $i = r+1, \ldots, n$. Therefore $\mathbf{AB} = \sum_{i=1}^{n} \mathbf{a}_i\mathbf{b}_i^T$ where $\mathbf{b}_i^T$ are the row vectors of $\mathbf{B}$, using partitioned multiplication. Hence

$$\mathbf{AB} = \sum_{i=1}^{r} \mathbf{a}_i\mathbf{b}_i^T + \sum_{i=r+1}^{n} \sum_{j=1}^{r} \alpha_{ij}\mathbf{a}_j\mathbf{b}_i^T = \sum_{i=1}^{r} \mathbf{a}_i \left(\mathbf{b}_i^T + \sum_{k=r+1}^{n} \alpha_{ki}\mathbf{b}_k^T \right)$$

so that all the column vectors of $\mathbf{AB}$ are made up of a linear combination of the r independent column vectors of $\mathbf{A}$, and therefore $\operatorname{rank} \mathbf{AB} \leq r$.

Furthermore, use of Theorem 3.15 gives $\operatorname{rank} \mathbf{B} = \operatorname{rank} \mathbf{B}^T$. Then use of the first part of this problem with $\mathbf{B}^T$ substituted for $\mathbf{A}$ and $\mathbf{A}^T$ for $\mathbf{B}$ gives $\operatorname{rank} \mathbf{B}^T\mathbf{A}^T \leq \operatorname{rank} \mathbf{B}$. Again, Theorem 3.15 gives $\operatorname{rank} \mathbf{AB} = \operatorname{rank} \mathbf{B}^T\mathbf{A}^T$, so that $\operatorname{rank} \mathbf{AB} \leq \operatorname{rank} \mathbf{B}$.

If $\mathbf{A}$ is nonsingular, $\operatorname{rank} \mathbf{B} = \operatorname{rank} \mathbf{A}^{-1}(\mathbf{AB}) \leq \operatorname{rank} \mathbf{AB}$, using $\mathbf{A}^{-1}$ for $\mathbf{A}$ and $\mathbf{AB}$ for $\mathbf{B}$ in the first result of this problem. But since $\operatorname{rank} \mathbf{AB} \leq \operatorname{rank} \mathbf{B}$, then $\operatorname{rank} \mathbf{AB} = \operatorname{rank} \mathbf{B}$ if $\mathbf{A}^{-1}$ exists. Similar reasoning can be used to prove the remaining statement.

3.14. Given n vectors $\mathbf{x}_1, \mathbf{x}_2, \ldots, \mathbf{x}_n$ in a generalized vector space possessing a scalar product. Define the Gram matrix $\mathbf{G}$ as the matrix whose elements $g_{ij} = (\mathbf{x}_i, \mathbf{x}_j)$. Prove that $\det \mathbf{G} = 0$ if and only if $\mathbf{x}_1, \mathbf{x}_2, \ldots, \mathbf{x}_n$ are linearly dependent. Note that $\mathbf{G}$ is a matrix whose elements are scalars, and that $\mathbf{x}_i$ might be a function of time.

Suppose $\det \mathbf{G} = 0$. Then from Theorem 3.7, $\beta_1\mathbf{g}_1 + \beta_2\mathbf{g}_2 + \cdots + \beta_n\mathbf{g}_n = \mathbf{0}$, where $\mathbf{g}$ is a column vector of $\mathbf{G}$. Then $0 = \sum_{i=1}^{n} \beta_i g_{ij} = \sum_{i=1}^{n} \beta_i(\mathbf{x}_i, \mathbf{x}_j)$.

Multiplying by a constant β_j^* and summing still gives zero, so that

$$0 = \sum_{j=1}^{n} \beta_j^* \sum_{i=1}^{n} \beta_i(\mathbf{x}_i, \mathbf{x}_j) = \left(\sum_{i=1}^{n} \beta_i^* \mathbf{x}_i, \sum_{j=1}^{n} \beta_j^* \mathbf{x}_j \right)$$

Use of property (2) of Definition 3.47 then gives $\sum_{i=1}^{n} \beta_i^* \mathbf{x}_i = \mathbf{0}$, which is the definition of linear dependence.

Now suppose the $\mathbf{x}_i$ are linearly dependent. Then there exist γ_i such that $\sum_{j=1}^{n} \gamma_j \mathbf{x}_j = \mathbf{0}$. Taking the inner product with any $\mathbf{x}_i$ gives $0 = \sum_{j=1}^{n} \gamma_j(\mathbf{x}_i, \mathbf{x}_j) = \sum_{j=1}^{n} \gamma_j g_{ij}$ for any i. Therefore $\sum_{j=1}^{n} \gamma_i \mathbf{g}_j = \mathbf{0}$ and the column vectors of $\mathbf{G}$ are linearly dependent so that $\det \mathbf{G} = 0$.

3.15. Given two linear transformations L_1 and L_2, both of whose domain and range are in $\mathcal{U}$. Show $\det(L_1L_2) = (\det L_1)(\det L_2)$ so that as a particular case $\det \mathbf{AB} = \det \mathbf{BA} = \det \mathbf{A} \det \mathbf{B}$.

Let $\mathcal{U}$ have a basis $\mathbf{b}_1, \mathbf{b}_2, \ldots, \mathbf{b}_n$. Using exterior products, from *(3.13)*,

$$\det(L_1L_2) = \frac{L_1L_2(\mathbf{b}_1) \wedge L_1L_2(\mathbf{b}_2) \wedge \cdots \wedge L_1L_2(\mathbf{b}_n)}{\mathbf{b}_1 \wedge \mathbf{b}_2 \wedge \cdots \wedge \mathbf{b}_n}$$

If L_2 is singular, the vectors $L_2(\mathbf{b}_i) = \mathbf{c}_i$ are linearly dependent, and so are the vectors $L_1L_2(\mathbf{b}_i) = L_1(\mathbf{c}_i)$. Then $\det(L_1L_2) = 0 = \det L_2$. If L_2 is nonsingular, the vectors $\mathbf{c}_i$ are linearly independent and form a basis of $\mathcal{U}$. Then $\mathbf{c}_1 \wedge \mathbf{c}_2 \wedge \cdots \wedge \mathbf{c}_n \neq 0$, so that

$$\begin{aligned} \det(L_1L_2) &= \frac{L_1(\mathbf{c}_1) \wedge L_1(\mathbf{c}_2) \wedge \cdots \wedge L_1(\mathbf{c}_n)}{\mathbf{c}_1 \wedge \mathbf{c}_2 \wedge \cdots \wedge \mathbf{c}_n} \, \frac{L_2(\mathbf{b}_1) \wedge L_2(\mathbf{b}_2) \wedge \cdots \wedge L_2(\mathbf{b}_n)}{\mathbf{b}_1 \wedge \mathbf{b}_2 \wedge \cdots \wedge \mathbf{b}_n} \\ &= (\det L_1)(\det L_2) \end{aligned}$$

3.16. Using exterior products, prove that $\det(\mathbf{I} + \mathbf{ab}^T) = 1 + \mathbf{b}^T\mathbf{a}$.

$$\begin{aligned} \det(\mathbf{I} + \mathbf{ab}^T) &= (\mathbf{e}_1 + b_1\mathbf{a}) \wedge (\mathbf{e}_2 + b_2\mathbf{a}) \wedge \cdots \wedge (\mathbf{e}_n + b_n\mathbf{a}) \\ &= \mathbf{e}_1 \wedge \mathbf{e}_2 \wedge \cdots \wedge \mathbf{e}_n + b_1\mathbf{a} \wedge \mathbf{e}_2 \wedge \cdots \wedge \mathbf{e}_n + b_2\mathbf{e}_1 \wedge \mathbf{a} \wedge \mathbf{e}_3 \wedge \cdots \wedge \mathbf{e}_n + \cdots \\ &\qquad + b_n\mathbf{e}_1 \wedge \cdots \wedge \mathbf{e}_{n-1} \wedge \mathbf{a} \end{aligned}$$

Use of Theorem 3.21 gives

$$\begin{aligned} \det(\mathbf{I} + \mathbf{ab}^T) &= 1 + b_1\mathbf{a}^T(\mathbf{e}_2 \wedge \cdots \wedge \mathbf{e}_n) - b_2\mathbf{a}^T(\mathbf{e}_1 \wedge \mathbf{e}_3 \wedge \cdots \wedge \mathbf{e}_n) \\ &\qquad + \cdots + (-1)^{n-1} b_n\mathbf{a}^T(\mathbf{e}_1 \wedge \cdots \wedge \mathbf{e}_{n-1}) \end{aligned}$$

but $\mathbf{e}_1 = (\mathbf{e}_2 \wedge \cdots \wedge \mathbf{e}_n)$, etc., so that

$$\det(\mathbf{I} + \mathbf{ab}^T) = 1 + \mathbf{a}^T b_1\mathbf{e}_1 + \mathbf{a}^T b_2\mathbf{e}_2 + \cdots + \mathbf{a}^T b_n\mathbf{e}_n = 1 + \mathbf{a}^T\mathbf{b}$$

Note use of this permits the definition of a projection matrix $\mathbf{P} = \mathbf{I} - \mathbf{ab}^T(\mathbf{a}^T\mathbf{b})^{-1}$ such that $\det \mathbf{P} = 0$, $\mathbf{Pa} = \mathbf{0}$, $\mathbf{P}^T\mathbf{b} = \mathbf{0}$, $\mathbf{P}^2 = \mathbf{P}$, and the transformation $\mathbf{y} = \mathbf{Px}$ leaves only the hyperplane $\mathbf{b}^T\mathbf{x}$ pointwise invariant, i.e. $\mathbf{b}^T\mathbf{y} = \mathbf{b}^T\mathbf{Px} = \mathbf{b}^T[\mathbf{I} - \mathbf{ab}^T(a^Tb)^{-1}]\mathbf{x} = (\mathbf{b}^T - \mathbf{b}^T)\mathbf{x} = \mathbf{0}$.

Supplementary Problems

3.17. Prove that an upper triangular matrix added to or multiplied by an upper triangular matrix results in an upper triangular matrix.

3.18. Using the formula given in Definition 3.13 for operations with elements, multiply the following matrices

$$\begin{pmatrix} a & 1 \\ 0 & b \end{pmatrix}\begin{pmatrix} 0 & 2 & j \\ \pi & x^2 & \sin t \end{pmatrix}$$

Next, partition in any compatible manner and verify the validity of partitioned multiplication.

3.19. Transpose $\begin{pmatrix} 0 & 2 & j \\ \pi & x^2 & \sin t \end{pmatrix}$, and then take the complex conjugate.

3.20. Prove $\mathbf{IA} = \mathbf{AI} = \mathbf{A}$ for any compatible matrix $\mathbf{A}$.

3.21. Prove all skew-symmetric matrices have all their diagonal elements equal to zero.

3.22. Prove $(\mathbf{AB})^\dagger = \mathbf{B}^\dagger\mathbf{A}^\dagger$.

3.23. Prove that matrix addition and multiplication are associative and distributive, and that matrix addition is commutative.

3.24. Find a nonzero matrix which, when multiplied by the matrix $\mathbf{B}$ of Example 3.5, page 40, results in the zero matrix. Hence conclude $\mathbf{AB} = \mathbf{0}$ does not necessarily mean $\mathbf{A} = \mathbf{0}$ or $\mathbf{B} = \mathbf{0}$.

3.25. How many times does one particular element appear in the sum for the determinant of an $n \times n$ matrix?

3.26. Prove $(\mathbf{AB})^{-1} = \mathbf{B}^{-1}\mathbf{A}^{-1}$ if the indicated inverses exist.

3.27. Prove $(\mathbf{A}^{-1})^T = (\mathbf{A}^T)^{-1}$.

3.28. Prove $\det \mathbf{A}^{-1} = (\det \mathbf{A})^{-1}$.

3.29. Verify

$$\det\begin{pmatrix} 2 & 2 \\ 1 & 3 \end{pmatrix} \det\begin{pmatrix} -1 & 0 \\ 1 & 1 \end{pmatrix} = \det\begin{pmatrix} 0 & 2 \\ 2 & 3 \end{pmatrix}$$

since $\begin{pmatrix} 2 & 2 \\ 1 & 3 \end{pmatrix}\begin{pmatrix} -1 & 0 \\ 1 & 1 \end{pmatrix} = \begin{pmatrix} 0 & 2 \\ 2 & 3 \end{pmatrix}$. Also verify

$$\begin{pmatrix} 0 & 2 \\ 2 & 3 \end{pmatrix}^{-1} = \begin{pmatrix} -1 & 0 \\ 1 & 1 \end{pmatrix}^{-1}\begin{pmatrix} 2 & 2 \\ 1 & 3 \end{pmatrix}^{-1}$$

3.30. Given $\mathbf{A} = \begin{pmatrix} 2 & 4 & 6 \\ 1 & 1 & 1 \\ 3 & 5 & 8 \end{pmatrix}$. Find $\mathbf{A}^{-1}$.

3.31. If both $\mathbf{A}^{-1}$ and $\mathbf{B}^{-1}$ exist, does $(\mathbf{A}+\mathbf{B})^{-1}$ exist in general?

3.32. Given a matrix $\mathbf{A}(t)$ whose elements are functions of time. Show $d\mathbf{A}^{-1}/dt = -\mathbf{A}^{-1}\dfrac{d\mathbf{A}}{dt}\mathbf{A}^{-1}$.

3.33. Let a nonsingular matrix $\mathbf{A}$ be partitioned into $\mathbf{A}_{11}$, $\mathbf{A}_{12}$, $\mathbf{A}_{21}$ and $\mathbf{A}_{22}$ such that $\mathbf{A}_{11}$ and $\mathbf{A}_{22} - \mathbf{A}_{21}\mathbf{A}_{11}^{-1}\mathbf{A}_{12}$ have inverses. Show that

$$\mathbf{A}^{-1} = \begin{pmatrix} \mathbf{I} & -\mathbf{A}_{11}^{-1}\mathbf{A}_{12} \\ \mathbf{0} & \mathbf{I} \end{pmatrix}\begin{pmatrix} \mathbf{A}_{11}^{-1} & \mathbf{0} \\ \mathbf{0} & (\mathbf{A}_{22} - \mathbf{A}_{21}\mathbf{A}_{11}^{-1}\mathbf{A}_{12})^{-1} \end{pmatrix}\begin{pmatrix} \mathbf{I} & \mathbf{0} \\ -\mathbf{A}_{21}\mathbf{A}_{11}^{-1} & \mathbf{I} \end{pmatrix}$$

and if $\mathbf{A}_{21} = \mathbf{0}$, then

$$\mathbf{A}^{-1} = \begin{pmatrix} \mathbf{A}_{11}^{-1} & -\mathbf{A}_{11}^{-1}\mathbf{A}_{12}\mathbf{A}_{22}^{-1} \\ \mathbf{0} & \mathbf{A}_{22}^{-1} \end{pmatrix}$$

3.34. Are the vectors $(2\ 0\ -1\ 3)$, $(1\ -3\ 4\ 0)$ and $(1\ 1\ -2\ 2)$ linearly independent?

3.35. Given the matrix equations

$$(a)\quad (\alpha_{11}\ \alpha_{12})\begin{pmatrix}\xi_1\\ \xi_2\end{pmatrix} = 0 \qquad (b)\quad \begin{pmatrix}\alpha_{11} & \alpha_{12}\\ \alpha_{21} & \alpha_{22}\end{pmatrix}\begin{pmatrix}\xi_1\\ \xi_2\end{pmatrix} = \begin{pmatrix}0\\ 0\end{pmatrix} \qquad (c)\quad \begin{pmatrix}\alpha_{11} & \alpha_{12}\\ \alpha_{21} & \alpha_{22}\\ \alpha_{31} & \alpha_{32}\end{pmatrix}\begin{pmatrix}\xi_1\\ \xi_2\end{pmatrix} = \begin{pmatrix}0\\ 0\\ 0\end{pmatrix}$$

Using algebraic manipulations on the scalar equations $\alpha_{i1}\xi_1 + \alpha_{i2}\xi_2 = 0$, find the conditions under which no solutions exist and the conditions under which many solutions exist, and thus verify Theorem 3.11 and the results of Problem 3.11.

3.36. Let $\mathbf{x}$ be in the null space of $\mathbf{A}$ and $\mathbf{y}$ be in the range space of $\mathbf{A}^T$. Show $\mathbf{x}^T\mathbf{y} = 0$.

3.37. Given matrix $\mathbf{A} = \begin{pmatrix}1 & 2 & 3 & 4\\ 1 & 1 & -1 & -1\end{pmatrix}$. Find a basis for the null space of $\mathbf{A}$.

3.38. For $\mathbf{A} = \begin{pmatrix}1 & -1\\ -2 & 2\end{pmatrix}$, show that (*a*) an arbitrary vector $\mathbf{z} = \begin{pmatrix}z_1\\ z_2\end{pmatrix}$ can be expressed as the sum of two vectors, $\mathbf{z} = \mathbf{x} + \mathbf{y}$, where $\mathbf{x}$ is in the range space of $\mathbf{A}$ and $\mathbf{y}$ is in the null space of the transpose of $\mathbf{A}$, and (*b*) this is true for any matrix $\mathbf{A}$.

3.39. Given $n \times k$ matrices $\mathbf{A}$ and $\mathbf{B}$ and an $m \times n$ matrix $\mathbf{X}$ such that $\mathbf{XA} = \mathbf{XB}$. Under what conditions can we conclude $\mathbf{A} = \mathbf{B}$?

3.40. Given $\mathbf{x}, \mathbf{y}$ in $\mathcal{U}$, where $\mathbf{b}_1, \mathbf{b}_2, \ldots, \mathbf{b}_n$ are an orthonormal basis of $\mathcal{U}$. Show that

$$(\mathbf{x}, \mathbf{y}) = \sum_{i=1}^{n} (\mathbf{x}, \mathbf{b}_i)(\mathbf{b}_i, \mathbf{y})$$

3.41. Given real vectors $\mathbf{x}$ and $\mathbf{y}$ such that $\|\mathbf{x}\|_2 = \|\mathbf{y}\|_2$. Show $(\mathbf{x} + \mathbf{y})$ is orthogonal to $(\mathbf{x} - \mathbf{y})$.

3.42. Show that $\operatorname{rank}(\mathbf{A} + \mathbf{B}) \le \operatorname{rank}\mathbf{A} + \operatorname{rank}\mathbf{B}$.

3.43. Define T as the operator that multiplies every vector in $\mathcal{V}_3$ by a constant α and then adds on a translation vector $\mathbf{t}_0$. Is T a linear operator?

3.44. Given the three vectors $\mathbf{a}_1 = (\sqrt{119}\ -4\ 3)$, $\mathbf{a}_2 = (\sqrt{119}\ -1\ 7)$ and $\mathbf{a}_3 = (\sqrt{119}\ -10\ -5)$, use the Gram-Schmit procedure on $\mathbf{a}_1$, $\mathbf{a}_2$ and $\mathbf{a}_3$ in the order given to find a set of orthonormal basis vectors.

3.45. Show that the exterior product $\boldsymbol{\phi}^p = \mathbf{a}_1 \wedge \cdots \wedge \mathbf{a}_p$ satisfies Definition 3.44, i.e. that is an element in a generalized vector space $\wedge^p\mathcal{U}$, the space of all linear combinations of p-fold exterior products.

3.46. Show $(\alpha_1\mathbf{e}_1 + \alpha_2\mathbf{e}_2 + \alpha_3\mathbf{e}_3) \wedge (\beta_1\mathbf{e}_1 + \beta_2\mathbf{e}_2 + \beta_3\mathbf{e}_3) = (\alpha_2\beta_3 - \alpha_3\beta_2)\mathbf{e}_1 + (\alpha_1\beta_3 - \alpha_3\beta_1)\mathbf{e}_2 + (\alpha_1\beta_2 - \alpha_2\beta_1)\mathbf{e}_3$, illustrating that $\mathbf{a} \wedge \mathbf{b}$ is the cross product in $\mathcal{V}_3$.

3.47. Given vectors $\mathbf{x}_1, \mathbf{x}_2, \ldots, \mathbf{x}_n$ and an $n \times n$ matrix $\mathbf{A}$ such that $\mathbf{y}_1, \mathbf{y}_2, \ldots, \mathbf{y}_n$ are linearly independent, where $\mathbf{y}_i = \mathbf{A}\mathbf{x}_i$. Prove that $\mathbf{x}_1, \mathbf{x}_2, \ldots, \mathbf{x}_n$ are linearly independent.

3.48. Prove that the dimension of $\wedge^p\mathcal{U} = \dfrac{n!}{(n-p)!\,p!}$

3.49. Show that the remainder for the Schwartz inequality

$$(\mathbf{a}, \mathbf{a})(\mathbf{b}, \mathbf{b}) - |(\mathbf{a}, \mathbf{b})|^2 = \frac{1}{2}\sum_{i=1}^{n}\sum_{j=1}^{n} |a_i b_j - a_j b_i|^2$$

for n-vectors. What is the remainder for the inner product defined as $(\mathbf{a}, \mathbf{b}) = \int a^*(t)\, b(t)\, dt$?

Answers to Supplementary Problems

3.18. $\begin{pmatrix} \pi & 2a + x^2 & aj + \sin t \\ b\pi & bx^2 & b \sin t \end{pmatrix}$

3.19. $\begin{pmatrix} 0 & \pi \\ 2 & x^2 \\ -j & \sin t \end{pmatrix}$

3.24. $\begin{pmatrix} -2\alpha & \alpha \\ -2\beta & \beta \end{pmatrix}$ for any α and β

3.25. $(n-1)!$ as is most easily seen by the Laplace expansion.

3.30. $\mathbf{A}^{-1} = \begin{pmatrix} -3/2 & 1 & 1 \\ 5/2 & 1 & -2 \\ -1 & -1 & 1 \end{pmatrix}$

3.31. No

3.34. No

3.37. $(5 \ -4 \ 1 \ 0)^T$ and $(6 \ -5 \ 0 \ 1)^T$ are a basis.

3.38. (*a*) $\mathbf{x} = \begin{pmatrix} 1 \\ -2 \end{pmatrix} \alpha$ and $\mathbf{y} = \begin{pmatrix} 2 \\ 1 \end{pmatrix} \theta$ where α and θ are arbitrary, and since $\mathbf{x}$ and $\mathbf{y}$ are independent they span $\mathcal{V}_2$.

3.39. The n column vectors of $\mathbf{X}$ must be linearly independent.

3.43. No, this is an *affine* transformation.

3.44. $\mathbf{b}_1 = (\sqrt{119} \ -4 \ 3)/12$, $\mathbf{b}_2 = (0 \ 3 \ 4)/5$ and $\mathbf{a}_3$ is coplanar with $\mathbf{a}_1$ and $\mathbf{a}_2$ so that only $\mathbf{b}_1$ and $\mathbf{b}_2$ are required.

3.48. One method of proof uses induction.

3.49. $\frac{1}{2} \int\int |a(t)\, b(\tau) - a(\tau)\, b(t)|^2 \, d\tau \, dt$

Chapter 4

Matrix Analysis

4.1 EIGENVALUES AND EIGENVECTORS

Definition 4.1: An *eigenvalue* of the $n \times n$ (square) matrix $\mathbf{A}$ is one of those scalars λ that permit a nontrivial ($\mathbf{x} \neq \mathbf{0}$) solution to the equation

$$\mathbf{A}\mathbf{x} = \lambda\mathbf{x} \tag{4.1}$$

Note this equation can be rewritten as $(\mathbf{A}-\lambda\mathbf{I})\mathbf{x} = \mathbf{0}$. Nontrivial solution vectors $\mathbf{x}$ exist only if $\det(\mathbf{A}-\lambda\mathbf{I}) = 0$.

Example 4.1.

Find the eigenvalues of $\begin{pmatrix} 3 & 4 \\ 1 & 3 \end{pmatrix}$. The eigenvalue equation is

$$\begin{pmatrix} 3 & 4 \\ 1 & 3 \end{pmatrix}\begin{pmatrix} x_1 \\ x_2 \end{pmatrix} = \lambda\begin{pmatrix} x_1 \\ x_2 \end{pmatrix}$$

Then

$$\left\{\begin{pmatrix} 3 & 4 \\ 1 & 3 \end{pmatrix} - \lambda\begin{pmatrix} 1 & 0 \\ 0 & 1 \end{pmatrix}\right\}\begin{pmatrix} x_1 \\ x_2 \end{pmatrix} = \begin{pmatrix} 0 \\ 0 \end{pmatrix} \quad \text{or} \quad \begin{pmatrix} 3-\lambda & 4 \\ 1 & 3-\lambda \end{pmatrix}\begin{pmatrix} x_1 \\ x_2 \end{pmatrix} = \begin{pmatrix} 0 \\ 0 \end{pmatrix}$$

The characteristic equation is $\det\begin{pmatrix} 3-\lambda & 4 \\ 1 & 3-\lambda \end{pmatrix} = 0$. Then $(3-\lambda)(3-\lambda)-4 = 0$ or $\lambda^2 - 6\lambda + 5 = 0$, a second-order polynomial equation whose roots are the eigenvalues $\lambda_1 = 1$, $\lambda_2 = 5$.

Definition 4.2: The *characteristic polynomial* of $\mathbf{A}$ is $\det(\mathbf{A}-\lambda\mathbf{I})$. Note the characteristic polynomial is an nth order polynomial. Then there are n eigenvalues $\lambda_1, \lambda_2, \ldots, \lambda_n$ that are the roots of this polynomial, although some might be repeated roots.

Definition 4.3: An eigenvalue of the square matrix $\mathbf{A}$ is said to be *distinct* if it is not a repeated root.

Definition 4.4: Associated with each eigenvalue λ_i of the $n \times n$ matrix $\mathbf{A}$ there is a nonzero solution vector $\mathbf{x}_i$ of the eigenvalue equation $\mathbf{A}\mathbf{x}_i = \lambda_i\mathbf{x}_i$. This solution vector is called an *eigenvector*.

Example 4.2.

In the previous example, the eigenvector assocated with the eigenvalue 1 is found as follows.

$$\begin{pmatrix} 3 & 4 \\ 1 & 3 \end{pmatrix}\begin{pmatrix} x_1 \\ x_2 \end{pmatrix} = (1)\begin{pmatrix} x_1 \\ x_2 \end{pmatrix} \quad \text{or} \quad \begin{pmatrix} 2 & 4 \\ 1 & 2 \end{pmatrix}\begin{pmatrix} x_1 \\ x_2 \end{pmatrix} = \begin{pmatrix} 0 \\ 0 \end{pmatrix}$$

Then $2x_1 + 4x_2 = 0$ and $x_1 + 2x_2 = 0$, from which $x_1 = -2x_2$. Thus the eigenvector $\mathbf{x}_1$ is $\mathbf{x}_1 = \begin{pmatrix} -2 \\ 1 \end{pmatrix} x_2$ where the scalar x_2 can be any number.

Note that eigenvectors have arbitrary length. This is true because for any scalar α, the equation $\mathbf{A}\mathbf{x} = \lambda\mathbf{x}$ has a solution vector $\alpha\mathbf{x}$ since $\mathbf{A}(\alpha\mathbf{x}) = \alpha\mathbf{A}\mathbf{x} = \alpha\lambda\mathbf{x} = \lambda(\alpha\mathbf{x})$.

Definition 4.5: An eigenvector is *normalized* to unity if its length is unity, i.e. $\|\mathbf{x}\| = 1$. Sometimes it is easier to normalize $\mathbf{x}$ such that one of the elements is unity.

Example 4.3.

The eigenvector normalized to unit length belonging to the eigenvalue 1 in the previous example is $\mathbf{x}_1 = \frac{1}{\sqrt{5}}\begin{pmatrix} -2 \\ 1 \end{pmatrix}$, whereas normalizing its first element to unity gives $\mathbf{x}_1 = \begin{pmatrix} 1 \\ -1/2 \end{pmatrix}$.

4.2 INTRODUCTION TO THE SIMILARITY TRANSFORMATION

Consider the general time-invariant continuous-time state equation with no inputs,

$$d\mathbf{x}(t)/dt = \mathbf{A}\mathbf{x}(t) \tag{4.2}$$

where $\mathbf{A}$ is a constant coefficient $n \times n$ matrix of real numbers. The initial condition is given as $\mathbf{x}(0) = \mathbf{x}_0$.

Example 4.4.

Written out, the state equations (*4.2*) are

$$\begin{aligned} dx_1(t)/dt &= a_{11}x_1(t) + a_{12}x_2(t) + \cdots + a_{1n}x_n(t) \\ dx_2(t)/dt &= a_{21}x_1(t) + a_{22}x_2(t) + \cdots + a_{2n}x_n(t) \\ &\cdots\cdots\cdots\cdots\cdots\cdots\cdots\cdots \\ dx_n(t)/dt &= a_{n1}x_1(t) + a_{n2}x_2(t) + \cdots + a_{nn}x_n(t) \end{aligned}$$

and the initial conditions are given, such as

$$\begin{pmatrix} x_1(0) \\ x_2(0) \\ \vdots \\ x_n(0) \end{pmatrix} = \begin{pmatrix} 7 \\ \pi \\ \vdots \\ y \end{pmatrix}$$

Now define a new variable, an n-vector $\mathbf{y}(t)$, by the one to one relationship

$$\mathbf{y}(t) = \mathbf{M}^{-1}\mathbf{x}(t) \tag{4.3}$$

It is required that $\mathbf{M}$ be an $n \times n$ nonsingular constant coefficient matrix so that the solution $\mathbf{x}$ can be determined from the solution for the new variable $\mathbf{y}(t)$. Putting $\mathbf{x}(t) = \mathbf{M}\mathbf{y}(t)$ into the system equation gives

$$\mathbf{M}\, d\mathbf{y}(t)/dt = \mathbf{A}\mathbf{M}\mathbf{y}(t)$$

Multiplying on the left by $\mathbf{M}^{-1}$ gives

$$d\mathbf{y}(t)/dt = \mathbf{M}^{-1}\mathbf{A}\mathbf{M}\mathbf{y}(t) \tag{4.4}$$

Definition 4.6: The transformation $\mathbf{T}^{-1}\mathbf{A}\mathbf{T}$, where $\mathbf{T}$ is an arbitrary matrix, is called a *similarity* transformation on $\mathbf{A}$. It is called similarity because the problem is similar to the original one but with a change of variables from $\mathbf{x}$ to $\mathbf{y}$.

Suppose $\mathbf{M}$ was chosen very cleverly to make $\mathbf{M}^{-1}\mathbf{A}\mathbf{M}$ a diagonal matrix $\mathbf{\Lambda}$. Then

$$\frac{d\mathbf{y}(t)}{dt} = \frac{d}{dt}\begin{pmatrix} y_1(t) \\ y_2(t) \\ \vdots \\ y_n(t) \end{pmatrix} = \begin{pmatrix} \lambda_1 & 0 & \ldots & 0 \\ 0 & \lambda_2 & \ldots & 0 \\ \ldots & \ldots & \ldots & \ldots \\ 0 & 0 & \ldots & \lambda_n \end{pmatrix}\begin{pmatrix} y_1(t) \\ y_2(t) \\ \vdots \\ y_n(t) \end{pmatrix} = \mathbf{\Lambda}\mathbf{y}(t)$$

Writing this equation out gives $dy_i/dt = \lambda_i y_i$ for $i = 1, 2, \ldots, n$. The solution can be expressed simply as $y_i(t) = y_i(0)e^{\lambda_i t}$. Therefore if an $\mathbf{M}$ such that $\mathbf{M}^{-1}\mathbf{AM} = \mathbf{\Lambda}$ can be found, solution of $d\mathbf{x}/dt = \mathbf{Ax}$ becomes easy. Although not always, such an $\mathbf{M}$ can usually be found. In cases where it cannot, a $\mathbf{T}$ can always be found where $\mathbf{T}^{-1}\mathbf{AT}$ is almost diagonal. Physically it must be the case that not all differential equations can be reduced to this simple form. Some differential equations have as solutions $te^{\lambda_i t}$, and there is no way to get this solution from the simple form.

The transformation $\mathbf{M}$ is constructed upon solution of the eigenvalue problem for all the eigenvectors $\mathbf{x}_i$, $i = 1, 2, \ldots, n$. Because $\mathbf{Ax}_i = \lambda_1 \mathbf{x}_i$ for $i = 1, 2, \ldots, n$, the equations can be "stacked up" using the rules of multiplication of partitioned matrices:

$$\begin{aligned} \mathbf{A}(\mathbf{x}_1 \mid \mathbf{x}_2 \mid \ldots \mid \mathbf{x}_n) &= (\mathbf{Ax}_1 \mid \mathbf{Ax}_2 \mid \ldots \mid \mathbf{Ax}_n) \\ &= (\lambda_1\mathbf{x}_1 \mid \lambda_2\mathbf{x}_2 \mid \ldots \mid \lambda_n\mathbf{x}_n) \\ &= (\mathbf{x}_1 \mid \mathbf{x}_2 \mid \ldots \mid \mathbf{x}_n)\begin{pmatrix} \lambda_1 & 0 & \ldots & 0 \\ 0 & \lambda_2 & \ldots & 0 \\ \ldots & \ldots & \ldots & \ldots \\ 0 & 0 & \ldots & \lambda_n \end{pmatrix} \\ &= (\mathbf{x}_1 \mid \mathbf{x}_2 \mid \ldots \mid \mathbf{x}_n)\mathbf{\Lambda} \end{aligned}$$

Therefore

$$\mathbf{M} = (\mathbf{x}_1 \mid \mathbf{x}_2 \mid \ldots \mid \mathbf{x}_n) \tag{4.5}$$

When $\mathbf{M}$ is singular, $\mathbf{\Lambda}$ cannot be found. Under a number of different conditions, it can be shown $\mathbf{M}$ is nonsingular. One of these conditions is stated as the next theorem, and other conditions will be found later.

Theorem 4.1: If the eigenvalues of an $n \times n$ matrix are distinct, then the eigenvectors are linearly independent.

Note that if the eigenvectors are linearly independent, $\mathbf{M}$ is nonsingular.

Proof: The proof is by contradiction. Let $\mathbf{A}$ have distinct eigenvalues. Let $\mathbf{x}_1, \mathbf{x}_2, \ldots, \mathbf{x}_n$ be the eigenvectors of $\mathbf{A}$, with $\mathbf{x}_1, \mathbf{x}_2, \ldots, \mathbf{x}_k$ independent and $\mathbf{x}_{k+1}, \ldots, \mathbf{x}_n$ dependent. Then $\mathbf{x}_j = \sum_{i=1}^{k} \beta_{ij}\mathbf{x}_i$ for $j = k+1, k+2, \ldots, n$ where not all $\beta_{ij} = 0$. Since $\mathbf{x}_j$ is an eigenvector,

$$\mathbf{Ax}_j = \lambda_j \mathbf{x}_j = \lambda_j \sum_{i=1}^{k} \beta_{ij}\mathbf{x}_i \quad \text{for } j = k+1, \ldots, n$$

Also,

$$\mathbf{Ax}_j = \mathbf{A}\left(\sum_{i=1}^{k} \beta_{ij}\mathbf{x}_i\right) = \sum_{i=1}^{k} \beta_{ij}\mathbf{Ax}_i = \sum_{i=1}^{k} \beta_{ij}\lambda_i\mathbf{x}_i$$

Subtracting this equation from the previous one gives

$$\mathbf{0} = \sum_{i=1}^{k} \beta_{ij}(\lambda_j - \lambda_i)\mathbf{x}_i$$

But the $\mathbf{x}_i$, $i = 1, 2, \ldots, k$, were assumed to be linearly independent. Because not all β_{ij} are zero, some $\lambda_i = \lambda_j$. This contradicts the assumption that $\mathbf{A}$ had distinct eigenvalues, and so all the eigenvectors of $\mathbf{A}$ must be linearly independent.

4.3 PROPERTIES OF SIMILARITY TRANSFORMATIONS

To determine when a $\mathbf{T}$ can be found such that $\mathbf{T}^{-1}\mathbf{AT}$ gives a diagonal matrix, the properties of a similarity transformation must be examined. Define

$$\mathbf{S} = \mathbf{T}^{-1}\mathbf{AT} \tag{4.6}$$

Then the eigenvalues of $\mathbf{S}$ are found as the roots of $\det(\mathbf{S}-\lambda\mathbf{I}) = 0$. But

$$\begin{aligned}\det(\mathbf{S}-\lambda\mathbf{I}) &= \det(\mathbf{T}^{-1}\mathbf{A}\mathbf{T}-\lambda\mathbf{I})\\ &= \det(\mathbf{T}^{-1}\mathbf{A}\mathbf{T}-\lambda\mathbf{T}^{-1}\mathbf{I}\mathbf{T})\\ &= \det[\mathbf{T}^{-1}(\mathbf{A}-\lambda\mathbf{I})\mathbf{T}]\end{aligned}$$

Using the product rule for determinants,

$$\det(\mathbf{S}-\lambda\mathbf{I}) = \det\mathbf{T}^{-1}\det(\mathbf{A}-\lambda\mathbf{I})\det\mathbf{T}$$

Since $\det\mathbf{T}^{-1} = (\det\mathbf{T})^{-1}$ from Problem 3.12, $\det(\mathbf{S}-\lambda\mathbf{I}) = \det(\mathbf{A}-\lambda\mathbf{I})$. Therefore we have proved

Theorem 4.2: All similar matrices have the same eigenvalues.

Corollary 4.3: All similar matrices have the same traces and determinants.

Proof of this corollary is given in Problem 4.1.

A useful fact to note here also is that all triangular matrices $\mathbf{B}$ display eigenvalues on the diagonal, because the determinant of the triangular matrix $(\mathbf{B}-\lambda\mathbf{I})$ is the product of its diagonal elements.

Theorem 4.4: A matrix $\mathbf{A}$ can be reduced to a diagonal matrix $\mathbf{\Lambda}$ by a similarity transformation if and only if a set of n linearly independent eigenvectors can be found.

Proof: By Theorem 4.2, the diagonal matrix $\mathbf{\Lambda}$ must have the eigenvalues of $\mathbf{A}$ appearing on the diagonal. If $\mathbf{AT} = \mathbf{T\Lambda}$, by partitioned matrix multiplication it is required that $\mathbf{A}\mathbf{t}_i = \lambda_i\mathbf{t}_i$, where $\mathbf{t}_i$ are the column vectors of $\mathbf{T}$. Therefore it is required that $\mathbf{T}$ have the eigenvectors of $\mathbf{A}$ as its column vectors, and $\mathbf{T}^{-1}$ exists if and only if its column vectors are linearly independent.

It has already been shown that when the eigenvalues are distinct, $\mathbf{T}$ is nonsingular. So consider what happens when the eigenvalues are not distinct. Theorem 4.4 says that the only way we can obtain a diagonal matrix is to find n linearly independent eigenvectors. Then there are two cases:

Case 1. For each root that is repeated k times, the space of eigenvectors belonging to that root is k-dimensional. In this case the matrix can still be reduced to a diagonal form.

Example 4.5.

Given the matrix $\mathbf{A} = \begin{pmatrix} 1 & 0 & 0\\ 1 & 1 & 1\\ -1 & 0 & 0\end{pmatrix}$. Then $\det(\mathbf{A}-\lambda\mathbf{I}) = -\lambda(1-\lambda)^2$ and the eigenvalues are 0, 1 and 1. For the zero eigenvalue, solution of $\mathbf{Ax} = \mathbf{0}$ gives $\mathbf{x} = (0\ \ 1\ \ -1)$. For the unity eigenvalue, the eigenvalue problem is

$$\begin{pmatrix} 1 & 0 & 0\\ 1 & 1 & 1\\ -1 & 0 & 0\end{pmatrix}\begin{pmatrix} x_1\\ x_2\\ x_3\end{pmatrix} = (1)\begin{pmatrix} x_1\\ x_2\\ x_3\end{pmatrix}$$

This gives the set of equations

$$\begin{aligned} 0 &= 0\\ x_1 + x_3 &= 0\\ -x_1 - x_3 &= 0\end{aligned}$$

Therefore all eigenvectors belonging to the eigenvalue 1 have the form

$$\begin{pmatrix} x_1 \\ x_2 \\ x_3 \end{pmatrix} = \begin{pmatrix} 0 \\ 1 \\ 0 \end{pmatrix} x_2 + \begin{pmatrix} 1 \\ 0 \\ -1 \end{pmatrix} x_1$$

where x_1 and x_2 are arbitrary. Hence any two linearly independent vectors in the space spanned by $(0\ 1\ 0)$ and $(1\ 0\ -1)$ will do. The transformation matrix is then

$$\mathbf{T} = \mathbf{M} = \begin{pmatrix} 0 & 0 & 1 \\ 1 & 1 & 0 \\ -1 & 0 & -1 \end{pmatrix}$$

and $\mathbf{T}^{-1}\mathbf{A}\mathbf{T} = \mathbf{\Lambda}$, where $\mathbf{\Lambda}$ has 0, 1 and 1 on the diagonal *in that order.*

Note that the occurrence of distinct eigenvalues falls into Case 1. Every distinct eigenvalue must have at least one eigenvector associated with it, and since there are n distinct eigenvalues there are n eigenvectors. By Theorem 4.1 these are linearly independent.

Case 2. The conditions of Case 1 do not hold. Then the matrix cannot be reduced to a diagonal form by a similarity transformation.

Example 4.6.

Given the matrix $\mathbf{A} = \begin{pmatrix} 1 & 1 \\ 0 & 1 \end{pmatrix}$. Since $\mathbf{A}$ is triangular, the eigenvalues are displayed as 1 and 1. Then the eigenvalue problem is

$$\begin{pmatrix} 1 & 1 \\ 0 & 1 \end{pmatrix}\begin{pmatrix} x_1 \\ x_2 \end{pmatrix} = (1)\begin{pmatrix} x_1 \\ x_2 \end{pmatrix}$$

which gives the set of equations $x_2 = 0$, $0 = 0$. All eigenvectors belonging to 1 have the form $(x_1\ 0)^T$. Two linearly independent eigenvectors are simply not available to form $\mathbf{M}$.

Because in Case 2 a diagonal matrix cannot be formed by a similarity transformation, there arises the question of what is the simplest matrix that is almost diagonal that can be formed by a similarity transformation. This is answered in the next section.

4.4 JORDAN FORM

The form closest to diagonal to which an arbitrary $n \times n$ matrix can be transformed by a similarity transformation is the Jordan form, denoted $\mathbf{J}$. Proof of its existence in all cases can be found in standard texts. In the interest of brevity we omit the lengthy development needed to show this form can always be obtained, and merely show how to obtain it. The Jordan form $\mathbf{J}$ is an upper triangular matrix and, as per the remarks of the preceding section, the eigenvalues of the $\mathbf{A}$ matrix must be displayed on the diagonal. If the $\mathbf{A}$ matrix has r linearly independent eigenvectors, the Jordan form has $n - r$ ones above the diagonal, and all other elements are zero. The general form is

$$\mathbf{J} = \begin{pmatrix} \mathbf{L}_{11}(\lambda_1) & & & & & & \mathbf{0} \\ & \mathbf{L}_{21}(\lambda_1) & & & & & \\ & & \ddots & & & & \\ & & & \mathbf{L}_{k1}(\lambda_1) & & & \\ & & & & \mathbf{L}_{12}(\lambda_2) & & \\ & & & & & \ddots & \\ \mathbf{0} & & & & & & \mathbf{L}_{mp}(\lambda_p) \end{pmatrix} \qquad (4.7)$$

Each $\mathbf{L}_{ji}(\lambda_i)$ is an upper triangular square matrix, called a Jordan block, on the diagonal of the Jordan form $\mathbf{J}$. Several $\mathbf{L}_{ji}(\lambda_i)$ can be associated with each value of λ_i, and may differ in dimension from one another. A general $\mathbf{L}_{ji}(\lambda_i)$ looks like

$$\mathbf{L}_{ji}(\lambda_i) = \begin{pmatrix} \lambda_i & 1 & 0 & \ldots & 0 \\ 0 & \lambda_i & 1 & \ldots & 0 \\ 0 & 0 & \lambda_i & \ldots & 0 \\ \ldots & \ldots & \ldots & \ldots & \ldots \\ 0 & 0 & 0 & \ldots & \lambda_i \end{pmatrix} \tag{4.8}$$

where λ_i are on the diagonal and ones occur in *all* places just above the diagonal.

Example 4.7.

Consider the Jordan form $\mathbf{J} = \begin{pmatrix} \lambda_1 & 1 & 0 & 0 \\ 0 & \lambda_1 & 0 & 0 \\ 0 & 0 & \lambda_1 & 0 \\ 0 & 0 & 0 & \lambda_2 \end{pmatrix}$. Because all ones must occur above the diagonal in a Jordan block, wherever a zero above the diagonal occurs in $\mathbf{J}$ there must occur a boundary between two Jordan blocks. Therefore this $\mathbf{J}$ contains three Jordan blocks,

$$\mathbf{L}_{11}(\lambda_1) = \begin{pmatrix} \lambda_1 & 1 \\ 0 & \lambda_1 \end{pmatrix}, \quad \mathbf{L}_{21}(\lambda_1) = \lambda_1, \quad \mathbf{L}_{12}(\lambda_2) = \lambda_2$$

There is one and only one linearly independent eigenvector associated with each Jordan block and vice versa. This leads to the calculation procedure for the other column vectors $\mathbf{t}_l$ of $\mathbf{T}$ called generalized eigenvectors associated with each Jordan block $L_{ji}(\lambda_i)$:

$$\begin{aligned} \mathbf{A}\mathbf{x}_i &= \lambda_i\mathbf{x}_i \\ \mathbf{A}\mathbf{t}_1 &= \lambda_i\mathbf{t}_1 + \mathbf{x}_i \\ \mathbf{A}\mathbf{t}_2 &= \lambda_i\mathbf{t}_2 + \mathbf{t}_1 \\ &\ldots\ldots\ldots\ldots \\ \mathbf{A}\mathbf{t}_l &= \lambda_i\mathbf{t}_l + \mathbf{t}_{l-1} \\ &\ldots\ldots\ldots\ldots \end{aligned} \tag{4.9}$$

Note the number of $\mathbf{t}_l$ equals the number of ones in the associated $\mathbf{L}_{ji}(\lambda_i)$. Then

$$\begin{aligned} \mathbf{A}(\mathbf{x}_i \mid \mathbf{t}_1 \mid \mathbf{t}_2 \mid \ldots \mid \mathbf{t}_l \mid \ldots) &= (\lambda_i\mathbf{x}_i \mid \lambda_i\mathbf{t}_1 + \mathbf{x}_i \mid \lambda_i\mathbf{t}_2 + \mathbf{t}_1 \mid \ldots \mid \lambda_i\mathbf{t}_l + \mathbf{t}_{l-1} \mid \ldots) \\ &= (\mathbf{x}_i \mid \mathbf{t}_1 \mid \mathbf{t}_2 \mid \ldots \mid \mathbf{t}_l \mid \ldots)\mathbf{L}_{ji}(\lambda_i) \end{aligned}$$

This procedure for calculating the $\mathbf{t}_l$ works very well as long as $\mathbf{x}_i$ is determined to within a multiplicative constant, because then each $\mathbf{t}_l$ is determined to within a multiplicative constant. However, difficulty is encountered whenever there is more than one Jordan block associated with a single value of an eigenvalue. Considerable background in linear algebra is required to find a construction procedure for the $\mathbf{t}_l$ in this case, which arises so seldom in practice that the general case will not be pursued here. If this case arises, a trial and error procedure along the lines of the next example can be used.

Example 4.8.

Find the transformation matrix $\mathbf{T}$ that reduces the matrix $\mathbf{A}$ to Jordan form, where

$$\mathbf{A} = \begin{pmatrix} 2 & 1 & 1 \\ 0 & 3 & 1 \\ 0 & -1 & 1 \end{pmatrix}$$

The characteristic equation is $(2-\lambda)(3-\lambda)(1-\lambda)+(2-\lambda)=0$. A factor $2-\lambda$ can be removed, and the remaining equation can be arranged so that the characteristic equation becomes $(2-\lambda)^3=0$. Solving for the eigenvectors belonging to the eigenvalue 2 results in

$$\begin{pmatrix} 0 & 1 & 1 \\ 0 & 1 & 1 \\ 0 & -1 & -1 \end{pmatrix}\begin{pmatrix} x_1 \\ x_2 \\ x_3 \end{pmatrix} = \begin{pmatrix} 0 \\ 0 \\ 0 \end{pmatrix}$$

Therefore any eigenvector can be expressed in a linear combination as

$$\mathbf{x} = \begin{pmatrix} 1 \\ 0 \\ 0 \end{pmatrix}\alpha + \begin{pmatrix} 0 \\ 1 \\ -1 \end{pmatrix}\beta$$

What combination should be tried to start the procedure described by equations *(4.9)*? Trying the general expression gives

$$\begin{pmatrix} 2 & 1 & 1 \\ 0 & 3 & 1 \\ 0 & -1 & 1 \end{pmatrix}\begin{pmatrix} \tau_1 \\ \tau_2 \\ \tau_3 \end{pmatrix} = 2\begin{pmatrix} \tau_1 \\ \tau_2 \\ \tau_3 \end{pmatrix} + \begin{pmatrix} 1 \\ 0 \\ 0 \end{pmatrix}\alpha + \begin{pmatrix} 0 \\ 1 \\ -1 \end{pmatrix}\beta$$

Then

$$\tau_2 + \tau_3 = \alpha$$
$$\tau_2 + \tau_3 = \beta$$
$$-\tau_2 - \tau_3 = -\beta$$

These equations are satisfied if $\alpha = \beta$ This gives the correct $\mathbf{x} = \alpha(1\ 1\ -1)^T$. Normalizing $\mathbf{x}$ by setting $\alpha = 1$ gives $\mathbf{t} = (\tau_1\ \tau_2\ 1-\tau_2)^T$. The transformation matrix is completed by any other linearly independent choice of $\mathbf{x}$, say $(0\ 1\ -1)^T$, and any choice of τ_1 and τ_2 such that $\mathbf{t}$ is linearly independent of the choices of all $\mathbf{x}$, say $\tau_1 = 0$ and $\tau_2 = 1$. This gives $\mathbf{AT} = \mathbf{TJ}$, or

$$\begin{pmatrix} 2 & 1 & 1 \\ 0 & 3 & 1 \\ 0 & -1 & 1 \end{pmatrix}\begin{pmatrix} 1 & 0 & 0 \\ 1 & 1 & 1 \\ -1 & 0 & -1 \end{pmatrix} = \begin{pmatrix} 1 & 0 & 0 \\ 1 & 1 & 1 \\ -1 & 0 & -1 \end{pmatrix}\begin{pmatrix} 2 & 1 & 0 \\ 0 & 2 & 0 \\ 0 & 0 & 2 \end{pmatrix}$$

4.5 QUADRATIC FORMS

Definition 4.7: A *quadratic form* $\mathcal{Q}$ is a real polynomial in the real variables $\xi_1, \xi_2, \ldots, \xi_n$ containing only terms of the form $\alpha_{ij}\xi_i\xi_j$, such that $\mathcal{Q} = \sum_{i=1}^{n}\sum_{j=1}^{n} \alpha_{ij}\xi_i\xi_j$, where α_{ij} is real for all i and j.

Example 4.9.

Some typical quadratic forms are

$$\mathcal{Q}_1 = 7\xi_1^2$$
$$\mathcal{Q}_2 = 3\xi_1^2 - 2\xi_1\xi_2 + \xi_2^2 + 5\xi_1\xi_3 - 7\xi_2\xi_1$$
$$\mathcal{Q}_3 = \alpha_{11}\xi_1^2 + \alpha_{12}\xi_1\xi_2 + \alpha_{21}\xi_2\xi_1 + \alpha_{22}\xi_2^2$$
$$\mathcal{Q}_4 = t\xi_1^2 + (1-t^2)\xi_1\xi_2 - e^t\xi_3^2$$

Theorem 4.5: All quadratic forms $\mathcal{Q}$ can be expressed as the inner product $(\mathbf{x}, \mathbf{Qx})$ and vice versa, where $\mathbf{Q}$ is an $n \times n$ Hermitian matrix, i.e. $\mathbf{Q}^\dagger = \mathbf{Q}$.

Proof: First $\mathcal{Q}$ to $(\mathbf{x}, \mathbf{Qx})$:

$$\mathcal{Q} = \sum_{i=1}^{n}\sum_{j=1}^{n} \alpha_{ij}\xi_i\xi_j \tag{4.10}$$

Let $\mathbf{Q} = \{q_{ij}\} = \frac{1}{2}\{\alpha_{ij} + \alpha_{ji}\}$. Then $q_{ij} = q_{ji}$, so $\mathbf{Q}$ is real and symmetric, and $\mathcal{Q} = \mathbf{x}^T\mathbf{Q}\mathbf{x}$.

Next, $(\mathbf{x}, \mathbf{Q}\mathbf{x})$ to $\mathcal{Q}$ (the problem is to prove the coefficients are real):

$$(\mathbf{x}, \mathbf{Q}\mathbf{x}) = \sum_{i=1}^{n}\sum_{j=1}^{n} q_{ij}\xi_i\xi_j \quad \text{and} \quad (\mathbf{x}, \mathbf{Q}^\dagger\mathbf{x}) = \sum_{i=1}^{n}\sum_{j=1}^{n} q_{ij}^*\xi_i\xi_j$$

Then

$$(\mathbf{x}, \mathbf{Q}\mathbf{x}) = \tfrac{1}{2}(\mathbf{x}, \mathbf{Q}\mathbf{x}) + \tfrac{1}{2}(\mathbf{x}, \mathbf{Q}^\dagger\mathbf{x}) = \tfrac{1}{2}\sum_{i=1}^{n}\sum_{j=1}^{n}(q_{ij} + q_{ij}^*)\xi_i\xi_j$$

So $(\mathbf{x}, \mathbf{Q}\mathbf{x}) = \sum_{i=1}^{n}\sum_{j=1}^{n} \operatorname{Re}(q_{ij})\xi_i\xi_j = \mathcal{Q}$ and the coefficients are real.

***Theorem 4.6*:** The eigenvalues of an $n \times n$ Hermitian matrix $\mathbf{Q} = \mathbf{Q}^\dagger$ are real, and the eigenvectors belonging to distinct eigenvalues are orthogonal.

The most important case of real symmetric $\mathbf{Q}$ is included in Theorem 4.6 because the set of real symmetric matrices is included in the set of Hermitian matrices.

Proof: The eigenvalue problems for specific λ_i and λ_j are

$$\begin{aligned} \mathbf{Q}\mathbf{x}_i &= \lambda_i\mathbf{x}_i \\ \mathbf{Q}\mathbf{x}_j &= \lambda_j\mathbf{x}_j \end{aligned} \tag{4.11}$$

Since $\mathbf{Q}$ is Hermitian,

$$\mathbf{Q}^\dagger\mathbf{x}_j = \lambda_j\mathbf{x}_j$$

Taking the complex conjugate transpose gives

$$\mathbf{x}_j^\dagger\mathbf{Q} = \lambda_j^*\mathbf{x}_j^\dagger \tag{4.12}$$

Multiplying (*4.12*) on the right by $\mathbf{x}_i$ and (*4.11*) on the left by $\mathbf{x}_j^\dagger$ gives

$$\mathbf{x}_j^\dagger\mathbf{Q}\mathbf{x}_i - \mathbf{x}_j^\dagger\mathbf{Q}\mathbf{x}_i = 0 = (\lambda_j^* - \lambda_i)\mathbf{x}_j^\dagger\mathbf{x}_i$$

If $j = i$, then $\sqrt{\mathbf{x}_i^\dagger\mathbf{x}_i}$ is a norm on $\mathbf{x}_i$ and cannot be zero, so that $\lambda_i = \lambda_i^*$, meaning each eigenvalue is real. Then if $j \neq i$, $\lambda_j^* - \lambda_i = \lambda_j - \lambda_i$. But for distinct eigenvalues, $\lambda_j - \lambda_i \neq 0$, so $\mathbf{x}_j^\dagger\mathbf{x}_i = \mathbf{0}$ and the eigenvectors are orthogonal.

***Theorem 4.7*:** Even if the eigenvalues are not distinct, a set of n orthonormal eigenvectors can be found for an $n \times n$ normal matrix $\mathbf{N}$.

The proof is left to the solved problems. Note both Hermitian and real symmetric matrices are normal so that Theorem 4.6 is a special case of this theorem.

Corollary 4.8: A Hermitian (or real symmetric) matrix $\mathbf{Q}$ can always be reduced to a diagonal matrix by a unitary transformation, where $\mathbf{U}^{-1}\mathbf{Q}\mathbf{U} = \mathbf{\Lambda}$ and $\mathbf{U}^{-1} = \mathbf{U}^\dagger$.

Proof: Since $\mathbf{Q}$ is Hermitian, it is also normal. Then by Theorem 4.7 there are n orthonormal eigenvectors and they are all independent. By Theorem 4.4 this is a necessary and sufficient condition for diagonalization. To show a transformation matrix is unitary, construct $\mathbf{U}$ with the orthonormal eigenvectors as column vectors. Then

$$\mathbf{U}^\dagger\mathbf{U} = \begin{pmatrix} \mathbf{x}_1^\dagger \\ \hline \mathbf{x}_2^\dagger \\ \hline \vdots \\ \hline \mathbf{x}_n^\dagger \end{pmatrix}(\mathbf{x}_1 \mid \mathbf{x}_2 \mid \ldots \mid \mathbf{x}_n) = \begin{pmatrix} \mathbf{x}_1^\dagger\mathbf{x}_1 & \mathbf{x}_1^\dagger\mathbf{x}_2 & \ldots & \mathbf{x}_1^\dagger\mathbf{x}_n \\ \mathbf{x}_2^\dagger\mathbf{x}_1 & \mathbf{x}_2^\dagger\mathbf{x}_2 & \ldots & \mathbf{x}_2^\dagger\mathbf{x}_n \\ \ldots & \ldots & \ldots & \ldots \\ \mathbf{x}_n^\dagger\mathbf{x}_1 & \mathbf{x}_n^\dagger\mathbf{x}_2 & \ldots & \mathbf{x}_n^\dagger\mathbf{x}_n \end{pmatrix}$$

But $\mathbf{x}_i^\dagger \mathbf{x}_j = (\mathbf{x}_i, \mathbf{x}_j) = \delta_{ij}$ because they are orthonormal. Then $\mathbf{U}^\dagger\mathbf{U} = \mathbf{I}$. Since the column vectors of $\mathbf{U}$ are linearly independent, $\mathbf{U}^{-1}$ exists, so multiplying on the right by $\mathbf{U}^{-1}$ gives $\mathbf{U}^\dagger = \mathbf{U}^{-1}$, which was to be proven.

Therefore if a quadratic form $\mathcal{Q} = \mathbf{x}^\dagger\mathbf{Q}\mathbf{x}$ is given, rotating coordinates by defining $\mathbf{x} = \mathbf{U}\mathbf{y}$ gives $\mathcal{Q} = \mathbf{y}^\dagger\mathbf{U}^\dagger\mathbf{Q}\mathbf{U}\mathbf{y} = \mathbf{y}^\dagger\mathbf{\Lambda}\mathbf{y}$. In other words, $\mathcal{Q}$ can be expressed as

$$\mathcal{Q} = \lambda_1|y_1|^2 + \lambda_2|y_2|^2 + \cdots + \lambda_n|y_n|^2$$

where the λ_i are the real eigenvalues of $\mathbf{Q}$. Note $\mathcal{Q}$ is always positive if the eigenvalues of $\mathbf{Q}$ are positive, unless $\mathbf{y}$, and hence $\mathbf{x}$, is identically the zero vector. Then the square root of $\mathcal{Q}$ is a norm of the $\mathbf{x}$ vector because an inner product can be defined as $(\mathbf{x}, \mathbf{y})_{\mathbf{Q}} = \mathbf{x}^\dagger\mathbf{Q}\mathbf{y}$.

Definition 4.8: An $n \times n$ Hermitian matrix $\mathbf{Q}$ is *positive definite* if its associated quadratic form $\mathcal{Q}$ is always positive except when $\mathbf{x}$ is identically the zero vector. Then $\mathbf{Q}$ is positive definite if and only if all its eigenvalues are > 0.

Definition 4.9: An $n \times n$ Hermitian matrix $\mathbf{Q}$ is *nonnegative definite* if its associated quadratic form $\mathcal{Q}$ is never negative. (It may be zero at times when $\mathbf{x}$ is not zero.) Then $\mathbf{Q}$ is nonnegative if and only if all its eigenvalues are ≥ 0.

Example 4.10.

$\mathcal{Q} = \xi_1^2 - 2\xi_1\xi_2 + \xi_2^2 = (\xi_1 - \xi_2)^2$ can be zero when $\xi_1 = \xi_2$, and so is nonnegative definite.

The geometric solution of constant $\mathcal{Q}$ when $\mathbf{Q}$ is positive definite is an ellipse in n-space.

***Theorem 4.9*:** A unique positive definite Hermitian matrix $\mathbf{R}$ exists such that $\mathbf{R}\mathbf{R} = \mathbf{Q}$, where $\mathbf{Q}$ is a Hermitian positive definite matrix. $\mathbf{R}$ is called the *square root* of $\mathbf{Q}$.

Proof: Let $\mathbf{U}$ be the unitary matrix that diagonalizes $\mathbf{Q}$. Then $\mathbf{Q} = \mathbf{U}\mathbf{\Lambda}\mathbf{U}^\dagger$. Since λ_{ii} is a positive diagonal element of $\mathbf{\Lambda}$, define $\mathbf{\Lambda}^{1/2}$ as the diagonal matrix of positive $\lambda_{ii}^{1/2}$.

$$\mathbf{Q} = \mathbf{U}\mathbf{\Lambda}^{1/2}\mathbf{\Lambda}^{1/2}\mathbf{U}^\dagger = \mathbf{U}\mathbf{\Lambda}^{1/2}\mathbf{U}^\dagger\mathbf{U}\mathbf{\Lambda}^{1/2}\mathbf{U}^\dagger$$

Now let $\mathbf{R} = \mathbf{U}\mathbf{\Lambda}^{1/2}\mathbf{U}^\dagger$ and it is symmetric, real and positive definite because its eigenvalues are positive. Uniqueness is proved in Problem 4.5.

One way to check if a Hermitian matrix is positive definite (or nonnegative definite) is to see if its eigenvalues are all positive (or nonnegative). Another way to check is to use Sylvester's criterion.

Definition 4.10: The mth leading *principal minor*, denoted $\det \mathbf{Q}_m$, of the $n \times n$ Hermitian matrix $\mathbf{Q}$ is the determinant of the matrix $\mathbf{Q}_m$ formed by deleting the last $n - m$ rows and columns of $\mathbf{Q}$.

***Theorem 4.10*:** A Hermitian matrix $\mathbf{Q}$ is positive definite (or nonnegative definite) if and only if all the leading principal minors of $\mathbf{Q}$ are positive (or nonnegative).

A proof is given in Problem 4.6.

Example 4.11.

Given $\mathbf{Q} = \{q_{ij}\}$. Then $\mathbf{Q}$ is positive definite if and only if

$$0 < \det \mathbf{Q}_1 = q_{11}; \quad 0 < \det \mathbf{Q}_2 = \det \begin{pmatrix} q_{11} & q_{12} \\ q_{12}^* & q_{22} \end{pmatrix}; \quad \ldots; \quad 0 < \det \mathbf{Q}_n = \det \mathbf{Q}$$

If $<$ is replaced by $\leq$, $\mathbf{Q}$ is nonnegative definite.

Rearrangement of the elements of $\mathbf{Q}$ sometimes leads to simpler algebraic inequalities.

Example 4.12.

The quadratic form

$$\mathcal{Q} = \mathbf{x}^\dagger \mathbf{Q} \mathbf{x} = (\xi_1 \ \xi_2)\begin{pmatrix} q_{11} & q_{12} \\ q_{12} & q_{22} \end{pmatrix}\begin{pmatrix} \xi_1 \\ \xi_2 \end{pmatrix} = q_{11}\xi_1^2 + 2q_{12}\xi_1\xi_2 + q_{22}\xi_2^2$$

is positive definite if $q_{11} > 0$ and $q_{11}q_{22} - q_{12}^2 > 0$. But $\mathcal{Q}$ can be written another way:

$$\mathcal{Q} = q_{11}\xi_1^2 + 2q_{12}\xi_1\xi_2 + q_{22}\xi_2^2 = (\xi_2 \ \xi_1)\begin{pmatrix} q_{22} & q_{12} \\ q_{12} & q_{11} \end{pmatrix}\begin{pmatrix} \xi_2 \\ \xi_1 \end{pmatrix}$$

which is positive definite if $q_{22} > 0$ and $q_{11}q_{22} - q_{12}^2 > 0$. The conclusion is that $\mathbf{Q}$ is positive definite if $\det \mathbf{Q} > 0$ and either q_{22} or q_{11} can be shown greater than zero.

4.6 MATRIX NORMS

Definition 4.11: A *norm of a matrix* $\mathbf{A}$, denoted $||\mathbf{A}||$, is the minimum value of κ such that $||\mathbf{Ax}|| \le \kappa ||\mathbf{x}||$ for all $\mathbf{x}$.

Geometrically, multiplication by a matrix $\mathbf{A}$ changes the length of a vector. Choose the vector $\mathbf{x}_0$ whose length is increased the most. Then $||\mathbf{A}||$ is the ratio of the length of $\mathbf{Ax}_0$ to the length of $\mathbf{x}_0$.

The matrix norm is understood to be the same kind of norm as the vector norm in its defining relationship. Vector norms have been covered in Section 3.10. Hence the matrix norm is to be taken in the sense of corresponding vector norm.

Example 4.13.

To find $||\mathbf{U}||_2$, where $\mathbf{U}^\dagger = \mathbf{U}^{-1}$, consider any nonzero vector $\mathbf{x}$.

$$||\mathbf{Ux}||_2^2 = \mathbf{x}^\dagger \mathbf{U}^\dagger \mathbf{U} \mathbf{x} = \mathbf{x}^\dagger \mathbf{x} = ||\mathbf{x}||_2^2 \quad \text{and so} \quad ||\mathbf{U}||_2 = 1$$

***Theorem 4.11*:** Properties of any matrix norm are:

(1) $||\mathbf{Ax}|| \le ||\mathbf{A}||\,||\mathbf{x}||$

(2) $||\mathbf{A}|| = \max ||\mathbf{Au}||$, where the maximum value of $||\mathbf{Au}||$ is to be taken over those $\mathbf{u}$ such that $||\mathbf{u}|| = 1$.

(3) $||\mathbf{A}+\mathbf{B}|| \le ||\mathbf{A}|| + ||\mathbf{B}||$

(4) $||\mathbf{AB}|| \le ||\mathbf{A}||\,||\mathbf{B}||$

(5) $|\lambda| \le ||\mathbf{A}||$ for any eigenvalue λ of $\mathbf{A}$.

(6) $||\mathbf{A}|| = 0$ if and only if $\mathbf{A} = \mathbf{0}$.

Proof: Since $||\mathbf{A}|| = \kappa_{\min}$, substitution into Definition 4.11 gives (1).

To show (2), consider the vector $\mathbf{u} = \mathbf{x}/\alpha$, where $\alpha = ||\mathbf{x}||$. Then

$$||\mathbf{Ax}|| = ||\alpha \mathbf{Au}|| = |\alpha|\,||\mathbf{Au}|| \le \kappa ||\mathbf{x}|| = \kappa |\alpha|\,||\mathbf{u}||$$

Division by $|\alpha|$ gives $||\mathbf{Au}|| \le \kappa ||\mathbf{u}||$, so that only unity length vectors need be considered instead of all $\mathbf{x}$ in Definition 4.11. Geometrically, since $\mathbf{A}$ is a linear operator its effect on the length of $\alpha\mathbf{x}$ is in direct proportion to its effect on $\mathbf{x}$, so that the selection of $\mathbf{x}_0$ should depend only on its direction.

To show (3),

$$||(\mathbf{A}+\mathbf{B})\mathbf{x}|| = ||\mathbf{Ax}+\mathbf{Bx}|| \le ||\mathbf{Ax}|| + ||\mathbf{Bx}|| \le (||\mathbf{A}|| + ||\mathbf{B}||)\,||\mathbf{x}||$$

for any $\mathbf{x}$. The first inequality results from the triangle inequality, Definition 3.46(4), for vector norms, as defined previously.

To show (4), use Definition 4.11 on the vectors **Bx** and **x**:

$$\|\mathbf{ABx}\| = \|\mathbf{A}(\mathbf{Bx})\| \leq \|\mathbf{A}\|\,\|\mathbf{Bx}\| \leq \|\mathbf{A}\|\,\|\mathbf{B}\|\,\|\mathbf{x}\|$$

for any **x**.

To show (5), consider the eigenvalue problem $\mathbf{Ax} = \lambda\mathbf{x}$. Then using the norm property of vectors Definition 3.46(5),

$$|\lambda|\,\|\mathbf{x}\| = \|\lambda\mathbf{x}\| = \|\mathbf{Ax}\| \leq \|\mathbf{A}\|\,\|\mathbf{x}\|$$

Since **x** is an eigenvector, $\|\mathbf{x}\| \neq 0$ and (5) follows.

To show (6), $\|\mathbf{A}\| = 0$ implies $\|\mathbf{Ax}\| = 0$ from Definition 4.11. Then $\mathbf{Ax} = \mathbf{0}$ so that $\mathbf{Ax} = \mathbf{0x}$ for any **x**. Therefore $\mathbf{A} = \mathbf{0}$. The converse $\|\mathbf{0}\| = 0$ is obvious.

Theorem 4.12: $\|\mathbf{A}\|_2 = \rho_{\max}$, where $\rho_{\max}^2$ is the maximum eigenvalue of $\mathbf{A}^\dagger\mathbf{A}$, and furthermore

$$0 \leq \rho_{\min} \leq \frac{\|\mathbf{Ax}\|_2}{\|\mathbf{x}\|_2} \leq \rho_{\max}$$

To calculate $\|\mathbf{A}\|_2$, find the maximum eigenvalue of $\mathbf{A}^\dagger\mathbf{A}$.

Proof: Consider the eigenvalue problem

$$\mathbf{A}^\dagger\mathbf{A}\mathbf{g}_i = \rho_i^2\mathbf{g}_i$$

Since $(\mathbf{x}, \mathbf{A}^\dagger\mathbf{Ax}) = (\mathbf{Ax}, \mathbf{Ax}) = \|\mathbf{Ax}\|_2^2 \geq 0$, then $\mathbf{A}^\dagger\mathbf{A}$ is nonnegative definite and $\rho_i^2 \geq 0$. Since $\mathbf{A}^\dagger\mathbf{A}$ is Hermitian, ρ_i^2 is real and the $\mathbf{g}_i$ can be chosen orthonormal. Express any **x** in $\mathcal{V}_n$ as $\mathbf{x} = \sum_{i=1}^{n} \xi_i\mathbf{g}_i$. Then

$$\frac{\|\mathbf{Ax}\|_2^2}{\|\mathbf{x}\|_2^2} = \frac{\sum_{i=1}^{n} \|\xi_i\mathbf{Ag}_i\|_2^2}{\sum_{i=1}^{n} \|\xi_i\mathbf{g}_i\|_2^2} = \frac{\sum_{i=1}^{n} \xi_i^2\rho_i^2}{\sum_{i=1}^{n} \xi_i^2}$$

Since $(\rho_i^2)_{\min}\sum_{i=1}^{n}\xi_i^2 \leq \sum_{i=1}^{n}\xi_i^2\rho_i^2 \leq (\rho_i^2)_{\max}\sum_{i=1}^{n}\xi_i^2$, taking square roots gives the inequality of the theorem. Note that $\|\mathbf{Ax}\|_2 = \rho_{\max}\|\mathbf{x}\|_2$ when **x** is the eigenvector $\mathbf{g}_i$ belonging to $(\rho_i^2)_{\max}$.

4.7 FUNCTIONS OF A MATRIX

Given an analytic scalar function $f(\alpha)$ of a scalar α, it can be uniquely expressed in a convergent Maclaurin series,

$$f(\alpha) = \sum_{k=0}^{\infty} f_k\alpha^k/k!$$

where $f_k = d^kf(\alpha)/d\alpha^k$ evaluated at $\alpha = 0$.

Definition 4.12: Given an analytic function $f(\alpha)$ of a scalar α, the *function of an* $n \times n$ *matrix* **A** is $f(\mathbf{A}) = \sum_{k=0}^{\infty} f_k\mathbf{A}^k/k!$.

Example 4.14.

Some functions of a matrix **A** are

$$\cos\mathbf{A} = (\cos 0)\mathbf{I} + (-\sin 0)\mathbf{A} + (-\cos 0)\mathbf{A}^2/2 + \cdots + (-1)^m\mathbf{A}^{2m}/(2m)! + \cdots$$

$$e^{\mathbf{A}t} = (e^0)\mathbf{I} + (e^0)\mathbf{A}t + (e^0)\mathbf{A}^2t^2/2 + \cdots + \mathbf{A}^kt^k/k! + \cdots$$

Theorem 4.13: If $\mathbf{T}^{-1}\mathbf{A}\mathbf{T} = \mathbf{J}$, then $f(\mathbf{A}) = \mathbf{T}f(\mathbf{J})\mathbf{T}^{-1}$.

Proof: Note $\mathbf{A}^k = \mathbf{A}\mathbf{A}\cdots\mathbf{A} = \mathbf{T}\mathbf{J}\mathbf{T}^{-1}\mathbf{T}\mathbf{J}\mathbf{T}^{-1}\cdots\mathbf{T}\mathbf{J}\mathbf{T}^{-1} = \mathbf{T}\mathbf{J}^k\mathbf{T}^{-1}$. Then

$$f(\mathbf{A}) = \sum_{k=0}^{\infty} f_k \mathbf{A}^k/k! = \sum_{k=0}^{\infty} f_k \mathbf{T}\mathbf{J}^k\mathbf{T}^{-1}/k!$$

$$= \mathbf{T}\left(\sum_{k=0}^{\infty} f_k \mathbf{J}^k/k!\right)\mathbf{T}^{-1} = \mathbf{T}f(\mathbf{J})\mathbf{T}^{-1}$$

Theorem 4.14: If $\mathbf{A} = \mathbf{T}\mathbf{\Lambda}\mathbf{T}^{-1}$, where $\mathbf{\Lambda}$ is the diagonal matrix of eigenvalues λ_i, then

$$f(\mathbf{A}) = f(\lambda_1)\mathbf{x}_1\mathbf{r}_1^\dagger + f(\lambda_2)\mathbf{x}_2\mathbf{r}_2^\dagger + \cdots + f(\lambda_n)\mathbf{x}_n\mathbf{r}_n^\dagger$$

Proof: From Theorem 4.13,

$$f(\mathbf{A}) = \mathbf{T}f(\mathbf{\Lambda})\mathbf{T}^{-1} \tag{4.13}$$

But

$$f(\mathbf{\Lambda}) = \sum_{k=0}^{\infty} f_k \mathbf{\Lambda}^k/k! = \sum_{k=0}^{\infty} \frac{f_k}{k!}\begin{pmatrix} \lambda_1^k & & & 0 \\ & \lambda_2^k & & \\ & & \ddots & \\ 0 & & & \lambda_n^k \end{pmatrix} = \begin{pmatrix} \sum_{k=0}^{\infty} f_k\lambda_1^k/k! & & & 0 \\ & \sum_{k=0}^{\infty} f_k\lambda_2^k/k! & & \\ & & \ddots & \\ 0 & & & \sum_{k=0}^{\infty} f_k\lambda_n^k/k! \end{pmatrix}$$

Therefore

$$f(\mathbf{\Lambda}) = \begin{pmatrix} f(\lambda_1) & & & 0 \\ & f(\lambda_2) & & \\ & & \ddots & \\ 0 & & & f(\lambda_n) \end{pmatrix} \tag{4.14}$$

Also, let $\mathbf{T} = (\mathbf{x}_1 \mid \mathbf{x}_2 \mid \ldots \mid \mathbf{x}_n)$ and $\mathbf{T}^{-1} = (\mathbf{r}_1 \mid \mathbf{r}_2 \mid \ldots \mid \mathbf{r}_n)^\dagger$, where $\mathbf{r}_i$ is the reciprocal basis vector. Then from (*4.13*) and (*4.14*),

$$f(\mathbf{A}) = (\mathbf{x}_1 \mid \mathbf{x}_2 \mid \ldots \mid \mathbf{x}_n)\begin{pmatrix} f(\lambda_1) & & & 0 \\ & f(\lambda_2) & & \\ & & \ddots & \\ 0 & & & f(\lambda_n) \end{pmatrix}\begin{pmatrix} \mathbf{r}_1^\dagger \\ \hline \mathbf{r}_2^\dagger \\ \hline \vdots \\ \hline \mathbf{r}_n^\dagger \end{pmatrix}$$

$$= (\mathbf{x}_1 \mid \mathbf{x}_2 \mid \ldots \mid \mathbf{x}_n)\begin{pmatrix} f(\lambda_1)\mathbf{r}_1^\dagger \\ \hline f(\lambda_2)\mathbf{r}_2^\dagger \\ \hline \vdots \\ \hline f(\lambda_n)\mathbf{r}_n^\dagger \end{pmatrix}$$

The theorem follows upon partitioned multiplication of these last two matrices.

The square root function $f(\alpha) = \alpha^{1/2}$ is not analytic upon substitution of any λ_i. Therefore the square root $\mathbf{R}$ of the positive definite matrix $\mathbf{Q}$ had to be adapted by always taking the positive square root of λ_i for uniqueness in Theorem 4.9.

Definition 4.13: Let $f(\alpha) = \alpha$ in Theorem 4.14 to get the *spectral representation* of $\mathbf{A} = \mathbf{T}\mathbf{\Lambda}\mathbf{T}^{-1}$:

$$\mathbf{A} = \sum_{i=1}^{n} \lambda_i \mathbf{x}_i \mathbf{r}_i^\dagger$$

Note that this is valid only for those $\mathbf{A}$ that can be diagonalized by a similarity transformation.

To calculate $f(\mathbf{A}) = \mathbf{T}f(\mathbf{J})\mathbf{T}^{-1}$, we must find $f(\mathbf{J})$. From equation (*4.7*),

$$\mathbf{J}^k = \begin{pmatrix} \mathbf{L}_{11}(\lambda_1) & & \mathbf{0} \\ & \ddots & \\ \mathbf{0} & & \mathbf{L}_{mn}(\lambda_n) \end{pmatrix}^k = \begin{pmatrix} \mathbf{L}_{11}^k(\lambda_1) & & \mathbf{0} \\ & \ddots & \\ \mathbf{0} & & \mathbf{L}_{mn}^k(\lambda_n) \end{pmatrix}$$

where the last equality follows from partitioned matrix multiplication. Then

$$f(\mathbf{J}) = \sum_{k=0}^{\infty} f_k \mathbf{J}^k/k! = \begin{pmatrix} f(\mathbf{L}_{11}) & & \mathbf{0} \\ & \ddots & \\ \mathbf{0} & & f(\mathbf{L}_{mn}) \end{pmatrix} \tag{4.15}$$

Hence, it is only necessary to find $f(\mathbf{L})$ and use (*4.15*). By calculation it can be found that for an $l \times l$ matrix $\mathbf{L}$,

$$\mathbf{L}^k(\lambda) = \begin{pmatrix} \lambda^k & k\lambda^{k-1} & \cdots & \binom{k}{l-1}\lambda^{k-(l-1)} \\ 0 & \lambda^k & \cdots & \binom{k}{l-2}\lambda^{k-(l-2)} \\ \cdots & \cdots & \cdots & \cdots \\ 0 & 0 & \cdots & \lambda^k \end{pmatrix} \tag{4.16}$$

where $\binom{n}{m} = \dfrac{n!}{(n-m)!\,m!}$, the number of combinations of n elements taken m at a time.

Then from $f(\mathbf{L}) = \sum_{n=0}^{\infty} f_k \mathbf{L}^k/k!$, the upper right hand terms α_{1l} are

$$\alpha_{1l} = \sum_{k=l-1}^{\infty} f_k \binom{k}{l-1} \lambda^{k-(l-1)}/k! = \sum_{k=l-1}^{\infty} f_k \lambda^{k-(l-1)}/[(l-1)!\,(k-l+1)!] \tag{4.17}$$

but

$$\frac{d^{l-1}f(\lambda)}{d\lambda^{l-1}} = \frac{d^{l-1}}{d\lambda^{l-1}} \sum_{k=0}^{\infty} f_k \lambda^k/k! = \sum_{k=l-1}^{\infty} f_k \lambda^{k-(l-1)}/(k-l+1)! \tag{4.18}$$

The series converge since $f(\lambda)$ is analytic, so that by comparing (*4.17*) and (*4.18*),

$$\alpha_{1l} = \frac{1}{(l-1)!} \frac{d^{l-1}f(\lambda)}{d\lambda^{l-1}}$$

Therefore

$$f(\mathbf{L}) = \begin{pmatrix} f(\lambda) & df/d\lambda & \cdots & [(l-1)!]^{-1} d^{l-1}f/d\lambda^{l-1} \\ 0 & f(\lambda) & \cdots & [(l-2)!]^{-1} d^{l-2}f/d\lambda^{l-2} \\ \cdots & \cdots & \cdots & \cdots \\ 0 & 0 & \cdots & f(\lambda) \end{pmatrix} \tag{4.19}$$

From (*4.19*) and (*4.15*) and Theorem 4.13, $f(\mathbf{A})$ can be computed.

Another almost equivalent method that can be used comes from the Cayley-Hamilton theorem.

***Theorem 4.15* (Cayley-Hamilton):** Given an arbitrary $n \times n$ matrix $\mathbf{A}$ with a characteristic polynomial $\phi(\lambda) = \det(\mathbf{A} - \lambda\mathbf{I})$. Then $\phi(\mathbf{A}) = \mathbf{0}$.

The proof is given in Problem 5.4.

Example 4.15.

Given $\mathbf{A} = \begin{pmatrix} 3 & 2 \\ 2 & 3 \end{pmatrix}$. Then

$$\det(\mathbf{A} - \lambda\mathbf{I}) = \phi(\lambda) = \lambda^2 - 6\lambda + 5 \tag{4.20}$$

By the Cayley-Hamilton theorem,

$$\phi(\mathbf{A}) = \mathbf{A}^2 - 6\mathbf{A} + 5\mathbf{I} = \begin{pmatrix} 13 & 12 \\ 12 & 13 \end{pmatrix} - 6\begin{pmatrix} 3 & 2 \\ 2 & 3 \end{pmatrix} + 5\begin{pmatrix} 1 & 0 \\ 0 & 1 \end{pmatrix} = \begin{pmatrix} 0 & 0 \\ 0 & 0 \end{pmatrix}$$

The Cayley-Hamilton theorem gives a means of expressing any power of a matrix in terms of a linear combination of $\mathbf{A}^m$ for $m = 0, 1, \ldots, n-1$.

Example 4.16.

From Example 4.15, the given $\mathbf{A}$ matrix satisfies $\mathbf{0} = \mathbf{A}^2 - 6\mathbf{A} + 5\mathbf{I}$. Then $\mathbf{A}^2$ can be expressed in terms of $\mathbf{A}$ and $\mathbf{I}$ by

$$\mathbf{A}^2 = 6\mathbf{A} - 5\mathbf{I} \tag{4.21}$$

Also $\mathbf{A}^3$ can be found by multiplying *(4.21)* by $\mathbf{A}$ and then using *(4.21)* again:

$$\mathbf{A}^3 = 6\mathbf{A}^2 - 5\mathbf{A} = 6(6\mathbf{A} - 5\mathbf{I}) - 5\mathbf{A} = 31\mathbf{A} - 30\mathbf{I}$$

Similarly any power of $\mathbf{A}$ can be found by this method, including $\mathbf{A}^{-1}$ if it exists, because *(4.21)* can be multiplied by $\mathbf{A}^{-1}$ to obtain

$$\mathbf{A}^{-1} = (6\mathbf{I} - \mathbf{A})/5$$

***Theorem 4.16*:** For an $n \times n$ matrix $\mathbf{A}$,

$$f(\mathbf{A}) = \gamma_1\mathbf{A}^{n-1} + \gamma_2\mathbf{A}^{n-2} + \cdots + \gamma_{n-1}\mathbf{A} + \gamma_n\mathbf{I}$$

where the scalars γ_i can be found from

$$f(\mathbf{J}) = \gamma_1\mathbf{J}^{n-1} + \gamma_2\mathbf{J}^{n-2} + \cdots + \gamma_{n-1}\mathbf{J} + \gamma_n\mathbf{I}$$

Here $f(\mathbf{J})$ is found from *(4.15)* and *(4.19)*, and very simply from *(4.14)* if $\mathbf{A}$ can be diagonalized.

This method avoids the calculation of $\mathbf{T}$ and $\mathbf{T}^{-1}$ at the expense of solving $f(\mathbf{J}) = \sum_{i=1}^{n} \gamma_i\mathbf{J}^{n-i}$ for the γ_i.

Proof: Since $f(\mathbf{A}) = \sum_{k=0}^{\infty} f_k\mathbf{A}^k/k!$ and $\mathbf{A}^k = \sum_{m=0}^{n-1} \alpha_{km}\mathbf{A}^m$ by the Cayley-Hamilton theorem, then

$$f(\mathbf{A}) = \sum_{k=0}^{\infty} f_k\left(\sum_{m=0}^{n-1} \alpha_{km}\mathbf{A}^m\right)\bigg/k! = \sum_{m=0}^{n-1} \mathbf{A}^m\left\{\sum_{k=0}^{\infty} f_k\alpha_{km}/k!\right\}$$

The quantity in brackets is γ_{n-m}.

Also, from Theorem 4.13,

$$f(\mathbf{J}) = \mathbf{T}^{-1}f(\mathbf{A})\mathbf{T} = \mathbf{T}^{-1}\sum_{i=1}^{n} \gamma_i\mathbf{A}^{n-i}\mathbf{T} = \sum_{i=1}^{n} \gamma_i\mathbf{T}^{-1}\mathbf{A}^{n-i}\mathbf{T} = \sum_{i=1}^{n} \gamma_i\mathbf{J}^{n-i}$$

Example 4.17.

For the $\mathbf{A}$ given in Example 4.15, $\cos\mathbf{A} = \gamma_1\mathbf{A} + \gamma_2\mathbf{I}$. Here $\mathbf{A}$ has eigenvalues $\lambda_1 = 1$, $\lambda_2 = 5$. From $\cos\mathbf{A} = \gamma_1\mathbf{A} + \gamma_2\mathbf{I}$ we obtain

$$\cos\lambda_1 = \gamma_1\lambda_1 + \gamma_2$$

$$\cos\lambda_2 = \gamma_1\lambda_2 + \gamma_2$$

Solving for γ_1 and γ_2 gives

$$\cos \mathbf{A} = \frac{\cos 1 - \cos 5}{1-5}\begin{pmatrix} 3 & 2 \\ 2 & 3 \end{pmatrix} + \frac{5\cos 1 - \cos 5}{5-1}\begin{pmatrix} 1 & 0 \\ 0 & 1 \end{pmatrix}$$

$$= \frac{\cos 1}{2}\begin{pmatrix} 1 & -1 \\ -1 & 1 \end{pmatrix} + \frac{\cos 5}{2}\begin{pmatrix} 1 & 1 \\ 1 & 1 \end{pmatrix}$$

Use of complex variable theory gives a very neat representation of $f(\mathbf{A})$ and leads to other computational procedures.

Theorem 4.17: If $f(\alpha)$ is analytic in a region containing the eigenvalues λ_i of $\mathbf{A}$, then

$$f(\mathbf{A}) = \frac{1}{2\pi j}\oint f(s)\,(s\mathbf{I}-\mathbf{A})^{-1}\,ds$$

where the contour integration is around the boundary of the region.

Proof: Since $f(\mathbf{A}) = \mathbf{T}f(\mathbf{J})\mathbf{T}^{-1} = \mathbf{T}\left[\frac{1}{2\pi j}\oint f(s)\,(s\mathbf{I}-\mathbf{J})^{-1}\,ds\right]\mathbf{T}^{-1}$, it suffices to show that $f(\mathbf{L}) = \frac{1}{2\pi j}\oint f(s)\,(s\mathbf{I}-\mathbf{L})^{-1}\,ds$. Since

$$s\mathbf{I} - \mathbf{L} = \begin{pmatrix} s-\lambda & -1 & \ldots & 0 \\ 0 & s-\lambda & \ldots & 0 \\ \ldots & \ldots & \ldots & \ldots \\ 0 & 0 & \ldots & s-\lambda \end{pmatrix}$$

then

$$(s\mathbf{I}-\mathbf{L})^{-1} = \frac{1}{(s-\gamma)^l}\begin{pmatrix} (s-\lambda)^{l-1} & (s-\lambda)^{l-2} & \ldots & 1 \\ 0 & (s-\lambda)^{l-1} & \ldots & s-\lambda \\ \ldots & \ldots & \ldots & \ldots \\ 0 & 0 & \ldots & (s-\lambda)^{l-1} \end{pmatrix}$$

The upper right hand term α_{1l} of $\frac{1}{2\pi j}\oint f(s)\,(s\mathbf{I}-\mathbf{L})^{-1}\,ds$ is

$$\alpha_{1l} = \frac{1}{2\pi j}\oint \frac{f(s)\,ds}{(s+\lambda)^l}$$

Because all the eigenvalues are within the contour, use of the Cauchy integral formula then gives

$$\alpha_{1l} = \frac{1}{(l-1)!}\,\frac{d^{l-1}f(\lambda)}{d\lambda^{l-1}}$$

which is identical with equation (*4.19*).

Example 4.18.

Using Theorem 4.17,

$$\cos \mathbf{A} = \frac{1}{2\pi j}\oint \cos s(s\mathbf{I}-\mathbf{A})^{-1}\,ds$$

For $\mathbf{A}$ as given in Example 4.15,

$$(s\mathbf{I}-\mathbf{A})^{-1} = \begin{pmatrix} s-3 & -2 \\ -2 & s-3 \end{pmatrix}^{-1} = \frac{1}{s^2-6s+5}\begin{pmatrix} s-3 & 2 \\ 2 & s-3 \end{pmatrix}$$

Then

$$\cos \mathbf{A} = \frac{1}{2\pi j}\oint \frac{\cos s}{(s-1)(s-5)}\begin{pmatrix} s-3 & 2 \\ 2 & s-3 \end{pmatrix} ds$$

Performing a *matrix* partial fraction expansion gives

$$\cos \mathbf{A} = \frac{1}{2\pi j}\oint \frac{\cos 1}{-4(s-1)}\begin{pmatrix} -2 & 2 \\ 2 & -2 \end{pmatrix} ds + \frac{1}{2\pi j}\oint \frac{\cos 5}{4(s-5)}\begin{pmatrix} 2 & 2 \\ 2 & 2 \end{pmatrix} ds$$

$$= \frac{\cos 1}{2}\begin{pmatrix} 1 & -1 \\ -1 & 1 \end{pmatrix} + \frac{\cos 5}{2}\begin{pmatrix} 1 & 1 \\ 1 & 1 \end{pmatrix}$$

4.8 PSEUDOINVERSE

When the determinant of an $n \times n$ matrix is zero, or even when the matrix is not square, there exists a way to obtain a solution "as close as possible" to the equation $\mathbf{Ax} = \mathbf{y}$. In this section we let $\mathbf{A}$ be an $m \times n$ real matrix and first examine the properties of the real symmetric square matrices $\mathbf{A}^T\mathbf{A}$ and $\mathbf{AA}^T$, which are $n \times n$ and $m \times m$ respectively.

Theorem 4.18: The matrix $\mathbf{B}^T\mathbf{B}$ is nonnegative definite.

Proof: Consider the quadratic form $\mathcal{Q} = \mathbf{y}^T\mathbf{y}$, which is never negative. Let $\mathbf{y} = \mathbf{Bx}$, where $\mathbf{B}$ can be $m \times n$. Then $\mathbf{x}^T\mathbf{B}^T\mathbf{Bx} \geq 0$, so that $\mathbf{B}^T\mathbf{B}$ is nonnegative definite.

From this theorem, Theorem 4.6 and Definition 4.9, the eigenvalues of $\mathbf{A}^T\mathbf{A}$ are either zero or positive and real. This is also true of the eigenvalues of $\mathbf{AA}^T$, in which case we let $\mathbf{B} = \mathbf{A}^T$ in Theorem 4.18.

Theorem 4.19: Let $\mathbf{A}$ be an $m \times n$ matrix of rank r, where $r \leq m$ and $r \leq n$; let $\mathbf{g}_i$ be an orthonormal eigenvector of $\mathbf{A}^T\mathbf{A}$; let $\mathbf{f}_i$ be an orthonormal eigenvector of $\mathbf{AA}^T$, and let ρ_i^2 be the nonzero eigenvalues of $\mathbf{A}^T\mathbf{A}$. Then

(1) ρ_i^2 are also the nonzero eigenvalues of $\mathbf{AA}^T$.

(2) $\mathbf{Ag}_i = \rho_i\mathbf{f}_i$ for $i = 1, 2, \ldots, r$

(3) $\mathbf{Ag}_i = \mathbf{0}$ for $i = r+1, \ldots, n$

(4) $\mathbf{A}^T\mathbf{f}_i = \rho_i\mathbf{g}_i$ for $i = 1, 2, \ldots, r$

(5) $\mathbf{A}^T\mathbf{f}_i = \mathbf{0}$ for $i = r+1, \ldots, m$

Proof: From Problem 3.12, there are exactly r nonzero eigenvalues of $\mathbf{A}^T\mathbf{A}$ and $\mathbf{AA}^T$. Then $\mathbf{A}^T\mathbf{Ag}_i = \rho_i^2\mathbf{g}_i$ for $i = 1, 2, \ldots, r$ and $\mathbf{A}^T\mathbf{Ag}_i = \mathbf{0}$ for $i = r+1, \ldots, n$. Define an m-vector $\mathbf{h}_i$ as $\mathbf{h}_i = \mathbf{Ag}_i/\rho_i$ for $i = 1, 2, \ldots, r$. Then

$$\mathbf{AA}^T\mathbf{h}_i = \mathbf{AA}^T\mathbf{Ag}_i/\rho_i = \mathbf{A}(\rho_i^2\mathbf{g}_i)/\rho_i = \rho_i^2\mathbf{h}_i$$

Furthermore,

$$\mathbf{h}_i^T\mathbf{h}_j = \mathbf{g}_i^T\mathbf{A}^T\mathbf{Ag}_j/\rho_i\rho_j = \rho_j\mathbf{g}_i^T\mathbf{g}_j/\rho_i = \delta_{ij}$$

Since for each i there is one normalized eigenvector, $\mathbf{h}_i$ can be taken equal to $\mathbf{f}_i$ and (2) is proven. Furthermore, since there are r of the ρ_i^2, these must be the eigenvalues of $\mathbf{AA}^T$ and (1) is proven. Also, we can find $\mathbf{f}_i$ for $i = r+1, \ldots, m$ such that $\mathbf{AA}^T\mathbf{f}_i = \mathbf{0}$ and are orthonormal. Hence the $\mathbf{f}_i$ are an orthonormal basis for $\mathcal{V}_m$ and the $\mathbf{g}_i$ are an orthonormal basis for $\mathcal{V}_n$.

To prove (3), $\mathbf{A}^T\mathbf{Ag}_i = \mathbf{0}$ for $i = r+1, \ldots, m$. Then $\|\mathbf{Ag}_i\|_2^2 = \mathbf{g}_i^T\mathbf{A}^T\mathbf{Ag}_i = 0$, so that $\mathbf{Ag}_i = \mathbf{0}$. Similarly, since $\mathbf{AA}^T\mathbf{f}_i = \mathbf{0}$ for $i = r+1, \ldots, m$, then $\mathbf{A}^T\mathbf{f}_i = \mathbf{0}$ and (5) is proven.

Finally, to prove (4),

$$\mathbf{A}^T\mathbf{f}_i = \mathbf{A}^T(\mathbf{Ag}_i/\rho_i) = (\mathbf{A}^T\mathbf{Ag}_i)/\rho_i = \rho_i^2\mathbf{g}_i/\rho_i = \rho_i\mathbf{g}_i$$

Example 4.19.

Let $\mathbf{A} = \begin{pmatrix} 6 & 0 & 4 & 0 \\ 6 & 1 & 0 & 6 \end{pmatrix}$. Then $m = r = 2$, $n = 4$.

$$\mathbf{A}\mathbf{A}^T = \begin{pmatrix} 52 & 36 \\ 36 & 73 \end{pmatrix} \qquad \mathbf{A}^T\mathbf{A} = \begin{pmatrix} 72 & 6 & 24 & 36 \\ 6 & 1 & 0 & 6 \\ 24 & 0 & 16 & 0 \\ 36 & 6 & 0 & 36 \end{pmatrix}$$

The eigenvalue of $\mathbf{AA}^T$ are $\rho_1^2 = 100$ and $\rho_2^2 = 25$. The eigenvalues of $\mathbf{A}^T\mathbf{A}$ are $\rho_1^2 = 100$, $\rho_2^2 = 25$, $\rho_3^2 = 0$, $\rho_4^2 = 0$. The eigenvectors of $\mathbf{A}^T\mathbf{A}$ and $\mathbf{AA}^T$ are

$$\begin{aligned}
\mathbf{g}_1 &= (\ 0.84 \quad 0.08 \quad 0.24 \quad 0.48)^T \\
\mathbf{g}_2 &= (\ 0.24 \quad -0.12 \quad 0.64 \quad -0.72)^T \\
\mathbf{g}_3 &= (-0.49 \quad 0.00 \quad 0.73 \quad 0.49)^T \\
\mathbf{g}_4 &= (\ 0.04 \quad -0.98 \quad -0.06 \quad -0.13)^T \\
\mathbf{f}_1 &= (\ 0.6 \quad 0.8)^T \\
\mathbf{f}_2 &= (\ 0.8 \quad -0.6)^T
\end{aligned}$$

From the above, propositions (1), (2), (3), (4) and (5) of Theorem 4.19 can be verified directly. Computationally it is easiest to find the eigenvalues ρ_1^2 and ρ_2^2 and eigenvectors $\mathbf{f}_1$ and $\mathbf{f}_2$ of $\mathbf{AA}^T$, and then obtain the $\mathbf{g}_i$ from propositions (4) and (3).

Theorem 4.20: Under the conditions of Theorem 4.19, $\mathbf{A} = \sum_{i=1}^{r} \rho_i \mathbf{f}_i \mathbf{g}_i^T$.

Proof: The $m \times n$ matrix $\mathbf{A}$ is a mapping from $\mathcal{V}_n$ to $\mathcal{V}_m$. Also, $\mathbf{g}_1, \mathbf{g}_2, \ldots, \mathbf{g}_n$ and $\mathbf{f}_1, \mathbf{f}_2, \ldots, \mathbf{f}_m$ form orthonormal bases for $\mathcal{V}_n$ and $\mathcal{V}_m$ respectively. For arbitrary $\mathbf{x}$ in $\mathcal{V}_n$,

$$\mathbf{x} = \sum_{i=1}^{n} \xi_i \mathbf{g}_i \qquad \text{where } \xi_i = \mathbf{g}_i^T \mathbf{x} \tag{4.22}$$

Use of properties (2) and (3) of Theorem 4.19 gives

$$\mathbf{Ax} = \sum_{i=1}^{n} \xi_i \mathbf{Ag}_i = \sum_{i=1}^{r} \xi_i \mathbf{Ag}_i + \sum_{i=r+1}^{n} \xi_i \mathbf{Ag}_i = \sum_{i=1}^{r} \xi_i \rho_i \mathbf{f}_i$$

From (*4.22*), $\xi_i = \mathbf{g}_i^T \mathbf{x}$ and so $\mathbf{Ax} = \sum_{i=1}^{r} \rho_i \mathbf{f}_i \mathbf{g}_i^T \mathbf{x}$. Since this holds for arbitrary $\mathbf{x}$, the theorem is proven.

Note that the representation of Theorem 4.20 holds even when $\mathbf{A}$ is rectangular, and has no spectral representation.

Example 4.20.

Using the $\mathbf{A}$ matrix and the results of Example 4.19,

$$\begin{pmatrix} 6 & 0 & 4 & 0 \\ 6 & 1 & 0 & 6 \end{pmatrix} = 10\begin{pmatrix} 0.6 \\ 0.8 \end{pmatrix}(0.84 \ \ 0.08 \ \ 0.24 \ \ 0.48) + 5\begin{pmatrix} 0.8 \\ -0.6 \end{pmatrix}(0.24 \ \ -0.12 \ \ 0.64 \ \ -0.72)$$

Definition 4.14: The *pseudoinverse*, denoted $\mathbf{A}^{-I}$, of the $m \times n$ real matrix $\mathbf{A}$ is the $n \times m$ real matrix $\mathbf{A}^{-I} = \sum_{i=1}^{r} \rho_i^{-1} \mathbf{g}_i \mathbf{f}_i^T$.

Example 4.21.

Again, for the $\mathbf{A}$ matrix of Example 4.19,

$$\mathbf{A}^{-I} = 0.1\begin{pmatrix} 0.84 \\ 0.08 \\ 0.24 \\ 0.48 \end{pmatrix}(0.6 \ \ 0.8) + 0.2\begin{pmatrix} 0.24 \\ -0.12 \\ 0.64 \\ -0.72 \end{pmatrix}(0.8 \ \ -0.6) = \begin{pmatrix} 0.0888 & 0.0384 \\ -0.0144 & 0.0208 \\ 0.1168 & -0.0576 \\ -0.0864 & 0.1248 \end{pmatrix}$$

Theorem 4.21: Given an $m \times n$ real matrix $\mathbf{A}$ and an arbitrary m-vector $\mathbf{y}$, consider the equation $\mathbf{Ax} = \mathbf{y}$. Define $\mathbf{x}_0 = \mathbf{A}^{-I}\mathbf{y}$. Then $\|\mathbf{Ax} - \mathbf{y}\|_2 \geq \|\mathbf{Ax}_0 - \mathbf{y}\|_2$ and for those $\mathbf{z} \neq \mathbf{x}_0$ such that $\|\mathbf{Az} - \mathbf{y}\|_2 = \|\mathbf{Ax}_0 - \mathbf{y}\|_2$, then $\|\mathbf{z}\|_2 > \|\mathbf{x}_0\|_2$.

In other words, if no solution to $\mathbf{Ax} = \mathbf{y}$ exists, $\mathbf{x}_0$ gives the closest possible solution. If the solution to $\mathbf{Ax} = \mathbf{y}$ is not unique, $\mathbf{x}_0$ gives the solution with the minimum norm.

Proof: Using the notation of Theorems 4.19 and 4.20, an arbitrary m-vector $\mathbf{y}$ and an arbitrary n-vector $\mathbf{x}$ can be written as

$$\mathbf{y} = \sum_{i=1}^{m} \eta_i \mathbf{f}_i, \qquad \mathbf{x} = \sum_{i=1}^{n} \xi_i \mathbf{g}_i \tag{4.23}$$

where $\eta_i = \mathbf{f}_i^T\mathbf{y}$ and $\xi_i = \mathbf{g}_i^T\mathbf{x}$. Then use of properties (2) and (3) of Theorem 4.19 gives

$$\mathbf{Ax} - \mathbf{y} = \sum_{i=1}^{n} \xi_i \mathbf{Ag}_i - \sum_{i=1}^{m} \eta_i \mathbf{f}_i = \sum_{i=1}^{r} (\xi_i \rho_i - \eta_i)\mathbf{f}_i + \sum_{i=r+1}^{m} \eta_i \mathbf{f}_i \tag{4.24}$$

Since the $\mathbf{f}_i$ are orthonormal,

$$\|\mathbf{Ax} - \mathbf{y}\|_2^2 = \sum_{i=1}^{r} (\xi_i \rho_i - \eta_i)^2 + \sum_{i=r+1}^{m} \eta_i^2$$

To minimize $\|\mathbf{Ax} - \mathbf{y}\|_2$ the best we can do is choose $\xi_i = \eta_i/\rho_i$ for $i = 1, 2, \ldots, r$. Then those vectors $\mathbf{z}$ in $\mathcal{V}_n$ that minimize $\|\mathbf{Ax} - \mathbf{y}\|_2$ can be expressed as

$$\mathbf{z} = \sum_{i=1}^{r} \mathbf{g}_i \eta_i/\rho_i + \sum_{i=r+1}^{n} \mathbf{g}_i \xi_i$$

where ξ_i for $i = r+1, \ldots, n$ is arbitrary. But

$$\|\mathbf{z}\|_2^2 = \sum_{i=1}^{r} (\eta_i/\rho_i)^2 + \sum_{r+1}^{n} \xi_i^2$$

The $\mathbf{z}$ with minimum norm must have $\xi_i = 0$ for $i = r+1, \ldots, n$. Then using $\eta_i = \mathbf{f}_i^T\mathbf{y}$ from (*4.23*) gives $\mathbf{z}$ with a minimum norm $= \sum_{i=1}^{r} \rho_i^{-1}\eta_i \mathbf{g}_i = \sum_{i=1}^{r} \rho_i^{-1}\mathbf{g}_i \mathbf{f}_i^T \mathbf{y} = \mathbf{A}^{-I}\mathbf{y} = \mathbf{x}_0$.

Example 4.22.

Solve the equations $6x_1 + 4x_3 = 1$ and $6x_1 + x_2 + 6x_4 = 10$.

This can be written $\mathbf{Ax} = \mathbf{y}$, where $\mathbf{y} = (1 \;\; 10)^T$ and $\mathbf{A}$ is the matrix of Example 4.19. Since the rank of $\mathbf{A}$ is 2 and these are four columns, the solution is not unique. The solution $\mathbf{x}_0$ with the minimum norm is

$$\mathbf{x}_0 = \mathbf{A}^{-I}\mathbf{y} = \begin{pmatrix} 0.0888 & 0.0384 \\ -0.0144 & 0.0208 \\ 0.1168 & -0.0576 \\ -0.0864 & 0.1248 \end{pmatrix} \begin{pmatrix} 1 \\ 10 \end{pmatrix} = \begin{pmatrix} 0.4728 \\ 0.1936 \\ -0.4592 \\ 1.1616 \end{pmatrix}$$

Theorem 4.22: If it exists, any solution to $\mathbf{Ax} = \mathbf{y}$ can be expressed as $\mathbf{x} = \mathbf{A}^{-I}\mathbf{y} + (\mathbf{I} - \mathbf{A}^{-I}\mathbf{A})\mathbf{z}$, where $\mathbf{z}$ is any arbitrary n-vector.

Proof: For a solution to exist, $\eta_i = 0$ for $i = r+1, \ldots, m$ in equation (*4.24*), and $\xi_i = \eta_i/\rho_i$ for $i = 1, 2, \ldots, r$. Then any solution $\mathbf{x}$ can be written as

$$\mathbf{x} = \sum_{i=1}^{r} \mathbf{g}_i \eta_i/\rho_i + \sum_{i=r+1}^{n} \mathbf{g}_i \zeta_i \tag{4.25}$$

where ζ_i for $i = r+1, \ldots, n$ are arbitrary scalars. Denote in $\mathcal{V}_n$ an arbitrary vector $\mathbf{z} = \sum_{i=1}^{n} \mathbf{g}_i \zeta_i$, where $\zeta_i = \mathbf{g}_i^T\mathbf{z}$. Note from Definition 4.14 and Theorem 4.20,

$$\mathbf{A}^{-I}\mathbf{A} = \sum_{k=1}^{r}\sum_{i=1}^{r} \rho_i^{-1}\mathbf{g}_i \mathbf{f}_i^T \mathbf{f}_k \mathbf{g}_k^T \rho_k = \sum_{k=1}^{r}\sum_{i=1}^{r} \rho_i^{-1}\rho_k \mathbf{g}_i \mathbf{g}_k^T \delta_{ik} = \sum_{i=1}^{r} \mathbf{g}_i \mathbf{g}_i^T \tag{4.26}$$

Furthermore, since the $\mathbf{g}_i$ are orthonormal basis vectors for $\mathcal{V}_n$, $\mathbf{I} = \sum_{i=1}^{n} \mathbf{g}_i\mathbf{g}_i^T$. Hence

$$(\mathbf{I}-\mathbf{A}^{-I}\mathbf{A})\mathbf{z} \quad = \quad \sum_{i=r+1}^{n} \mathbf{g}_i\mathbf{g}_i^T\mathbf{z} \quad = \quad \sum_{i=r+1}^{n} \mathbf{g}_i\zeta_i \tag{4.27}$$

From equation (*4.23*), $\eta_i = \mathbf{f}_i^T\mathbf{y}$ so that substitution of (*4.27*) into (*4.25*) and use of Definition 4.14 for the pseudoinverse gives

$$\mathbf{x} = \mathbf{A}^{-I}\mathbf{y} + (\mathbf{I}-\mathbf{A}^{-I}\mathbf{A})\mathbf{z}$$

Some further properties of the pseudoinverse are:

1. If $\mathbf{A}$ is nonsingular, $\mathbf{A}^{-I} = \mathbf{A}^{-1}$.
2. $\mathbf{A}^{-I}\mathbf{A} \neq \mathbf{A}\mathbf{A}^{-I}$ in general.
3. $\mathbf{A}\mathbf{A}^{-I}\mathbf{A} = \mathbf{A}$
4. $\mathbf{A}^{-I}\mathbf{A}\mathbf{A}^{-I} = \mathbf{A}^{-I}$
5. $(\mathbf{A}\mathbf{A}^{-I})^T = \mathbf{A}\mathbf{A}^{-I}$
6. $(\mathbf{A}^{-I}\mathbf{A})^T = \mathbf{A}^{-I}\mathbf{A}$
7. $\mathbf{A}^{-I}\mathbf{A}\mathbf{w} = \mathbf{w}$ for all $\mathbf{w}$ in the range space of $\mathbf{A}^T$.
8. $\mathbf{A}^{-I}\mathbf{x} = \mathbf{0}$ for all $\mathbf{x}$ in the null space of $\mathbf{A}^T$.
9. $\mathbf{A}^{-I}(\mathbf{y}+\mathbf{z}) = \mathbf{A}^{-I}\mathbf{y} + \mathbf{A}^{-I}\mathbf{z}$ for all $\mathbf{y}$ in the range space of $\mathbf{A}$ and all $\mathbf{z}$ in the null space of $\mathbf{A}^T$.
10. Properties 3–6 and also properties 7–9 completely define a unique $\mathbf{A}^{-I}$ and are sometimes used as definitions of $\mathbf{A}^{-I}$.
11. Given a diagonal matrix $\mathbf{\Lambda} = \operatorname{diag}(\lambda_1, \lambda_2, \ldots, \lambda_n)$ where some λ_i may be zero. Then $\mathbf{\Lambda}^{-I} = \operatorname{diag}(\lambda_1^{-1}, \lambda_2^{-1}, \ldots, \lambda_n^{-1})$ where 0^{-1} is taken to be 0.
12. Given a Hermitian matrix $\mathbf{H} = \mathbf{H}^\dagger$. Let $\mathbf{H} = \mathbf{U}\mathbf{\Lambda}\mathbf{U}^\dagger$ where $\mathbf{U}^\dagger = \mathbf{U}^{-1}$. Then $\mathbf{H}^{-I} = \mathbf{U}\mathbf{\Lambda}^{-I}\mathbf{U}^\dagger$ where $\mathbf{\Lambda}^{-I}$ can be found from 11.
13. Given an arbitrary $m \times n$ matrix $\mathbf{A}$. Let $\mathbf{H} = \mathbf{A}^\dagger\mathbf{A}$. Then $\mathbf{A}^{-I} = \mathbf{H}^{-I}\mathbf{A}^\dagger = (\mathbf{A}\mathbf{H}^{-I})^\dagger$ where $\mathbf{H}^{-I}$ can be computed from 12.
14. $(\mathbf{A}^{-I})^{-I} = \mathbf{A}$
15. $(\mathbf{A}^T)^{-I} = (\mathbf{A}^{-I})^T$
16. The rank of $\mathbf{A}$, $\mathbf{A}^\dagger\mathbf{A}$, $\mathbf{A}\mathbf{A}^\dagger$, $\mathbf{A}^{-I}$, $\mathbf{A}^{-I}\mathbf{A}$ and $\mathbf{A}\mathbf{A}^{-I}$ equals $\operatorname{tr}(\mathbf{A}\mathbf{A}^{-I})_{\mathbf{Q}} = r$.
17. If $\mathbf{A}$ is square, there exists a unique polar decomposition $\mathbf{A} = \mathbf{U}\mathbf{H}$, where $\mathbf{H}^2 = \mathbf{A}^\dagger\mathbf{A}$, $\mathbf{U} = \mathbf{A}\mathbf{A}^{-I}$, $\mathbf{U}^\dagger = \mathbf{U}^{-I}$. If and only if $\mathbf{A}$ is nonsingular, $\mathbf{H}$ is positive definite real symmetric and $\mathbf{U}$ is nonsingular. When $\mathbf{A}$ is singular, $\mathbf{H}$ becomes nonnegative definite and $\mathbf{U}$ becomes singular.
18. If $\mathbf{A}(t)$ is a general matrix of continuous time functions, $\mathbf{A}^{-I}(t)$ may have discontinuities in the time functions of its elements.

Proofs of these properties are left to the solved and supplementary problems.

Solved Problems

4.1. Show that all similar matrices have the same determinants and traces.

To show this, we show that the determinant of a matrix equals the product of its eigenvalues and that the trace of a matrix equals the sum of its eigenvalues, and then use Theorem 4.2.

Factoring the characteristic polynomial gives

$$\det(\mathbf{A}-\lambda\mathbf{I}) = (\lambda_1-\lambda)(\lambda_2-\lambda)\cdots(\lambda_n-\lambda) = \lambda_1\lambda_2\cdots\lambda_n + \cdots + (\lambda_1+\lambda_2+\cdots+\lambda_n)(-\lambda)^{n-1} + (-\lambda)^n$$

Setting $\lambda = 0$ gives $\det \mathbf{A} = \lambda_1\lambda_2\cdots\lambda_n$. Furthermore,

$$\begin{aligned}\det(\mathbf{A}-\lambda\mathbf{I}) &= (\mathbf{a}_1-\lambda\mathbf{e}_1)\wedge(\mathbf{a}_2-\lambda\mathbf{e}_2)\wedge\cdots\wedge(\mathbf{a}_n-\lambda\mathbf{e}_n)\\ &= \mathbf{a}_1\wedge\mathbf{a}_2\wedge\cdots\wedge\mathbf{a}_n + (-\lambda)[\mathbf{e}_1\wedge\mathbf{a}_2\wedge\cdots\wedge\mathbf{a}_n + \mathbf{a}_1\wedge\mathbf{e}_2\wedge\cdots\wedge\mathbf{a}_n + \cdots + \mathbf{a}_1\wedge\mathbf{a}_2\wedge\cdots\wedge\mathbf{e}_n]\\ &\quad + \cdots + (-\lambda)^{n-1}[\mathbf{a}_1\wedge\mathbf{e}_2\wedge\cdots\wedge\mathbf{e}_n + \mathbf{e}_1\wedge\mathbf{a}_2\wedge\cdots\wedge\mathbf{e}_n + \cdots + \mathbf{e}_1\wedge\mathbf{e}_2\wedge\cdots\wedge\mathbf{a}_n]\\ &\quad + (-\lambda)^n\mathbf{e}_1\wedge\mathbf{e}_2\wedge\cdots\wedge\mathbf{e}_n\end{aligned}$$

Comparing coefficients of $-\lambda$ again gives $\lambda_1\lambda_2\cdots\lambda_n = \mathbf{a}_1\wedge\mathbf{a}_2\wedge\cdots\wedge\mathbf{a}_n$, and also

$$\lambda_1+\lambda_2+\cdots+\lambda_n = \mathbf{a}_1\wedge\mathbf{e}_2\wedge\cdots\wedge\mathbf{e}_n + \mathbf{e}_1\wedge\mathbf{a}_2\wedge\cdots\wedge\mathbf{e}_n + \cdots + \mathbf{e}_1\wedge\mathbf{e}_2\wedge\cdots\wedge\mathbf{a}_n$$

However,

$$\mathbf{a}_1\wedge\mathbf{e}_2\wedge\cdots\wedge\mathbf{e}_n = (a_{11}\mathbf{e}_1+a_{21}\mathbf{e}_2+\cdots+a_{n1}\mathbf{e}_n)\wedge\mathbf{e}_2\wedge\cdots\wedge\mathbf{e}_n = a_{11}\mathbf{e}_1\wedge\mathbf{e}_2\wedge\cdots\wedge\mathbf{e}_n = a_{11}$$

and similarly $\mathbf{e}_1\wedge\mathbf{a}_2\wedge\cdots\wedge\mathbf{e}_n = a_{22}$, etc. Therefore

$$\lambda_1+\lambda_2+\cdots+\lambda_n = a_{11}+a_{22}+\cdots+a_{nn} = \operatorname{tr}\mathbf{A}$$

4.2. Reduce the matrix **A** to Jordan form, where

$$(a)\quad \mathbf{A} = \begin{pmatrix} 8 & -8 & -2\\ 4 & -3 & -2\\ 3 & -4 & 1\end{pmatrix} \qquad (c)\quad \mathbf{A} = \begin{pmatrix} -1 & 2 & -1\\ 0 & -1 & 0\\ 0 & 0 & -1\end{pmatrix}$$

$$(b)\quad \mathbf{A} = \begin{pmatrix} -1 & 1 & -1\\ 0 & 1 & -4\\ 0 & 1 & -3\end{pmatrix} \qquad (d)\quad \mathbf{A} = \begin{pmatrix} -1 & 0 & 0\\ 0 & -1 & 0\\ 0 & 0 & -1\end{pmatrix}$$

(*a*) Calculation of $\det\begin{pmatrix} 8-\lambda & -8 & -2\\ 4 & -3-\lambda & -2\\ 3 & -4 & 1-\lambda\end{pmatrix} = 0$ gives $\lambda_1 = 1$, $\lambda_2 = 2$, $\lambda_3 = 3$. The eigenvector $\mathbf{x}_i$ is solved for from the equations

$$\begin{pmatrix} 8-\lambda_i & -8 & -2\\ 4 & -3-\lambda_i & -2\\ 3 & -4 & 1-\lambda_i\end{pmatrix}\mathbf{x}_i = \begin{pmatrix}0\\0\\0\end{pmatrix} \quad \text{giving} \quad \mathbf{x}_1 = \begin{pmatrix}4\\3\\2\end{pmatrix}, \quad \mathbf{x}_2 = \begin{pmatrix}3\\2\\1\end{pmatrix}, \quad \mathbf{x}_3 = \begin{pmatrix}2\\1\\1\end{pmatrix}$$

where the third element has been normalized to one in $\mathbf{x}_2$ and $\mathbf{x}_3$ and to two in $\mathbf{x}_1$. Then

$$\begin{pmatrix} 8 & -8 & -2\\ 4 & -3 & -2\\ 3 & -4 & 1\end{pmatrix}\begin{pmatrix} 4 & 3 & 2\\ 3 & 2 & 1\\ 2 & 1 & 1\end{pmatrix} = \begin{pmatrix} 4 & 3 & 2\\ 3 & 2 & 1\\ 2 & 1 & 1\end{pmatrix}\begin{pmatrix} 1 & 0 & 0\\ 0 & 2 & 0\\ 0 & 0 & 3\end{pmatrix}$$

(*b*) Calculation of $\det\begin{pmatrix} -1-\lambda & 1 & -1\\ 0 & 1-\lambda & -4\\ 0 & 1 & -3-\lambda\end{pmatrix} = 0$ gives $\lambda_1 = \lambda_2 = \lambda_3 = -1$.

Solution of the eigenvalue problem $(\mathbf{A}-(-1)\mathbf{I})\mathbf{x}=\mathbf{0}$ gives

$$\begin{pmatrix} 0 & 1 & -1 \\ 0 & 2 & -4 \\ 0 & 1 & -2 \end{pmatrix}\begin{pmatrix} x_1 \\ x_2 \\ x_3 \end{pmatrix} = \begin{pmatrix} 0 \\ 0 \\ 0 \end{pmatrix}$$

and only one vector, $(\alpha \;\; 0 \;\; 0)^T$ where α is arbitrary. Therefore it can be concluded there is only one Jordan block $\mathbf{L}_{11}(-1)$, so

$$\mathbf{J} \;=\; \mathbf{L}_{11}(-1) \;=\; \begin{pmatrix} -1 & 1 & 0 \\ 0 & -1 & 1 \\ 0 & 0 & -1 \end{pmatrix}$$

Solving,

$$\begin{pmatrix} 0 & 1 & -1 \\ 0 & 2 & -4 \\ 0 & 1 & -2 \end{pmatrix}\mathbf{t}_1 \;=\; \begin{pmatrix} \alpha \\ 0 \\ 0 \end{pmatrix} \;=\; \mathbf{x}_1$$

gives $\mathbf{t}_1 = (\beta \;\; 2\alpha \;\; \alpha)^T$ where β is arbitrary. Finally, from

$$\begin{pmatrix} 0 & 1 & -1 \\ 0 & 2 & -4 \\ 0 & 1 & -2 \end{pmatrix}\mathbf{t}_2 \;=\; \begin{pmatrix} \beta \\ 2\alpha \\ \alpha \end{pmatrix} \;=\; \mathbf{t}_1$$

we find $\mathbf{t}_2 = (\gamma \;\; 2\beta-\alpha \;\; \beta-\alpha)^T$ where γ is arbitrary. Choosing α, β and γ to be 1, 0 and 0 respectively gives a nonsingular $(\mathbf{x}_1 \mid \mathbf{t}_1 \mid \mathbf{t}_2)$, so that

$$\begin{pmatrix} 1 & 0 & 0 \\ 0 & 1 & -1 \\ 0 & 1 & -2 \end{pmatrix}\begin{pmatrix} -1 & 1 & -1 \\ 0 & 1 & -4 \\ 0 & 1 & -3 \end{pmatrix}\begin{pmatrix} 1 & 0 & 0 \\ 0 & 2 & -1 \\ 0 & 1 & -1 \end{pmatrix} \;=\; \begin{pmatrix} -1 & 1 & 0 \\ 0 & -1 & 1 \\ 0 & 0 & -1 \end{pmatrix}$$

(*c*) Since $\mathbf{A}$ is triangular, it exhibits its eigenvalues on the diagonal, so $\lambda_1=\lambda_2=\lambda_3=-1$. The solution of $\mathbf{Ax}=(-1)\mathbf{x}$ is $\mathbf{x}=(\alpha \;\; \beta \;\; 2\beta)^T$, so there are two linearly independent eigenvectors $\alpha(1 \;\; 0 \;\; 0)^T$ and $\beta(0 \;\; 1 \;\; 2)^T$. Therefore there are two Jordan blocks $\mathbf{L}_1(-1)$ and $\mathbf{L}_2(-1)$. These can form two different Jordan matrices $\mathbf{J}_1$ or $\mathbf{J}_2$:

$$\mathbf{J}_1 \;=\; \begin{pmatrix} \mathbf{L}_1(-1) & \mathbf{0} \\ \mathbf{0} & \mathbf{L}_2(-1) \end{pmatrix} \;=\; \begin{pmatrix} -1 & 1 & 0 \\ 0 & -1 & 0 \\ 0 & 0 & -1 \end{pmatrix} \quad \text{or} \quad \mathbf{J}_2 \;=\; \begin{pmatrix} \mathbf{L}_2(-1) & \mathbf{0} \\ \mathbf{0} & \mathbf{L}_1(-1) \end{pmatrix} \;=\; \begin{pmatrix} -1 & 0 & 0 \\ 0 & -1 & 1 \\ 0 & 0 & -1 \end{pmatrix}$$

It makes no difference whether we choose $\mathbf{J}_1$ or $\mathbf{J}_2$ because we merely reorder the eigenvectors and generalized eigenvector in the $\mathbf{T}$ matrix, i.e.,

$$\mathbf{A}(\mathbf{x}_1 \mid \mathbf{t}_1 \mid \mathbf{x}_2) \;=\; (\mathbf{x}_1 \mid \mathbf{t}_1 \mid \mathbf{x}_2)\mathbf{J}_1 \qquad \mathbf{A}(\mathbf{x}_2 \mid \mathbf{x}_1 \mid \mathbf{t}_1) \;=\; (\mathbf{x}_2 \mid \mathbf{x}_1 \mid \mathbf{t}_1)\mathbf{J}_2$$

How do we choose α and β to get the correct $\mathbf{x}_1$ to solve for $\mathbf{t}_1$? From (*4.19*)

$$(\mathbf{A}-\lambda\mathbf{I})\mathbf{t}_1 \;=\; \begin{pmatrix} 0 & 2 & -1 \\ 0 & 0 & 0 \\ 0 & 0 & 0 \end{pmatrix}\mathbf{t}_1 \;=\; \begin{pmatrix} \alpha \\ \beta \\ 2\beta \end{pmatrix} \;=\; \mathbf{x}_1$$

from which $\beta=0$ so that $\mathbf{x}_1=(\alpha \;\; 0 \;\; 0)^T$ and $\mathbf{t}_1 = (\gamma \;\; \delta \;\; 2\delta-\alpha)^T$ where γ and δ are also arbitrary. From Problem 4.41 we can always take $\alpha=1$ and $\gamma, \delta, \text{etc.}=0$, but in general

$$\begin{pmatrix} -1 & 2 & -1 \\ 0 & -1 & 0 \\ 0 & 0 & -1 \end{pmatrix}\begin{pmatrix} \alpha & \gamma & \alpha \\ 0 & \delta & \beta \\ 0 & 2\delta-\alpha & 2\beta \end{pmatrix} \;=\; \begin{pmatrix} \alpha & \gamma & \alpha \\ 0 & \delta & \beta \\ 0 & 2\delta-\alpha & 2\beta \end{pmatrix}\begin{pmatrix} -1 & 1 & 0 \\ 0 & -1 & 0 \\ 0 & 0 & -1 \end{pmatrix}$$

Any choice of α, β, γ and δ such that the inverse of the $\mathbf{T}$ matrix exists will give a similarity transformation to a Jordan form.

(*d*) The $\mathbf{A}$ matrix is already in Jordan form. Any nonsingular matrix $\mathbf{T}$ will transform it, since $\mathbf{A}=-\mathbf{I}$ and $\mathbf{T}^{-1}(-\mathbf{I})\mathbf{T}=-\mathbf{I}$. This can also be seen from the eigenvalue problem

$$(\mathbf{A}-\lambda\mathbf{I})\mathbf{x} \;=\; (-\mathbf{I}-(-1)\mathbf{I})\mathbf{x} \;=\; \mathbf{0x} \;=\; \mathbf{0}$$

so that any 3-vector $\mathbf{x}$ is an eigenvector. The space of eigenvectors belonging to -1 is three dimensional, so there are three Jordan blocks $\mathbf{L}_{11}(\lambda)=-1$, $\mathbf{L}_{12}(\lambda)=-1$ and $\mathbf{L}_{13}(\lambda)=-1$ on the diagonal for $\lambda=-1$.

4.3. Show that a general normal matrix $\mathbf{N}$ (i.e. $\mathbf{NN}^\dagger = \mathbf{N}^\dagger\mathbf{N}$), not necessarily with distinct eigenvalues, can be diagonalized by a similarity transformation $\mathbf{U}$ such that $\mathbf{U}^\dagger = \mathbf{U}^{-1}$.

The proof is by induction. First, it is true for a 1×1 matrix, because it is already a diagonal matrix and $\mathbf{U} = \mathbf{I}$. Now assume it is true for a $k-1 \times k-1$ matrix and prove it is true for a $k \times k$ matrix; i.e. assume that for $\mathbf{U}_{k-1}^\dagger = \mathbf{U}_{k-1}^{-1}$

$$\mathbf{U}_{k-1}^\dagger \mathbf{N}_{k-1}\mathbf{U}_{k-1} = \begin{pmatrix} \lambda_1 & & 0 \\ & \ddots & \\ 0 & & \lambda_{k-1} \end{pmatrix}$$

Let
$$\mathbf{N}_k = \begin{pmatrix} \mathbf{n}_1^\dagger \\ \hline \mathbf{n}_2^\dagger \\ \hline \vdots \\ \hline \mathbf{n}_k^\dagger \end{pmatrix}$$

Form $\mathbf{T}$ with the first column vector equal to the eigenvector $\mathbf{x}_1$ belonging to λ_1, an eigenvalue of $\mathbf{N}_k$. Then form $k-1$ other orthonormal vectors $\mathbf{x}_2, \mathbf{x}_3, \ldots, \mathbf{x}_k$ from $\mathcal{V}_k$ using the Gram-Schmit process, and make $\mathbf{T} = (\mathbf{x}_1 \mid \mathbf{x}_2 \mid \ldots \mid \mathbf{x}_k)$. Note $\mathbf{T}^\dagger\mathbf{T} = \mathbf{I}$. Then,

$$\mathbf{N}_k\mathbf{T} = \begin{pmatrix} \mathbf{n}_1^\dagger \\ \hline \mathbf{n}_2^\dagger \\ \hline \vdots \\ \hline \mathbf{n}_k^\dagger \end{pmatrix} (\mathbf{x}_1 \mid \mathbf{x}_2 \mid \ldots \mid \mathbf{x}_k) = \begin{pmatrix} \lambda_1 x_{11} & \mathbf{n}_1^\dagger\mathbf{x}_2 & \ldots & \mathbf{n}_1^\dagger\mathbf{x}_k \\ \lambda_1 x_{21} & \mathbf{n}_2^\dagger\mathbf{x}_2 & \ldots & \mathbf{n}_2^\dagger\mathbf{x}_k \\ \cdots & \cdots & \cdots & \cdots \\ \lambda_1 x_{k1} & \mathbf{n}_k^\dagger\mathbf{x}_2 & \ldots & \mathbf{n}_k^\dagger\mathbf{x}_k \end{pmatrix}$$

and
$$\mathbf{T}^\dagger\mathbf{N}_k\mathbf{T} = \begin{pmatrix} \mathbf{x}_1^\dagger \\ \hline \mathbf{x}_2^\dagger \\ \hline \vdots \\ \hline \mathbf{x}_k^\dagger \end{pmatrix} \left(\lambda_1\mathbf{x}_1 \;\middle|\; \begin{matrix} \mathbf{n}_1^\dagger\mathbf{x}_2 & \ldots & \mathbf{n}_1^\dagger\mathbf{x}_k \\ \cdots & \cdots & \cdots \\ \mathbf{n}_k^\dagger\mathbf{x}_2 & \ldots & \mathbf{n}_k^\dagger\mathbf{x}_k \end{matrix} \right) = \begin{pmatrix} \lambda_1 & \alpha_{12} & \ldots & \alpha_{1k} \\ 0 & \alpha_{22} & \ldots & \alpha_{2k} \\ \cdots & \cdots & \cdots & \cdots \\ 0 & \alpha_{k2} & \ldots & \alpha_{kk} \end{pmatrix}$$

where the α_{ij} are some numbers.

But $\mathbf{T}^\dagger\mathbf{N}_k\mathbf{T}$ is normal, because

$$\mathbf{T}^\dagger\mathbf{N}_k\mathbf{T}(\mathbf{T}^\dagger\mathbf{N}_k\mathbf{T})^\dagger = \mathbf{T}^\dagger\mathbf{N}_k\mathbf{T}\mathbf{T}^\dagger\mathbf{N}_k^\dagger\mathbf{T} = \mathbf{T}^\dagger\mathbf{N}_k\mathbf{N}_k^\dagger\mathbf{T} = (\mathbf{T}^\dagger\mathbf{N}_k\mathbf{T})^\dagger(\mathbf{T}^\dagger\mathbf{N}_k\mathbf{T})$$

Therefore

$$\begin{pmatrix} \lambda_1 & \alpha_{12} & \ldots & \alpha_{1k} \\ 0 & \alpha_{22} & \ldots & \alpha_{2k} \\ \cdots & \cdots & \cdots & \cdots \\ 0 & \alpha_{k2} & \ldots & \alpha_{kk} \end{pmatrix} \begin{pmatrix} \lambda_1^* & 0 & \ldots & 0 \\ \alpha_{12}^* & \alpha_{22}^* & \ldots & \alpha_{2k}^* \\ \cdots & \cdots & \cdots & \cdots \\ \alpha_{1k}^* & \alpha_{2k}^* & \ldots & \alpha_{kk}^* \end{pmatrix} = \begin{pmatrix} \lambda_1^* & 0 & \ldots & 0 \\ \alpha_{12}^* & \alpha_{22}^* & \ldots & \alpha_{k2}^* \\ \cdots & \cdots & \cdots & \cdots \\ \alpha_{1k}^* & \alpha_{2k}^* & \ldots & \alpha_{kk}^* \end{pmatrix} \begin{pmatrix} \lambda_1 & \alpha_{12} & \ldots & \alpha_{1k} \\ 0 & \alpha_{12} & \ldots & \alpha_{2k} \\ \cdots & \cdots & \cdots & \cdots \\ 0 & \alpha_{k2} & \ldots & \alpha_{kk} \end{pmatrix}$$

Equating the first element of the matrix product on the left with the first element of the product on the right gives

$$|\lambda_1|^2 + |\alpha_{12}|^2 + \cdots + |\alpha_{1k}|^2 = |\lambda_1|^2$$

Therefore $\alpha_{12}, \alpha_{13}, \ldots, \alpha_{1k}$ must all be zero so that

$$\mathbf{T}^\dagger\mathbf{N}_k\mathbf{T} = \left(\begin{array}{c|c} \lambda_1 & 0 \;\ldots\; 0 \\ \hline 0 & \\ \vdots & \mathbf{A}_{k-1} \\ 0 & \end{array} \right) \quad \text{where} \quad \mathbf{A}_{k-1} = \begin{pmatrix} \alpha_{22} & \ldots & \alpha_{2,k-1} \\ \cdots & \cdots & \cdots \\ \alpha_{k-1,2} & \ldots & \alpha_{k-1,k-1} \end{pmatrix}$$

and $\mathbf{A}_{k-1}$ is normal. Since $\mathbf{A}_{k-1}$ is $k-1 \times k-1$ and normal, by the inductive hypothesis there exists a $\mathbf{U}_{k-1}^\dagger = \mathbf{U}_{k-1}^{-1}$ such that $\mathbf{U}_{k-1}^\dagger\mathbf{A}_{k-1}\mathbf{U}_{k-1} = \mathbf{D}$, where $\mathbf{D}$ is a diagonal matrix.

Define $\mathbf{S}_k$ such that $\mathbf{S}_k = \left(\begin{array}{c|ccc} 1 & 0 & \cdots & 0 \\ \hline 0 & & & \\ \vdots & & \mathbf{U}_{k-1} & \\ 0 & & & \end{array}\right)$. Then $\mathbf{S}_k^\dagger \mathbf{S}_k = \mathbf{I}$ and

$$\mathbf{S}_k^\dagger \mathbf{T}^\dagger \mathbf{N}_k \mathbf{T}\mathbf{S}_k = \left(\begin{array}{c|ccc} 1 & 0 & \cdots & 0 \\ \hline 0 & & & \\ \vdots & & \mathbf{U}_{k-1}^\dagger & \\ 0 & & & \end{array}\right)\left(\begin{array}{c|ccc} \lambda_1 & 0 & \cdots & 0 \\ \hline 0 & & & \\ \vdots & & \mathbf{A}_{k-1} & \\ 0 & & & \end{array}\right)\left(\begin{array}{c|ccc} 1 & 0 & \cdots & 0 \\ \hline 0 & & & \\ \vdots & & \mathbf{U}_{k-1} & \\ 0 & & & \end{array}\right) = \left(\begin{array}{c|ccc} \lambda_1 & 0 & \cdots & 0 \\ \hline 0 & & & \\ \vdots & & \mathbf{D} & \\ 0 & & & \end{array}\right)$$

Therefore the matrix $\mathbf{TS}_k$ diagonalizes $\mathbf{N}_k$, and by Theorem 4.2, $\mathbf{D}$ has the other eigenvalues $\lambda_2, \lambda_3, \ldots, \lambda_k$ of $\mathbf{N}_k$ on the diagonal.

Finally, to show $\mathbf{TS}_k$ is unitary,

$$\mathbf{I} = \mathbf{S}_k^\dagger \mathbf{S}_k = \mathbf{S}_k^\dagger \mathbf{I} \mathbf{S}_k = \mathbf{S}_k^\dagger \mathbf{T}^\dagger \mathbf{T} \mathbf{S}_k = (\mathbf{TS}_k)^\dagger (\mathbf{TS}_k)$$

4.4. Prove that two $n \times n$ Hermitian matrices $\mathbf{A} = \mathbf{A}^\dagger$ and $\mathbf{B} = \mathbf{B}^\dagger$ can be simultaneously diagonalized by an orthonormal matrix $\mathbf{U}$ (i.e. $\mathbf{U}^\dagger\mathbf{U} = \mathbf{I}$) if and only if $\mathbf{AB} = \mathbf{BA}$.

If $\mathbf{A} = \mathbf{U}\mathbf{\Lambda}\mathbf{U}^\dagger$ and $\mathbf{B} = \mathbf{U}\mathbf{D}\mathbf{U}^\dagger$, then

$$\mathbf{AB} = \mathbf{U\Lambda U^\dagger UDU^\dagger} = \mathbf{U\Lambda DU^\dagger} = \mathbf{UD\Lambda U^\dagger} = \mathbf{UDU^\dagger U\Lambda U^\dagger} = \mathbf{BA}$$

Therefore all matrices that can be simultaneously diagonalized by an orthonormal $\mathbf{U}$ commute.

To show the converse, start with $\mathbf{AB} = \mathbf{BA}$. Assume $\mathbf{A}$ has distinct eigenvalues. Then $\mathbf{Ax}_i = \lambda_i \mathbf{x}_i$, so that $\mathbf{ABx}_i = \mathbf{BAx}_i = \lambda_i \mathbf{Bx}_i$. Hence if $\mathbf{x}_i$ is an eigenvector of $\mathbf{A}$, so is $\mathbf{Bx}_i$. For distinct eigenvalues, the eivenvectors are proportional, so that $\mathbf{Bx}_i = \rho_i \mathbf{x}_i$ where ρ_i is a constant of proportionality. But then ρ_i is also an eigenvalue of $\mathbf{B}$, and $\mathbf{x}_i$ is an eigenvector of $\mathbf{B}$. By normalizing the $\mathbf{x}_i$ so that $\mathbf{x}_i^\dagger \mathbf{x}_i = 1$, $\mathbf{U} = (\mathbf{x}_1 | \ldots | \mathbf{x}_n)$ simultaneously diagonalizes $\mathbf{A}$ and $\mathbf{B}$.

If neither $\mathbf{A}$ nor $\mathbf{B}$ have distinct eigenvalues, the proof is slightly more complicated. Let λ be an eigenvalue of $\mathbf{A}$ having multiplicity m. For nondistinct eigenvalues, all eigenvectors of λ belong in the m dimensional null space of $\mathbf{A} - \lambda\mathbf{I}$ spanned by orthonormal $\mathbf{x}_1, \mathbf{x}_2, \ldots, \mathbf{x}_m$. Therefore $\mathbf{Bx}_i = \sum_{i=1}^{m} c_{ij}\mathbf{x}_j$, where the constants c_{ij} can be determined by $c_{ij} = \mathbf{x}_j^\dagger \mathbf{Bx}_i$. Then for $\mathbf{C} = \{c_{ij}\}$ and $\mathbf{X} = (\mathbf{x}_1 | \mathbf{x}_2 | \ldots | \mathbf{x}_n)$, $\mathbf{C} = \mathbf{X}^\dagger\mathbf{BX} = \mathbf{X}^\dagger\mathbf{B}^\dagger\mathbf{X} = \mathbf{C}^\dagger$, so $\mathbf{C}$ is an $m \times m$ Hermitian matrix. Then $\mathbf{C} = \mathbf{U}_m\mathbf{D}_m\mathbf{U}_m^\dagger$ where $\mathbf{D}_m$ and $\mathbf{U}_m$ are $m \times m$ diagonal and unitary matrices respectively. Now $\mathbf{A}(\mathbf{XU}_m) = \lambda(\mathbf{XU}_m)$ since linear combinations of eigenvectors are still eigenvectors, and $\mathbf{D}_m = \mathbf{U}_m^\dagger\mathbf{X}^\dagger\mathbf{BXU}_m$. Therefore the set of m column vectors of $\mathbf{XU}_m$ together with all other normalized eigenvectors of $\mathbf{A}$ can diagonalize both $\mathbf{A}$ and $\mathbf{B}$. Finally, $(\mathbf{XU}_m)^\dagger(\mathbf{XU}_m) = (\mathbf{U}_m^\dagger\mathbf{X}^\dagger\mathbf{XU}_m) = (\mathbf{U}_m^\dagger\mathbf{I}_m\mathbf{U}_m) = \mathbf{I}_m$, so that the column vectors $\mathbf{XU}_m$ are orthonormal.

4.5. Show the positive definite Hermitian square root $\mathbf{R}$, such that $\mathbf{R}^2 = \mathbf{Q}$, is unique.

Since $\mathbf{R}$ is Hermitian, $\mathbf{U\Lambda}_1\mathbf{U}^\dagger = \mathbf{R}$ where $\mathbf{U}$ is orthonormal. Also, $\mathbf{R}^2$ and $\mathbf{R}$ commute, so that both $\mathbf{R}$ and $\mathbf{Q}$ can be simultaneously reduced to diagonal form by Problem 4.4, and $\mathbf{Q} = \mathbf{UDU}^\dagger$. Therefore $\mathbf{D} = \mathbf{\Lambda}_1^2$. Suppose another matrix $\mathbf{S}^2 = \mathbf{Q}$ such that $\mathbf{S} = \mathbf{V\Lambda}_2\mathbf{V}^\dagger$. By similar reasoning, $\mathbf{D} = \mathbf{\Lambda}_2^2$. Since a number > 0 has a unique positive square root, $\mathbf{\Lambda}_2 = \mathbf{\Lambda}_1$ and $\mathbf{V}$ and $\mathbf{U}$ are matrices of orthonormal eigenvectors. The normalized eigenvectors corresponding to distinct eigenvalues are unique. For any nondistinct eigenvalue with orthonormal eigenvectors $\mathbf{x}_1, \mathbf{x}_2, \ldots, \mathbf{x}_n$,

$$(\mathbf{x}_1 | \ldots | \mathbf{x}_m)\begin{pmatrix} \lambda & & 0 \\ & \ddots & \\ 0 & & \lambda \end{pmatrix}\begin{pmatrix} \mathbf{x}_1^\dagger \\ \hline \vdots \\ \hline \mathbf{x}_m^\dagger \end{pmatrix} = \lambda\mathbf{XX}^\dagger$$

and for any other linear combination of orthonormal eigenvectors, $\mathbf{y}_1, \mathbf{y}_2, \ldots, \mathbf{y}_m$,

$$(\mathbf{y}_1 \mid \ldots \mid \mathbf{y}_m) = (\mathbf{x}_1 \mid \ldots \mid \mathbf{x}_m)\mathbf{T}_m$$

where $\mathbf{T}_m^\dagger = \mathbf{T}_m^{-1}$. Then

$$(\mathbf{y}_1 \mid \ldots \mid \mathbf{y}_m)\begin{pmatrix} \lambda & & 0 \\ & \ddots & \\ 0 & & \lambda \end{pmatrix}\begin{pmatrix} \mathbf{y}_1^\dagger \\ \vdots \\ \mathbf{y}_m^\dagger \end{pmatrix} = (\mathbf{x}_1 \mid \ldots \mid \mathbf{x}_m)\mathbf{T}_m\mathbf{T}_m^\dagger\begin{pmatrix} \mathbf{x}_1^\dagger \\ \vdots \\ \mathbf{x}_m^\dagger \end{pmatrix}$$

$$= (\mathbf{x}_1 \mid \ldots \mid \mathbf{x}_m)\begin{pmatrix} \lambda & & 0 \\ & \ddots & \\ 0 & & \lambda \end{pmatrix}\begin{pmatrix} \mathbf{x}_1^\dagger \\ \vdots \\ \mathbf{x}_m^\dagger \end{pmatrix}$$

Hence $\mathbf{R}$ and $\mathbf{S}$ are equal even though $\mathbf{U}$ and $\mathbf{V}$ may differ slightly when $\mathbf{Q}$ has nondistinct eigenvalues.

4.6. Prove Sylvester's theorem: A Hermitian matrix $\mathbf{Q}$ is positive definite if and only if all principal minors $\det \mathbf{Q}_m > 0$.

If $\mathbf{Q}$ is positive definite, $\mathcal{Q} = (\mathbf{x}, \mathbf{Q}\mathbf{x}) \geqq 0$ for any $\mathbf{x}$. Let $\mathbf{x}_m$ be the vector of the first m elements of $\mathbf{x}$. For those $\mathbf{x}^0$ whose last $m-n$ elements are zero, $(\mathbf{x}_m, \mathbf{Q}_m\mathbf{x}_m) = (\mathbf{x}^0, \mathbf{Q}\mathbf{x}^0) \geqq 0$. Therefore $\mathbf{Q}_m$ is positive definite, and all its eigenvalues are positive. From Problem 4.1, the determinant of any matrix equals the product of its eigenvalues, so $\det \mathbf{Q}_m > 0$.

If $\det \mathbf{Q}_m > 0$ for $m = 1, 2, \ldots, n$, we proceed by induction. For $n = 1$, $\det \mathbf{Q} = \lambda_1 > 0$.

Assume now that if $\det \mathbf{Q}_1 > 0, \ldots, \det \mathbf{Q}_{n-1} > 0$, then $\mathbf{Q}_{n-1}$ is positive definite and must possess an inverse. Partition $\mathbf{Q}_n$ as

$$\mathbf{Q}_n = \left(\begin{array}{c|c} \mathbf{Q}_{n-1} & \mathbf{q} \\ \hline \mathbf{q}^\dagger & q_{nn} \end{array}\right) = \left(\begin{array}{c|c} \mathbf{I} & \mathbf{0} \\ \hline \mathbf{q}^\dagger\mathbf{Q}_{n-1}^{-1} & 1 \end{array}\right)\left(\begin{array}{c|c} \mathbf{Q}_{n-1} & \mathbf{0} \\ \hline \mathbf{0} & q_{nn} - \mathbf{q}^\dagger\mathbf{Q}_{n-1}^{-1}\mathbf{q} \end{array}\right)\left(\begin{array}{c|c} \mathbf{I} & \mathbf{Q}_{n-1}^{-1}\mathbf{q} \\ \hline \mathbf{0} & 1 \end{array}\right) \tag{4.28}$$

We are also given $\det \mathbf{Q}_n > 0$. Then use of Problem 3.5 gives $\det \mathbf{Q}_n = (q_{nn} - \mathbf{q}^\dagger\mathbf{Q}_{n-1}^{-1}\mathbf{q}) \det \mathbf{Q}_{n-1}$, so that $q_{nn} - \mathbf{q}^\dagger\mathbf{Q}_{n-1}^{-1}\mathbf{q} > 0$. Hence

$$(\mathbf{x}_{n-1}^\dagger \mid x_n^*)\left(\begin{array}{c|c} \mathbf{Q}_{n-1} & \mathbf{0} \\ \hline \mathbf{0} & q_{nn} - \mathbf{q}\mathbf{Q}_{n-1}^{-1}\mathbf{q} \end{array}\right)\left(\begin{array}{c} \mathbf{x}_{n-1} \\ \hline x_n \end{array}\right) > 0$$

for any vector $(\mathbf{x}_{n-1}^\dagger \mid x_n^*)$. Then for any vectors $\mathbf{y}$ defined by

$$\mathbf{y} = \left(\begin{array}{c|c} \mathbf{I} & \mathbf{Q}_{n-1}^{-1}\mathbf{q} \\ \hline \mathbf{0} & 1 \end{array}\right)^{-1}\left(\begin{array}{c} \mathbf{x}_{n-1} \\ \hline x_n \end{array}\right)$$

substitution into $(\mathbf{x}, \mathbf{Q}\mathbf{x})$ and use of *(4.28)* will give $(\mathbf{y}, \mathbf{Q}_n\mathbf{y}) > 0$.

4.7. Show that if $\|\mathbf{A}\| < 1$, then $(\mathbf{I} - \mathbf{A})^{-1} = \sum_{n=0}^{\infty} \mathbf{A}^n$.

Let $\mathbf{S}_k = \sum_{n=0}^{k} \mathbf{A}^n$ and $\mathbf{S} = \sum_{n=0}^{\infty} \mathbf{A}^n$. Then

$$\|\mathbf{S} - \mathbf{S}_k\| = \left\| \sum_{n=k+1}^{\infty} \mathbf{A}^k \right\| \leq \sum_{n=k+1}^{\infty} \|\mathbf{A}\|^k$$

by properties (3) and (4) of Theorem 4.11. Using property (6) and $||\mathbf{A}|| < 1$ gives $\mathbf{S} = \lim_{k\to\infty} \mathbf{S}_k$. Note $||\mathbf{A}^{k+1}|| \leq ||\mathbf{A}||^{k+1}$ so that $\lim_{k\to\infty} \mathbf{A}^{k+1} = \mathbf{0}$. Since $(\mathbf{I}-\mathbf{A})\mathbf{S}_k = \mathbf{I} - \mathbf{A}^{k+1}$, taking limits as $k \to \infty$ gives $(\mathbf{I}-\mathbf{A})\mathbf{S} = \mathbf{I}$. Since $\mathbf{S}$ exists, it is $(\mathbf{I}-\mathbf{A})^{-1}$. This is called a contraction mapping, because $||(\mathbf{I}-\mathbf{A})\mathbf{x}|| \leq (1-||\mathbf{A}||)||\mathbf{x}|| \leq ||\mathbf{x}||$.

4.8. Find the spectral representation of $\mathbf{A} = \begin{pmatrix} 1 & 1 & 2 \\ -1 & 1 & -2 \\ 0 & 0 & 1 \end{pmatrix}$.

The spectral representation of $\mathbf{A} = \sum_{i=1}^{3} \lambda_i \mathbf{x}_i \mathbf{r}_i^\dagger$. The matrix $\mathbf{A}$ has eigenvalues $\lambda_1 = 1$, $\lambda_2 = 1-j$ and $\lambda_3 = 1+j$, and eigenvectors $\mathbf{x}_1 = (1\ 1\ -0.5)^T$, $\mathbf{x}_2 = (j\ 1\ 0)^T$ and $\mathbf{x}_3 = (-j\ 1\ 0)^T$. The reciprocal basis $\mathbf{r}_i$ can be found as

$$\begin{pmatrix} \mathbf{r}_1^\dagger \\ \mathbf{r}_2^\dagger \\ \mathbf{r}_3^\dagger \end{pmatrix} = \begin{pmatrix} 1 & j & -j \\ 1 & 1 & 1 \\ -0.5 & 0 & 0 \end{pmatrix}^{-1} = 0.5\begin{pmatrix} 0 & 0 & -4 \\ -j & 1 & 2-2j \\ j & 1 & 2+2j \end{pmatrix}$$

Then

$$\mathbf{A} = (1)\begin{pmatrix} 1 \\ 1 \\ -0.5 \end{pmatrix}(0\ \ 0\ \ -2) + (1-j)\begin{pmatrix} j \\ 1 \\ 0 \end{pmatrix}(-0.5j\ \ 0.5\ \ 1-j) + (1+j)\begin{pmatrix} -j \\ 1 \\ 0 \end{pmatrix}(0.5j\ \ 0.5\ \ 1+j)$$

4.9. Show that the relations $\mathbf{A}\mathbf{A}^{-I}\mathbf{A} = \mathbf{A}$, $\mathbf{A}^{-I}\mathbf{A}\mathbf{A}^{-I} = \mathbf{A}^{-I}$, $(\mathbf{A}\mathbf{A}^{-I})^T = \mathbf{A}\mathbf{A}^{-I}$ and $(\mathbf{A}^{-I}\mathbf{A})^T = \mathbf{A}^{-I}\mathbf{A}$ define a unique matrix $\mathbf{A}^{-I}$, that can also be expressed as in Definition 4.15.

Represent $\mathbf{A} = \sum_{i=1}^{r} \rho_i \mathbf{f}_i \mathbf{g}_i^T$ and $\mathbf{A}^{-I} = \sum_{k=1}^{r} \rho_k^{-1} \mathbf{g}_k \mathbf{f}_k^T$. Then

$$\begin{aligned} \mathbf{A}\mathbf{A}^{-I}\mathbf{A} &= \Big(\sum_{i=1}^{r} \rho_i \mathbf{f}_i \mathbf{g}_i^T\Big)\Big(\sum_{j=1}^{r} \rho_j^{-1} \mathbf{g}_j \mathbf{f}_j^T\Big)\Big(\sum_{k=1}^{r} \rho_k \mathbf{f}_k \mathbf{g}_k^T\Big) \\ &= \sum_{i=1}^{r}\sum_{j=1}^{r}\sum_{k=1}^{r} \rho_i \rho_j^{-1} \rho_k \mathbf{f}_i \mathbf{g}_i^T \mathbf{g}_j \mathbf{f}_j^T \mathbf{f}_k \mathbf{g}_k^T \end{aligned}$$

Since $\mathbf{g}_i^T \mathbf{g}_j = \delta_{ij}$ and $\mathbf{f}_j^T \mathbf{f}_k = \delta_{jk}$,

$$\mathbf{A}\mathbf{A}^{-I}\mathbf{A} = \sum_{i=1}^{r} \rho_i \mathbf{f}_i \mathbf{g}_i^T = \mathbf{A}$$

Similarly $\mathbf{A}^{-I}\mathbf{A}\mathbf{A}^{-I} = \mathbf{A}^{-I}$, and from equation (*4.26*),

$$\mathbf{A}^{-I}\mathbf{A} = \sum_{i=1}^{r} \mathbf{g}_i \mathbf{g}_i^T = \Big(\sum_{i=1}^{r} \mathbf{g}_i \mathbf{g}_i^T\Big)^T = (\mathbf{A}^{-I}\mathbf{A})^T$$

Similarly $(\mathbf{A}\mathbf{A}^{-I})^T = \Big(\sum_{i=1}^{r} \mathbf{f}_i \mathbf{f}_i^T\Big)^T = \mathbf{A}\mathbf{A}^{-I}$.

To show uniqueness, assume two solutions $\mathbf{X}$ and $\mathbf{Y}$ satisfy the four relations. Then

1. $\mathbf{AXA} = \mathbf{A}$	3. $(\mathbf{AX})^T = \mathbf{AX}$	5. $\mathbf{AYA} = \mathbf{A}$	7. $(\mathbf{AY})^T = \mathbf{AY}$
2. $\mathbf{XAX} = \mathbf{X}$	4. $(\mathbf{XA})^T = \mathbf{XA}$	6. $\mathbf{YAY} = \mathbf{Y}$	8. $(\mathbf{YA})^T = \mathbf{YA}$

and transposing 1 and 5 gives

9. $\mathbf{A}^T\mathbf{X}^T\mathbf{A}^T = \mathbf{A}^T$ 10. $\mathbf{A}^T\mathbf{Y}^T\mathbf{A}^T = \mathbf{A}^T$

The following chain of equalities can be established by using the equation number above the equals sign as the justification for that step.

$$\mathbf{X} \overset{2}{=} \mathbf{XAX} \overset{4}{=} \mathbf{A}^T\mathbf{X}^T\mathbf{X} \overset{10}{=} \mathbf{A}^T\mathbf{Y}^T\mathbf{A}^T\mathbf{X}^T\mathbf{X} \overset{4}{=} \mathbf{A}^T\mathbf{Y}^T\mathbf{XAX} \overset{2}{=} \mathbf{A}^T\mathbf{Y}^T\mathbf{X} \overset{8}{=} \mathbf{YAX}$$
$$\overset{6}{=} \mathbf{YAYAX} \overset{3}{=} \mathbf{YAYX}^T\mathbf{A}^T \overset{7}{=} \mathbf{YY}^T\mathbf{A}^T\mathbf{X}^T\mathbf{A}^T \overset{9}{=} \mathbf{YY}^T\mathbf{A}^T \overset{7}{=} \mathbf{YAY} \overset{6}{=} \mathbf{Y}$$

Therefore the four relations given form a definition for the pseudoinverse that is equivalent to Definition 4.15.

4.10. The outcome of y of a certain experiment is thought to depend linearly upon a parameter x, such that $y = \alpha x + \beta$. The experiment is repeated three times, during which x assumes values $x_1 = 1$, $x_2 = -1$ and $x_3 = 0$, and the corresponding outcomes $y_1 = 2$, and $y_2 = -2$ and $y_3 = 3$. If the linear relation is true,

$$\begin{aligned} 2 &= \alpha(1) + \beta \\ -2 &= \alpha(-1) + \beta \\ 3 &= \alpha(0) + \beta \end{aligned}$$

However, experimental uncertainties are such that the relations are not quite satisfied in each case, so α and β are to be chosen such that

$$\sum_{i=1}^{3} (y_i - \alpha x_i - \beta)^2$$

is minimum. Explain why the pseudoinverse can be used to select the best α and β, and then calculate the best α and β using the pseudoinverse.

The equations can be written in the form $\mathbf{y} = \mathbf{Ax}$ as

$$\begin{pmatrix} 2 \\ -2 \\ 3 \end{pmatrix} = \begin{pmatrix} 1 & 1 \\ -1 & 1 \\ 0 & 1 \end{pmatrix} \begin{pmatrix} \alpha \\ \beta \end{pmatrix}$$

Defining $\mathbf{x}_0 = \mathbf{A}^{-I}\mathbf{y}$, by Theorem 4.21, $\|\mathbf{y} - \mathbf{Ax}_0\|_2^2 = \sum_{i=1}^{3} (y_i - \alpha_0 x_i - \beta_0)^2$ is minimized. Since $\mathbf{A}^T\mathbf{A} = \begin{pmatrix} 2 & 0 \\ 0 & 3 \end{pmatrix}$, then $\rho_1 = \sqrt{2}$ and $\rho_2 = \sqrt{3}$, and $\mathbf{g}_1 = (1\ 0)$ and $\mathbf{g}_2 = (0\ 1)$. Since $\mathbf{f}_i = \mathbf{Ag}_i/\rho_i$, then $\mathbf{f}_1 = (1\ -1\ 0)/\sqrt{2}$ and $\mathbf{f}_2 = (1\ 1\ 1)/\sqrt{3}$. Now $\mathbf{A}^{-I}$ can be calculated from Definition 4.15 to be

$$\mathbf{A}^{-I} = \begin{pmatrix} \frac{1}{2} & -\frac{1}{2} & 0 \\ \frac{1}{3} & \frac{1}{3} & \frac{1}{3} \end{pmatrix}$$

so that the best $\alpha_0 = 2$ and $\beta_0 = 1$.

Note this procedure can be applied to $\sum_{i=1}^{n} (y_i - \alpha x_i^2 - \beta x_i - \gamma)^2$, etc.

4.11. Show that if an $n \times n$ matrix $\mathbf{C}$ is nonsingular, then a matrix $\mathbf{B}$ exists such that $\mathbf{C} = e^{\mathbf{B}}$, or $\mathbf{B} = \ln \mathbf{C}$.

Reduce $\mathbf{C}$ to Jordan form, so that $\mathbf{C} = \mathbf{TJT}^{-1}$. Then $\mathbf{B} = \ln \mathbf{C} = \mathbf{T} \ln \mathbf{JT}^{-1}$, so that the problem is to find $\ln \mathbf{L}(\lambda)$ where $\mathbf{L}(\lambda)$ is an $l \times l$ Jordan block, because

$$\ln \mathbf{J} = \begin{pmatrix} \ln \mathbf{L}_{11}(\lambda_1) & & \mathbf{0} \\ & \ddots & \\ \mathbf{0} & & \ln \mathbf{L}_{km}(\lambda_m) \end{pmatrix}$$

Using the Maclaurin series for the logarithm of $\mathbf{L}(\lambda) - \lambda\mathbf{I}$,

$$\ln \mathbf{L}(\lambda) = \ln[\lambda\mathbf{I} + \mathbf{L}(\lambda) - \lambda\mathbf{I}] = \mathbf{I}\ln\lambda - \sum_{i=1}^{\infty} (-i\lambda)^{-i}[\mathbf{L}(\lambda) - \lambda\mathbf{I}]^i$$

Note all the eigenvalues ξ of $\mathbf{L}(\lambda) - \lambda\mathbf{I}$ are zero, so that the characteristic equation is $\xi^l = 0$. Then by the Cayley-Hamilton theorem, $[\mathbf{L}(\lambda) - \lambda\mathbf{I}]^l = \mathbf{0}$, so that

$$\ln \mathbf{L}(\lambda) = \mathbf{I}\ln\lambda - \sum_{i=1}^{l-1} (-i\lambda)^{-i}[\mathbf{L}(\lambda) - \lambda\mathbf{I}]^i$$

Since $\lambda \neq 0$ because $\mathbf{C}^{-1}$ exists, $\ln \mathbf{L}(\lambda)$ exists and can be calculated, so that $\ln \mathbf{J}$ and hence $\ln \mathbf{C}$ can be found.

Note $\mathbf{B}$ may be complex, because in the 1×1 case, $\ln(-1) = j\pi$. Also, in the converse case where $\mathbf{B}$ is given, $\mathbf{C}$ is always nonsingular for arbitrary $\mathbf{B}$ because $\mathbf{C}^{-1} = e^{-\mathbf{B}}$.

Supplementary Problems

4.12. Why does at least one nonzero eigenvector belong to each distinct eigenvalue?

4.13. Find the eigenvalues and eigenvectors of $\mathbf{A}$ where $\mathbf{A} = \begin{pmatrix} 1 & 0 & -4 \\ 0 & 3 & 0 \\ -2 & 0 & -1 \end{pmatrix}$.

4.14. Suppose all the eigenvalues of $\mathbf{A}$ are zero. Can we conclude that $\mathbf{A} = \mathbf{0}$?

4.15. Prove by induction that the generalized eigenvector $\mathbf{t}_l$ of equation (*4.9*) lies in the null space of $(\mathbf{A} - \lambda_i\mathbf{I})^{l+1}$.

4.16. Let $\mathbf{x}$ be an eigenvector of both $\mathbf{A}$ and $\mathbf{B}$. Is $\mathbf{x}$ also an eigenvector of $(\mathbf{A} - 2\mathbf{B})$?

4.17. Let $\mathbf{x}_1$ and $\mathbf{x}_2$ be eigenvectors of a matrix $\mathbf{A}$ corresponding to the eigenvalues of λ_1 and λ_2, where $\lambda_1 \neq \lambda_2$. Show that $\alpha\mathbf{x}_1 + \beta\mathbf{x}_2$ is not an eigenvector of $\mathbf{A}$ if $\alpha \neq 0$ and $\beta \neq 0$.

4.18. Using a similarity transformation to a diagonal matrix, solve the set of difference equations

$$\begin{pmatrix} x_1(n+1) \\ x_2(n+1) \end{pmatrix} = \begin{pmatrix} 0 & 2 \\ 2 & -3 \end{pmatrix}\begin{pmatrix} x_1(n) \\ x_2(n) \end{pmatrix} \quad \text{where} \quad \begin{pmatrix} x_1(0) \\ x_2(0) \end{pmatrix} = \begin{pmatrix} 2 \\ 1 \end{pmatrix}$$

4.19. Show that all the eigenvalues of the unitary matrix $\mathbf{U}$, where $\mathbf{U}^\dagger\mathbf{U} = \mathbf{I}$, have an absolute value of one.

4.20. Find the unitary matrix $\mathbf{U}$ and the diagonal matrix $\mathbf{\Lambda}$ such that

$$\mathbf{U}^\dagger \begin{pmatrix} \frac{3}{2} & 0 & -\frac{1}{2} \\ 0 & 1 & 0 \\ -\frac{1}{2} & 0 & \frac{3}{2} \end{pmatrix} \mathbf{U} = \mathbf{\Lambda}$$

Check your work.

4.21. Reduce to the matrix $\mathbf{A} = \begin{pmatrix} 1 & 3/5 & -4/5 \\ 0 & 1 & 0 \\ 0 & 0 & 1 \end{pmatrix}$ to Jordan form.

4.22. Given a 3×3 real matrix $\mathbf{P} \neq \mathbf{0}$ such that $\mathbf{P}^2 = \mathbf{0}$. Find its Jordan form.

4.23. Given the matrix $\mathbf{A} = \begin{pmatrix} 0 & 1 & 0 \\ -1 & 2 & 0 \\ 1 & 0 & 1 \end{pmatrix}$. Find the transformation matrix $\mathbf{T}$ that reduces $\mathbf{A}$ to Jordan form, and identify the Jordan blocks $\mathbf{L}_{ij}(\lambda_i)$.

4.24. Find the eigenvalues and eigenvectors of $\mathbf{A} = \begin{pmatrix} 1+\epsilon/2 & 1 \\ 0 & 1-\epsilon/2 \end{pmatrix}$. What happens as $\epsilon \to 0$?

4.25. Find the square root of $\begin{pmatrix} 5 & 4 \\ 4 & 5 \end{pmatrix}$.

4.26. Given the quadratic form $\mathcal{Q} = \xi_1^2 + 2\xi_1\xi_2 + 4\xi_2^2 + 4\xi_2\xi_3 + 2\xi_3^2$. Is it positive definite?

4.27. Show that $\sum_{n=0}^{\infty} \mathbf{A}^n/n!$ converges for any $\mathbf{A}$.

4.28. Show that the coefficient α_n of $\mathbf{I}$ in the Cayley-Hamilton theorem $\mathbf{A}^n + \alpha_1\mathbf{A}^{n-1} + \cdots + \alpha_n\mathbf{I} = \mathbf{0}$ is zero if and only if $\mathbf{A}$ is singular.

4.29. Does $\mathbf{A}^2 f(\mathbf{A}) = [f(\mathbf{A})]\mathbf{A}^2$?

4.30. Let the matrix $\mathbf{A}$ have distinct eigenvalues λ_1 and λ_2. Does $\mathbf{A}^3(\mathbf{A} - \lambda_1\mathbf{I})(\mathbf{A} - \lambda_2\mathbf{I}) = \mathbf{0}$?

4.31. Given a real vector $\mathbf{x} = (x_1 \ x_2 \ \ldots \ x_n)^T$ and a scalar α. Define the vector

$$\text{grad}_{\mathbf{x}}\,\alpha = (\partial\alpha/\partial x_1 \ \partial\alpha/\partial x_2 \ \ldots \ \partial\alpha/\partial x_n)^T$$

Show $\text{grad}_{\mathbf{x}}\,\mathbf{x}^T\mathbf{Q}\mathbf{x} = 2\mathbf{Q}\mathbf{x}$ if $\mathbf{Q}$ is symmetric, and evaluate $\text{grad}_{\mathbf{x}}\,\mathbf{x}^T\mathbf{A}\mathbf{x}$ for a nonsymmetric $\mathbf{A}$.

4.32. Given a basis $\mathbf{x}_1, \mathbf{x}_2, \ldots, \mathbf{x}_n$ of $\mathcal{V}_n$, and its reciprocal basis $\mathbf{r}_1, \mathbf{r}_2, \ldots, \mathbf{r}_n$. Show that $\mathbf{I} = \sum_{i=1}^{n} \mathbf{x}_i\mathbf{r}_i^\dagger$.

4.33. Suppose $\mathbf{A}^2 = \mathbf{0}$. Can $\mathbf{A}$ be nonsingular?

4.34. Find $e^{\mathbf{A}t}$, where $\mathbf{A}$ is the matrix of Problem 4.2(*a*).

4.35. Find $e^{\mathbf{A}t}$ for $\mathbf{A} = \begin{pmatrix} \sigma & \omega \\ -\omega & \sigma \end{pmatrix}$.

4.36. Find the pseudoinverse of $\begin{pmatrix} 4 & -3 \\ 0 & 0 \end{pmatrix}$.

4.37. Find the pseudoinverse of $\mathbf{A} = \begin{pmatrix} 1 & 2 & 0 \\ 3 & 6 & 0 \end{pmatrix}$.

4.38. Prove that the listed properties 1-18 of the pseudoinverse are true.

4.39. Given an $m \times n$ real matrix $\mathbf{A}$ and scalars θ_i such that

$$\mathbf{A}\mathbf{g}_i = \theta_i^n \mathbf{f}_i \qquad \mathbf{A}\mathbf{f}_i = \theta_i^{2-n}\mathbf{g}_i$$

Show that only $n = 1$ gives $\mathbf{f}_i^T\mathbf{f}_j = \delta_{ij}$ if $\mathbf{g}_i^T\mathbf{g}_j = \delta_{ij}$.

4.40. Given a real $m \times n$ matrix $\mathbf{A}$. Starting with the eigenvalues and eigenvectors of the real symmetric $(n+m) \times (n+m)$ matrix $\left(\begin{array}{c|c} \mathbf{0} & \mathbf{A}^T \\ \hline \mathbf{A} & \mathbf{0} \end{array}\right)$, derive the conclusions of Theorems 4.19 and 4.20.

4.41. Show that the $\mathbf{T}$ matrix of $\mathbf{J} = \mathbf{T}^{-1}\mathbf{AT}$ is arbitrary to within n constants if $\mathbf{A}$ is an $n \times n$ matrix in which each Jordan block has distinct eigenvalues. Specifically, show $\mathbf{T} = \mathbf{T}_0\mathbf{K}$ where $\mathbf{T}_0$ is fixed and

$$\mathbf{K} = \begin{pmatrix} \mathbf{K}_1 & \mathbf{0} & \cdots & \mathbf{0} \\ \mathbf{0} & \mathbf{K}_2 & \cdots & \mathbf{0} \\ \cdots & \cdots & \cdots & \cdots \\ \mathbf{0} & \mathbf{0} & \cdots & \mathbf{K}_m \end{pmatrix}$$

where $\mathbf{K}_j$ is an $l \times l$ matrix corresponding to the jth Jordan block $\mathbf{L}_j$ of the form

$$\mathbf{K}_j = \begin{pmatrix} \alpha_1 & \alpha_2 & \alpha_3 & \cdots & \alpha_l \\ 0 & \alpha_1 & \alpha_2 & \cdots & \alpha_{l-1} \\ 0 & 0 & \alpha_1 & \cdots & \alpha_{l-2} \\ \cdots & \cdots & \cdots & \cdots & \cdots \\ 0 & 0 & 0 & \cdots & \alpha_1 \end{pmatrix}$$

where α_i is an arbitrary constant.

4.42. Show that for any partitioned matrix $(\mathbf{A} \mid \mathbf{0})$, $(\mathbf{A} \mid \mathbf{0})^{-I} = \left(\dfrac{\mathbf{A}^{-I}}{\mathbf{0}}\right)$.

4.43. Show that $|\mathbf{x}^\dagger\mathbf{A}\mathbf{x}| \leq \|\mathbf{A}\|_2 \, \|\mathbf{x}\|_2^2$.

4.44. Show that another definition of the pseudoinverse is $\mathbf{A}^{-I} = \lim_{\epsilon \to 0} (\mathbf{A}^T\mathbf{A} + \epsilon\mathbf{P})^{-1}\mathbf{A}^T$, where $\mathbf{P}$ is any positive definite symmetric matrix that commutes with $\mathbf{A}^T\mathbf{A}$.

4.45. Show $\|\mathbf{A}^{-I}\| = 0$ if and only if $\mathbf{A} = \mathbf{0}$.

Answers to Supplementary Problems

4.12. Because $\det(\mathbf{A} - \lambda_i\mathbf{I}) = 0$, the column vectors of $\mathbf{A} - \lambda_i\mathbf{I}$ are linearly dependent, giving a null space of at least one dimension. The eigenvector of λ_i lies in the null space of $\mathbf{A} - \lambda_i\mathbf{I}$.

4.13. $\lambda = 3, 3, -3, \quad x_3 = \alpha(-2 \;\; 0 \;\; 1) + \beta(0 \;\; 1 \;\; 0); \quad x_{-3} = \gamma(1 \;\; 0 \;\; 1)$

4.14. No

4.16. Yes

4.18. $x_1(n) = 2$ and $x_2(n) = 1$ for all n.

4.20. $\mathbf{A} = \begin{pmatrix} 1 & 0 & 0 \\ 0 & 1 & 0 \\ 0 & 0 & 2 \end{pmatrix}, \quad \mathbf{U} = \dfrac{1}{\sqrt{2}} \begin{pmatrix} 1 & 0 & 1 \\ 0 & \sqrt{2} & 0 \\ 1 & 0 & -1 \end{pmatrix}$

4.21. $\mathbf{T} = \begin{pmatrix} 1 & 0 & 0 \\ 0 & 0.6 & 0.8 \\ 0 & -0.8 & 0.6 \end{pmatrix}, \quad \mathbf{J} = \begin{pmatrix} 1 & 1 & 0 \\ 0 & 1 & 0 \\ 0 & 0 & 1 \end{pmatrix}$

4.22. $\mathbf{J} = \begin{pmatrix} 0 & 1 & 0 \\ 0 & 0 & 0 \\ 0 & 0 & 0 \end{pmatrix}$ or $\mathbf{J} = \begin{pmatrix} 0 & 0 & 0 \\ 0 & 0 & 1 \\ 0 & 0 & 0 \end{pmatrix}$

4.23. $\mathbf{T} = \begin{pmatrix} 0 & \alpha & \tau \\ 0 & \alpha & \tau+\alpha \\ \alpha & \tau & \theta \end{pmatrix}$ for α, τ, θ arbitrary, and $\mathbf{J} = \begin{pmatrix} 1 & 1 & 0 \\ 0 & 1 & 1 \\ 0 & 0 & 1 \end{pmatrix}$, all one big Jordan block.

4.24. $\lambda_1 = 1 + \varepsilon/2$, $\lambda_2 = 1 - \epsilon/2$, $\mathbf{x}_1 = (1\ 0)$, $\mathbf{x}_2 = (1\ \epsilon)$; as $\epsilon \to 0$, $\lambda_1 = \lambda_2$, $\mathbf{x}_1 = \mathbf{x}_2$. Slight perturbations on the eigenvalues break the multiplicity.

4.25. $\begin{pmatrix} 2 & 1 \\ 1 & 2 \end{pmatrix}$

4.26. Yes

4.27. Use matrix norms.

4.28. If $\mathbf{A}^{-1}$ exists, $\mathbf{A}^{-1} = \alpha_n^{-1}[\mathbf{A}^{n-1} + \alpha_1\mathbf{A}^{n-2} + \cdots]$. If $\alpha_n = \lambda_1\lambda_2\ldots\lambda_n = 0$, then at least one eigenvalue is zero and $\mathbf{A}$ is singular.

4.29. Yes

4.30. Yes

4.31. $(\mathbf{A}^T + \mathbf{A})\mathbf{x}$

4.33. No

4.34. $$e^{\mathbf{A}t} = \begin{pmatrix} -4e^t + 6e^{2t} - e^{3t} & -3e^t + 4e^{2t} - e^{3t} & -e^t + 2e^{2t} - e^{3t} \\ 8e^t - 9e^{2t} + e^{3t} & 6e^t - 6e^{2t} + e^{3t} & 2e^t - 3e^{2t} + e^{3t} \\ -4e^t + 3e^{2t} + e^{3t} & -3e^t + 2e^{2t} + e^{3t} & -e^t + e^{2t} + e^{3t} \end{pmatrix}$$

4.35. $$e^{\mathbf{A}t} = e^{\sigma t}\begin{pmatrix} \cos\omega t & \sin\omega t \\ -\sin\omega t & \cos\omega t \end{pmatrix}$$

4.36. $\frac{1}{25}\begin{pmatrix} 4 & 0 \\ -3 & 0 \end{pmatrix}$

4.37. $$\mathbf{A}^{-I} = \frac{1}{50}\begin{pmatrix} 1 & 3 \\ 2 & 6 \\ 0 & 0 \end{pmatrix}$$

4.40. There are r nonzero positive eigenvalues ρ_i and r nonzero negative eigenvalues $-\rho_i$. Corresponding to the eigenvalues ρ_i are the eigenvectors $\begin{pmatrix} \mathbf{g}_i \\ \mathbf{f}_i \end{pmatrix}$ and to $-\rho_i$ are $\begin{pmatrix} \mathbf{g}_i \\ -\mathbf{f}_i \end{pmatrix}$. Spectral representation of $\left(\begin{array}{c|c} \mathbf{0} & \mathbf{A}^T \\ \mathbf{A} & \mathbf{0} \end{array}\right)$ then gives the desired result.

4.43. $|\mathbf{x}^\dagger\mathbf{A}\mathbf{x}| \leq \|\mathbf{x}\|_2 \|\mathbf{A}\mathbf{x}\|_2$ by Schwartz' inequality.

Chapter 5

Solutions to the Linear State Equation

5.1 TRANSITION MATRIX

From Section 1.3, a solution to a nonlinear state equation with an input $\mathbf{u}(t)$ and an initial condition $\mathbf{x}_0$ can be written in terms of its trajectory in state space as $\mathbf{x}(t) = \boldsymbol{\phi}(t; \mathbf{u}(\tau), \mathbf{x}_0, t_0)$. Since the state of a zero-input system does not depend on $\mathbf{u}(\tau)$, it can be written $\mathbf{x}(t) = \boldsymbol{\phi}(t; \mathbf{x}_0, t_0)$. Furthermore, if the system is linear, then it is linear in the initial condition so that from Theorem 3.20 we obtain the

Definition 5.1: The *transition matrix*, denoted $\boldsymbol{\Phi}(t, t_0)$, is the $n \times n$ matrix such that $\mathbf{x}(t) = \boldsymbol{\phi}(t; \mathbf{x}_0, t_0) = \boldsymbol{\Phi}(t, t_0)\mathbf{x}_0$.

This is true for any t_0, i.e. $\mathbf{x}(t) = \boldsymbol{\Phi}(t, \tau)\,\mathbf{x}(\tau)$ for $\tau > t$ as well as $\tau \leq t$. Substitution of $\mathbf{x}(t) = \boldsymbol{\Phi}(t, t_0)\mathbf{x}_0$ for arbitrary $\mathbf{x}_0$ in the zero-input linear state equation $d\mathbf{x}/dt = \mathbf{A}(t)\mathbf{x}$ gives the matrix equation for $\boldsymbol{\Phi}(t, t_0)$,

$$\partial\boldsymbol{\Phi}(t, t_0)/\partial t = \mathbf{A}(t)\,\boldsymbol{\Phi}(t, t_0) \tag{5.1}$$

Since for any $\mathbf{x}_0$, $\mathbf{x}_0 = \mathbf{x}(t_0) = \boldsymbol{\Phi}(t_0, t_0)\mathbf{x}_0$, the initial condition on $\boldsymbol{\Phi}(t, t_0)$ is

$$\boldsymbol{\Phi}(t_0, t_0) = \mathbf{I} \tag{5.2}$$

Notice that if the transition matrix can be found, we have the solution to a time-varying linear differential equation. Also, analogous to the continuous time case, the discrete time transition matrix obeys

$$\boldsymbol{\Phi}(k+1, m) = \mathbf{A}(k)\,\boldsymbol{\Phi}(k, m) \tag{5.3}$$

with

$$\boldsymbol{\Phi}(m, m) = \mathbf{I} \tag{5.4}$$

so that $\mathbf{x}(k) = \boldsymbol{\Phi}(k, m)\,\mathbf{x}(m)$.

***Theorem 5.1*:** Properties of the continuous time transition matrix for a linear, time-varying system are

(1) transition property

$$\boldsymbol{\Phi}(t_2, t_0) = \boldsymbol{\Phi}(t_2, t_1)\,\boldsymbol{\Phi}(t_1, t_0) \tag{5.5}$$

(2) inversion property

$$\boldsymbol{\Phi}(t_0, t_1) = \boldsymbol{\Phi}^{-1}(t_1, t_0) \tag{5.6}$$

(3) separation property

$$\boldsymbol{\Phi}(t_1, t_0) = \boldsymbol{\theta}(t_1)\,\boldsymbol{\theta}^{-1}(t_0) \tag{5.7}$$

(4) determinant property

$$\det \boldsymbol{\Phi}(t_1, t_0) = e^{\int_{t_0}^{t_1} [\operatorname{tr} \mathbf{A}(\tau)]\, d\tau} \tag{5.8}$$

and properties of the discrete time transition matrix are

(5) transition property

$$\boldsymbol{\Phi}(k, m) = \boldsymbol{\Phi}(k, l)\,\boldsymbol{\Phi}(l, m) \tag{5.9}$$

(6) inversion property

$$\Phi(m,k) = \Phi^{-1}(k,m) \tag{5.10}$$

(7) separation property

$$\Phi(m,k) = \theta(m)\theta^{-1}(k) \tag{5.11}$$

(8) determinant property

$$\det \Phi(k,m) = [\det \mathbf{A}(k-1)][\det \mathbf{A}(k-2)]\cdots[\det \mathbf{A}(m)] \quad \text{for } k > m \tag{5.12}$$

In the continuous time case, $\Phi^{-1}(t,t_0)$ always exists. However, in rather unusual circumstances, $\mathbf{A}(k)$ may be singular for some k, so there is no guarantee that the inverses in equations (*5.10*) and (*5.11*) exist.

Proof of Theorem 5.1: Because we have a linear zero-input dynamical system, the transition relations (*1.6*) and (*1.7*) become $\Phi(t,t_0)\mathbf{x}(t_0) = \Phi(t,t_1)\mathbf{x}(t_1)$ and $\mathbf{x}(t_1) = \Phi(t_1,t_0)\mathbf{x}(t_0)$. Combining these relations gives $\Phi(t,t_0)\mathbf{x}(t_0) = \Phi(t,t_1)\Phi(t_1,t_0)\mathbf{x}(t_0)$. Since $\mathbf{x}(t_0)$ is an arbitrary initial condition, the transition property is proven. Setting $t_2 = t_0$ in equation (*5.5*) and using (*5.2*) gives $\Phi(t_0,t_1)\Phi(t_1,t_0) = \mathbf{I}$, so that if $\det \Phi(t_0,t_1) \neq 0$ the inversion property is proven. Furthermore let $\theta(t) = \Phi(t,0)$ and set $t_1 = 0$ in equation (*5.5*) so that $\Phi(t_2,t_0) = \theta(t_2)\Phi(0,t_0)$. Use of (*5.6*) gives $\Phi(0,t_0) = \Phi^{-1}(t_0,0) = \theta^{-1}(t_0)$ so that the separation property is proven.

To prove the determinant property, partition Φ into its row vectors $\phi_1, \phi_2, \ldots, \phi_n$. Then

$$\det \Phi = \phi_1 \wedge \phi_2 \wedge \cdots \wedge \phi_n$$

and

$$d(\det \Phi)/dt = d\phi_1/dt \wedge \phi_2 \wedge \cdots \wedge \phi_n + \phi_1 \wedge d\phi_2/dt \wedge \cdots \wedge \phi_n + \cdots + \phi_1 \wedge \phi_2 \wedge \cdots \wedge d\phi_n/dt \tag{5.13}$$

From the differential equation (*5.1*) for Φ, the row vectors are related by

$$d\phi_i/dt = \sum_{k=1}^{n} a_{ik}(t)\,\phi_k \qquad \text{for } i = 1,2,\ldots,n$$

Because this is a linear, time-varying dynamical system, each element $a_{ik}(t)$ is continuous and single-valued, so that this uniquely represents $d\phi_i/dt$ for each t.

$$\phi_1 \wedge \cdots \wedge d\phi_i/dt \wedge \cdots \wedge \phi_n = \phi_1 \wedge \cdots \wedge \sum_{k=1}^{n} a_{ik}\phi_k \wedge \cdots \wedge \phi_n = a_{ii}\phi_1 \wedge \cdots \wedge \phi_i \wedge \cdots \wedge \phi_n$$

Then from equation (*5.13*),

$$\begin{aligned} d(\det \Phi)/dt &= a_{11}\phi_1 \wedge \phi_2 \wedge \cdots \wedge \phi_n + a_{22}\phi_1 \wedge \phi_2 \wedge \cdots \wedge \phi_n + \cdots + a_{nn}\phi_1 \wedge \phi_2 \wedge \cdots \wedge \phi_n \\ &= [\operatorname{tr} \mathbf{A}(t)]\phi_1 \wedge \phi_2 \wedge \cdots \wedge \phi_n = [\operatorname{tr} \mathbf{A}(t)] \det \Phi \end{aligned}$$

Separating variables gives

$$d(\det \Phi)/\det \Phi = \operatorname{tr} \mathbf{A}(t)\,dt$$

Integrating and taking antilogarithms results in

$$\det \Phi(t,t_0) = \gamma e^{\int_{t_0}^{t} [\operatorname{tr} \mathbf{A}(\tau)]\,d\tau}$$

where γ is the constant of integration. Setting $t = t_0$ gives $\det \Phi(t_0,t_0) = \det \mathbf{I} = 1 = \gamma$, so that the determinant property is proven. Since $e^{f(t)} = 0$ if and only if $f(t) = -\infty$, the inverse of $\Phi(t,t_0)$ always exists because the elements of $\mathbf{A}(t)$ are bounded.

The proof of the properties for the discrete time transition matrix is quite similar, and the reader is referred to the supplementary problems.

5.2 CALCULATION OF THE TRANSITION MATRIX FOR TIME-INVARIANT SYSTEMS

***Theorem 5.2*:** The transition matrix for a time-invariant linear differential system is

$$\mathbf{\Phi}(t, \tau) = e^{\mathbf{A}(t-\tau)} \tag{5.14}$$

and for a time-invariant linear difference system is

$$\mathbf{\Phi}(k, m) = \mathbf{A}^{k-m} \tag{5.15}$$

Proof: The Maclaurin series for $e^{\mathbf{A}t} = \sum_{k=0}^{\infty} \mathbf{A}^k t^k/k!$, which is uniformly convergent as shown in Problem 4.27. Differentiating with respect to t gives $de^{\mathbf{A}t}/dt = \sum_{k=0}^{\infty} \mathbf{A}^{k+1} t^k/k!$, so substitution into equation (*5.1*) verifies that $e^{\mathbf{A}(t-\tau)}$ is a solution. Furthermore, for $t = \tau$, $e^{\mathbf{A}(t-\tau)} = \mathbf{I}$, so this is the unique solution starting from $\mathbf{\Phi}(\tau, \tau) = \mathbf{I}$.

Also, substitution of $\mathbf{\Phi}(k, m) = \mathbf{A}^{k-m}$ into equation (*5.3*) verifies that it is a solution, and for $k = m$, $\mathbf{A}^{k-m} = \mathbf{I}$. Note that $e^{\mathbf{A}} e^{\mathbf{B}} \neq e^{\mathbf{B}} e^{\mathbf{A}}$ in general, but $e^{\mathbf{A}t_0} e^{\mathbf{A}t_1} = e^{\mathbf{A}t_1} e^{\mathbf{A}t_0} = e^{\mathbf{A}(t_0+t_1)}$ and $\mathbf{A} e^{\mathbf{A}t} = e^{\mathbf{A}t} \mathbf{A}$, as is easily shown using the Maclaurin series.

Since time-invariant linear systems are the most important, numerical calculation of $e^{\mathbf{A}t}$ is often necessary. However, sometimes only $\mathbf{x}(t)$ for $t \geq t_0$ is needed. Then $\mathbf{x}(t)$ can be found by some standard differential equation routine such as Runge-Kutta or Adams, etc., on the digital computer or by simulation on the analog computer.

When $e^{\mathbf{A}t}$ must be found, a number of methods for numerical calculation are available. No one method has yet been found that is the easiest in all cases. Here we present four of the most useful, based on the methods of Section 4.7.

1. Series method:

$$e^{\mathbf{A}t} = \sum_{k=0}^{\infty} \mathbf{A}^k t^k/k! \tag{5.16}$$

2. Eigenvalue method:

$$e^{\mathbf{A}t} = \mathbf{T} e^{\mathbf{J}t} \mathbf{T}^{-1} \tag{5.17}$$

and, if the eigenvalues are distinct,

$$e^{\mathbf{A}t} = \sum_{i=1}^{n} e^{\lambda_i t} \mathbf{x}_i \mathbf{r}_i^\dagger$$

3. Cayley-Hamilton:

$$e^{\mathbf{A}t} = \sum_{i=0}^{n-1} \gamma_i(t) \mathbf{A}^i \tag{5.18}$$

where the $\gamma_i(t)$ are evaluated from $e^{\mathbf{J}t} = \sum_{i=0}^{n-1} \gamma_i(t) \mathbf{J}^i$. Note that from (*4.15*),

$$e^{\mathbf{J}t} = \begin{pmatrix} e^{\mathbf{L}_{11}(\lambda_1)t} & \mathbf{0} & \cdots & \mathbf{0} & \mathbf{0} & \cdots & \mathbf{0} \\ \mathbf{0} & e^{\mathbf{L}_{21}(\lambda_1)t} & \cdots & \mathbf{0} & \mathbf{0} & \cdots & \mathbf{0} \\ \cdots & \cdots & \cdots & \cdots & \cdots & \cdots & \cdots \\ \mathbf{0} & \mathbf{0} & \cdots & e^{\mathbf{L}_{j1}(\lambda_1)t} & \mathbf{0} & \cdots & \mathbf{0} \\ \mathbf{0} & \mathbf{0} & \cdots & \mathbf{0} & e^{\mathbf{L}_{12}(\lambda_2)t} & \cdots & \mathbf{0} \\ \cdots & \cdots & \cdots & \cdots & \cdots & \cdots & \cdots \\ \mathbf{0} & \mathbf{0} & \cdots & \mathbf{0} & \mathbf{0} & \cdots & e^{\mathbf{L}_{mk}(\lambda_k)t} \end{pmatrix} \tag{5.19}$$

where if $\mathbf{L}_{ji}(\lambda_i)$ is $l \times l$,

$$\mathbf{L}_{ji}(\lambda_i) = \overbrace{\begin{pmatrix} \lambda_i & 1 & \cdots & 0 \\ 0 & \lambda_i & \cdots & 0 \\ \cdots & \cdots & \cdots & \cdots \\ 0 & 0 & \cdots & \lambda_i \end{pmatrix}}^{l} \Big\updownarrow l \tag{5.20}$$

then

$$e^{\mathbf{L}_{ji}(\lambda_i)t} = \begin{pmatrix} e^{\lambda_i t} & te^{\lambda_i t} & \cdots & t^{l-1}e^{\lambda_i t}/(l-1)! \\ 0 & e^{\lambda_i t} & \cdots & t^{l-2}e^{\lambda_i t}/(l-2)! \\ \cdots & \cdots & \cdots & \cdots \\ 0 & 0 & \cdots & e^{\lambda_i t} \end{pmatrix} \tag{5.21}$$

4. Resolvent matrix:

$$e^{\mathbf{A}t} = \mathcal{L}^{-1}\{\mathbf{R}(s)\} \tag{5.22}$$

where $\mathbf{R}(s) = (s\mathbf{I} - \mathbf{A})^{-1}$.

The hard part of this method is computing the inverse of $(s\mathbf{I} - \mathbf{A})$, since it is a polynomial in s. For matrices with many zero elements, substitution and elimination is about the quickest method. For the general case up to about third order, Cramer's rule can be used. Somewhat higher order systems can be handled from the flow diagram of the Laplace transformed system. The elements $r_{ij}(s)$ of $\mathbf{R}(s)$ are the response of the ith state (integrator) to a unit impulse input of the jth state (integrator). For higher order systems, Leverrier's algorithm might be faster.

***Theorem 5.3*:** *Leverrier's algorithm.* Define the $n \times n$ real matrices $\mathbf{F}_1, \mathbf{F}_2, \ldots, \mathbf{F}_n$ and scalars $\theta_1, \theta_2, \ldots, \theta_n$ as follows:

$$\begin{aligned} \mathbf{F}_1 &= \mathbf{I} & \theta_1 &= -\operatorname{tr} \mathbf{AF}_1/1 \\ \mathbf{F}_2 &= \mathbf{AF}_1 + \theta_1\mathbf{I} & \theta_2 &= -\operatorname{tr} \mathbf{AF}_2/2 \\ &\cdots\cdots & &\cdots\cdots \\ \mathbf{F}_n &= \mathbf{AF}_{n-1} + \theta_{n-1}\mathbf{I} & \theta_n &= -\operatorname{tr} \mathbf{AF}_n/n \end{aligned}$$

Then

$$(s\mathbf{I} - \mathbf{A})^{-1} = \frac{s^{n-1}\mathbf{F}_1 + s^{n-2}\mathbf{F}_2 + \cdots + s\mathbf{F}_{n-1} + \mathbf{F}_n}{s^n + \theta_1 s^{n-1} + \cdots + \theta_{n-1}s + \theta_n} \tag{5.23}$$

Also, $\mathbf{AF}_n + \theta_n\mathbf{I} = \mathbf{0}$, to check the method. Proof is given in Problem 5.4.

Having $\mathbf{R}(s)$, a matrix partial fraction expansion can be performed. First, factor $\det \mathbf{R}(s)$ as

$$\det \mathbf{R}(s) = s^n + \theta_1 s^{n-1} + \cdots + \theta_{n-1}s + \theta_n = (s-\lambda_1)(s-\lambda_2)\cdots(s-\lambda_n) \tag{5.24}$$

where the λ_i are the eigenvalues of $\mathbf{A}$ and the poles of the system. Next, expand $\mathbf{R}(s)$ in matrix partial fractions. If the eigenvalues are distinct, this has the form

$$\mathbf{R}(s) = \frac{1}{s-\lambda_1}\mathbf{R}_1 + \frac{1}{s-\lambda_2}\mathbf{R}_2 + \cdots + \frac{1}{s-\lambda_n}\mathbf{R}_n \tag{5.25}$$

where $\mathbf{R}_k$ is the matrix-valued residue

$$\mathbf{R}_k = (s-\lambda_k)\mathbf{R}(s)\,|_{s=\lambda_k} \tag{5.26}$$

For mth order roots λ, the residue of $(s-\lambda)^{-i}$ is

$$\mathbf{R}_i = \frac{1}{(m-i)!}\frac{d^{m-i}}{ds^{m-i}}[(s-\lambda)^m\mathbf{R}(s)]\bigg|_{s=\lambda} \tag{5.27}$$

Then $e^{\mathbf{A}t}$ is easily found by taking the inverse Laplace transform. In the case of distinct roots, equation (*5.25*) becomes

$$e^{\mathbf{A}t} = e^{\lambda_1 t}\mathbf{R}_1 + e^{\lambda_2 t}\mathbf{R}_2 + \cdots + e^{\lambda_n t}\mathbf{R}_n \tag{5.28}$$

Note, from the spectral representation equation (*5.17*),

$$\mathbf{R}_i = \mathbf{x}_i\mathbf{r}_i^\dagger$$

so that the eigenvectors $\mathbf{x}_i$ and their reciprocal $\mathbf{r}_i$ can easily be found from the $\mathbf{R}_i$.

In the case of repeated roots,

$$\mathcal{L}^{-1}\{(s-\lambda_i)^{-m}\} = t^{m-1}e^{\lambda_i t}/(m-1)! \tag{5.29}$$

To find $\mathbf{A}^k$, methods similar to those discussed to find $e^{\mathbf{A}t}$ are available.

1. Series method:

$$\mathbf{A}^k = \mathbf{A}^k \tag{5.30}$$

2. Eigenvalue method:

$$\mathbf{A}^k = \mathbf{T}\mathbf{J}^k\mathbf{T}^{-1} \tag{5.31}$$

   and for distinct eigenvalues

$$\mathbf{A}^k = \sum_{i=1}^{n} \lambda_i^k \mathbf{x}_i \mathbf{r}_i^\dagger$$

3. Cayley-Hamilton:

$$\mathbf{A}^k = \sum_{i=0}^{n-1} \gamma_i(k)\mathbf{A}^i \tag{5.32}$$

where the $\gamma_i(k)$ are evaluated from $\mathbf{J}^k = \sum_{i=1}^{n-1} \gamma_i(k)\mathbf{J}^i$ where from equation (*4.15*),

$$\mathbf{J}^k = \begin{pmatrix} \mathbf{L}_{11}^k(\lambda_1) & \mathbf{0} & \dots & \mathbf{0} & \mathbf{0} & \dots & \mathbf{0} \\ \mathbf{0} & \mathbf{L}_{21}^k(\lambda_1) & \dots & \mathbf{0} & \mathbf{0} & \dots & \mathbf{0} \\ \dots & \dots & \dots & \dots & \dots & \dots & \dots \\ \mathbf{0} & \mathbf{0} & \dots & \mathbf{L}_{j1}^k(\lambda_1) & \mathbf{0} & \dots & \mathbf{0} \\ \mathbf{0} & \mathbf{0} & \dots & \mathbf{0} & \mathbf{L}_{12}^k(\lambda_2) & \dots & \mathbf{0} \\ \dots & \dots & \dots & \dots & \dots & \dots & \dots \\ \mathbf{0} & \mathbf{0} & \dots & \mathbf{0} & \mathbf{0} & \dots & \mathbf{L}_{mi}^k(\lambda_i) \end{pmatrix} \tag{5.33}$$

and if $\mathbf{L}_{ji}(\lambda_i)$ is $l \times l$ as in equation (*5.20*),

$$\mathbf{L}_{ji}^k(\lambda_i) = \begin{pmatrix} \lambda_i^k & k\lambda_i^{k-1} & \dots & (k!\,\lambda_i^{k+1-l})[(l-1)!\,(k-l+1)!]^{-1} \\ 0 & \lambda_i^k & \dots & (k!\,\lambda_i^{k+2-l})[(l-2)!\,(k-l+2)!]^{-1} \\ \dots & \dots & \dots & \dots \\ 0 & 0 & \dots & \lambda_i^k \end{pmatrix} \tag{5.34}$$

4. Resolvent matrix:

$$\mathbf{A}^k = \mathcal{Z}^{-1}\{z\mathbf{R}(z)\} \tag{5.35}$$

   where $\mathbf{R}(z) = (z\mathbf{I}-\mathbf{A})^{-1}$.

Since $\mathbf{R}(z)$ is exactly the same form as $\mathbf{R}(s)$ except with z for s, the inversion procedures given previously are exactly the same.

The series method is useful if $\mathbf{A}^k = 0$ for some $k = k_0$. Then the series truncates at $k_0 - 1$. Because the eigenvalue problem $\mathbf{Ax} = \lambda\mathbf{x}$ can be multiplied by $\mathbf{A}^{k-1}$ to obtain $\mathbf{0} = \mathbf{A}^k\mathbf{x} = \lambda\mathbf{A}^{k-1}\mathbf{x} = \lambda^k\mathbf{x}$, then $\lambda = 0$. Therefore the series method is useful only for systems with k_0 poles only at the origin. Otherwise it suffers from slow convergence, roundoff, and difficulties in recognizing the resulting infinite series.

The eigenvalue method is not very fast because each eigenvector must be computed. However, at the 1968 Joint Automatic Control Conference it was the general consensus that this was the only method that anyone had any experience with that could compute $e^{\mathbf{A}t}$ up to twentieth order.

The Cayley-Hamilton method is very similar to the eigenvalue method, and usually involves a few more multiplications.

The resolvent matrix method is usually simplest for systems of less than tenth order. This is the extension to matrix form of the usual Laplace transform techniques for single input–single output that has worked so successfully in the past. For very high order systems, Leverrier's algorithm involves very high powers of **A**, which makes the spread of the eigenvalues very large unless **A** is scaled properly. However, it involves no matrix inversions, and gives a means of checking the amount of roundoff in that $\mathbf{AF}_n + \theta_n\mathbf{I}$ should equal **0**. In the case of distinct roots, $\mathbf{R}_i = \mathbf{x}_i\mathbf{r}_i^\dagger$ so that the eigenvectors can easily be obtained. Perhaps a combination of both Leverrier's algorithm and the eigenvalue method might be useful for very high order systems.

5.3 TRANSITION MATRIX FOR TIME-VARYING DIFFERENTIAL SYSTEMS

There is NO general solution for the transition matrix of a time-varying linear system such as there is for the time-invariant case.

Example 5.1.

We found that the transformation $\mathbf{A} = \mathbf{TJT}^{-1}$ gave a general solution

$$\mathbf{\Phi}(t, t_0) = \mathbf{\Phi}(t - t_0) = e^{\mathbf{A}(t-t_0)} = \mathbf{T}\, e^{\mathbf{J}(t-t_0)}\,\mathbf{T}^{-1}$$

for the time-invariant case. For the time-varying case,

$$d\mathbf{x}/dt = \mathbf{A}(t)\mathbf{x}$$

Then $\mathbf{A}(t) = \mathbf{T}(t)\,\mathbf{J}(t)\mathbf{T}^{-1}(t)$, where the elements of **T** and **J** must be functions of t. Attempting a change of variable $\mathbf{x} = \mathbf{T}(t)\mathbf{y}$ results in

$$d\mathbf{y}/dt = \mathbf{J}(t)\mathbf{y} - \mathbf{T}^{-1}(t)\,(d\mathbf{T}(t)/dt)\mathbf{y}$$

which does not simplify unless $d\mathbf{T}(t)/dt = \mathbf{0}$ or some very fortunate combination of elements.

We may conclude that knowledge of the time-varying eigenvalues of a time-varying system usually does *not* help.

The behavior of a time-varying system depends on the behavior of the coefficients of the $\mathbf{A}(t)$ matrix.

Example 5.2.

Given the time-varying scalar system $d\xi/dt = \xi \operatorname{sgn}(t - t_1)$ where sgn is the signum function, so that $\operatorname{sgn}(t - t_1) = -1$ for $t < t_1$ and $\operatorname{sgn}(t - t_1) = +1$ for $t > t_1$. This has a solution $\xi(t) = \xi(t_0)e^{-(t-t_0)}$ for $t < t_1$ and $\xi(t) = \xi(t_1)e^{(t-t_1)}$ for $t > t_1$. For times $t < t_1$, the system appears stable, but actually the solution grows without bound as $t \to \infty$. We shall see in Chapter 9 that the concept of stability must be carefully defined for a time-varying system.

Also, the phenomenon of finite escape time can arise in a time-varying linear system, whereas this is impossible in a time-invariant linear system.

Example 5.3.

Consider the time-varying scalar system

$$d\xi/dt = \xi/(t-t_1)^2 \quad \text{with} \quad \xi(t_0) = \xi_0, \ t_0 \leq t < t_1$$

Then

$$\xi(t) = \xi_0 e^{[(t-t_1)^{-1}-(t_0-t_1)^{-1}]}$$

and the solution is represented in Fig. 5-1. The solution goes to infinity in a finite time.

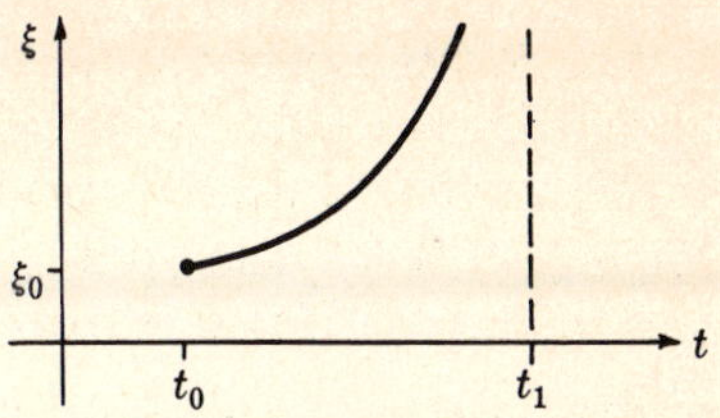

Fig. 5-1

These and other peculiarities make the analysis of time-varying linear systems relatively more difficult than the analysis of time-invariant linear systems. However, the analysis of time-varying systems is of considerable practical importance. For instance, a time-varying linear system usually results from the linearization of a nonlinear system about a nominal trajectory (see Section 1.6). Since a control system is usually designed to keep the variations from the nominal small, the time-varying linear system is a good approximation.

Since there is no general solution for the transition matrix, what can be done? In certain special cases a closed-form solution is available. A computer can almost always find a numerical solution, and with the use of the properties of the transition matrix (Theorem 5.1) this makes a powerful tool for analysis. Finally and perhaps most importantly, solutions for systems with an input can be expressed in terms of the transition matrix.

5.4 CLOSED FORMS FOR SPECIAL CASES OF TIME-VARYING LINEAR DIFFERENTIAL SYSTEMS

***Theorem 5.4*:** A general *scalar* time-varying linear differential system $d\xi/dt = \alpha(t)\xi$ has the scalar transition matrix

$$\phi(t,\tau) = e^{\int_\tau^t \alpha(\eta)\,d\eta}$$

Proof: Separating variables in the original equation, $d\xi/\xi = \alpha(t)dt$. Integrating and taking antilogarithms gives $\xi(t) = \xi_0 e^{\int_\tau^t \alpha(\eta)\,d\eta}$.

***Theorem 5.5*:** If $\mathbf{A}(t)\,\mathbf{A}(\tau) = \mathbf{A}(\tau)\,\mathbf{A}(t)$ for all t, τ, the time-varying linear differential system $d\mathbf{x}/dt = \mathbf{A}(t)\mathbf{x}$ has the transition matrix $\mathbf{\Phi}(t,\tau) = e^{\int_\tau^t \mathbf{A}(\eta)\,d\eta}$.

This is a severe requirement on $\mathbf{A}(t)$, and is usually met only on final examinations.

Proof: Use of the series form for the exponential gives

$$e^{\int_\tau^t \mathbf{A}(\eta)\,d\eta} = \mathbf{I} + \int_\tau^t \mathbf{A}(\eta)\,d\eta + \frac{1}{2!}\int_\tau^t \mathbf{A}(\eta)\,d\eta \int_\tau^t \mathbf{A}(\theta)\,d\theta + \cdots \tag{5.36}$$

Taking derivatives,

$$\frac{\partial}{\partial t} e^{\int_\tau^t \mathbf{A}(\eta)\,d\eta} = \mathbf{A}(t) + \frac{1}{2}\mathbf{A}(t)\int_\tau^t \mathbf{A}(\eta)\,d\eta + \frac{1}{2}\int_\tau^t \mathbf{A}(\eta)\,d\eta\,\mathbf{A}(t) + \cdots \tag{5.37}$$

But from equation (*5.36*),

$$\mathbf{A}(t)e^{\int_\tau^t \mathbf{A}(\eta)\,d\eta} = \mathbf{A}(t) + \mathbf{A}(t)\int_\tau^t \mathbf{A}(\eta)\,d\eta + \cdots$$

This equation and (*5.37*) are equal if and only if

$$\mathbf{A}(t)\int_\tau^t \mathbf{A}(\eta)\,d\eta = \int_\tau^t \mathbf{A}(\eta)\,d\eta\,\mathbf{A}(t)$$

Differentiating with respect to τ and multiplying by -1 gives the requirement

$$\mathbf{A}(t)\,\mathbf{A}(\tau) = \mathbf{A}(\tau)\,\mathbf{A}(t)$$

Only $\mathbf{A}(t)\,\mathbf{A}(\tau) = \mathbf{G}(t,\tau)$ need be multiplied in the application of this test. Substitution of τ for t and t for τ will then indicate if $\mathbf{G}(t,\tau) = \mathbf{G}(\tau,t)$.

Example 5.4.

Given $\mathbf{A}(t) = \begin{pmatrix} t^2 & t \\ 0 & 1 \end{pmatrix}$. Then from $\mathbf{A}(t)\,\mathbf{A}(\tau) = \mathbf{G}(t,\tau) = \begin{pmatrix} t^2\tau^2 & t^2\tau + t \\ 0 & 1 \end{pmatrix}$, we see immediately $\mathbf{G}(\tau,t) = \begin{pmatrix} \tau^2 t^2 & \tau^2 t + \tau \\ 0 & 1 \end{pmatrix} \neq \mathbf{G}(t,\tau)$.

Theorem 5.6: A piecewise time-invariant system, in which $\mathbf{A}(t) = \mathbf{A}_i$ for $t_i \leq t \leq t_{i+1}$ for $i = 0, 1, 2, \ldots$ where each $\mathbf{A}_i$ is a constant matrix, has the transition matrix

$$\mathbf{\Phi}(t,t_0) = e^{\mathbf{A}_i(t-t_i)}\mathbf{\Phi}(t_i,t_0) \quad \text{for} \quad t_i \leq t \leq t_{i+1}$$

Proof: Use of the continuity property of dynamical systems, the transition property (equation (5.5)) and the transition matrix for time-invariant systems gives this proof.

Successive application of this theorem gives

$$\mathbf{\Phi}(t,t_0) = e^{\mathbf{A}_0(t-t_0)} \qquad \text{for } t_0 \leq t \leq t_1$$

$$\mathbf{\Phi}(t,t_0) = e^{\mathbf{A}_1(t-t_1)}e^{\mathbf{A}_0(t_1-t_0)} \qquad \text{for } t_1 \leq t \leq t_2$$

etc.

Example 5.5.

Given the flow diagram of Fig. 5-2 with a switch S that switches from the lower position to the upper position at time t_1. Then $dx_1/dt = x_1$ for $t_0 \leq t < t_1$ and $dx_1/dt = 2x_1$ for $t_1 \leq t$. The solutions during each time interval are $x_1(t) = x_{10}e^{t-t_0}$ for $t_0 \leq t < t_1$ and $x_1(t) = x_1(t_1)e^{2(t-t_1)}$ for $t_1 \leq t$, where $x_1(t_1) = x_{10}e^{t_1-t_0}$ by continuity.

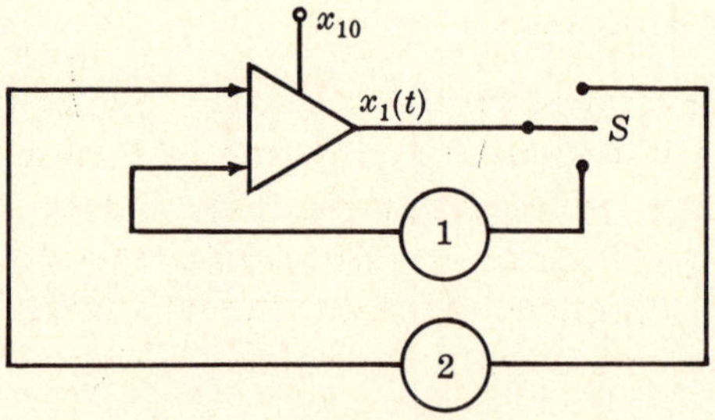

Fig. 5-2

It is common practice to approximate slowly varying coefficients by piecewise constants. This can be dangerous because errors tend to accumulate, but often suggests means of system design that can be checked by simulation with the original system.

Another special case is that the nth order time-varying equation

$$t^n\frac{d^ny}{dt^n} + \alpha_1 t^{n-1}\frac{d^{n-1}y}{dt^{n-1}} + \cdots + \alpha_{n-1}t\frac{dy}{dt} + \alpha_n y = 0$$

can be solved by assuming $y = t^\lambda$. Then a scalar polynomial results for λ, analogous to the characteristic equation. If there are multiplicities of order m in the solution of this polynomial, $y = (\ln t)^{m-1}t^\lambda$ is a solution for $i = 0, 1, 2, \ldots, m-1$.

A number of "classical" second order linear equations have closed form solution in the sense that the properties of the solutions have been investigated.

Bessel's equation:

$$t^2\ddot{y} + (1-2\alpha)t\dot{y} + [\beta\gamma t^{2\gamma} + (\alpha^2 - p^2\gamma^2)]y = 0 \tag{5.38}$$

Associated Legendre equation:

$$(1-t^2)\ddot{y} - 2t\dot{y} + [n(n+1) - m^2/(1-t^2)]y = 0$$

Hermite equation:

$$\ddot{y} - 2t\dot{y} + 2\alpha y = 0$$

or, with $z = e^{t^2/2}y$,

$$\ddot{z} + (1 - t^2 + 2\alpha)z = 0$$

Laguerre equation:

$$t\ddot{y} + (1-t)\dot{y} + \alpha y = 0$$

with solution $L_n(t)$, or

$$t\ddot{y} + (k+1-t)\dot{y} + (\alpha - k)y = 0$$

with solution $d^kL_n(t)/dt^k$.

Hypergeometric equation:

Ordinary:

$$t(1-t)\ddot{y} + [\gamma - (\alpha+\beta+1)t]\dot{y} - \alpha\beta y = 0$$

Confluent:

$$t\ddot{y} + (\gamma - t)\dot{y} - \alpha y = 0$$

Mathieu equation:

$$\ddot{y} + (\alpha + \beta\cos t)y = 0 \tag{5.39}$$

or, with $\tau = \cos^2 t$,

$$4\tau(1-\tau)\ddot{y} + 2(1-2\tau)\dot{y} + [\alpha + \beta(2\tau - 1)]y = 0$$

The solutions and details on their behavior are available in standard texts on engineering mathematics and mathematical physics.

Also available in the linear time-varying case are a number of methods to give $\mathbf{\Phi}(t,\tau)$ as an infinite series. Picard iteration, Peano-Baker integration, perturbation techniques, etc., can be used, and sometimes give quite rapid convergence. However, even only three or four terms in a series representation greatly complicate any sort of design procedure, so discussion of these series techniques is left to standard texts. Use of a digital or analog computer is recommended for those cases in which a closed form solution is not readily found.

5.5 PERIODICALLY-VARYING LINEAR DIFFERENTIAL SYSTEMS

Floquet theory is applicable to time-varying linear systems whose coefficients are constant or vary periodically. Floquet theory does not help find the solution, but instead gives insight into the general behavior of periodically-varying systems.

***Theorem 5.7*:** (Floquet). Given the dynamical linear time-varying system $d\mathbf{x}/dt = \mathbf{A}(t)\mathbf{x}$, where $\mathbf{A}(t) = \mathbf{A}(t+\omega)$. Then

$$\mathbf{\Phi}(t,\tau) = \mathbf{P}(t,\tau)e^{\mathbf{R}(t-\tau)}$$

where $\mathbf{P}(t,\tau) = \mathbf{P}(t+\omega,\tau)$ and $\mathbf{R}$ is a constant matrix.

Proof: The transition matrix satisfies

$$\partial\mathbf{\Phi}(t,\tau)/\partial t = \mathbf{A}(t)\,\mathbf{\Phi}(t,\tau) \quad \text{with} \quad \mathbf{\Phi}(\tau,\tau) = \mathbf{I}$$

Setting $t = t+\omega$, and using $\mathbf{A}(t) = \mathbf{A}(t+\omega)$ gives

$$\partial\mathbf{\Phi}(t+\omega,\tau)/\partial t = \mathbf{A}(t+\omega)\,\mathbf{\Phi}(t+\omega,\tau) = \mathbf{A}(t)\,\mathbf{\Phi}(t+\omega,\tau) \tag{5.40}$$

It was shown in Example 3.25 that the solutions to $d\mathbf{x}/dt = \mathbf{A}(t)\mathbf{x}$ for any initial condition form a generalized vector space. The column vectors $\boldsymbol{\phi}_i(t,\tau)$ of $\mathbf{\Phi}(t,\tau)$ span this vector space, and since $\det\mathbf{\Phi}(t,\tau) \neq 0$, the $\boldsymbol{\phi}_i(t,\tau)$ are a basis. But equation *(5.40)* states that the $\boldsymbol{\phi}_i(t+\omega,\tau)$ are solutions to $d\mathbf{x}/dt = \mathbf{A}(t)\mathbf{x}$, so that

$$\boldsymbol{\phi}_i(t+\omega,\tau) = \sum_{j=1}^{n} c_{ji}\boldsymbol{\phi}_j(t,\tau) \quad \text{for } i = 1,2,\ldots,n$$

Rewriting this in matrix form, where $\mathbf{C} = \{c_{ji}\}$,

$$\mathbf{\Phi}(t+\omega,\tau) = \mathbf{\Phi}(t,\tau)\mathbf{C} \tag{5.41}$$

Then

$$\mathbf{C} = \mathbf{\Phi}(\tau,t)\,\mathbf{\Phi}(t+\omega,\tau)$$

Note that $\mathbf{C}^{-1}$ exists, since it is the product of two nonsingular matrices. Therefore by Problem 4.11 the logarithm of $\mathbf{C}$ exists and will be written in the form

$$\mathbf{C} = e^{\omega\mathbf{R}} \tag{5.42}$$

If $\mathbf{P}(t,\tau)$ can be any matrix, it is merely a change of variables to write

$$\mathbf{\Phi}(t,\tau) = \mathbf{P}(t,\tau)e^{\mathbf{R}(t-\tau)} \tag{5.43}$$

But from equations *(5.43)*, *(5.41)* and *(5.42)*,

$$\begin{aligned}\mathbf{P}(t+\omega,\tau) &= \mathbf{\Phi}(t+\omega,\tau)e^{-\mathbf{R}(t+\omega-\tau)} = \mathbf{\Phi}(t,\tau)e^{\omega\mathbf{R}}e^{-\mathbf{R}(t+\omega-\tau)} \\ &= \mathbf{\Phi}(t,\tau)e^{-\mathbf{R}(t-\tau)} = \mathbf{P}(t,\tau)\end{aligned}$$

From *(5.41)*–*(5.43)*, $\mathbf{R} = \omega^{-1}\ln[\mathbf{\Phi}(\tau,t)\,\mathbf{\Phi}(t+\omega,\tau)]$ and $\mathbf{P}(t,\tau) = \mathbf{\Phi}(t,\tau)e^{-\mathbf{R}(t-\tau)}$, so that to find $\mathbf{R}$ and $\mathbf{P}(t,\tau)$ the solution $\mathbf{\Phi}(t,\tau)$ must already be known. It may be concluded that Floquet's theorem does not give the solution, but rather shows the form of the solution. The matrix $\mathbf{P}(t,\tau)$ gives the periodic part of the solution, and $e^{\mathbf{R}(t-\tau)}$ gives the envelope of the solution. Since $e^{\mathbf{R}(t-\tau)}$ is the transition matrix to $d\mathbf{z}/dt = \mathbf{R}\mathbf{z}$, this is the *constant coefficient* equation for the envelope $\mathbf{z}(t)$. If the system $d\mathbf{z}/dt = \mathbf{R}\mathbf{z}$ has all poles in the left half plane, the original time-varying system $\mathbf{x}(t)$ is stable. If $\mathbf{R}$ has all eigenvalues in the left half plane except for some on the imaginary axis, the steady state of $\mathbf{z}(t)$ is periodic with the frequency of its imaginary eigenvalues. To have a periodic envelope in the sense that no element of $\mathbf{z}(t)$ behaves exponentially, *all* the eigenvalues of $\mathbf{R}$ must be on the imaginary axis. If any eigenvalues of $\mathbf{R}$ are in the right half plane, then $\mathbf{z}(t)$ and $\mathbf{x}(t)$ are unstable. In particular, if the coefficients of the $\mathbf{A}(t)$ matrix are continuous functions of some parameter α, the eigenvalues of $\mathbf{R}$ are also continuous functions of α, so that periodic solutions form the stability boundaries of the system.

Example 5.6.

Consider the Mathieu equation $d^2x/dt^2 + (\alpha + \beta\cos t)x = 0$ *(5.39)*. Its periodic solutions are called Mathieu functions, which exist only for certain combinations of α and β. The values of α and β for which these periodic solutions exist are given by the curves in Fig. 5.3 below. These curves then form the boundary for regions of stability.

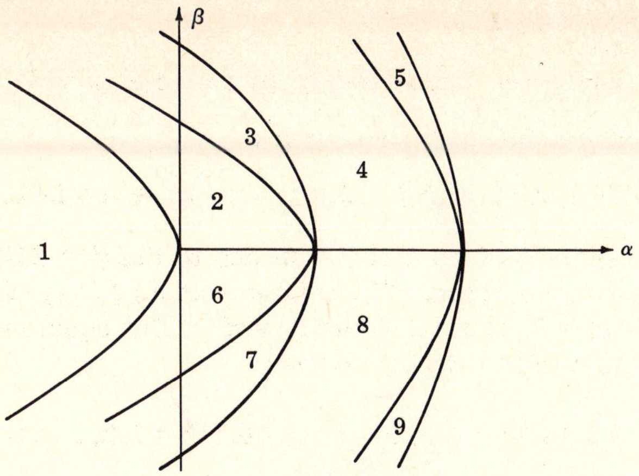

Fig. 5-3

Whether the regions are stable or unstable can be determined by considering the point $\beta = 0$ and $\alpha < 0$ in region 1. This is known to be unstable, so the whole region 1 is unstable. Since the curves are stability boundaries, regions 2 and 6 are stable. Similarly all the odd numbered regions are unstable and all the even numbered regions are stable. The line $\beta = 0$, $\alpha \geqq 0$ represents a degenerate case, which agrees with physical intuition.

It is interesting to note from the example above that an originally unstable system might be stabilized by the introduction of a periodically-varying parameter, and vice versa.

Another use of Floquet theory is in simulation of $\mathbf{\Phi}(t, \tau)$. Only $\mathbf{\Phi}(t, \tau)$ for one period ω need be calculated numerically and then Floquet's theorem can be used to generate the solution over the whole time span.

5.6 SOLUTION OF THE LINEAR STATE EQUATIONS WITH INPUT

Knowledge of the transition matrix gives the solution to the linear state equation with input, even in time-varying systems.

Theorem 5.8: Given the linear differential system with input

$$d\mathbf{x}/dt = \mathbf{A}(t)\mathbf{x} + \mathbf{B}(t)\mathbf{u}$$

$$\mathbf{y} = \mathbf{C}(t)\mathbf{x} + \mathbf{D}(t)\mathbf{u} \tag{2.39}$$

with transition matrix $\mathbf{\Phi}(t, \tau)$ obeying $\partial\mathbf{\Phi}(t, \tau)/\partial t = \mathbf{A}(t)\,\mathbf{\Phi}(t, \tau)$ [equation (*5.1*)]. Then

$$\mathbf{x}(t) = \mathbf{\Phi}(t, t_0)\,\mathbf{x}(t_0) + \int_{t_0}^{t} \mathbf{\Phi}(t, \tau)\,\mathbf{B}(\tau)\,\mathbf{u}(\tau)\,d\tau$$

$$\mathbf{y}(t) = \mathbf{C}(t)\,\mathbf{\Phi}(t, t_0)\,\mathbf{x}(t_0) + \int_{t_0}^{t} \mathbf{C}(t)\,\mathbf{\Phi}(t, \tau)\,\mathbf{B}(\tau)\,\mathbf{u}(\tau)\,d\tau + \mathbf{D}(t)\mathbf{u} \tag{5.44}$$

The integral is the superposition integral, and in the time-invariant case it becomes a convolution integral.

Proof: Since the equation $d\mathbf{x}/dt = \mathbf{A}(t)\mathbf{x}$ has a solution $\mathbf{x}(t) = \mathbf{\Phi}(t, t_0)$, in accordance with the method of variation of parameters, we change variables to $\mathbf{k}(t)$ where

$$\mathbf{x}(t) = \mathbf{\Phi}(t, t_0)\,\mathbf{k}(t) \tag{5.45}$$

Substituting into equation (*2.39*),

$$d\mathbf{x}/dt = (\partial\mathbf{\Phi}/\partial t)\mathbf{k} + \mathbf{\Phi} d\mathbf{k}/dt = \mathbf{A}(t)\mathbf{\Phi}\mathbf{k} + \mathbf{B}(t)\mathbf{u}$$

Use of equation (*5.1*) and multiplication by $\mathbf{\Phi}(t_0, t)$ gives

$$d\mathbf{k}/dt = \mathbf{\Phi}(t_0, t)\,\mathbf{B}(t)\,\mathbf{u}(t)$$

Integrating from t_0 to t,

$$\mathbf{k}(t) = \mathbf{k}(t_0) + \int_{t_0}^{t} \mathbf{\Phi}(t_0, \tau)\,\mathbf{B}(\tau)\,\mathbf{u}(\tau)\,d\tau \tag{5.46}$$

Since equation (*5.45*) evaluated at $t = t_0$ gives $\mathbf{x}(t_0) = \mathbf{k}(t_0)$, use of (*5.45*) in (*5.46*) yields

$$\mathbf{\Phi}(t_0, t)\,\mathbf{x}(t) = \mathbf{x}(t_0) + \int_{t_0}^{t} \mathbf{\Phi}(t_0, \tau)\,\mathbf{B}(\tau)\,\mathbf{u}(\tau)\,d\tau$$

Multiplying by $\mathbf{\Phi}(t, t_0)$ completes the proof for $\mathbf{x}(t)$. Substituting into $\mathbf{y}(t) = \mathbf{C}(t)\,\mathbf{x}(t) + \mathbf{D}(t)\,\mathbf{u}(t)$ gives $\mathbf{y}(t)$.

In the constant coefficient case, use of equation (*5.14*) gives

$$\mathbf{x}(t) = e^{\mathbf{A}(t-t_0)}\mathbf{x}(t_0) + \int_{t_0}^{t} e^{\mathbf{A}(t-\tau)}\mathbf{B}\mathbf{u}(\tau)\,d\tau \tag{5.47}$$

and

$$\mathbf{y}(t) = \mathbf{C}e^{\mathbf{A}(t-t_0)}\mathbf{x}(t_0) + \int_{t_0}^{t} \mathbf{C}e^{\mathbf{A}(t-\tau)}\mathbf{B}\mathbf{u}(\tau)\,d\tau + \mathbf{D}\mathbf{u}(t) \tag{5.48}$$

This is the vector convolution integral.

***Theorem 5.9*:** Given the linear difference equation

$$\mathbf{x}(k+1) = \mathbf{A}(k)\,\mathbf{x}(k) + \mathbf{B}(k)\,\mathbf{u}(k)$$
$$\mathbf{y}(k) = \mathbf{C}(k)\,\mathbf{x}(k) + \mathbf{D}(k)\,\mathbf{u}(k) \tag{2.40}$$

with transition matrix $\mathbf{\Phi}(k, m)$ obeying $\mathbf{\Phi}(k+1, m) = \mathbf{A}(k)\,\mathbf{\Phi}(k, m)$ [equation (*5.3*)]. Then

$$\mathbf{x}(k) = \prod_{i=m}^{k-1} \mathbf{A}(i)\,\mathbf{x}(m) + \sum_{j=m}^{k-2} \prod_{i=j+1}^{k-1} \mathbf{A}(i)\,\mathbf{B}(j)\,\mathbf{u}(j) + \mathbf{B}(k-1)\,\mathbf{u}(k-1) \tag{5.49}$$

where the order of multiplication starts with the largest integer, i.e. $\mathbf{A}(k-1)\,\mathbf{A}(k-2)\cdots$.

Proof: Stepping equation (*2.40*) up one gives

$$\mathbf{x}(m+2) = \mathbf{A}(m+1)\,\mathbf{x}(m+1) + \mathbf{B}(m+1)\,\mathbf{u}(m+1)$$

Substituting in equation (*2.40*) for $\mathbf{x}(m+1)$,

$$\mathbf{x}(m+2) = \mathbf{A}(m+1)\,\mathbf{A}(m)\,\mathbf{x}(m) + \mathbf{A}(m+1)\,\mathbf{B}(m)\,\mathbf{u}(m) + \mathbf{B}(m+1)\,\mathbf{u}(m+1)$$

Repetitive stepping and substituting gives

$$\begin{aligned}\mathbf{x}(k) = {} & \mathbf{A}(k-1)\cdots\mathbf{A}(m)\,\mathbf{x}(m) + \mathbf{A}(k-1)\cdots\mathbf{A}(m+1)\,\mathbf{B}(m)\,\mathbf{u}(m) \\ & + \mathbf{A}(k-1)\cdots\mathbf{A}(m+2)\,\mathbf{B}(m+1)\,\mathbf{u}(m+1) + \cdots + \mathbf{A}(k-1)\,\mathbf{B}(k-2)\,\mathbf{u}(k-2) \\ & + \mathbf{B}(k-1)\,\mathbf{u}(k-1)\end{aligned}$$

This is equation (*5.49*) with the sums and products written out.

5.7 TRANSITION MATRIX FOR TIME-VARYING DIFFERENCE EQUATIONS

Setting $\mathbf{B} = \mathbf{0}$ in equation (5.49) gives the transition matrix for time-varying difference equations as

$$\mathbf{\Phi}(k,m) = \prod_{i=m}^{k-1} \mathbf{A}(i) \quad \text{for } k > m \tag{5.50}$$

For $k = m$, $\mathbf{\Phi}(m,m) = \mathbf{I}$, and if $\mathbf{A}^{-1}(i)$ exists for all i, $\mathbf{\Phi}(k,m) = \prod_{i=k}^{m-1} \mathbf{A}^{-1}(i)$ for $k<m$. Then equation (5.49) can be written as

$$\mathbf{x}(k) = \mathbf{\Phi}(k,m)\,\mathbf{x}(m) + \sum_{j=m}^{k-1} \mathbf{\Phi}(k, j+1)\,\mathbf{B}(j)\,\mathbf{u}(j) \tag{5.51}$$

This is very similar to the corresponding equation (5.44) for differential equations except the integral is replaced by a sum.

Often difference equations result from periodic sampling and holding inputs to differential systems.

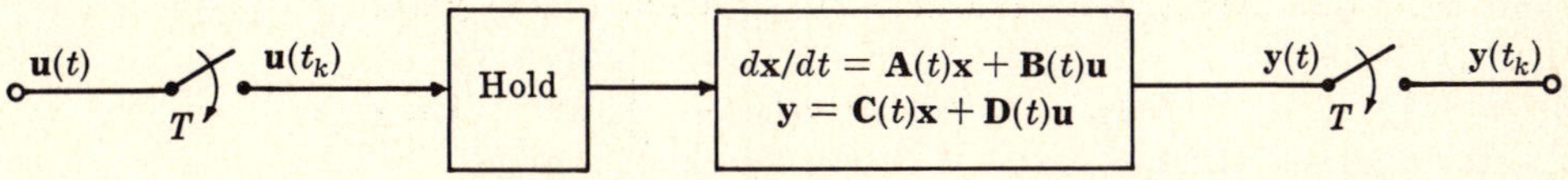

Fig. 5-4

In Fig. 5-4 the output of the hold element is $\mathbf{u}(k) = \mathbf{u}(t_k)$ for $t_k \leq t < t_{k+1}$, where $t_{k+1} - t_k = T$ for all k. Use of (5.44) at time $t = t_{k+1}$ and $t_0 = t_k$ gives

$$\mathbf{x}(k+1) = \mathbf{x}(t_{k+1}) = \mathbf{\Phi}(t_{k+1}, t_k)\,\mathbf{x}(k) + \int_{t_k}^{t_{k+1}} \mathbf{\Phi}(t_{k+1}, \tau)\,\mathbf{B}(\tau)\,\mathbf{u}(k)\,d\tau \tag{5.52}$$

Comparison with the difference equations (2.40) results in

$$\mathbf{A}_s(k) = \mathbf{\Phi}(t_{k+1}, t_k) \qquad \mathbf{B}_s(k) = \int_{t_k}^{t_{k+1}} \mathbf{\Phi}(t_{k+1}, \tau)\,\mathbf{B}(\tau)\,d\tau$$

$$\mathbf{C}_s(k) = \mathbf{C}(t_k) \qquad \mathbf{D}_s(k) = \mathbf{D}(t_k)$$

where the subscript s refers to the difference equations of the sampled system. For time-invariant differential systems, $\mathbf{A}_s = e^{\mathbf{A}T}$, $\mathbf{B}_s = \int_0^T e^{\mathbf{A}(T-\tau)}\mathbf{B}\,d\tau$, $\mathbf{C}_s = \mathbf{C}$ and $\mathbf{D}_s = \mathbf{D}$. Since in this case $\mathbf{A}_s$ is a matrix exponential, it is nonsingular no matter what $\mathbf{A}$ is (see the comment after Problem 4.11).

Although equation (5.50) is always a representation for $\mathbf{\Phi}(k,m)$, its behavior is not usually displayed. Techniques corresponding to the differential case can be used to show this behavior. For instance, Floquet's theorem becomes $\mathbf{\Phi}(k,m) = \mathbf{P}(k,m)\mathbf{R}^{k-m}$, where $\mathbf{P}(k,m) = \mathbf{P}(k+\omega, m)$ if $\mathbf{A}(k) = (k+\omega)$. Also

$$\frac{(k+n)!}{k!}\,y(k+n) + \alpha_1 \frac{(k+n-1)!}{k!}\,y(k+n-1) + \cdots + \alpha_{n-1}(k+1)\,y(k+1) + \alpha_n y(k) = 0$$

has solutions of the form $\lambda^k/k!$. Piecewise time-invariant, classical second order linear, and series solutions also have a corresponding discrete time form.

5.8 IMPULSE RESPONSE MATRICES

With zero initial condition $\mathbf{x}(t_0) = \mathbf{0}$, from *(5.44)* the output $\mathbf{y}(t)$ is

$$\mathbf{y}(t) = \int_{t_0}^{t} \mathbf{C}(t)\,\mathbf{\Phi}(t,\tau)\,\mathbf{B}(\tau)\,\mathbf{u}(\tau)\,d\tau + \mathbf{D}(t)\,\mathbf{u}(t) \tag{5.53}$$

This suggests that $\mathbf{y}(t)$ can be written as a matrix generalization of the superposition integral,

$$\mathbf{y}(t) = \int_{-\infty}^{t} \mathbf{H}(t,\tau)\,\mathbf{u}(\tau)\,d\tau \tag{5.54}$$

where $\mathbf{H}(t,\tau)$ is the impulse response matrix, i.e. $h_{ij}(t,\tau)$ is the response of the ith output at time t due to an impulse at the jth input at time τ. Comparison of equations *(5.53)* and *(5.54)* gives

$$\mathbf{H}(t,\tau) = \begin{cases} \mathbf{C}(t)\,\mathbf{\Phi}(t,\tau)\,\mathbf{B}(\tau) + \delta(t-\tau)\,\mathbf{D}(t) & t \geqq \tau \\ \mathbf{0} & t < \tau \end{cases} \tag{5.55}$$

In the time-invariant case the Laplace transformation of $\mathbf{H}(t,0)$ gives the transfer function matrix

$$\mathcal{L}\{\mathbf{H}(t,0)\} = \mathbf{C}(s\mathbf{I}-\mathbf{A})^{-1}\mathbf{B} + \mathbf{D} \tag{5.56}$$

Similarly, for discrete-time systems $\mathbf{y}(k)$ can be expressed as

$$\mathbf{y}(k) = \sum_{m=-\infty}^{k} \mathbf{H}(k,m)\,\mathbf{u}(m) \tag{5.57}$$

where

$$\mathbf{H}(k,m) = \begin{cases} \mathbf{C}(k)\,\mathbf{\Phi}(k,m+1)\,\mathbf{B}(m) & k > m \\ \mathbf{D}(k) & k = m \\ \mathbf{0} & k < m \end{cases} \tag{5.58}$$

Also, the $\mathcal{Z}$ transfer function matrix in the time-invariant case is

$$\mathcal{Z}\{\mathbf{H}(k,0)\} = \mathbf{C}(z\mathbf{I}-\mathbf{A})^{-1}\mathbf{B} + \mathbf{D} \tag{5.59}$$

5.9 THE ADJOINT SYSTEM

The concept of the adjoint occurs quite frequently, especially in optimization problems.

Definition 5.2: The *adjoint*, denoted L_a, of a linear operator L is defined by the relation

$$(\mathbf{p}, L\mathbf{x}) = (L_a\mathbf{p}, \mathbf{x}) \qquad \text{for all } \mathbf{x} \text{ and } \mathbf{p} \tag{5.60}$$

We are concerned with the system $d\mathbf{x}/dt = \mathbf{A}(t)\mathbf{x}$. Defining $L = \mathbf{A}(t) - \mathbf{I}d/dt$, this becomes $L\mathbf{x} = \mathbf{0}$. Using the scalar product $(\mathbf{p},\mathbf{x}) = \int_{t_0}^{t_1} \mathbf{p}^\dagger \mathbf{x}\,dt$, the adjoint system is found from equation *(5.60)* using integration by parts.

$$\begin{aligned}(\mathbf{p}, L\mathbf{x}) &= \int_{t_0}^{t_1} \mathbf{p}^\dagger \mathbf{A}(t)\mathbf{x}\,dt - \int_{t_0}^{t_1} \mathbf{p}^\dagger \frac{d\mathbf{x}}{dt}\,dt \\ &= \int_{t_0}^{t_1} \left[\mathbf{x}^\dagger\mathbf{A}^\dagger\mathbf{p} + \mathbf{x}^\dagger \frac{d\mathbf{p}}{dt}\right]^* dt + \mathbf{p}^\dagger(t_0)\,\mathbf{x}(t_0) - \mathbf{p}^\dagger(t_1)\,\mathbf{x}(t_0)\end{aligned}$$

For the case $\mathbf{p}(t_0) = \mathbf{0} = \mathbf{p}(t_1)$, we find $L_a = \mathbf{A}^\dagger(t) + \mathbf{I}d/dt$.

Since $L\mathbf{x} = \mathbf{0}$ for all $\mathbf{x}$, then $(\mathbf{p}, L\mathbf{x}) = 0$ for all $\mathbf{x}$ and $\mathbf{p}$. Using (*5.60*) it can be concluded $L_a\mathbf{p} = \mathbf{0}$, so that the *adjoint system* is defined by the relation

$$d\mathbf{p}/dt = -\mathbf{A}^\dagger(t)\mathbf{p} \tag{5.61}$$

Denote the transition matrix of this adjoint system as $\Psi(t, t_0)$, i.e.,

$$\partial\Psi(t, t_0)/\partial t = -\mathbf{A}^\dagger(t)\,\Psi(t, t_0) \quad \text{with} \quad \Psi(t_0, t_0) = \mathbf{I} \tag{5.62}$$

Theorem 5.10: Given the system $d\mathbf{x}/dt = \mathbf{A}(t)\mathbf{x} + \mathbf{B}(t)\mathbf{u}$ and its adjoint system $d\mathbf{p}/dt = -\mathbf{A}^\dagger(t)\mathbf{p}$. Then

$$\mathbf{p}^\dagger(t_1)\,\mathbf{x}(t_1) = \mathbf{p}^\dagger(t_0)\,\mathbf{x}(t_0) + \int_{t_0}^{t_1} \mathbf{p}^\dagger(t)\,\mathbf{B}(t)\,\mathbf{u}(t)\,dt \tag{5.63}$$

and

$$\Psi^\dagger(t, t_0) = \Phi^{-1}(t, t_0) \tag{5.64}$$

The column vectors $\psi_i(t, t_0)$ of $\Psi(t, t_0)$ are the reciprocal basis to the column vectors $\phi_i(t, t_0)$ of $\Phi(t, t_0)$. Also, if $\mathbf{u}(t) = \mathbf{0}$, then $\mathbf{p}^\dagger(t)\,\mathbf{x}(t)$ = scalar constant for any t.

Proof: Differentiate $\mathbf{p}^\dagger(t)\,\mathbf{x}(t)$ to obtain

$$d(\mathbf{p}^\dagger\mathbf{x})/dt = (d\mathbf{p}^\dagger/dt)\mathbf{x} + \mathbf{p}^\dagger\,d\mathbf{x}/dt$$

Using the system equation $d\mathbf{x}/dt = \mathbf{A}(t)\mathbf{x} + \mathbf{B}(t)\mathbf{u}$ and equation (*5.61*) gives

$$d(\mathbf{p}^\dagger\mathbf{x})/dt = \mathbf{p}^\dagger\mathbf{B}\mathbf{u}$$

Integration from t_0 to t_1 then yields (*5.63*). Furthermore if $\mathbf{u}(t) = \mathbf{0}$,

$$\mathbf{p}^\dagger(t_0)\,\mathbf{x}(t_0) = \mathbf{p}^\dagger(t)\,\mathbf{x}(t)$$

From the transition relation, $\mathbf{x}(t) = \Phi(t, t_0)\,\mathbf{x}(t_0)$ and $\mathbf{p}(t) = \Psi(t, t_0)\,\mathbf{p}(t_0)$ so that

$$\mathbf{p}^\dagger(t_0)\,\mathbf{I}\mathbf{x}(t_0) = \mathbf{p}^\dagger(t_0)\,\Psi^\dagger(t, t_0)\,\Phi(t, t_0)\,\mathbf{x}(t_0)$$

for any $\mathbf{p}(t_0)$ and $\mathbf{x}(t_0)$. Therefore (*5.64*) must hold.

The adjoint system transition matrix gives another way to express the forced solution of (*5.44*):

$$\mathbf{x}(t) = \Phi(t, t_0)\,\mathbf{x}(t_0) + \int_{t_0}^{t} \Phi(t, \tau)\,\mathbf{B}(\tau)\,\mathbf{u}(\tau)\,d\tau$$

The variable of integration τ is the second argument of $\Phi(t, \tau)$, which sometimes poses simulation difficulties. Since $\Phi(t_0, \tau) = \Phi^{-1}(\tau, t_0) = \Psi^\dagger(\tau, t_0)$, this becomes

$$\mathbf{x}(t) = \Phi(t, t_0)\left[\mathbf{x}(t_0) + \int_{t_0}^{t} \Psi^\dagger(\tau, t_0)\,\mathbf{B}(\tau)\,\mathbf{u}(\tau)\,d\tau\right] \tag{5.65}$$

in which the variable of integration τ is the first argument of $\Psi(\tau, t_0)$.

The adjoint often can be used conveniently when a final value is given and the system motion backwards in time must be found.

Example 5.7.

Given $d\mathbf{x}/dt = \begin{pmatrix} t & 2t \\ 3t & 2t \end{pmatrix}\mathbf{x}$. Use the adjoint system to find the set of states $(x_1(1)\ x_2(1))$ that permit the system to pass through the point $x_1(2) = 1$.

The adjoint system is $d\mathbf{p}/dt = -t\begin{pmatrix} 1 & 3 \\ 2 & 2 \end{pmatrix}\mathbf{p}$. This has a transition matrix

$$\Psi(t, \tau) = 0.2e^{(t^2/2)-(\tau^2/2)}\begin{pmatrix} 3 & -3 \\ -2 & 2 \end{pmatrix} + 0.2e^{2\tau^2-2t^2}\begin{pmatrix} 2 & 3 \\ 2 & 3 \end{pmatrix}$$

Since $\mathbf{p}^\dagger(2)\,\mathbf{x}(2) = \mathbf{p}^\dagger(1)\,\mathbf{x}(1)$, if we choose $\mathbf{p}^\dagger(2) = (1\ \ 0)$, then $(1\ \ 0)\mathbf{x}(2) = x_1(2) = 1 = \mathbf{p}^\dagger(1)\,\mathbf{x}(1) = p_1(1)\,x_1(1) + p_2(1)\,x_2(1)$. But $\mathbf{p}(1) = \mathbf{\Psi}(1,2)\,\mathbf{p}(2)$, so that

$$\mathbf{p}^\dagger(1) \;=\; 0.2(3e^{-1.5} + 2e^6 \quad 2e^6 - 2e^{-1.5})$$

The set of states $\mathbf{x}(1)$ that gives $x_1(2) = 1$ is determined by

$$1 \;=\; (0.6e^{-1.5} + 0.4e^6)x_1(1) + (0.4e^6 - 0.4e^{-1.5})x_2(1)$$

Solved Problems

5.1. Given $\mathbf{\Phi}(t,t_0)$, find $\mathbf{A}(t)$.

Using (5.2) in (5.1) at time $t_0 = t$, $\mathbf{A}(t) = \partial\mathbf{\Phi}(t,t_0)/\partial t$ evaluated at $t_0 = t$. This is a quick check on any solution.

5.2. Find the transition matrix for the system

$$\frac{d\mathbf{x}}{dt} \;=\; \begin{pmatrix} -1 & 0 & 0 \\ 0 & -4 & 4 \\ 0 & -1 & 0 \end{pmatrix}\mathbf{x}$$

by (a) series method, (b) eigenvalue method, (c) Cayley-Hamilton method, (d) resolvent matrix method. In the resolvent matrix method, find $(s\mathbf{I} - \mathbf{A})^{-1}$ by (1) substitution and elimination, (2) Cramer's rule, (3) flow diagram, (4) Leverrier's algorithm.

(a) Series method. From (5.16), $e^{\mathbf{A}t} = \mathbf{I} + \mathbf{A}t/1! + \mathbf{A}^2 t^2/2! + \cdots$. Substituting for $\mathbf{A}$,

$$e^{\mathbf{A}t} = \begin{pmatrix} 1 & 0 & 0 \\ 0 & 1 & 0 \\ 0 & 0 & 1 \end{pmatrix} + \begin{pmatrix} -t & 0 & 0 \\ 0 & -4t & 4t \\ 0 & -t & 0 \end{pmatrix} + \begin{pmatrix} t^2/2 & 0 & 0 \\ 0 & 6t^2 & -8t^2 \\ 0 & 2t^2 & -2t^2 \end{pmatrix} + \cdots$$

$$= \begin{pmatrix} 1 - t + t^2/2 + \cdots & 0 & 0 \\ 0 & 1 - 2t + 4t^2/2 - 2t(1-2t) + \cdots & 4t(1 - 2t + 4t^2/2 + \cdots) \\ 0 & -t(1 - 2t + 4t^2/2 + \cdots) & 1 - 2t + 4t^2/2 + 2t(1-2t) + \cdots \end{pmatrix}$$

Recognizing the series expression for e^{-t} and e^{-2t} gives

$$e^{\mathbf{A}t} = \begin{pmatrix} e^{-t} & 0 & 0 \\ 0 & (1-2t)e^{-2t} & 4te^{-2t} \\ 0 & -te^{-2t} & (1+2t)e^{-2t} \end{pmatrix}$$

(b) The eigenvalues of $\mathbf{A}$ are $-1, -2$ and -2, with corresponding eigenvectors $(1\ 0\ 0)^T$ and $(0\ 2\ 1)^T$. The generalized eigenvector corresponding to -2 is $(0\ 1\ 1)^T$, so that

$$\begin{pmatrix} -1 & 0 & 0 \\ 0 & -4 & 4 \\ 0 & -1 & 0 \end{pmatrix} = \begin{pmatrix} 1 & 0 & 0 \\ 0 & 2 & 1 \\ 0 & 1 & 1 \end{pmatrix}\begin{pmatrix} -1 & 0 & 0 \\ 0 & -2 & 1 \\ 0 & 0 & -2 \end{pmatrix}\begin{pmatrix} 1 & 0 & 0 \\ 0 & 1 & -1 \\ 0 & -1 & 2 \end{pmatrix}.$$

Using equation (5.17),

$$e^{\mathbf{A}t} = \begin{pmatrix} 1 & 0 & 0 \\ 0 & 2 & 1 \\ 0 & 1 & 1 \end{pmatrix}\begin{pmatrix} e^{-t} & 0 & 0 \\ 0 & e^{-2t} & te^{-2t} \\ 0 & 0 & e^{-2t} \end{pmatrix}\begin{pmatrix} 1 & 0 & 0 \\ 0 & 1 & -1 \\ 0 & -1 & 2 \end{pmatrix}$$

Multiplying out the matrices gives the answer obtained in (a).

(c) Again, the eigenvalues of **A** are calculated to be -1, -2 and -2. To find the $\gamma_i(t)$ in equation *(5.18)*,

$$e^{\mathbf{J}t} = \begin{pmatrix} e^{-t} & 0 & 0 \\ 0 & e^{-2t} & te^{-2t} \\ 0 & 0 & e^{-2t} \end{pmatrix} = \gamma_0 \begin{pmatrix} 1 & 0 & 0 \\ 0 & 1 & 0 \\ 0 & 0 & 1 \end{pmatrix} + \gamma_1 \begin{pmatrix} -1 & 0 & 0 \\ 0 & -2 & 1 \\ 0 & 0 & -2 \end{pmatrix} + \gamma_2 \begin{pmatrix} 1 & 0 & 0 \\ 0 & 4 & -4 \\ 0 & 0 & 4 \end{pmatrix}$$

which gives the equations

$$\begin{aligned} e^{-t} &= \gamma_0 - \gamma_1 + \gamma_2 \\ e^{-2t} &= \gamma_0 - 2\gamma_1 + 4\gamma_2 \\ te^{-2t} &= \gamma_1 - 4\gamma_2 \end{aligned}$$

Solving for the γ_i,

$$\begin{aligned} \gamma_0 &= 4e^{-t} - 3e^{-2t} - 2te^{-2t} \\ \gamma_1 &= 4e^{-t} - 4e^{-2t} - 3te^{-2t} \\ \gamma_2 &= e^{-t} - e^{-2t} - te^{-2t} \end{aligned}$$

Using *(5.18)* then gives

$$e^{\mathbf{A}t} = (4e^{-t} - 3e^{-2t} - 2te^{-2t}) \begin{pmatrix} 1 & 0 & 0 \\ 0 & 1 & 0 \\ 0 & 0 & 1 \end{pmatrix} + (4e^{-t} - 4e^{-2t} - 3te^{-2t}) \begin{pmatrix} -1 & 0 & 0 \\ 0 & -4 & 4 \\ 0 & -1 & 0 \end{pmatrix}$$
$$+ (e^{-t} - e^{-2t} - te^{-2t}) \begin{pmatrix} 1 & 0 & 0 \\ 0 & 12 & -16 \\ 0 & 4 & -4 \end{pmatrix}$$

Summing these matrices again gives the answer obtained in (*a*).

(*d*1) Taking the Laplace transform of the original equation,

$$\begin{aligned} s\mathcal{L}(x_1) - x_{10} &= -\mathcal{L}(x_1) \\ s\mathcal{L}(x_2) - x_{20} &= -4\mathcal{L}(x_2) + 4\mathcal{L}(x_3) \\ s\mathcal{L}(x_3) - x_{30} &= -\mathcal{L}(x_2) \end{aligned}$$

Solving these equations by substitution and elimination,

$$\begin{aligned} \mathcal{L}(x_1) &= \frac{x_{10}}{s+1} \\ \mathcal{L}(x_2) &= \left[\frac{1}{s+2} - \frac{2}{(s+2)^2}\right] x_{20} + \frac{4x_{30}}{(s+2)^2} \\ \mathcal{L}(x_3) &= -\frac{x_{20}}{(s+2)^2} + \left[\frac{1}{s+2} + \frac{2}{(s+2)^2}\right] x_{30} \end{aligned}$$

Putting this in matrix form $\mathcal{L}(x) = \mathbf{R}(s)x_0$,

$$\mathbf{R}(s) = \begin{pmatrix} \frac{1}{s+1} & 0 & 0 \\ 0 & \frac{1}{s+2} - \frac{2}{(s+2)^2} & \frac{4}{(s+2)^2} \\ 0 & \frac{1}{(s+2)^2} & \frac{1}{s+2} + \frac{2}{(s+2)^2} \end{pmatrix}$$

Inverse Laplace transformation gives $e^{\mathbf{A}t}$ as found in (*a*).

(*d*2) From *(5.22)*,

$$\mathbf{R}^{-1}(s) = \begin{pmatrix} s+1 & 0 & 0 \\ 0 & s+4 & -4 \\ 0 & 1 & s \end{pmatrix}$$

Using Cramer's rule,

$$\mathbf{R}(s) = \frac{1}{(s+1)(s+2)^2}\begin{pmatrix} (s+2)^2 & 0 & 0 \\ 0 & s(s+1) & 4(s+1) \\ 0 & -(s+1) & (s+1)(s+4) \end{pmatrix}$$

Performing a partial fraction expansion,

$$\mathbf{R}(s) = \frac{1}{s+1}\begin{pmatrix} 1 & 0 & 0 \\ 0 & 0 & 0 \\ 0 & 0 & 0 \end{pmatrix} + \frac{1}{s+2}\begin{pmatrix} 0 & 0 & 0 \\ 0 & 1 & 0 \\ 0 & 0 & 1 \end{pmatrix} + \frac{1}{(s+2)^2}\begin{pmatrix} 0 & 0 & 0 \\ 0 & -2 & 4 \\ 0 & -1 & 2 \end{pmatrix}$$

Addition will give $\mathbf{R}(s)$ as in ($d1$).

($d3$) The flow diagram of the Laplace transformed system is shown in Fig. 5-5.

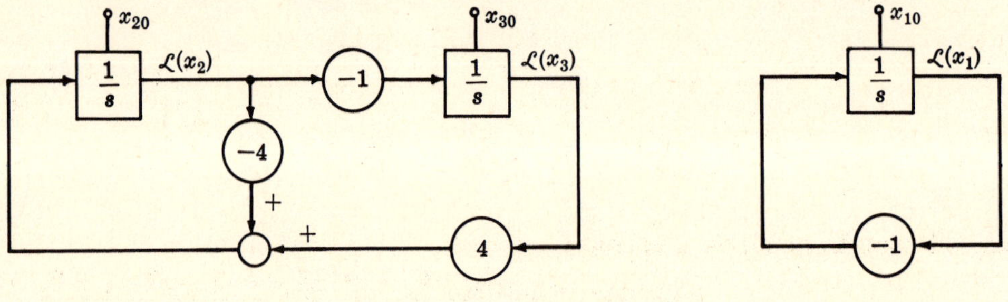

Fig. 5-5

For $x_{10} = 1$, $\mathcal{L}(x_1) = 1/(s+1)$ and $\mathcal{L}(x_2) = \mathcal{L}(x_3) = 0$.

For $x_{20} = 1$, $\mathcal{L}(x_1) = 0$, $\mathcal{L}(x_2) = s/(s+2)^2$ and $\mathcal{L}(x_3) = 4/(s+2)^2$.

For $x_{30} = 1$, $\mathcal{L}(x_1) = 0$, $\mathcal{L}(x_2) = -1/(s+2)^2$ and $\mathcal{L}(x_3) = (s+4)/(s+2)^2$.

Therefore,

$$\mathbf{R}(s) = \begin{pmatrix} \dfrac{1}{s+1} & 0 & 0 \\ 0 & \dfrac{s}{(s+2)^2} & \dfrac{4}{(s+2)^2} \\ 0 & \dfrac{-1}{(s+2)^2} & \dfrac{s+4}{(s+2)^2} \end{pmatrix}$$

Again a partial fraction expansion can be performed to obtain the previous result.

($d4$) Using Theorem 5.3 for Leverrier's algorithm,

$$\mathbf{F}_1 = \mathbf{I} \qquad \theta_1 = 5$$

$$\mathbf{F}_2 = \begin{pmatrix} -1 & 0 & 0 \\ 0 & -4 & 4 \\ 0 & -1 & 0 \end{pmatrix} + 5\mathbf{I} \qquad \theta_2 = 8$$

$$\mathbf{F}_3 = \begin{pmatrix} -4 & 0 & 0 \\ 0 & -8 & 4 \\ 0 & -1 & -4 \end{pmatrix} + 8\mathbf{I} \qquad \theta_3 = 4$$

Using equation (*5.23*),

$$\mathbf{R}(s) = \frac{s^2\begin{pmatrix} 1 & 0 & 0 \\ 0 & 1 & 0 \\ 0 & 0 & 1 \end{pmatrix} + s\begin{pmatrix} 4 & 0 & 0 \\ 0 & 1 & 4 \\ 0 & -1 & 5 \end{pmatrix} + \begin{pmatrix} 4 & 0 & 0 \\ 0 & 0 & 4 \\ 0 & -1 & 4 \end{pmatrix}}{s^3 + 5s^2 + 8s + 4}$$

A partial fraction expansion of this gives the previous result.

5.3. Using (*a*) the eigenvalue method and then (*b*) the resolvent matrix, find the transition matrix for

$$\mathbf{x}(k+1) = \begin{pmatrix} -1 & 1 \\ 0 & -2 \end{pmatrix} \mathbf{x}(k)$$

(*a*) The eigenvalues are -1 and -2, with eigenvectors $(1\ 0)^T$ and $(1\ -1)^T$ respectively. The reciprocal basis is $(1\ 1)$ and $(0\ -1)$. Using the spectral representation equation (*5.31*),

$$\mathbf{A}^k = (-1)^k \begin{pmatrix} 1 \\ 0 \end{pmatrix} (1\ \ 1) + (-2)^k \begin{pmatrix} 1 \\ -1 \end{pmatrix} (0\ \ -1)$$

(*b*) From equation (*5.35*),

$$\mathbf{R}(z) = \begin{pmatrix} z+1 & -1 \\ 0 & z+2 \end{pmatrix}^{-1} = \frac{1}{(z+1)(z+2)} \begin{pmatrix} z+2 & 1 \\ 0 & z+1 \end{pmatrix}$$

so that

$$\mathbf{A}^k = \mathcal{Z}^{-1} \begin{pmatrix} \dfrac{z}{z+1} & \dfrac{z}{z+1} - \dfrac{z}{z+2} \\ 0 & \dfrac{z}{z+2} \end{pmatrix} = \begin{pmatrix} (-1)^k & (-1)^k - (-2)^k \\ 0 & (-2)^k \end{pmatrix}$$

5.4. Prove Leverrier's algorithm,

$$(s\mathbf{I} - \mathbf{A})^{-1} = \frac{s^{n-1}\mathbf{F}_1 + s^{n-2}\mathbf{F}_2 + \cdots + s\mathbf{F}_{n-1} + \mathbf{F}_n}{s^n + \theta_1 s^{n-1} + \cdots + \theta_{n-1}s + \theta_n} \tag{5.23}$$

and the Cayley-Hamilton theorem (Theorem 4.15).

Let $\det(s\mathbf{I} - \mathbf{A}) = \phi(s) = s^n + \theta_1 s^{n-1} + \cdots + \theta_{n-1}s + \theta_n$. Use of Cramer's rule gives $\phi(s)(s\mathbf{I} - \mathbf{A})^{-1} = \mathbf{F}(s)$ where $\mathbf{F}(s)$ is the adjugate matrix, i.e. the matrix of signed cofactors transposed of $(s\mathbf{I} - \mathbf{A})$.

An intermediate result must be proven before proceeding, namely that $\operatorname{tr}\mathbf{F}(s) = d\phi/ds$. In the proof of Theorem 3.21, it was shown the cofactor c_{1j} of a general matrix $\mathbf{B}$ can be represented as $c_{1j} = \mathbf{e}_j \wedge \mathbf{b}_2 \wedge \cdots \wedge \mathbf{b}_n$ and similarly it can be shown $c_{ij} = \mathbf{b}_1 \wedge \cdots \wedge \mathbf{b}_{i-1} \wedge \mathbf{e}_j \wedge \mathbf{b}_{i+1} \wedge \cdots \wedge \mathbf{b}_n$, so that letting $\mathbf{B} = s\mathbf{I} - \mathbf{A}$ and using $\operatorname{tr}\mathbf{F}(s) = c_{11}(s) + c_{22}(s) + \cdots + c_{nn}(s)$ gives

$$\operatorname{tr}\mathbf{F}(s) = \mathbf{e}_1 \wedge (s\mathbf{e}_2 - \mathbf{a}_2) \wedge \cdots \wedge (s\mathbf{e}_n - \mathbf{a}_n) + (s\mathbf{e}_1 - \mathbf{a}_1) \wedge \mathbf{e}_2 \wedge \cdots \wedge (s\mathbf{e}_1 - \mathbf{a}_n)$$
$$+ \cdots + (s\mathbf{e}_1 - \mathbf{a}_1) \wedge \cdots \wedge (s\mathbf{e}_{n-1} - \mathbf{a}_{n-1}) \wedge \mathbf{e}_n$$

But

$$\phi(s) = \det(s\mathbf{I} - \mathbf{A}) = (s\mathbf{e}_1 - \mathbf{a}_1) \wedge (s\mathbf{e}_2 - \mathbf{a}_2) \wedge \cdots \wedge (s\mathbf{e}_n - \mathbf{a}_n)$$

and

$$d\phi/ds = \mathbf{e}_1 \wedge (s\mathbf{e}_2 - \mathbf{a}_2) \wedge \cdots \wedge (s\mathbf{e}_n - \mathbf{a}_n) + \cdots + (s\mathbf{e}_1 - \mathbf{a}_1) \wedge \cdots \wedge \mathbf{e}_n = \operatorname{tr}\mathbf{F}(s)$$

and so the intermediate result is established.

Substituting the definitions for $\phi(s)$ and $\mathbf{F}(s)$ into the intermediate result,

$$\operatorname{tr}\mathbf{F}(s) = \operatorname{tr}(s^{n-1}\mathbf{F}_1 + s^{n-2}\mathbf{F}_2 + \cdots + s\mathbf{F}_{n-1} + \mathbf{F}_n)$$
$$= s^{n-1}\operatorname{tr}\mathbf{F}_1 + s^{n-2}\operatorname{tr}\mathbf{F}_2 + \cdots + s\operatorname{tr}\mathbf{F}_{n-1} + \operatorname{tr}\mathbf{F}_n$$
$$d\phi/ds = ns^{n-1} + (n-1)\theta_1 s^{n-2} + \cdots + \theta_{n-1}$$

Equating like powers of s,

$$\operatorname{tr}\mathbf{F}_{k+1} = (n-k)\theta_k \tag{5.66}$$

for $k = 1, 2, \ldots, n-1$, and $\operatorname{tr}\mathbf{F}_1 = n$. Rearranging (*5.23*) as

$$(s^n + \theta_1 s^{n-1} + \cdots + \theta_n)\mathbf{I} = (s\mathbf{I} - \mathbf{A})(s^{n-1}\mathbf{F}_1 + s^{n-2}\mathbf{F}_2 + \cdots + \mathbf{F}_n)$$

and equating the powers of s gives $\mathbf{I} = \mathbf{F}_1$, $\theta_n\mathbf{I} = -\mathbf{A}\mathbf{F}_n$, and

$$\theta_k\mathbf{I} = -\mathbf{A}\mathbf{F}_k + \mathbf{F}_{k+1} \tag{5.67}$$

for $k = 1, 2, \ldots, n-1$. These are half of the relationships required. The other half are obtained by taking their trace

$$n\theta_k = -\operatorname{tr} \mathbf{AF}_k + \operatorname{tr} \mathbf{F}_{k+1}$$

and substituting into (*5.66*) to get

$$k\theta_k = -\operatorname{tr} \mathbf{AF}_k$$

which are the other half of the relationships needed for the proof.

To prove the Cayley-Hamilton theorem, successively substitute for $\mathbf{F}_{k+1}$, i.e.,

$$\begin{aligned}
\mathbf{F}_1 &= \mathbf{I} \\
\mathbf{F}_2 &= \theta_1\mathbf{I} + \mathbf{AF}_1 = \theta_1\mathbf{I} + \mathbf{A} \\
\mathbf{F}_3 &= \theta_2\mathbf{I} + \mathbf{AF}_2 = \theta_2\mathbf{I} + \theta_1\mathbf{A} + \mathbf{A}^2 \\
&\cdots\cdots\cdots\cdots\cdots\cdots\cdots\cdots \\
\mathbf{F}_n &= \theta_{n-1}\mathbf{I} + \theta_{n-2}\mathbf{A} + \cdots + \theta_1\mathbf{A}^{n-2} + \mathbf{A}^{n-1}
\end{aligned}$$

Using the last relation $\theta_n\mathbf{I} = -\mathbf{AF}_n$ then gives

$$\mathbf{0} = \theta_n\mathbf{I} + \theta_{n-1}\mathbf{A} + \cdots + \theta_1\mathbf{A}^{n-1} + \mathbf{A}^n = \phi(\mathbf{A})$$

which is the Cayley-Hamilton theorem.

5.5. Given the time-varying system

$$\frac{d\mathbf{x}}{dt} = \begin{pmatrix} \alpha & e^{-t} \\ -e^{-t} & \alpha \end{pmatrix} \mathbf{x}$$

Find the transition matrix and verify the transition properties [equations (*5.5*)-(*5.8*)].

Note that $\mathbf{A}(t)$ commutes with $\mathbf{A}(\tau)$, i.e.,

$$\mathbf{A}(t)\,\mathbf{A}(\tau) = \begin{pmatrix} \alpha^2 - e^{-(t+\tau)} & \alpha(e^{-\tau} + e^{-t}) \\ -\alpha(e^{-\tau} + e^{-t}) & \alpha^2 - e^{-(t+\tau)} \end{pmatrix} = \mathbf{A}(\tau)\,\mathbf{A}(t)$$

It can be shown similar to Problem 4.4 that two $n \times n$ matrices $\mathbf{B}$ and $\mathbf{C}$ with n independent eigenvectors can be simultaneously diagonalized by a nonsingular matrix $\mathbf{T}$ if and only if $\mathbf{B}$ commutes with $\mathbf{C}$. Identifying $\mathbf{A}(t)$ and $\mathbf{A}(\tau)$ for fixed t and τ with $\mathbf{B}$ and $\mathbf{C}$, then $\mathbf{A}(t) = \mathbf{T}(t)\mathbf{\Lambda}(t)\mathbf{T}^{-1}(t)$ and $\mathbf{A}(\tau) = \mathbf{T}(\tau)\mathbf{\Lambda}(\tau)\mathbf{T}^{-1}(\tau)$ means that $\mathbf{T}(t) = \mathbf{T}(\tau)$ for all t and τ. This implies the matrix of eigenvectors $\mathbf{T}$ is constant, so $d\mathbf{T}/dt = \mathbf{0}$. Referring to Example 5.1, when $d\mathbf{T}(t)/dt = \mathbf{0}$ then a general solution exists for this special case of $\mathbf{A}(t)\mathbf{A}(\tau) = \mathbf{A}(\tau)\mathbf{A}(t)$.

For the given time-varying system, $\mathbf{A}(t)$ has the eigenvalues $\lambda_1 = \alpha + je^{-t}$ and $\lambda_2 = \alpha - je^{-t}$. Since $\mathbf{A}(t)$ commutes with $\mathbf{A}(\tau)$, the eigenvectors are constant, $(j \;\; -1)^T$ and $(j \;\; 1)^T$. Consequently

$$\mathbf{T} = \begin{pmatrix} j & j \\ -1 & 1 \end{pmatrix} \qquad \mathbf{\Lambda}(t) = \begin{pmatrix} \alpha - je^{-t} & 0 \\ 0 & \alpha + je^{-t} \end{pmatrix} \qquad \mathbf{T}^{-1} = \frac{1}{2}\begin{pmatrix} -j & -1 \\ -j & 1 \end{pmatrix}$$

From Theorem 5.5,

$$\mathbf{\Phi}(t, \tau) = \mathbf{T}e^{\int_\tau^t \mathbf{\Lambda}(\eta)\,d\eta}\,\mathbf{T}^{-1}$$

Substituting the numerical values, integrating and multiplying out gives

$$\mathbf{\Phi}(t, \tau) = e^{\alpha(t-\tau)} \begin{pmatrix} \cos(e^{-\tau} - e^{-t}) & \sin(e^{-\tau} - e^{-t}) \\ -\sin(e^{-\tau} - e^{-t}) & \cos(e^{-\tau} - e^{-t}) \end{pmatrix}$$

To check this, use Problem 5.1:

$$\frac{\partial \Phi}{\partial t} = \alpha e^{\alpha(t-\tau)}\begin{pmatrix} \cos(e^{-\tau}-e^{-t}) & \sin(e^{-\tau}-e^{-t}) \\ -\sin(e^{-\tau}-e^{-t}) & \cos(e^{-\tau}-e^{-t}) \end{pmatrix} - e^{\alpha(t-\tau)+t}\begin{pmatrix} \sin(e^{-\tau}-e^{-t}) & -\cos(e^{-\tau}-e^{-t}) \\ \cos(e^{-\tau}-e^{-t}) & \sin(e^{-\tau}-e^{-t}) \end{pmatrix}$$

Setting $\tau = t$ in this gives $\mathbf{A}(t)$.

To verify $\Phi(t_2, t_0) = \Phi(t_2, t_1)\,\Phi(t_1, t_0)$, note

$$\begin{aligned} \cos(e^{-t_0}-e^{-t_2}) &= \cos(e^{-t_0}-e^{-t_1}+e^{-t_1}-e^{-t_2}) \\ &= \cos(e^{-t_0}-e^{-t_1})\cos(e^{-t_1}-e^{-t_2}) - \sin(e^{-t_0}-e^{-t_1})\sin(e^{-t_1}-e^{-t_2}) \end{aligned}$$

and

$$\sin(e^{-t_0}-e^{-t_2}) = \sin(e^{-t_0}-e^{-t_1})\cos(e^{-t_1}-e^{-t_2}) + \cos(e^{-t_0}-e^{-t_1})\sin(e^{-t_1}-e^{-t_2})$$

so that

$$\begin{aligned} e^{-\alpha(t_2-t_0)}\begin{pmatrix} \cos(e^{-t_0}-e^{-t_2}) & \sin(e^{-t_0}-e^{-t_2}) \\ -\sin(e^{-t_0}-e^{-t_2}) & \cos(e^{-t_0}-e^{-t_2}) \end{pmatrix} &= e^{-\alpha(t_2-t_1)}\begin{pmatrix} \cos(e^{-t_1}-e^{-t_2}) & \sin(e^{-t_1}-e^{-t_2}) \\ -\sin(e^{-t_1}-e^{-t_2}) & \cos(e^{-t_1}-e^{-t_2}) \end{pmatrix} \\ &\quad \times e^{-\alpha(t_1-t_0)}\begin{pmatrix} \cos(e^{-t_0}-e^{-t_1}) & \sin(e^{-t_0}-e^{-t_1}) \\ -\sin(e^{-t_0}-e^{-t_1}) & \cos(e^{-t_0}-e^{-t_1}) \end{pmatrix} \end{aligned}$$

To verify $\Phi^{-1}(t_1, t_0) = \Phi(t_0, t_1)$, calculate $\Phi^{-1}(t_1, t_0)$ by Cramer's rule:

$$\Phi^{-1}(t_1, t_0) = \frac{e^{\alpha(t_0-t_1)}}{\cos^2(e^{-t_0}-e^{-t_1}) + \sin^2(e^{-t_0}-e^{-t_1})}\begin{pmatrix} \cos(e^{-t_0}-e^{-t_1}) & -\sin(e^{-t_0}-e^{-t_1}) \\ \sin(e^{-t_0}-e^{-t_1}) & \cos(e^{-t_0}-e^{-t_1}) \end{pmatrix}$$

Since $\cos^2(e^{-t_0}-e^{-t_1}) + \sin^2(e^{-t_0}-e^{-t_1}) = 1$ and $\sin(e^{-t_0}-e^{-t_1}) = -\sin(e^{-t_1}-e^{t_0})$, this equals $\Phi(t_0, t_1)$.

To verify $\Phi(t_1, t_0) = \boldsymbol{\theta}(t_1)\boldsymbol{\theta}^{-1}(t_0)$, set

$$\boldsymbol{\theta}(t) = e^{\alpha t}\begin{pmatrix} \cos(1-e^{-t}) & \sin(1-e^{-t}) \\ -\sin(1-e^{-t}) & \cos(1-e^{-t}) \end{pmatrix}$$

Then

$$\boldsymbol{\theta}^{-1}(t) = e^{-\alpha t}\begin{pmatrix} \cos(1-e^{-t}) & -\sin(1-e^{-t}) \\ \sin(1-e^{-t}) & \cos(1-e^{-t}) \end{pmatrix}$$

Since we have

$$\cos(e^{-t_0}-e^{-t_1}) = \cos(e^{-t_0}-1)\cos(1-e^{-t_1}) - \sin(e^{-t_0}-1)\sin(1-e^{-t_1})$$

and a similar formula for the sine, multiplication of $\boldsymbol{\theta}(t_1)$ and $\boldsymbol{\theta}^{-1}(t_0)$ will verify this.

Finally, $\det \Phi(t, \tau) = e^{2\alpha(t-\tau)}$ and $\operatorname{tr} \mathbf{A}(t) = 2\alpha$, so that integration shows that the determinant property holds.

5.6. Find the transition matrix for the system

$$d\mathbf{x}/dt = \begin{pmatrix} 0 & 1 \\ -1-(\alpha^2+\alpha)/t & -2\alpha - 1/t \end{pmatrix}\mathbf{x}$$

Writing out the matrix equations, we find $dx_1/dt = x_2$ and so

$$d^2x_1/dt^2 + (2\alpha + 1/t)\,dx_1/dt + [1 + (\alpha^2+\alpha)/t]x_1 = 0$$

Multiplying by the integrating factor $te^{\alpha t}$, which was found by much trial and error with the equation, gives

$$t(e^{\alpha t}\,d^2x_1/dt^2 + 2\alpha e^{\alpha t}\,dx_1/dt + \alpha^2 e^{\alpha t}x_1) + (e^{\alpha t}\,dx_1/dt + \alpha e^{\alpha t}x_1) + te^{\alpha t}x_1 = 0$$

which can be rewritten as

$$td^2(e^{\alpha t}x_1)/dt^2 + d(e^{\alpha t}x_1)/dt + te^{\alpha t}x_1 = 0$$

This has the same form as Bessel's equation (*5.38*), so the solution is

$$\begin{aligned} x_1(t) &= c_1e^{-\alpha t}J_0(t) + c_2e^{-\alpha t}Y_0(t) \\ x_2(t) &= dx_1/dt = -\alpha x_1(t) + e^{-\alpha t}[-c_1J_1(t) + c_2dY_0/dt] \end{aligned}$$

To solve for the constants c_1 and c_2,

$$e^{\alpha\tau}\begin{pmatrix} x_1(\tau) \\ x_2(\tau) + \alpha x_1(\tau) \end{pmatrix} = \begin{pmatrix} J_0(\tau) & Y_0(\tau) \\ -J_1(\tau) & dY_0/dt \end{pmatrix}\begin{pmatrix} c_1 \\ c_2 \end{pmatrix}$$

so that $\mathbf{x}(t) = \mathbf{F}\mathbf{G}(t)\,\mathbf{G}^{-1}(\tau)\,\mathbf{F}^{-1}\mathbf{x}(\tau)$, where

$$\mathbf{F} = \begin{pmatrix} 1 & 0 \\ -\alpha & 1 \end{pmatrix} \qquad \mathbf{G}(t) = \begin{pmatrix} J_0(t) & Y_0(t) \\ -J_1(t) & dY_0/dt \end{pmatrix}$$

and so $\mathbf{\Phi}(t,\tau) = \mathbf{F}\mathbf{G}(t)\,\mathbf{G}^{-1}(\tau)\mathbf{F}^{-1}$. This is true only for t and $\tau > 0$ or t and $\tau < 0$, because at the point $t = 0$, the elements of the original $\mathbf{A}(t)$ matrix blow up. This accounts for $Y_0(0) = \infty$.

Admittedly this problem was contrived, and in practice a man-made system would only accidentally have this form. However, Bessel's equation often occurs in nature, and knowledge that as $t \to \infty$, $J_0(t) \sim \sqrt{2/\pi t}\ \cos(t - \pi/4)$ and $Y_0(t) \sim \sqrt{2/\pi t}\ \sin(t - \pi/4)$ gives great insight into the behavior of the system.

5.7. Given the time-varying difference equation $\mathbf{x}(n+1) = \mathbf{A}(n)\,\mathbf{x}(n)$, where $\mathbf{A}(n) = \mathbf{A}_0$ if n is even and $\mathbf{A}(n) = \mathbf{A}_1$ if n is odd. Find the fundamental matrix, analyze by Floquet theory, and give the conditions for stability if $\mathbf{A}_0$ and $\mathbf{A}_1$ are nonsingular.

From equation (5.50),

$$\mathbf{\Phi}(k,m) = \begin{cases} \mathbf{A}_1\mathbf{A}_0\mathbf{A}_1\cdots\mathbf{A}_1\mathbf{A}_0 & \text{if } k \text{ is even and } m \text{ is even} \\ \mathbf{A}_0\mathbf{A}_1\mathbf{A}_0\cdots\mathbf{A}_1\mathbf{A}_0 & \text{if } k \text{ is odd and } m \text{ is even} \\ \mathbf{A}_1\mathbf{A}_0\mathbf{A}_1\cdots\mathbf{A}_0\mathbf{A}_1 & \text{if } k \text{ is odd and } m \text{ is odd} \\ \mathbf{A}_0\mathbf{A}_1\mathbf{A}_0\cdots\mathbf{A}_0\mathbf{A}_1 & \text{if } k \text{ is even and } m \text{ is odd} \end{cases}$$

For m even, $\mathbf{\Phi}(k,m) = \mathbf{P}(k,m)\,(\mathbf{A}_1\mathbf{A}_0)^{(k-m)/2}$, where $\mathbf{P}(k,m) = \mathbf{I}$ if k is even and $\mathbf{P}(k,m) = (\mathbf{A}_0\mathbf{A}_1^{-1})^{1/2}$ if k is odd. For m odd, $\mathbf{\Phi}(k,m) = \mathbf{P}(k,m)\,(\mathbf{A}_0\mathbf{A}_1)^{(k-m)/2}$, where $\mathbf{P}(k,m) = \mathbf{I}$ if k is odd and $\mathbf{P}(k,m) = (\mathbf{A}_1\mathbf{A}_0^{-1})^{1/2}$ if k is even. For instability, the eigenvalues of $\mathbf{R} = (\mathbf{A}_1\mathbf{A}_0)^{1/2}$ must be outside the unit circle. Since the eigenvalues of $\mathbf{B}^2$ are the squares of the eigenvalues of $\mathbf{B}$, it is enough to find the eigenvalues of $\mathbf{A}_1\mathbf{A}_0$. This agrees with the stability analysis of the equation $\mathbf{x}(n+2) = \mathbf{A}_1\mathbf{A}_0\mathbf{x}(n)$.

5.8. Find the impulse response of the system

$$d^2y/dt^2 + (1-2\alpha)\,dy/dt + (\alpha^2 - \alpha + e^{-2t})y = u$$

Choose $x_1 = y$ and $x_2 = e^t(dx_1/dt - \alpha x_1)$ to find a state representation in which $\mathbf{A}(\tau)$ commutes with $\mathbf{A}(t)$. Then in matrix form the system is

$$\frac{d}{dt}\begin{pmatrix} x_1 \\ x_2 \end{pmatrix} = \begin{pmatrix} \alpha & e^{-t} \\ -e^{-t} & \alpha \end{pmatrix}\begin{pmatrix} x_1 \\ x_2 \end{pmatrix} + \begin{pmatrix} 0 \\ e^t \end{pmatrix} u$$

$$y = (1 \ \ 0)\mathbf{x}$$

From equation (5.55) and $\mathbf{\Phi}(t,\tau)$ obtained from Problem 5.5,

$$\mathbf{H}(t,\tau) = e^{\tau + \alpha(t-\tau)} \sin(e^{-\tau} - e^{-t})$$

This is the response $y(t)$ to an input $u(t) = \delta(t-\tau)$.

5.9. In the system of Problem 5.8, let $u(t) = e^{(\alpha-2)t}$, $y(t_0) = y_0$ and $(dy/dt)(t_0) = \alpha y_0$. Find the complete response.

From equations (5.44) and Problem 5.5,

$$y(t) = e^{\alpha(t-t_0)} \cos(e^{-t_0} - e^{-t})y_0 + e^{\alpha t}\int_{t_0}^{t} e^{-\tau} \sin(e^{-\tau} - e^{-t})\,d\tau$$

Changing variables from τ to η in the integral, where $e^{-\tau} - e^{-t} = \eta$, gives

$$y(t) = e^{\alpha(t-t_0)} \cos(e^{-t_0} - e^{-t}) y_0 + e^{-\alpha t}[1 - \cos(e^{-t} - e^{-t_0})]$$

Notice this problem cannot be solved by Laplace transformation in one variable.

5.10. Given a step input $U(s) = 6/s$ into a system with a transfer function

$$H(s) = \frac{s+1}{s^2 + 5s + 6}$$

Find the output $y(t)$ assuming zero initial conditions.

The easiest way to do this is by using classical techniques.

$$\mathcal{L}\{y(t)\} = U(s)H(s) = \frac{6(s+1)}{s^3 + 5s^2 + 6s} = \frac{1}{s} + \frac{3}{s+2} - \frac{4}{s+3}$$

Taking the inverse Laplace transform determines $y = 1 + 3e^{-2t} - 4e^{-3t}$.

Doing this by state space techniques shows how it corresponds with the classical techniques. From Problem 2.3 the state equations are

$$\frac{d}{dt}\begin{pmatrix} x_1 \\ x_2 \end{pmatrix} = \begin{pmatrix} -2 & 0 \\ 0 & -3 \end{pmatrix}\begin{pmatrix} x_1 \\ x_2 \end{pmatrix} + \begin{pmatrix} 1 \\ 1 \end{pmatrix} u$$

$$y = (-1 \quad 2)\begin{pmatrix} x_1 \\ x_2 \end{pmatrix}$$

The transition matrix is obviously

$$\boldsymbol{\Phi}(t, \tau) = \begin{pmatrix} e^{-2(t-\tau)} & 0 \\ 0 & e^{-3(t-\tau)} \end{pmatrix}$$

The response can be expressed directly in terms of (*5.44*).

$$\begin{aligned} y(t) &= \boldsymbol{\Phi}(t, t_0)\mathbf{0} + \int_{t_0}^{t} (-1 \quad 2)\begin{pmatrix} e^{-2(t-\tau)} & 0 \\ 0 & e^{-3(t-\tau)} \end{pmatrix}\begin{pmatrix} 1 \\ 1 \end{pmatrix} 6u(\tau - t_0)\, d\tau \\ &= 1 + 3e^{-2t} - 4e^{-3t} \end{aligned} \qquad (5.68)$$

This integral is usually very complicated to solve analytically, although it is easy for a computer. Instead, we shall use the transfer function matrix of equation (*5.56*).

$$\begin{aligned} \mathcal{L}\{\mathbf{H}(t, 0)\} &= (-1 \quad 2)\begin{pmatrix} (s+2)^{-1} & 0 \\ 0 & (s+3)^{-1} \end{pmatrix}\begin{pmatrix} 1 \\ 1 \end{pmatrix} \\ &= \frac{2}{s+3} - \frac{1}{s+2} = \frac{s+1}{s^2 + 5s + 6} = H(s) \end{aligned}$$

This is indeed our original transfer function, and the integral (*5.68*) is a convolution and its Laplace transform is

$$\mathcal{L}\{y(t)\} = \left(\frac{s+1}{s^2 + 5s + 6}\right)\left(\frac{1}{s}\right)$$

whose inverse Laplace transform gives $y(t)$ as before.

5.11. Using the adjoint matrix, synthesize a form of control law for use in guidance.

We desire to guide a vehicle to a final state $\mathbf{x}(t_f)$, which is known. From (*5.44*),

$$\mathbf{x}(t_f) = \boldsymbol{\Phi}(t_f, t)\,\mathbf{x}(t) + \int_{t}^{t_f} \boldsymbol{\Phi}(t_f, \tau)\,\mathbf{B}(\tau)\,\mathbf{u}(\tau)\, d\tau$$

Choose $\mathbf{u}(t) = \mathbf{U}(t)\mathbf{c}$ where $\mathbf{U}(t)$ is a prespecified matrix of time functions that are easily mechanized, such as polynomials in t. The vector $\mathbf{c}$ is constant, except that at intervals of time it is recomputed as knowledge of $\mathbf{x}(t)$ becomes better. Then $\mathbf{c}$ can be computed as

$$\mathbf{c} = \left[\int_{t_f}^{t} \mathbf{\Phi}(t_f, \tau)\,\mathbf{B}(\tau)\,\mathbf{U}(\tau)\,d\tau\right]^{-1} [\mathbf{\Phi}(t_f, t)\,\mathbf{x}(t) - \mathbf{x}(t_f)]$$

However, this involves finding $\mathbf{\Phi}(t_f, t)$ as the transition matrix of $d\mathbf{x}/dt = \mathbf{A}(t)\mathbf{x}$ with $\mathbf{x}(t)$ as the initial condition going to $\mathbf{x}(t_f)$. Therefore $\mathbf{\Phi}(t_f, t)$ would have to be computed at each recomputation of $\mathbf{c}$, starting with the best estimates of $\mathbf{x}(t)$. To avoid this, the adjoint transition matrix $\mathbf{\Psi}(\tau, t_f)$ can be found starting with the final time t_f, and be stored and used for all recomputations of $\mathbf{c}$ because, from equation (*5.64*), $\mathbf{\Psi}^\dagger(\tau, t_f) = \mathbf{\Phi}(t_f, \tau)$ and $\mathbf{c}$ is found from

$$\mathbf{c} = \left[\int_{t_f}^{t} \mathbf{\Psi}^\dagger(\tau, t_f)\,\mathbf{B}(\tau)\,\mathbf{U}(\tau)\,d\tau\right]^{-1} [\mathbf{\Psi}^\dagger(t, t_f)\,\mathbf{x}(t) - \mathbf{x}(t_f)]$$

Supplementary Problems

5.12. Prove equations (*5.9*), (*5.10*), (*5.11*), and (*5.12*).

5.13. Given $\mathbf{\Phi}(k, m)$, how can $\mathbf{A}(k)$ be found?

5.14. Prove that $\mathbf{A}e^{\mathbf{A}t} = e^{\mathbf{A}t}\mathbf{A}$ and then find the conditions on $\mathbf{A}$ and $\mathbf{B}$ such that $e^{\mathbf{A}}e^{\mathbf{B}} = e^{\mathbf{B}}e^{\mathbf{A}}$.

5.15. Verify that $\mathbf{\Phi}(t, \tau) = e^{\mathbf{A}(t-\tau)}$ and $\mathbf{\Phi}(k, m) = \mathbf{A}^{k-m}$ satisfy the properties of a transition matrix given in equations (*5.5*)-(*5.12*).

5.16. Given the fundamental matrix

$$\mathbf{\Phi}(t, \tau) = \frac{1}{2}\begin{pmatrix} e^{-4(t-\tau)} + 1 & e^{-4(t-\tau)} - 1 \\ e^{-4(t-\tau)} - 1 & e^{-4(t-\tau)} + 1 \end{pmatrix}$$

What is the state equation corresponding to this fundamental matrix?

5.17. Find the transition matrix to the system $d\mathbf{x}/dt = \begin{pmatrix} 0 & 2 \\ 2 & -3 \end{pmatrix}\mathbf{x}$.

5.18. Calculate the transition matrix $\mathbf{\Phi}(t, 0)$ for $d\mathbf{x}/dt = \begin{pmatrix} 1 & 0 \\ 0 & 0 \end{pmatrix}\mathbf{x}$ using (*a*) reduction to Jordan form, (*b*) the Maclaurin series, (*c*) the resolvent matrix.

5.19. Find $e^{\mathbf{A}t}$ by the series method, where $\mathbf{A} = \begin{pmatrix} 0 & 1 & 1 \\ 0 & 0 & 1 \\ 0 & 0 & 0 \end{pmatrix}$. This shows a case where the series method is the easiest.

5.20. Find $e^{\mathbf{A}t}$ using the resolvent matrix and Leverrier's algorithm, where $\mathbf{A} = \begin{pmatrix} -3 & 1 & 0 \\ 1 & -3 & 0 \\ 0 & 0 & -3 \end{pmatrix}$.

5.21. Find $e^{\mathbf{A}t}$ using the Cayley-Hamilton method, where $\mathbf{A}$ is the matrix given in Problem 5.20.

5.22. Use the eigenvalue method to find $e^{\mathbf{A}t}$ for

$$\mathbf{A} = \begin{pmatrix} -1 & -2 & 3 \\ 0 & 1 & -1 \\ -1 & -1 & 2 \end{pmatrix}$$

5.23. Use the resolvent matrix and Cramer's rule to find $e^{\mathbf{A}t}$ for $\mathbf{A}$ as given in Problem 5.22.

5.24. Use the resolvent matrix and Cramer's rule to find $\mathbf{A}^k$ for $\mathbf{A}$ as given in Problem 5.22.

5.25. Find $e^{\mathbf{A}t}$ by using the Maclaurin series, Cayley-Hamilton and resolvent matrix methods when $\mathbf{A} = \begin{pmatrix} 2 & -2 \\ 0 & 0 \end{pmatrix}$.

5.26. Find the fundamental, or transition, matrix for the system

$$\frac{d\mathbf{x}}{dt} = \begin{pmatrix} \sigma & \omega & 0 \\ -\omega & \sigma & 0 \\ 0 & 0 & \lambda \end{pmatrix}\mathbf{x}$$

using the matrix Laplace transform method.

5.27. Given the continuous time system

$$d\mathbf{x}/dt = \begin{pmatrix} -\frac{3}{4} & -\frac{1}{4} \\ -\frac{1}{4} & -\frac{3}{4} \end{pmatrix}\mathbf{x} + \begin{pmatrix} 1 \\ 0 \end{pmatrix}u \qquad \mathbf{x}(0) = \begin{pmatrix} 0 \\ 0 \end{pmatrix}$$

$$y = (1 \quad 0)\mathbf{x} + 4u$$

Compute $y(t)$ using the transition matrix if u is a unit step function. Compare this with the solution obtained by finding the 1×1 transfer function matrix for the input to the output.

5.28. Given the discrete time system

$$\mathbf{x}(n+1) = \begin{pmatrix} -\frac{3}{4} & -\frac{1}{4} \\ -\frac{1}{4} & -\frac{3}{4} \end{pmatrix}\mathbf{x}(n) + \begin{pmatrix} 1 \\ 0 \end{pmatrix}u(n) \qquad \mathbf{x}(0) = \begin{pmatrix} 0 \\ 0 \end{pmatrix}$$

$$y(n) = (1 \quad 0)x(n) + 4u(n)$$

Compute $y(n)$ using the transition matrix if u is the series of ones $1, 1, 1, \ldots, 1, \ldots$.

5.29. (*a*) Calculate $\Phi(t, t_0)$ for the system $d\mathbf{x}/dt = \begin{pmatrix} -1 & 2 \\ 0 & 1 \end{pmatrix}\mathbf{x}$ using Laplace transforms.

(*b*) Calculate $\Phi(k, m)$ for the system $\mathbf{x}(k+1) = \begin{pmatrix} -1 & 2 \\ 0 & 1 \end{pmatrix}\mathbf{x}(k)$ using $\mathcal{Z}$ transforms.

5.30. How does the spectral representation for $e^{\mathbf{A}t}$ extend to the case where the eigenvalues of $\mathbf{A}$ are not distinct?

5.31. In the Cayley-Hamilton method of finding $e^{\mathbf{A}t}$, show that the equation $e^{\mathbf{J}t} = \sum_{i=0}^{n-1} \gamma_i(t)\mathbf{A}^i$ can always be solved for the $\gamma_i(t)$. For simplicity, consider only the case of distinct eigenvalues.

5.32. Show that the column vectors of $\Phi(t, \tau)$ span the vector space of solutions to $d\mathbf{x}/dt = \mathbf{A}(t)\mathbf{x}$.

5.33. Show $\mathbf{A}(t)\,\mathbf{A}(\tau) = \mathbf{A}(\tau)\,\mathbf{A}(t)$ when $\mathbf{A}(t) = \alpha(t)\mathbf{C}$, where $\mathbf{C}$ is a constant $n \times n$ matrix and $\alpha(t)$ is a scalar function of t. Also, find the conditions on $a_{ij}(t)$ such that $\mathbf{A}(t)\,\mathbf{A}(\tau) = \mathbf{A}(\tau)\,\mathbf{A}(t)$ for a 2×2 $\mathbf{A}(t)$ matrix.

5.34. Given the time-varying system

$$\frac{d\mathbf{x}}{dt} = \begin{pmatrix} 0 & 1 \\ -\alpha^{-2}(t) & -\dot{\alpha}(t)\,\alpha^{-1}(t) \end{pmatrix}\mathbf{x}$$

Find the transition matrix. *Hint:* Find an integrating factor.

5.35. Prove Floquet's theorem for discrete time systems, $\mathbf{\Phi}(k, m) = \mathbf{P}(k, m)\mathbf{R}^{k-m}$ where $\mathbf{P}(k, m) = \mathbf{P}(k + \omega, m)$ if $\mathbf{A}(k) = \mathbf{A}(k + \omega)$.

5.36. Given the time-varying periodic system $d\mathbf{x}/dt = \begin{pmatrix} \sin t & \sin t \\ \sin t & \sin t \end{pmatrix}\mathbf{x}$. Find the transition matrix $\mathbf{\Phi}(t, t_0)$ and verify it satisfies Floquet's result $\mathbf{\Phi}(t, t_0) = \mathbf{P}(t, t_0)e^{\mathbf{R}(t-t_0)}$ where $\mathbf{P}$ is periodic and $\mathbf{R}$ is constant. Also find the fundamental matrix of the adjoint system.

5.37. The linear system shown in Fig. 5-6 is excited by a square wave $s(t)$ with period 2 and amplitude $|s(t)| = 1$. The system equation is $\ddot{y} + [\beta + \alpha \operatorname{sgn}(\sin \pi t)]y = 0$.

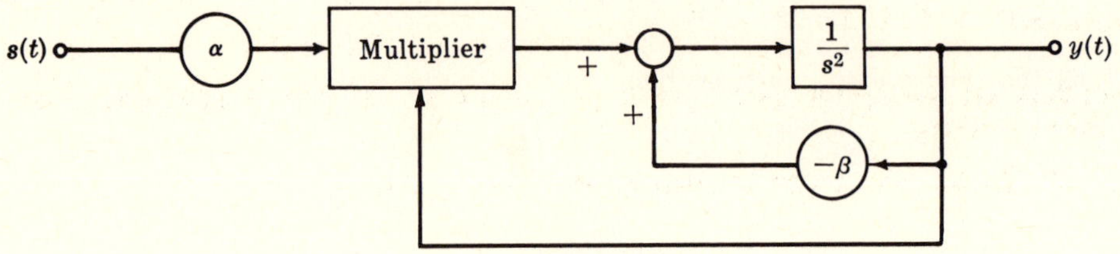

Fig. 5-6

It is found experimentally that the relationship between α and β that permits a periodic solution can be plotted as shown in Fig. 5-7.

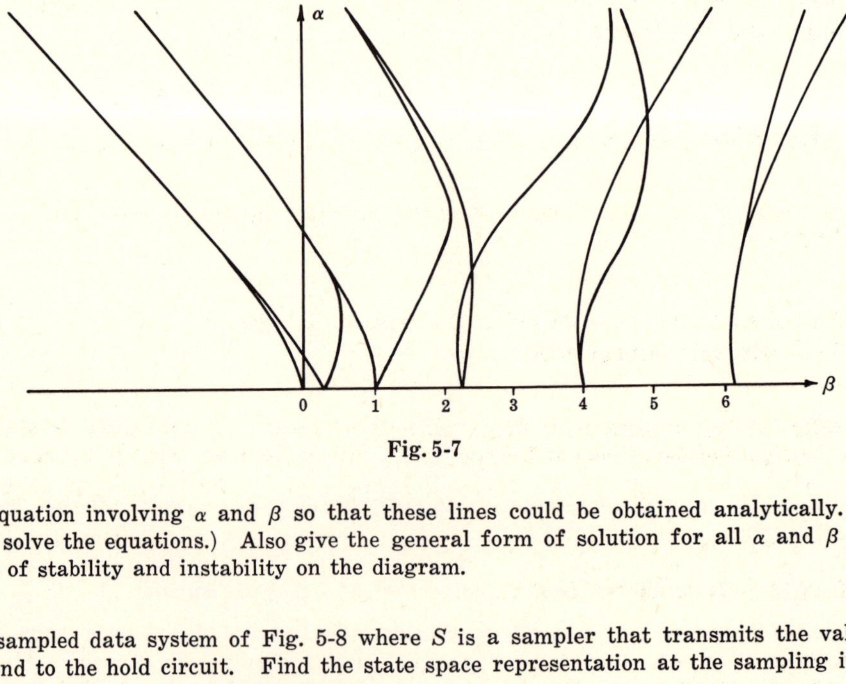

Fig. 5-7

Find the equation involving α and β so that these lines could be obtained analytically. (Do not attempt to solve the equations.) Also give the general form of solution for all α and β and mark the regions of stability and instability on the diagram.

5.38. Given the sampled data system of Fig. 5-8 where S is a sampler that transmits the value of $e(t)$ once a second to the hold circuit. Find the state space representation at the sampling instants of the closed loop system. Use Problem 5.10.

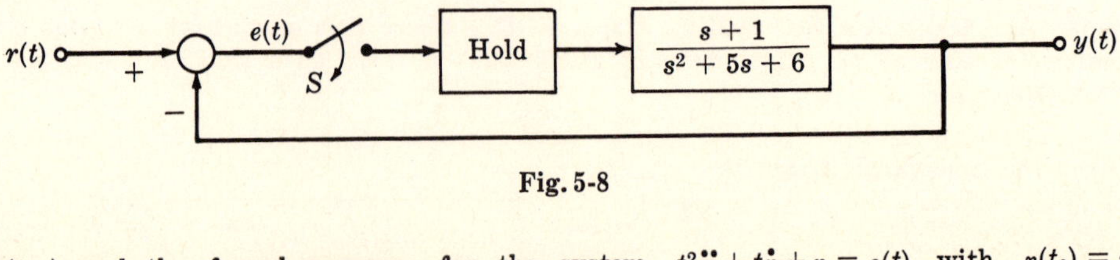

Fig. 5-8

5.39. Find $\mathbf{\Phi}(t, \tau)$ and the forced response for the system $t^2\ddot{\eta} + t\dot{\eta} + \eta = \rho(t)$ with $\eta(t_0) = \eta_0$ and $\dot{\eta}(t_0) = \dot{\eta}_0$.

5.40. Consider the system

$$\frac{d}{dt}\begin{pmatrix} x_1 \\ x_2 \end{pmatrix} = \begin{pmatrix} a & b \\ c & d \end{pmatrix}\begin{pmatrix} x_1 \\ x_2 \end{pmatrix} + \begin{pmatrix} e \\ f \end{pmatrix} u \qquad \text{or} \qquad \dot{\mathbf{x}} = \mathbf{Ax} + \mathbf{Bu}$$

$$\begin{pmatrix} y_1 \\ y_2 \end{pmatrix} = \begin{pmatrix} g & h \\ l & m \end{pmatrix}\begin{pmatrix} x_1 \\ x_2 \end{pmatrix} \qquad \text{or} \qquad \mathbf{y} = \mathbf{Cx} + \mathbf{Du}$$

Find the transfer functions $\mathcal{L}\{y_1\}/\mathcal{L}\{u_1\}$ and $\mathcal{L}\{y_2\}/\mathcal{L}\{u_2\}$ using the relation

$$\mathcal{L}\{y\}/\mathcal{L}\{u\} = [\mathbf{C}(\mathbf{I}s - \mathbf{A})^{-1}\mathbf{B} + \mathbf{D}]$$

5.41. The steady state response $\mathbf{x}_{ss}(t)$ of an asymptotically stable linear differential system satisfies the equation

$$d\mathbf{x}_{ss}/dt = \mathbf{A}(t)\mathbf{x}_{ss} + \mathbf{B}(t)\,\mathbf{u}(t)$$

but does not satisfy the initial condition $\mathbf{x}_{ss}(t_0) = \mathbf{x}_0$ and has no reference to the initial time t_0. $\mathbf{\Phi}(t,\tau)$ is known.

(*a*) Verify, by substitution into the given equation, that

$$\mathbf{x}_{ss}(t) = \int^t \mathbf{\Phi}(t,\tau)\,\mathbf{B}(\tau)\,\mathbf{u}(\tau)\,d\tau$$

where $\int^t$ is the indefinite integral evaluated at $\tau = t$. *Hint:* For an arbitrary vector function $\mathbf{f}(t,\tau)$,

$$\frac{d}{dt}\int_{\alpha(t)}^{\beta(t)} \mathbf{f}(t,\tau)\,d\tau = \mathbf{f}(t,\beta(t))\frac{d\beta}{dt} - \mathbf{f}(t,\alpha(t))\frac{d\alpha}{dt} + \int_{\alpha(t)}^{\beta(t)} \frac{\partial}{\partial t}\mathbf{f}(t,\tau)\,d\tau$$

(*b*) Suppose $\mathbf{A}(t) = \mathbf{A}(t+T)$, $\mathbf{B}(t) = \mathbf{B}(t+T)$, $\mathbf{u}(t) = \mathbf{u}(t+T)$, and the system is stable. Find an expression for a periodic $\mathbf{x}_{ss}(t) = \mathbf{x}_{ss}(t+T)$ in the form

$$\mathbf{x}_{ss}(t) = \mathbf{K}(T)\int_t^{t+T} \mathbf{\Phi}(t,\tau)\,\mathbf{B}(\tau)\,\mathbf{u}(\tau)\,d\tau$$

where $\mathbf{K}(T)$ is an $n \times n$ matrix to be found depending on T and independent of t.

5.42. Check that $h(t) = e^{t+\alpha(t-\tau)}\sin(e^{-\tau} - e^{-t})$ satisfies the system equation of Problem 5.8 with $u(t) = \delta(t-\tau)$.

5.43. Find in closed form the response of the system $(1-t^2)\dfrac{d^2y}{dt^2} - \dfrac{1}{t}\dfrac{dy}{dt} = u$ to the input $u(t) = t\sqrt{t^2-1}$ with zero initial conditions.

5.44. Consider the scalar system $dy/dt = -(1+t)y + (1+t)u$. If the initial condition is $y(0) = 10.0$, find the sign and magnitude of the impulse in $u(t)$ required at $t = 1.0$ to make $y(2) = 1.0$.

5.45. Given the system $d\mathbf{x}/dt = \begin{pmatrix} 0 & -3/t^2 \\ -1 & 2/t \end{pmatrix}\mathbf{x}$. Find the relationship between $x_1(t)$ and $x_2(t)$ such that $x_1(t_f) = 1$, using the adjoint.

5.46. Show that if an $n \times n$ nonsingular matrix solution $\mathbf{T}(t)$ to the equation $d\mathbf{T}/dt = \mathbf{A}(t)\mathbf{T} - \mathbf{TD}(t)$ is known explicitly, where $\mathbf{D}(t)$ is an $n \times n$ diagonal matrix, then an explicit solution to $d\mathbf{x}/dt = \mathbf{A}(t)\mathbf{x}$ is also known.

Answers to Supplementary Problems

5.13. $\mathbf{A}(k) = \mathbf{\Phi}(k+1, k)$

5.14. If and only if $\mathbf{AB} = \mathbf{BA}$ does $e^{\mathbf{A}}e^{\mathbf{B}} = e^{\mathbf{B}}e^{\mathbf{A}}$.

5.15. $\det e^{\mathbf{A}(t-\tau)} = \det e^{\mathbf{J}(t-\tau)} = e^{(t-\tau)\Sigma\lambda_i} = e^{(t-\tau)\operatorname{tr}\mathbf{A}}$

5.16. $\mathbf{A} = \begin{pmatrix} -2 & -2 \\ -2 & -2 \end{pmatrix}$

5.17. $\mathbf{\Phi}(t,\tau) = \frac{1}{4}\left[e^{t-\tau}\begin{pmatrix} 2 \\ 1 \end{pmatrix}(1 \;\; 2) + e^{4(\tau-t)}\begin{pmatrix} -2 \\ 1 \end{pmatrix}(-1 \;\; 2)\right]$

5.18. $\begin{pmatrix} e^t & 0 \\ 0 & 1 \end{pmatrix}$

5.19. $e^{\mathbf{A}t} = \begin{pmatrix} 1 & t & t+t^2 \\ 0 & 1 & t \\ 0 & 0 & 1 \end{pmatrix}$

5.20. $\gamma_1 = 9, \;\; \gamma_2 = 26, \;\; \gamma_3 = 24$

$$e^{\mathbf{A}t} = 0.5e^{-2t}\begin{pmatrix} 1 & 1 & 0 \\ 1 & 1 & 0 \\ 0 & 0 & 0 \end{pmatrix} + e^{-3t}\begin{pmatrix} 0 & 0 & 0 \\ 0 & 0 & 0 \\ 0 & 0 & 1 \end{pmatrix} + 0.5e^{-4t}\begin{pmatrix} 1 & -1 & 0 \\ -1 & 1 & 0 \\ 0 & 0 & 0 \end{pmatrix}$$

5.21. $e^{\mathbf{A}t} = (6e^{-2t} - 8e^{-3t} + 3e^{-4t})\mathbf{I} + 0.5(7e^{-2t} - 12e^{-3t} + 5e^{-4t})\mathbf{A} + 0.5(e^{-2t} - 2e^{-3t} + e^{-4t})\mathbf{A}^2$

5.22. $e^{\mathbf{A}t} = \begin{pmatrix} 1 & 0 & -1 \\ -1 & -1 & 2 \\ 1 & 1 & 1 \end{pmatrix}\begin{pmatrix} e^t & te^t & 0 \\ 0 & e^t & 0 \\ 0 & 0 & 1 \end{pmatrix}\begin{pmatrix} 1 & 1 & 1 \\ -1 & 0 & 1 \\ 0 & 1 & 1 \end{pmatrix}$

5.23. $(s\mathbf{I} - \mathbf{A})^{-1} = \mathbf{F}(s)/s(s-1)^2 = \mathbf{D}/s + \mathbf{B}/(s-1) + \mathbf{C}/(s-1)^2$ where

$$\mathbf{D} = \begin{pmatrix} 1 & 1 & -1 \\ 1 & 1 & -1 \\ 1 & 1 & -1 \end{pmatrix} \qquad \mathbf{C} = \begin{pmatrix} -1 & -1 & 2 \\ 1 & 1 & -2 \\ 0 & 0 & 0 \end{pmatrix} \qquad \mathbf{B} = \begin{pmatrix} 0 & -1 & 1 \\ -1 & 0 & 1 \\ -1 & -1 & 0 \end{pmatrix}$$

5.24. $\mathbf{A}^k = \begin{pmatrix} -k & -k-1 & 2k+1 \\ k-1 & k & -2k+1 \\ -1 & -1 & 2 \end{pmatrix}$

5.25. $e^{\mathbf{A}t} = \begin{pmatrix} e^{2t} & 1-e^{2t} \\ 0 & 1 \end{pmatrix}$

5.27. $y(t) = \frac{1}{2}[11 - 2e^{-(t-t_0)/2} - e^{-(t-t_0)}]U(t-t_0)$

5.28. $y(k) = 4 + [7 - 3(-1)^k - 4(-1/2)^k]/12$

5.29. $\mathbf{\Phi}(t,t_0) = \begin{pmatrix} e^{t_0-t} & e^{t-t_0} - e^{t_0-t} \\ 0 & e^{t-t_0} \end{pmatrix}, \qquad \mathbf{\Phi}(k,m) = \begin{pmatrix} (-1)^{k-m} & 1-(-1)^{k-m} \\ 0 & 1 \end{pmatrix}$

5.30. Let the generalized eigenvector $\mathbf{t}_1$ of $\mathbf{A}$ have a reciprocal basis vector $\mathbf{s}_1$, etc. Then

$$e^{\mathbf{A}t} = e^{\lambda t}\mathbf{x}_1\mathbf{r}_1^\dagger + e^{\lambda t}\mathbf{t}_1\mathbf{s}_1^\dagger + te^{\lambda t}\mathbf{x}_1\mathbf{s}_1^\dagger + \cdots$$

5.31. A Vandermonde matrix in the eigenvalues results, which is then always invertible.

5.33. Requires only $a_{12}(t)\,a_{21}(\tau) = a_{12}(\tau)\,a_{21}(t)$ and $[a_{22}(t) - a_{11}(t)]\,a_{21}(\tau) = a_{21}(t)[a_{22}(\tau) - a_{11}(\tau)]$.

5.34. $$\Phi(t,t_0) = \begin{pmatrix} \cos\int_{t_0}^{t}\alpha^{-1}(\tau)\,d\tau & \alpha(t_0)\sin\int_{t_0}^{t}\alpha^{-1}(\tau)\,d\tau \\ -\alpha^{-1}(t)\sin\int_{t_0}^{t}\alpha^{-1}(\tau)\,d\tau & \cos\int_{t_0}^{t}\alpha^{-1}(\tau)\,d\tau \end{pmatrix}$$

5.36. $$\Phi(t,t_0) = e^{\tau}\begin{pmatrix} \cosh\tau & \sinh\tau \\ \sinh\tau & \cosh\tau \end{pmatrix}$$

where $\tau = \cos t_0 - \cos t$, so $\Phi(t,t_0) = \mathbf{P}(t,t_0)$ and $\mathbf{R} = \mathbf{0}$. Also $\Psi^{\dagger}(t,t_0) = \Phi^{-1}(t,t_0) = \Phi(t_0,t)$.

5.37. Let $\gamma^2 = \beta + \alpha$, $\delta^2 = \beta - \alpha$. Then $e^{2\mathbf{R}} = \Phi(2,0) = \Phi(2,1)\,\Phi(1,0)$ where

$$\Phi(2,1) = \begin{pmatrix} \cos\delta & (\sin\delta)/\delta \\ -\delta\sin\delta & \cos\delta \end{pmatrix} \qquad \Phi(1,0) = \begin{pmatrix} \cos\gamma & (\sin\gamma)/\gamma \\ -\gamma\sin\gamma & \cos\gamma \end{pmatrix}$$

For periodicity of the envelope $\mathbf{z}(t+4\pi/\theta) = \mathbf{z}(t)$, eigenvalues λ of $e^{2\mathbf{R}} = e^{\pm j\theta}$.

$$\det(\lambda I - e^{2\mathbf{R}}) = \lambda^2 - \lambda 2\cos\theta + 1 = \lambda^2 - \lambda\,\mathrm{tr}\,e^{2\mathbf{R}} + \det e^{2\mathbf{R}}$$

$$\det\Phi(2,1)\det\Phi(1,0) = 1$$

$$2\cos\theta = 2\cos\gamma\cos\delta - (\gamma/\delta + \delta/\gamma)\sin\gamma\sin\delta$$

The stability boundaries are then determined by

$$\pm 2 = 2\cos\gamma\cos\delta - (\gamma/\delta + \delta/\gamma)\sin\gamma\sin\delta$$

The solution is of the form $\Phi(t,\tau) = \mathbf{P}(t,\tau)e^{\mathbf{R}(t-\tau)}$ and the given curves form the stability boundaries between unstable regions and periodic regions.

Reference: B. Van Der Pol and M. J. O. Strutt, On the Stability of the Solutions of Mathieu's Equation, *Philosophical Magazine*, 7th series, vol. V, January-June 1928, pp. 18-38.

5.38. $$\mathbf{x}(k+1) = \begin{pmatrix} (1+e^{-2})/2 & e^{-2}-1 \\ (1-e^{-3})/3 & (5e^{-3}-2)/3 \end{pmatrix}\mathbf{x}(k) + \begin{pmatrix} 3-3e^{-2} \\ 2-2e^{-3} \end{pmatrix}\frac{r(k)}{6}$$

5.39. $$\eta(t) = \eta_0\cos(\ln t/t_0) + \dot{\eta}_0 t_0 \sin(\ln t/t_0) - \int_{t_0}^{t}\sin(\ln\tau/t)\,\rho(\tau)/\tau\,d\tau$$

5.41. $\mathbf{K}(T) = (e^{-\mathbf{R}T} - \mathbf{I})^{-1}$

5.42. Since $h(t)$ is an element of the transition matrix,

$$d^2h/dt^2 + (1-2)\,dh/dt + (\alpha^2 - \alpha + e^{-2t})h = 0 \qquad \text{for } t \neq \tau$$

Also, since dh/dt and h are continuous in t,

$$\lim_{\epsilon\to 0}\int_{\tau-\epsilon}^{\tau+\epsilon} d^2h/dt^2\,dt = \int_{\tau-\epsilon}^{\tau+\epsilon}\delta(t-\tau)\,dt = 1$$

5.43. $y(t) = (1/4 - \sqrt{t^2-1})(t-t_0) + (\sin^{-1}t - \sin^{-1}t_0)/2$

5.44. $$u(t) = \frac{1-10e^{-4}}{2e^{-5/2}}\,\delta(t-1.0)$$

5.45. $4 = (3/\tau + \tau^3)x_1(t_0) + 3(1-\tau^4)x_2(t_0)/t_f$ where $\tau = t_f/t_0$

5.46. Let $\mathbf{x} = \mathbf{T}(t)\mathbf{z}$.

Chapter 6

Controllability and Observability

6.1 INTRODUCTION TO CONTROLLABILITY AND OBSERVABILITY

Can all the states of a system be controlled and/or observed? This fundamental question arises surprisingly often in both practical and theoretical investigations and is most easily investigated using state space techniques.

Definition 6.1: A state $\mathbf{x}_1$ of a system is *controllable* if all initial conditions $\mathbf{x}_0$ at any previous time t_0 can be transferred to $\mathbf{x}_1$ in a finite time by some control function $\mathbf{u}(t, \mathbf{x}_0)$.

If all states $\mathbf{x}_1$ are controllable, the system is called *completely controllable* or simply *controllable*. If controllability is restricted to depend on t_0, the state is said to be *controllable at time* t_0. If the *state* can be transferred from $\mathbf{x}_0$ to $\mathbf{x}_1$ as quickly as desired independent of t_0, instead of in some finite time, that state is *totally controllable*. The *system* is totally controllable if all states are totally controllable. Finally, we may talk about the output $\mathbf{y}$ instead of the state $\mathbf{x}$ and give similar definitions for *output controllable*, e.g. an output controllable at time t_0 means that a particular output $\mathbf{y}$ can be attained starting from any arbitrary $\mathbf{x}_0$ at t_0.

To determine complete controllability at time t_0 for linear systems, it is necessary and sufficient to investigate whether the zero state instead of all initial states can be transferred to all final states. Writing the complete solution for the linear case,

$$\mathbf{x}(t_1) \quad = \quad \mathbf{\Phi}(t_1, t_0)\,\mathbf{x}(t_0) + \int_{t_0}^{t_1} \mathbf{\Phi}(t_1, \tau)\,\mathbf{B}(\tau)\,\mathbf{u}(\tau)\,d\tau$$

which is equivalent to starting from the zero state and going to a final state $\hat{\mathbf{x}}(t_1) = \mathbf{x}(t_1) - \mathbf{\Phi}(t_1, t_0)\,\mathbf{x}(t_0)$. Therefore if we can show the linear system can go from $\mathbf{0}$ to any $\hat{\mathbf{x}}(t_1)$, then it can go from any $\mathbf{x}(t_0)$ to any $\mathbf{x}(t_1)$.

The concept of observability will turn out to be the dual of controllability.

Definition 6.2: A state $\mathbf{x}(t)$ at some given t of a system is *observable* if knowledge of the input $\mathbf{u}(\tau)$ and output $\mathbf{y}(\tau)$ over a finite time segment $t_0 < \tau \leq t$ completely determines $\mathbf{x}(t)$.

If all states $\mathbf{x}(t)$ are observable, the system is called *completely observable*. If observability depends on t_0, the state is said to be *observable at* t_0. If the state can be determined for τ in any arbitrarily small time segment independent of t_0, it is *totally observable*. Finally, we may talk about observability when $\mathbf{u}(\tau) = \mathbf{0}$, and give similar definitions for *zero-input observable*.

To determine complete observability for linear systems, it is necessary and sufficient to see if the initial state $\mathbf{x}(t_0)$ of the zero-input system can be completely determined from $\mathbf{y}(\tau)$, because knowledge of $\mathbf{x}(t_0)$ and $\mathbf{u}(\tau)$ permits $\mathbf{x}(t)$ to be calculated from the complete solution equation (*5.44*).

We have already encountered uncontrollable and unobservable states in Example 1.7, page 3. These states were physically disconnected from the input or the output. By physically disconnected we mean that for all time the flow diagram shows no connection, i.e. the control passes through a scalor with zero gain. Then it follows that any state vectors having elements that are disconnected from the input or output will be uncontrollable or unobservable. However, there exist uncontrollable and unobservable systems in which the flow diagram is not always disconnected.

Example 6.1.

Consider the time-varying system

$$\frac{d}{dt}\begin{pmatrix} x_1 \\ x_2 \end{pmatrix} = \begin{pmatrix} \alpha & 0 \\ 0 & \beta \end{pmatrix}\begin{pmatrix} x_1 \\ x_2 \end{pmatrix} + \begin{pmatrix} e^{\alpha t} \\ e^{\beta t} \end{pmatrix} u$$

From the flow diagram of Fig. 6-1 it can be seen that $u(t)$ passes through scalors that are never zero.

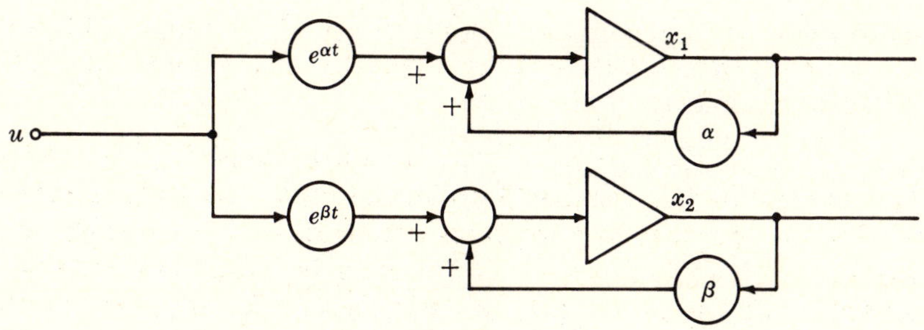

Fig. 6-1

For zero initial conditions,

$$x_1(t_1)e^{-\alpha t_1} = \int_{t_0}^{t_1} u(\tau)\,d\tau = x_2(t_1)e^{-\beta t_1}$$

Only those $x_2(t_1) = x_1(t_1)e^{t_1(\beta-\alpha)}$ can be reached at t_1, so that $x_2(t_1)$ is fixed after $x_1(t_1)$ is chosen. Therefore the system is not controllable.

6.2 CONTROLLABILITY IN TIME-INVARIANT LINEAR SYSTEMS

For time-invariant systems $d\mathbf{x}/dt = \mathbf{A}\mathbf{x} + \mathbf{B}\mathbf{u}$ in the case where $\mathbf{A}$ has distinct eigenvalues, connectedness between the input and all elements of the state vector becomes equivalent to the strongest form of controllability, totally controllable. We shall first consider the case of a scalar input to avoid complexity of notation, and consider distinct eigenvalues before going on to the general case.

***Theorem 6.1*:** Given a scalar input $u(t)$ to the time-invariant system $d\mathbf{x}/dt = \mathbf{A}\mathbf{x} + \mathbf{b}u$, where $\mathbf{A}$ has distinct eigenvalues λ_i. Then the system is totally controllable if and only if the vector $\mathbf{f} = \mathbf{M}^{-1}\mathbf{b}$ has no zero elements. $\mathbf{M}$ is the modal matrix with eigenvectors of $\mathbf{A}$ as its column vectors.

Proof: *Only if* part: Change variables by $\mathbf{x} = \mathbf{M}\mathbf{z}$. Then the system becomes $d\mathbf{z}/dt = \mathbf{\Lambda}\mathbf{z} + \mathbf{f}u$, where $\mathbf{\Lambda}$ is the diagonal matrix of distinct eigenvalues λ_i. A flow diagram of this system is shown in Fig. 6-2 below.

If any element f_i of the $\mathbf{f}$ vector is zero, the element of the state vector z_i is disconnected from the control. Consequently any $\mathbf{x}$ made up of a linear combination of the z's involving z_i will be uncontrollable. Therefore if the system is totally controllable, then all elements f_i must be nonzero.

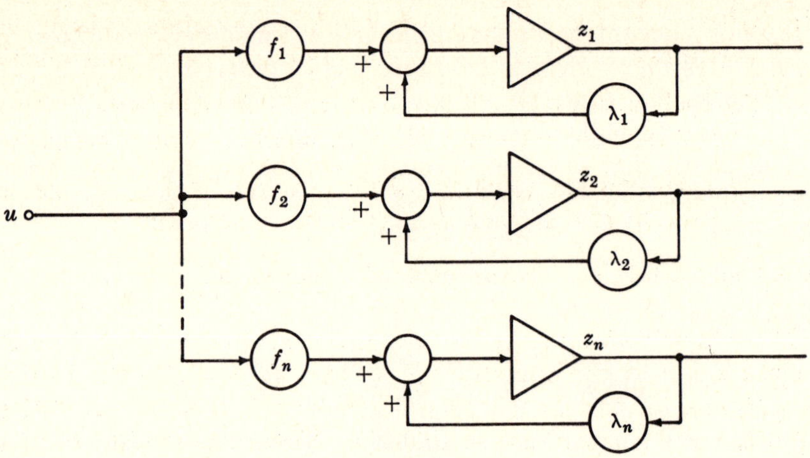

Fig. 6-2

If part: Now we assume all f_i are nonzero, and from the remarks following Definition 6.1 we need investigate only whether the transformed system can be transferred to an arbitrary $\mathbf{z}(t_1)$ from $\mathbf{z}(t_0) = \mathbf{0}$, where t_1 can be arbitrarily close to t_0. To do this, note

$$z_i(t_1) \quad = \quad \int_{t_0}^{t_1} e^{\lambda_i(t_1-\tau)} f_i u(\tau)\, d\tau \qquad i = 1, 2, \ldots, n \tag{6.1}$$

It is true, but yet unproven, that if the f_i are nonzero, many different $u(\tau)$ can transfer $\mathbf{0}$ to $\mathbf{z}(t_1)$.

Now we construct a particular $u(\tau)$ that will always do the job. Prescribe $u(\tau)$ as

$$u(\tau) \quad = \quad \sum_{k=1}^{n} \mu_k e^{-\lambda_k^*(\tau - t_1)} \tag{6.2}$$

where the μ_k are constants to be chosen. Substituting the construction (*6.2*) into equation (*6.1*) gives

$$z_i(t_1) \quad = \quad \sum_{k=1}^{n} f_i \mu_k (e^{\lambda_k(t_1-\tau)}, e^{\lambda_i(t_1-\tau)}) \qquad \text{for all } i \tag{6.3}$$

where the inner product is defined as

$$(\theta(\tau), \phi(\tau)) \quad = \quad \int_{t_0}^{t_1} \theta^*(\tau)\, \phi(\tau)\, d\tau$$

Equation (*6.3*) can be written in matrix notation as

$$\begin{pmatrix} z_1(t_1)/f_1 \\ z_2(t_1)/f_2 \\ \vdots \\ z_n(t_1)/f_n \end{pmatrix} = \begin{pmatrix} g_{11} & g_{12} & \cdots & g_{1n} \\ g_{21} & g_{22} & \cdots & g_{2n} \\ \cdots & \cdots & \cdots & \cdots \\ g_{1n} & g_{2n} & \cdots & g_{nn} \end{pmatrix} \begin{pmatrix} \mu_1 \\ \mu_2 \\ \vdots \\ \mu_n \end{pmatrix} \tag{6.4}$$

where $g_{ik} = (e^{\lambda_k(t_1-\tau)}, e^{\lambda_i(t_1-\tau)})$. Note that because $f_i \neq 0$ by assumption, division by f_i is permitted. Since the time functions $e^{\lambda_i(t_1-\tau)}$ are obviously linearly independent, the Gram matrix $\{g_{ik}\}$ is nonsingular (see Problem 3.14). Hence we can always solve equation (*6.4*) for $\mu_1, \mu_2, \ldots, \mu_n$, which means the control (*6.2*) will always work.

Now we can consider what happens when $\mathbf{A}$ is not restricted to have distinct eigenvalues.

Theorem 6.2: Given a scalar input $u(t)$ to the time-invariant system $d\mathbf{x}/dt = \mathbf{Ax} + \mathbf{b}u$, where $\mathbf{A}$ is arbitrary. Then the system is totally controllable if and only if:

(1) each eigenvalue λ_i associated with a Jordan block $\mathbf{L}_{ji}(\lambda_i)$ is distinct from an eigenvalue associated with another Jordan block, and

(2) each element f_i of $\mathbf{f} = \mathbf{T}^{-1}\mathbf{b}$, where $\mathbf{T}^{-1}\mathbf{AT} = \mathbf{J}$, associated with the bottom row of each Jordan block is nonzero.

Note that Theorem 6.1 is a special case of Theorem 6.2.

Proof: *Only if* part: The system is assumed controllable. The flow diagram for one $l \times l$ Jordan block $\mathbf{L}_{ji}(\lambda_i)$ of the transformed system $d\mathbf{z}/dt = \mathbf{Jz} + \mathbf{f}u$ is shown in Fig. 6-3. The control u is connected to $z_1, z_2, \ldots, z_{l-1}$ and z_l only if f_l is nonzero. It does not matter if $f_1, f_2, \ldots$ and f_{l-1} are zero or not, so that the controllable system requires condition (2) to hold. Furthermore suppose condition (1) did not hold. Then the bottom rows of two different Jordan blocks with the same eigenvalues $[\mathbf{L}_{\nu i}(\lambda_i)$ and $\mathbf{L}_{\eta i}(\lambda_i)]$ could be written as

$$dz_\nu/dt = \lambda_i z_\nu + f_\nu u$$

$$dz_\eta/dt = \lambda_i z_\eta + f_\eta u$$

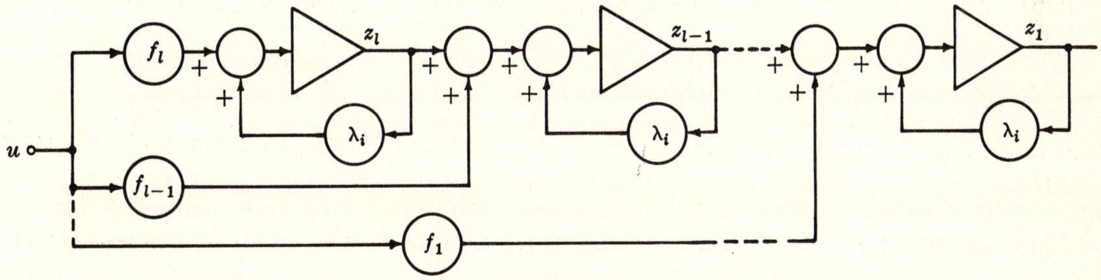

Fig. 6-3

Consider the particular state having one element equal to $f_\eta z_\nu - f_\nu z_\eta$. Then

$$d(f_\eta z_\nu - f_\nu z_\eta)/dt = f_\eta(\lambda_i z_\nu + f_\nu u) - f_\nu(\lambda_i z_\eta + f_\eta u)$$

$$= \lambda_i(f_\eta z_\nu - f_\nu z_\eta)$$

Therefore $f_\eta z_\nu(t) - f_\nu z_\eta(t) = [f_\eta z_\nu(0) - f_\nu z_\eta(0)]e^{\lambda_i t}$ and is independent of the control. We have found a particular state that is not controllable, so if the system is controllable, condition (1) must hold.

If part: Again, a control can be constructed in similar manner to equation (*6.2*) to show the system is totally controllable if conditions (1) and (2) of Theorem 6.2 hold.

Example 6.2.

To illustrate why condition (1) of Theorem 6.2 is important, consider the system

$$\frac{d}{dt}\begin{pmatrix} x_1 \\ x_2 \end{pmatrix} = \begin{pmatrix} 2 & 0 \\ 0 & 2 \end{pmatrix}\begin{pmatrix} x_1 \\ x_2 \end{pmatrix} + \begin{pmatrix} 3 \\ 1 \end{pmatrix} u$$

$$y = (1 \quad -3)\begin{pmatrix} x_1 \\ x_2 \end{pmatrix}$$

Then $\mathcal{L}\{x_1\} = (3\mathcal{L}\{u\} + x_{10})/(s-2)$ and $\mathcal{L}\{x_2\} = (\mathcal{L}\{u\} + x_{20})/(s-2)$ so that $y(t) = (x_{10} - 3x_{20})e^{2t}$ regardless of the action of the control. The input is physically connected to the state and the state is physically connected to the output, but the output cannot be controlled.

For discrete-time systems, an analogous theorem holds.

Theorem 6.3: Given a scalar input $u(m)$ to the time-invariant system $\mathbf{x}(m+1) = \mathbf{A}\mathbf{x} + \mathbf{b}u(m)$, where $\mathbf{A}$ is arbitrary. Then the system is completely controllable if and only if conditions (1) and (2) of Theorem 6.2 hold.

Proof: *Only if* part: This is analogous to the *only if* part of Theorem 6.2, in that the flow diagram shows the control is disconnected from at least one element of the state vector if condition (2) does not hold, and a particular state vector with an element equal to $f_\eta z_\nu - f_\nu z_\eta$ is uncontrollable if condition (1) does not hold.

If part: Consider the transformed systems $\mathbf{z}(m+1) = \mathbf{J}\mathbf{z}(m) + \mathbf{f}u(m)$, and for simplicity assume distinct roots so that $\mathbf{J} = \mathbf{\Lambda}$. Then for zero initial condition,

$$z_i(m) \;=\; [\lambda^{m-1}u(0) + \lambda^{m-2}u(1) + \cdots + u(m-1)]f_i$$

For an nth order system, the desired state can be reached on the nth step because

$$\begin{pmatrix} z_1(n)/f_1 \\ z_2(n)/f_2 \\ \vdots \\ z_n(n)/f_n \end{pmatrix} = \begin{pmatrix} \lambda_1^{n-1} & \lambda_1^{n-2} & \ldots & 1 \\ \lambda_2^{n-1} & \lambda_2^{n-2} & \ldots & 1 \\ \cdots & \cdots & \cdots & \cdots \\ \lambda_n^{n-1} & \lambda_n^{n-2} & \ldots & 1 \end{pmatrix} \begin{pmatrix} u(0) \\ u(1) \\ \vdots \\ u(n-1) \end{pmatrix} \tag{6.5}$$

Note that a Vandermonde matrix with distinct elements results, and so it is nonsingular. Therefore we can solve (*6.5*) for a control sequence $u(0), u(1), \ldots, u(n-1)$ to bring the system to a desired state in n steps if the conditions of the theorem hold.

For discrete-time systems with a scalar input, it takes at least n steps to transfer to an arbitrary desired state. The corresponding control can be found from equation (*6.5*), called dead beat control. Since it takes n steps, only complete controllability was stated in the theorem. We could (but will not) change the definition of total controllability to say that in the case of discrete-time systems, transfer in n steps is total control.

The phenomenon of hidden oscillations in sampled data systems deserves some mention here. Given a periodic function, such as $\sin \omega t$, if we sample it at a multiple of its period it will be undetectable. Referring to Fig. 6-4, it is impossible to tell from the sample points whether the dashed straight line or the sine wave is being sampled. This has nothing to do with controllability or observability, because it represents a failure of the abstract object (the difference equation) to represent a physical object. In this case, a differential-difference equation can be used to represent behavior between sampling instants.

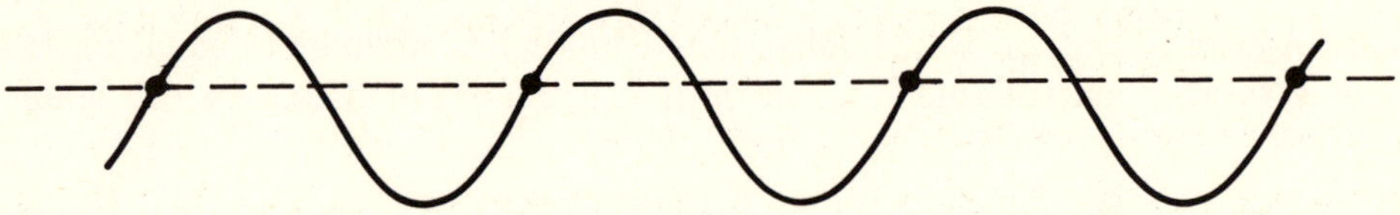

Fig. 6-4

6.3 OBSERVABILITY IN TIME-INVARIANT LINEAR SYSTEMS

Analogous to Theorem 6.1, connectedness between state and output becomes equivalent to total observability for $d\mathbf{x}/dt = \mathbf{A}\mathbf{x} + \mathbf{B}\mathbf{u}$, $y = \mathbf{c}^\dagger\mathbf{x} + \mathbf{d}^T\mathbf{u}$, when the system is stationary and $\mathbf{A}$ has distinct eigenvalues. To avoid complexity, first we consider scalar outputs and distinct eigenvalues.

Theorem 6.4: Given a scalar output $y(t)$ to the time-invariant system $d\mathbf{x}/dt = \mathbf{Ax} + \mathbf{Bu}$, $y = \mathbf{c}^\dagger\mathbf{x} + \mathbf{d}^T\mathbf{u}$, where $\mathbf{A}$ has distinct eigenvalues λ_i. Then the system is totally observable if and only if the vector $\mathbf{g}^\dagger = \mathbf{c}^\dagger\mathbf{M}$ has no zero elements. $\mathbf{M}$ is the modal matrix with eigenvectors of $\mathbf{A}$ as its column vectors.

Proof: From the remarks following Definition 6.2 we need to see if $\mathbf{x}(t_0)$ can be reconstructed from measurement of $\mathbf{y}(\tau)$ over $t_0 < \tau \leq t$ in the case where $\mathbf{u}(\tau) = \mathbf{0}$. We do this by changing variables as $\mathbf{x} = \mathbf{Mz}$. Then the system becomes $d\mathbf{z}/dt = \mathbf{\Lambda z}$ and $y = \mathbf{c}^\dagger\mathbf{Mz} = \mathbf{g}^\dagger\mathbf{z}$. The flow diagram for this system is given in Fig. 6.5.

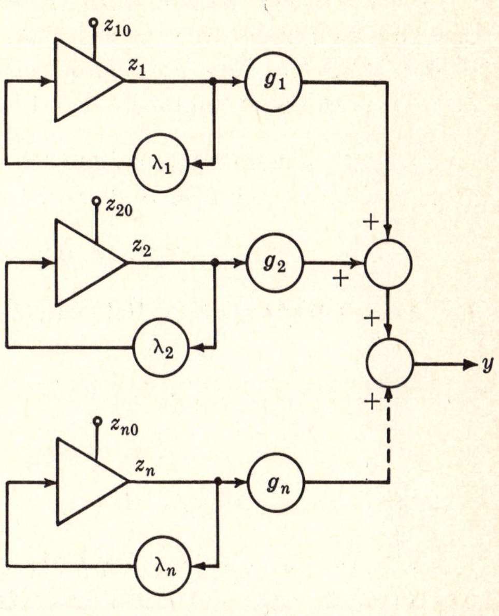

Fig. 6-5

Each $z_i(t) = z_i(t_0)e^{\lambda_i(t-t_0)}$ can be determined by taking in measurements of $y(t)$ at times $\tau_k = t_0 + (t_1 - t_0)k/n$ for $k = 1, 2, \ldots, n$ and solving the set of equations

$$y(\tau_k) = \sum_{i=1}^{n} g_i z_i(t_0) e^{\lambda_i k(t_1 - t_0)/n}$$

for $g_i z_i(t_0)$. When written in matrix form this set of equations gives a Vandermonde matrix which is always nonsingular if the λ_i are distinct. If all $g_i \neq 0$, then all $z_i(t_0)$ can be found. To find $\mathbf{x}(t)$, use $\mathbf{x}(t_0) = \mathbf{Mz}(t_0)$ and $d\mathbf{x}/dt = \mathbf{Ax} + \mathbf{Bu}$. Only if $g_i \neq 0$ is each state connected to y.

The extension to a general $\mathbf{A}$ matrix is similar to the controllability Theorem 6.2.

Theorem 6.5: Given a scalar output $y(t)$ from the time-invariant system $d\mathbf{x}/dt = \mathbf{Ax} + \mathbf{Bu}$, $y = \mathbf{c}^\dagger\mathbf{x} + \mathbf{d}^T\mathbf{u}$, where $\mathbf{A}$ is arbitrary. Then the system is totally observable if and only if:

(1) each eigenvalue λ_i associated with a Jordan block $\mathbf{L}_{ji}(\lambda_i)$ is distinct from an eigenvalue associated with another Jordan block, and

(2) each element g_i of $\mathbf{g}^\dagger = \mathbf{c}^\dagger\mathbf{T}$, where $\mathbf{T}^{-1}\mathbf{AT} = \mathbf{J}$, associated with the top row of each Jordan block is nonzero.

The proof is similar to that of Theorem 6.2.

Theorem 6.6: Given a scalar output $y(m)$ from the time-invariant system $\mathbf{x}(m+1) = \mathbf{Ax}(m) + \mathbf{Bu}(m)$, $y(m) = \mathbf{c}^\dagger\mathbf{x}(m) + \mathbf{d}^T\mathbf{u}(m)$, where $\mathbf{A}$ is arbitrary. Then the system is completely observable if and only if conditions (1) and (2) of Theorem 6.5 hold.

The proof is similar to that of Theorem 6.3.

Now we can classify the elements of the state vectors of $d\mathbf{x}/dt = \mathbf{Ax} + \mathbf{Bu}$, $\mathbf{y} = \mathbf{Cx} + \mathbf{Du}$ and of $\mathbf{x}(m+1) = \mathbf{Ax}(m) + \mathbf{Bu}(m)$, $\mathbf{y} = \mathbf{Cx} + \mathbf{Du}$ according to whether they are controllable or not, and whether they are observable or not. In particular, in the single input–single output case when $\mathbf{A}$ has distinct eigenvalues, those elements of the state z_i that have nonzero f_i and g_i are both controllable and observable, those elements z_j that have zero f_j but nonzero g_j are uncontrollable but observable, etc. When $\mathbf{A}$ has repeated eigenvalues, a glance at Fig. 6-3 shows that z_k is controllable if and only if not all of $f_k, f_{k+1}, \ldots, f_l$ are zero, and the eigenvalues associated with individual Jordan blocks are distinct.

Unobservable and uncontrollable elements of a state vector cancel out of the transfer function of a single input–single output system.

***Theorem 6.7*:** For the single input–single output system $d\mathbf{x}/dt = \mathbf{A}\mathbf{x} + \mathbf{b}u$, $y = \mathbf{c}^\dagger\mathbf{x} + du$, the transfer function $\mathbf{c}^\dagger(s\mathbf{I} - \mathbf{A})^{-1}\mathbf{b}$ has poles that are canceled by zeros if and only if some states are uncontrollable and/or unobservable. A similar statement holds for discrete-time systems.

Proof: First, note that the Jordan flow diagram (see Section 2.4) to represent the transfer function cannot be drawn with repeated eigenvalues associated with different Jordan blocks. (Try it! The elements belonging to a particular eigenvalue must be combined.) Furthermore, if the **J** matrix has repeated eigenvalues associated with different Jordan blocks, an immediate cancellation occurs. This can be shown by considering the bottom rows of two Jordan blocks with identical eigenvalues λ.

$$\frac{d}{dt}\begin{pmatrix} z_1 \\ z_2 \end{pmatrix} = \begin{pmatrix} \lambda & 0 \\ 0 & \lambda \end{pmatrix}\begin{pmatrix} z_1 \\ z_2 \end{pmatrix} + \begin{pmatrix} b_1 \\ b_2 \end{pmatrix} u$$

$$y = (c_1 \ \ c_2)\begin{pmatrix} z_1 \\ z_2 \end{pmatrix} + du$$

Then for zero intial conditions $\mathcal{L}\{z_1\} = (s-\lambda)^{-1}b_1\mathcal{L}\{u\}$ and $\mathcal{L}\{z_2\} = (s-\lambda)^{-1}b_2\mathcal{L}\{u\}$, so that

$$\mathcal{L}\{y\} = [c_1(s-\lambda)^{-1}b_1 + c_2(s-\lambda)^{-1}b_2 + d]\mathcal{L}\{u\}$$

Combining terms gives

$$\mathcal{L}\{y\} = \left(d + \frac{c_1b_1 + c_2b_2}{s-\lambda}\right)\mathcal{L}\{u\}$$

This is a first-order transfer function representing a second-order system, so a cancellation has occurred. Starting from the transfer function gives a system representation of $dz/dt = \lambda z + u$, $y = (c_1b_1 + c_2b_2)z + du$, which illustrates why the Jordan flow diagram cannot be drawn for systems with repeated eigenvalues.

Now consider condition (2) of Theorems 6.2 and 6.5. Combining the flow diagrams of Figs. 6-2 and 6-5 to represent the element z_i of the bottom row of a Jordan block gives Fig. 6-6.

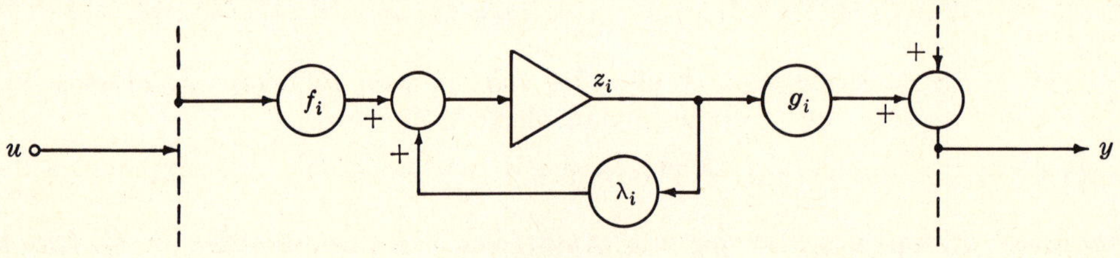

Fig. 6-6

Comparing this figure with Fig. 2-12 shows that $f_ig_i = \rho_i$, the residue of λ_i. If and only if $\rho_i = 0$, a cancellation occurs, and $\rho_i = 0$ if and only if f_i and/or $g_i = 0$, which occurs when the system is uncontrollable and/or unobservable.

Note that it is the uncontrollable and/or unobservable element of the state vector that is canceled from the transfer function.

6.4 DIRECT CRITERIA FROM A, B, AND C

If we need not determine which elements of the state vector are controllable and observable, but merely need to investigate the controllability and/or observability of the whole system, calculation of the Jordan form is not necessary. Criteria using only **A**, **B** and **C** are available and provide an easy and general way to determine if the system is completely controllable and observable.

Theorem 6.8: The time-invariant system $d\mathbf{x}/dt = \mathbf{A}\mathbf{x} + \mathbf{B}\mathbf{u}$ is totally controllable if and only if the $n \times nm$ matrix $\mathbf{Q}$ has rank n, where

$$\mathbf{Q} = (\mathbf{B} \mid \mathbf{AB} \mid \ldots \mid \mathbf{A}^{n-1}\mathbf{B})$$

Note that this criterion is valid for vector $\mathbf{u}$, and so is more general than Theorem 6.2. One method of determining if $\operatorname{rank}\mathbf{Q} = n$ is to see if $\det(\mathbf{Q}\mathbf{Q}^T) \neq 0$, since $\operatorname{rank}\mathbf{Q} = \operatorname{rank}\mathbf{Q}\mathbf{Q}^T$ from property 16, page 87. However, there exist faster machine methods for determining the rank.

Proof: To reach an arbitrary $\mathbf{x}(t_1)$ from the zero initial state, we must find a control $\mathbf{u}(\tau)$ such that

$$\mathbf{x}(t_1) = \int_{t_0}^{t_1} e^{\mathbf{A}(t_1-\tau)}\mathbf{B}\mathbf{u}(\tau)\,d\tau \tag{6.6}$$

Use of Theorem 4.16 gives $e^{\mathbf{A}(t_1-\tau)} = \sum_{i=1}^{n} \gamma_i(\tau)\mathbf{A}^{n-i}$ so that substitution into (*6.6*) gives

$$\mathbf{x}(t_1) = (\mathbf{B}\mathbf{w}_1 + \mathbf{AB}\mathbf{w}_2 + \cdots + \mathbf{A}^{n-1}\mathbf{B}\mathbf{w}_n) = \mathbf{Q}\begin{pmatrix} \mathbf{w}_1 \\ \vdots \\ \mathbf{w}_n \end{pmatrix}$$

where $\mathbf{w}_k = \int_{t_0}^{t_1} \gamma_{n+1-k}(\tau)\mathbf{u}(\tau)\,d\tau$. Hence $\mathbf{x}(t_1)$ lies in the range space of $\mathbf{Q}$, so that $\mathbf{Q}$ must have rank n to reach an arbitrary vector in $\mathcal{V}_n$. Therefore if the system is totally controllable, $\mathbf{Q}$ has rank n.

Now we assume $\mathbf{Q}$ has rank n and show that the system is totally controllable. This time we construct $\mathbf{u}(\tau)$ as

$$\mathbf{u}(\tau) = \boldsymbol{\mu}_1\delta(\tau - t_0) + \boldsymbol{\mu}_2\delta^{(1)}(\tau - t_0) + \cdots + \boldsymbol{\mu}_n\delta^{(n-1)}(\tau - t_0) \tag{6.7}$$

where $\boldsymbol{\mu}_k$ are constant m-vectors to be found and $\delta^{(k)}(\tau)$ is the kth derivative of the Dirac delta function. Substituting this construction into equation (*6.6*) gives

$$\mathbf{x}(t_1) = e^{\mathbf{A}(t_1-t_0)}\mathbf{B}\boldsymbol{\mu}_1 + e^{\mathbf{A}(t_1-t_0)}\mathbf{AB}\boldsymbol{\mu}_2 + \cdots + e^{\mathbf{A}(t_1-t_0)}\mathbf{A}^{n-1}\mathbf{B}\boldsymbol{\mu}_n \tag{6.8}$$

since $\int_{t_0}^{t_1} e^{\mathbf{A}(t_1-\tau)}\mathbf{B}\delta(\tau - t_0)\,d\tau = e^{\mathbf{A}(t_1-t_0)}\mathbf{B}$ and the defining relation for $\delta^{(k)}$ is

$$\int_{-\infty}^{\infty} \delta^{(k)}(t-\xi)\,\mathbf{g}(\xi)\,d\xi = d^k\mathbf{g}/dt^k$$

Using the inversion property of transition matrices and the definition of $\mathbf{Q}$, equation (*6.8*) can be rewritten

$$e^{\mathbf{A}(t_0-t_1)}\mathbf{x}(t_1) = \mathbf{Q}\begin{pmatrix} \boldsymbol{\mu}_1 \\ \hline \vdots \\ \hline \boldsymbol{\mu}_n \end{pmatrix}$$

From Problem 3.11, page 63, a solution for the $\boldsymbol{\mu}_i$ always exists if $\operatorname{rank}\mathbf{Q} = n$. Hence some $\boldsymbol{\mu}_i$ (perhaps not unique) always exists such that the control (*6.7*) will drive the system to $\mathbf{x}(t_1)$.

The construction for the control (*6.7*) gives some insight as to why completely controllable stationary linear systems can be transferred to any desired state as quickly as possible. No restrictions are put on the magnitude or shape of $\mathbf{u}(\tau)$. If the magnitude of the control is bounded, the set of states to which the system can be transferred by t_1 are called the *reachable* states at t_1, which has the dual concept in observability as *recoverable* states at t_1. Any further discussion of this point is beyond the scope of this text.

A proof can be given involving a construction for a bounded $\mathbf{u}(t)$ similar to equation (*6.2*), instead of the unbounded $\mathbf{u}(t)$ of (*6.7*). However, as $t_1 \to t_0$, any control must become unbounded to introduce a jump from $\mathbf{x}(t_0)$ to $\mathbf{x}(t_1)$.

The dual theorem to the one just proven is

Theorem 6.9: The time-invariant system $d\mathbf{x}/dt = \mathbf{Ax} + \mathbf{Bu}$, $\mathbf{y} = \mathbf{Cx} + \mathbf{Du}$ is totally observable if and only if the $kn \times n$ matrix $\mathbf{P}$ has rank n where

$$\mathbf{P} = \begin{pmatrix} \mathbf{C} \\ \hline \mathbf{CA} \\ \hline \vdots \\ \hline \mathbf{CA}^{n-1} \end{pmatrix}$$

Theorems 6.8 and 6.9 are also true (replacing *totally* by *completely*) for the discrete-time system $\mathbf{x}(m+1) = \mathbf{Ax}(m) + \mathbf{Bu}(m)$, $\mathbf{y}(m) = \mathbf{Cx}(m) + \mathbf{Du}(m)$. Since the proof of Theorem 6.9 is quite similar to that of Theorem 6.8 in the continuous-time case, we give a proof for the discrete-time case. It is sufficient to see if the initial state $\mathbf{x}(l)$ can be reconstructed from knowledge of $\mathbf{y}(m)$ *for* $l \leq m < \infty$, in the case where $\mathbf{u}(m) = \mathbf{0}$. From the state equation,

$$\begin{aligned} \mathbf{y}(l) &= \mathbf{Cx}(l) \\ \mathbf{y}(l+1) &= \mathbf{Cx}(l+1) = \mathbf{CAx}(l) \\ &\cdots\cdots\cdots\cdots \\ \mathbf{y}(l+n-1) &= \mathbf{CA}^{n-1}\mathbf{x}(l) \end{aligned}$$

This can be rewritten as

$$\begin{pmatrix} \mathbf{y}(l) \\ \hline \vdots \\ \hline \mathbf{y}(l+n-1) \end{pmatrix} = \mathbf{Px}(l)$$

and a unique solution exists if and only if $\mathbf{P}$ has rank n, as shown in Problem 3.11, page 63.

6.5 CONTROLLABILITY AND OBSERVABILITY OF TIME-VARYING SYSTEMS

In time-varying systems, the difference between totally and completely controllable becomes important.

Example 6.3.

Consider the time-varying scalar system $dx/dt = -x + b(t)u$ and $y = x$, where $b(t)$ is either zero or unity as shown in Fig. 6-7.

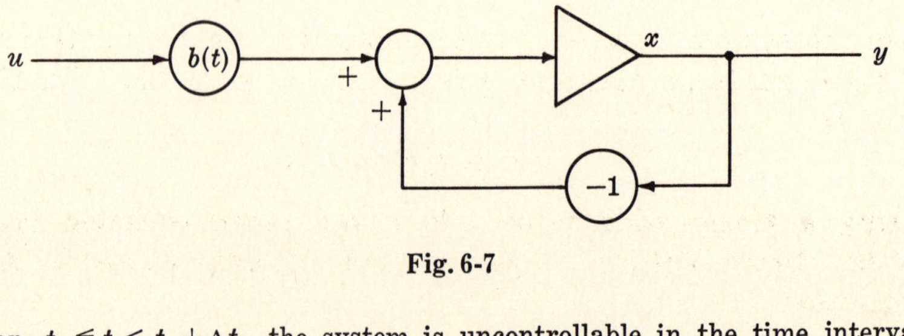

Fig. 6-7

If $b(t) = 0$ for $t_0 \leq t < t_0 + \Delta t$, the system is uncontrollable in the time interval $[t_0, t_0 + \Delta t)$. If $b(t) = 1$ for $t_0 + \Delta t \leq t \leq t_1$, the system is totally controllable in the time interval $t_0 + \Delta t \leq t \leq t_1$. However, the system is not totally controllable, but is completely controllable over the time interval $t_0 \leq t \leq t_1$ to reach a desired final state $x(t_1)$.

Now suppose $b(t) = 0$ for $t_1 < t \leq t_2$, and we wish to reach a final state $x(t_2)$. The state $x(t_2)$ can be reached by controlling the system such that the state at time t_1 is $x(t_1) = e^{(t_2-t_1)}x(t_2)$. Then with zero input the free system will coast to the desired state $x(t_2) = \Phi(t_2, t_1)\, x(t_1)$. Therefore if the system is totally controllable for any time interval $t_c \leq t \leq t_c + \Delta t$, it is completely controllable for all $t \geq t_c$.

For the time-varying system, a criterion analogous to Theorem 6.8 can be formed.

***Theorem 6.10*:** The time-varying system $d\mathbf{x}/dt = \mathbf{A}(t)\mathbf{x} + \mathbf{B}(t)\mathbf{u}$ is totally controllable if and only if the matrix $\mathbf{Q}(t)$ has rank n for times t everywhere dense in $[t_0, t_1]$, where $\mathbf{Q}(t) = (\mathbf{Q}_1 \mid \mathbf{Q}_2 \mid \ldots \mid \mathbf{Q}_n)$, in which $\mathbf{Q}_1 = \mathbf{B}(t)$ and $\mathbf{Q}_{k+1} = -\mathbf{A}(t)\mathbf{Q}_k + d\mathbf{Q}_k/dt$ for $k = 1, 2, \ldots, n-1$. Here $\mathbf{A}(t)$ and $\mathbf{B}(t)$ are assumed piecewise differentiable at least $n-2$ and $n-1$ times, respectively.

The phrase "for times t everywhere dense in $[t_0, t_1]$" essentially means that there can exist only isolated points in t in which $\operatorname{rank}\mathbf{Q} < n$. Because this concept occurs frequently, we shall abbreviate it to "$\mathbf{Q}(t)$ has rank n(e.d.)".

Proof: First we assume $\operatorname{rank}\mathbf{Q} = n$ in an interval containing time η such that $t_0 < \eta < t_1$ and show the system is totally controllable.

Construct $\mathbf{u}(\tau)$ as

$$\mathbf{u}(\tau) = \sum_{k=1}^{n} \mu_k \delta^{(k-1)}(\tau - \eta) \tag{6.9}$$

To attain an arbitrary $\mathbf{x}(t_1)$ starting from $\mathbf{x}(t_0) = \mathbf{0}$ we must have

$$\mathbf{x}(t_1) = \int_{t_0}^{t_1} \mathbf{\Phi}(t_1, \tau)\, \mathbf{B}(\tau)\, \mathbf{u}(\tau)\, d\tau = \sum_{k=1}^{n} \frac{d^{k-1}}{d\tau^{k-1}} [\mathbf{\Phi}(t_1, \tau)\, \mathbf{B}(\tau)] \bigg|_{\tau=\eta} \mu_k \tag{6.10}$$

But

$$\frac{d}{d\tau}[\mathbf{\Phi}(t_1, \tau)\, \mathbf{B}(\tau)] = \mathbf{\Phi}(t_1, \tau)\frac{d\mathbf{B}}{d\tau} + \left[\frac{d}{d\tau}\mathbf{\Phi}^{-1}(\tau, t_1)\right]\mathbf{B}(\tau) \tag{6.11}$$

Note $\mathbf{0} = d\mathbf{I}/dt = d(\mathbf{\Phi\Phi}^{-1})/dt = \mathbf{A\Phi\Phi}^{-1} + \mathbf{\Phi} d\mathbf{\Phi}^{-1}/dt$ so that $d\mathbf{\Phi}^{-1}/dt = -\mathbf{\Phi}^{-1}\mathbf{A}$ and equation (*6.11*) becomes

$$\frac{d}{d\tau}[\mathbf{\Phi}(t_1, \tau)\, \mathbf{B}(\tau)] = \mathbf{\Phi}(t_1, \tau)\, \mathbf{Q}_2(\tau)$$

Similarly,

$$\frac{d^{k-1}}{d\tau^{k-1}}[\mathbf{\Phi}(t_1, \tau)\, \mathbf{B}(\tau)] = \mathbf{\Phi}(t_1, \tau)\, \mathbf{Q}_k(\tau) \tag{6.12}$$

Therefore (*6.10*) becomes

$$\mathbf{x}(t_1) = \mathbf{\Phi}(t_1, \eta)\, \mathbf{Q}(\eta) \begin{pmatrix} \mu_1 \\ \hline \vdots \\ \hline \mu_n \end{pmatrix}$$

A solution always exists for the μ_k because $\operatorname{rank}\mathbf{Q} = n$ and $\mathbf{\Phi}$ is nonsingular.

Now assume the system is totally controllable and show $\operatorname{rank}\mathbf{Q} = n$. From Problem 6.8, page 142, there exists no constant n-vector $\mathbf{z} \neq \mathbf{0}$ such that, for times t everywhere dense in $t_0 \leq t \leq t_1$,

$$\mathbf{z}^\dagger \mathbf{\Phi}(t_0, t)\, \mathbf{B}(t) = \mathbf{0}$$

By differentiating k times with respect to t and using equation (*6.12*), this becomes $\mathbf{z}^\dagger \mathbf{\Phi}(t_0, t)\, \mathbf{Q}_k(t) = \mathbf{0}$. Since $\mathbf{\Phi}(t_0, t)$ is always nonsingular, there is no n-vector $\mathbf{y}^\dagger = \mathbf{z}^\dagger \mathbf{\Phi}(t_0, t) \neq \mathbf{0}$ such that

$$\mathbf{0} = (\mathbf{y}^\dagger \mathbf{Q}_1 \mid \mathbf{y}^\dagger \mathbf{Q}_2 \mid \ldots \mid \mathbf{y}^\dagger \mathbf{Q}_n) = \mathbf{y}^\dagger \mathbf{Q} = y_1 \mathbf{q}_1 + y_2 \mathbf{q}_2 + \cdots + y_n \mathbf{q}_n$$

where $\mathbf{q}_i$ are the row vectors of $\mathbf{Q}$. Since the n row vectors are then linearly independent, the rank of $\mathbf{Q}(t)$ is n(e.d.).

Theorem 6.11: The time-varying system $d\mathbf{x}/dt = \mathbf{A}(t)x + \mathbf{B}(t)\mathbf{u}$, $\mathbf{y} = \mathbf{C}(t)\mathbf{x} + \mathbf{D}(t)\mathbf{u}$ is totally observable if and only if the matrix $\mathbf{P}(t)$ has rank n(e.d.), where $\mathbf{P}^T(t) = (\mathbf{P}_1^T \mid \mathbf{P}_2^T \mid \ldots \mid \mathbf{P}_n^T)$ in which $\mathbf{P}_1 = \mathbf{C}(t)$ and $\mathbf{P}_{k+1} = \mathbf{P}_k\mathbf{A}(t) + d\mathbf{P}_k/dt$ for $k = 1, 2, \ldots, n-1$. Again $\mathbf{A}(t)$ and $\mathbf{B}(t)$ are assumed piecewise differentiable at least $n-2$ and $n-1$ times, respectively.

Again, the proof is somewhat similar to Theorem 6.10 and will not be given. The situation is somewhat different for the discrete-time case, because generalizing the proof following Theorem 6.9 leads to the criterion $\operatorname{rank} \mathbf{P} = n$, where for this case

$$\mathbf{P}_1 = \mathbf{C}(l), \quad \mathbf{P}_2 = \mathbf{C}(l+1)\,\mathbf{A}(l), \quad \ldots, \quad \mathbf{P}_k = \mathbf{C}(l+k-1)\,\mathbf{A}(l+k-2)\cdots\mathbf{A}(l)$$

The situation changes somewhat when only complete controllability is required. Since any system that is totally controllable must be completely controllable, if rank $\mathbf{Q}(t)$ has rank n for some $t > t_0$ [not rank n(e.d)] then $\mathbf{x}(t_0)$ can be transferred to $\mathbf{x}(t_1)$ for $t_1 \geqq t$. On the other hand, systems for which $\operatorname{rank} \mathbf{Q}(t) < n$ for all t might be completely controllable (but not totally controllable).

Example 6.4.

Given the system

$$\frac{d}{dt}\begin{pmatrix} x_1 \\ x_2 \end{pmatrix} = -\begin{pmatrix} \alpha & 0 \\ 0 & \beta \end{pmatrix}\begin{pmatrix} x_1 \\ x_2 \end{pmatrix} + \begin{pmatrix} f_1(t) \\ f_2(t) \end{pmatrix} u$$

where

$$f_1(t) = \begin{cases} \sin t & 2k\pi \leqq t < (2k+1)\pi \\ 0 & (2k+1)\pi \leqq t < 2(k+1)\pi \end{cases} \qquad k = 0, \pm 1, \pm 2, \ldots$$

and $f_2(t) = f_1(t+\pi)$. Then

$$\mathbf{Q}(t) = \begin{pmatrix} f_1(t) & \alpha f_1(t) + df_1/dt \\ f_2(t) & \beta f_2(t) + df_2/dt \end{pmatrix}$$

At each instant of time, one row of $\mathbf{Q}(t)$ is zero, so $\operatorname{rank} \mathbf{Q}(t) = 1$ for all t. However,

$$\mathbf{\Phi}(t_0, t)\,\mathbf{B}(t) = \begin{pmatrix} e^{\alpha(t-t_0)} & 0 \\ 0 & e^{\beta(t-t_0)} \end{pmatrix}\begin{pmatrix} f_1(t) \\ f_2(t) \end{pmatrix} = \begin{pmatrix} f_1(t)e^{\alpha(t-t_0)} \\ f_2(t)e^{\beta(t-t_0)} \end{pmatrix}$$

If $t - t_0 > \pi$, the rows of $\mathbf{\Phi}(t_0, t)\,\mathbf{B}(t)$ are linearly independent time functions, and from Problem 6.30, page 145, the system is completely controllable for $t_1 - t_0 > \pi$. The system is not totally controllable because for every t_0, if $t_2 - t_0 < \pi$, either $f_1(\tau)$ or $f_2(\tau)$ is zero for $t_0 \leqq \tau \leqq t_2$.

However, for systems with analytic $\mathbf{A}(t)$ and $\mathbf{B}(t)$, it can be shown that complete controllability implies total controllability. Therefore $\operatorname{rank} \mathbf{Q} = n$ is necessary and sufficient for complete controllability also. (Note $f_1(t)$ and $f_2(t)$ are not analytic in Example 6.4.) For complete controllability in a nonanalytic system with $\operatorname{rank} \mathbf{Q}(t) < n$, the rank of $\mathbf{\Phi}(t, \tau)\,\mathbf{B}(\tau)$ must be found.

6.6 DUALITY

In this chapter we have repeatedly used the same kind of proof for observability as was used for controllability. Kalman first remarked on this duality between observing a dynamic system and controlling it. He notes the determinant of the **W** matrix of Problem 6.8, page 142, is analogous to Shannon's definition of information. The dual of the optimal control problem of Chapter 10 is the Kalman filter. This duality is manifested by the following two systems:

System #1:

$$\begin{aligned} d\mathbf{x}/dt &= \mathbf{A}(t)\mathbf{x} + \mathbf{B}(t)\mathbf{u} \\ \mathbf{y} &= \mathbf{C}(t)\mathbf{x} + \mathbf{D}(t)\mathbf{u} \end{aligned}$$

System #2:

$$\begin{aligned} d\mathbf{w}/dt &= -\mathbf{A}^\dagger(t)\mathbf{w} + \mathbf{C}^\dagger(t)\mathbf{v} \\ \mathbf{z} &= \mathbf{B}^\dagger(t)\mathbf{w} + \mathbf{D}^\dagger(t)\mathbf{v} \end{aligned}$$

Then system #1 is totally controllable (observable) if and only if system #2 is totally observable (controllable), which can be shown immediately from Theorems 6.10 and 6.11.

Solved Problems

6.1. Given the system

$$\frac{d\mathbf{x}}{dt} = \begin{pmatrix} 1 & 2 & -1 \\ 0 & 1 & 0 \\ 1 & -4 & 3 \end{pmatrix}\mathbf{x} + \begin{pmatrix} 0 \\ 0 \\ 1 \end{pmatrix}u, \qquad y = (1\ \ -1\ \ 1)\mathbf{x}$$

find the controllable and uncontrollable states and then find the observable and unobservable states.

Following the standard procedure for transforming a matrix to Jordan form gives $\mathbf{A} = \mathbf{TJT}^{-1}$ as

$$\begin{pmatrix} 1 & 2 & -1 \\ 0 & 1 & 0 \\ 1 & -4 & 3 \end{pmatrix} = \begin{pmatrix} -1 & 0 & 0 \\ 0 & 0 & 1 \\ 1 & 1 & 2 \end{pmatrix}\begin{pmatrix} 2 & 1 & 0 \\ 0 & 2 & 0 \\ 0 & 0 & 1 \end{pmatrix}\begin{pmatrix} -1 & 0 & 0 \\ 1 & -2 & 1 \\ 0 & 1 & 0 \end{pmatrix}$$

Then $\mathbf{f} = \mathbf{T}^{-1}\mathbf{b} = (0\ 1\ 0)^T$ and $\mathbf{g}^T = \mathbf{c}^T\mathbf{T} = (0\ 1\ 1)$. The flow diagram of the Jordan system is shown in Fig. 6-8.

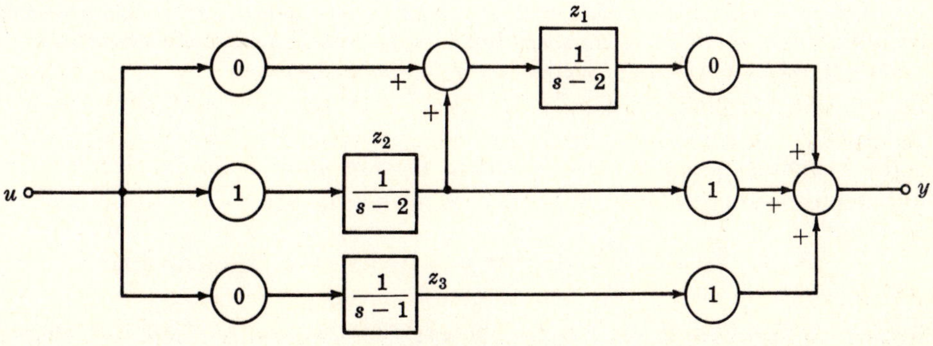

Fig. 6-8

The element z_1 of the state vector $\mathbf{z}$ is controllable (through z_2) but unobservable. The element z_2 is both controllable and observable. The element z_3 is uncontrollable but observable.

Note Theorem 6.1 is inapplicable.

6.2. Find the elements of the state vector $\mathbf{z}$ that are both controllable and observable for the system of Problem 6.1.

Taking the Laplace transform of the system with zero initial conditions gives the transfer function. Using the same symbols for original and transformed variables, we have

$$sx_1 = x_1 + 2x_2 - x_3 \tag{6.13}$$

$$sx_2 = x_2 \tag{6.14}$$

$$sx_3 = x_1 - 4x_2 + 3x_3 + u \tag{6.15}$$

$$y = x_1 - x_2 + x_3 \tag{6.16}$$

From (6.15), $x_1 = 4x_2 + (s-3)x_3 - u$. Putting this in (6.13) gives $(4s-6)x_2 = (s-1)u - (s-2)^2 x_3$. Substituting this in (6.14), $(s-1)(s-2)^2 x_3 = (s-1)^2 u$. Then from (6.16),

$$y = \left[\frac{(s-1)^2}{(s-2)^3} - 0 + \frac{-1}{(s-2)^2}\right]u = \frac{(s-1)(s-2)}{(s-1)(s-2)^2}u$$

Thus the transfer function $h(s) = (s-2)^{-1}$, and from Theorem 6.7 the only observable and controllable element of the state vector is z_2 as defined in Problem 6.1.

6.3. Given the system of Problem 6.1. Is it totally observable and totally controllable?

Forming the $\mathbf{P}$ and $\mathbf{Q}$ matrices of Theorems 6.9 and 6.8 gives

$$\mathbf{P} = \begin{pmatrix} 1 & -1 & 1 \\ \hline 2 & -3 & 2 \\ \hline 4 & -7 & 4 \end{pmatrix} \qquad \mathbf{Q} = \left(\begin{array}{c|c|c} 0 & -1 & -4 \\ 0 & 0 & 0 \\ 1 & 3 & 8 \end{array}\right)$$

Then $\operatorname{rank}\mathbf{P} = 2$, the dimension of the observable state space; and $\operatorname{rank}\mathbf{Q} = 2$, the dimension of the controllable state space. Hence the system is neither controllable nor observable.

6.4. Given the time-invariant system

$$\frac{d}{dt}\begin{pmatrix} x_1 \\ x_2 \end{pmatrix} = \begin{pmatrix} 0 & \alpha \\ 0 & -1 \end{pmatrix}\begin{pmatrix} x_1 \\ x_2 \end{pmatrix} + \begin{pmatrix} 0 \\ 1 \end{pmatrix}u, \qquad y = (1 \;\; 0)\begin{pmatrix} x_1 \\ x_2 \end{pmatrix}$$

and that $u(t) = e^{-t}$ and $y(t) = 2 - \alpha te^{-t}$. Find $x_1(t)$ and $x_2(t)$. Find $x_1(0)$ and $x_2(0)$. What happens when $\alpha = 0$?

Since $y = x_1$, then $x_1(t) = 2 - \alpha te^{-t}$. Also, $dx_1/dt = \alpha x_2$, so differentiating the output gives $x_2(t) = -e^{-t} + te^{-t}$. Then $x_1(0) = 2$ and $x_2(0) = -1$. When $\alpha = 0$, this procedure does not work because $dx_1/dt = 0$. There is no way to find $x_2(t)$, because x_2 is unobservable as can be verified from Theorem 6.5. (For $\alpha = 0$, the system is in Jordan form.)

6.5. The normalized equations of vertical motion $y(r, \theta, t)$ for a circular drumhead being struck by a force $u(t)$ at a point $r = r_0$, $\theta = \theta_0$ are

$$\frac{\partial^2 y}{\partial t^2} = \nabla^2 y + 2\pi r\,\delta(r - r_0)\,\delta(\theta - \theta_0)\,u(t) \tag{6.17}$$

where $y(r_1, \theta, t) = 0$ at $r = r_1$, the edge of the drum. Can this force excite all the modes of vibration of the drum?

The solution for the mode shapes is

$$y(r, \theta, t) = \sum_{m=1}^{\infty}\sum_{n=0}^{\infty} J_n(\kappa_m r/r_1)[x_{2n,m}(t)\cos 2n\pi\theta + x_{2n+1,m}(t)\sin 2n\pi\theta]$$

where κ_m is the mth zero of the nth order Bessel function $J_n(r)$. Substituting the motion of the first harmonic $m = 1$, $n = 1$ into equation (6.17) gives

$$d^2x_{21}/dt^2 = \lambda x_{21} + \gamma \cos 2\pi\theta_0 u$$

$$d^2x_{31}/dt^2 = \lambda x_{31} + \gamma \sin 2\pi\theta_0 u$$

where $\lambda = \kappa_m^2 + (2\pi)^2$ and $\gamma = r_0 J_1(\kappa_1 r_0/r_1)$. Using the controllability criteria, it can be found that one particular state that is not influenced by $u(t)$ is the first harmonic rotated so that its node line is at angle θ_0. This is illustrated in Fig. 6-9. A noncolinear point of application of another force is needed to excite, or damp out, this particular uncontrollable mode.

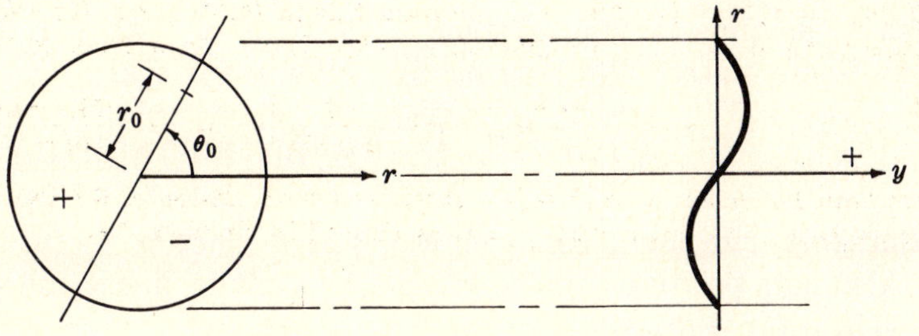

Fig. 6-9

6.6. Consider the system

$$\frac{d\mathbf{x}}{dt} = \begin{pmatrix} 1 & 1 & 0 \\ 0 & 1 & 0 \\ 0 & 0 & 1 \end{pmatrix}\mathbf{x} + \begin{pmatrix} 0 \\ 1 \\ 1 \end{pmatrix} u_1 + \mathbf{b}u_2$$

where u_1 and u_2 are two separate scalar controls. Determine whether the system is totally controllable if

(1) $\mathbf{b} = (0\ 0\ 0)^T$

(2) $\mathbf{b} = (0\ 0\ 1)^T$

(3) $\mathbf{b} = (1\ 0\ 0)^T$

For each case, we investigate the controllability matrix $\mathbf{Q} = (\mathbf{B} \mid \mathbf{AB} \mid \mathbf{A}^2\mathbf{B})$ for

$$\mathbf{A} = \begin{pmatrix} 1 & 1 & 0 \\ 0 & 1 & 0 \\ 0 & 0 & 1 \end{pmatrix} \quad \text{and} \quad \mathbf{B} = \left(\begin{array}{c|c} 0 & \\ 1 & \mathbf{b} \\ 1 & \end{array}\right)$$

For $\mathbf{b} = \mathbf{0}$ it is equivalent to scalar control, and by condition (1) of Theorem 6.2 the system is uncontrollable. For case (2),

$$\mathbf{Q} = \left(\begin{array}{cc|cc|cc} 0 & 0 & 1 & 0 & 2 & 0 \\ 1 & 0 & 1 & 0 & 1 & 0 \\ 1 & 1 & 1 & 1 & 1 & 1 \end{array}\right)$$

The first three columns are linearly independent, so $\mathbf{Q}$ has rank 3 and the system is controllable. For case (3),

$$\mathbf{Q} = \begin{pmatrix} 0 & 1 & 1 & 1 & 2 & 1 \\ 1 & 0 & 1 & 0 & 1 & 0 \\ 1 & 0 & 1 & 0 & 1 & 0 \end{pmatrix}$$

The bottom two rows are identical and so the system is uncontrollable.

6.7. Investigate total controllability of the time-varying system

$$\frac{d\mathbf{x}}{dt} = \begin{pmatrix} t & 0 \\ 1 & 0 \end{pmatrix} x + \begin{pmatrix} e^t \\ 2 \end{pmatrix} u$$

The $\mathbf{Q}(t)$ matrix of Theorem 6.10 is

$$\mathbf{Q}(t) = \begin{pmatrix} e^t & e^t(1-t) \\ 2 & -e^t \end{pmatrix}$$

Then $\det \mathbf{Q}(t) = e^t(e^t + 2t - 2)$. Since $e^t + 2t = 2$ only at one instant of time, $\operatorname{rank} \mathbf{Q}(t) = 2$(e.d.).

6.8. Show that the time-varying system $d\mathbf{x}/dt = \mathbf{A}(t)\mathbf{x} + \mathbf{B}(t)\mathbf{u}$ is totally controllable if and only if the matrix $\mathbf{W}(t, \tau)$ is postive definite for every τ and every $t > \tau$, where

$$\mathbf{W}(t, \tau) = \int_\tau^t \mathbf{\Phi}(\tau, \eta)\, \mathbf{B}(\eta)\, \mathbf{B}^\dagger(\eta)\, \mathbf{\Phi}^\dagger(\tau, \eta)\, d\eta$$

Note that this criterion depends on $\mathbf{\Phi}(t, \tau)$ and is not as useful as Theorems 6.10 and 6.11. Also note positive definite $\mathbf{W}$ is equivalent to linear independence of the rows of $\mathbf{\Phi}(\tau, \eta)\, \boldsymbol{B}(\eta)$ for $\tau \leq \eta \leq t$.

If $\mathbf{W}(t, \tau)$ is positive definite, $\mathbf{W}^{-1}$ exists. Then choose

$$\mathbf{u}(\tau) = -\mathbf{B}^\dagger(\tau)\, \mathbf{\Phi}^\dagger(t_1, \tau)\, \mathbf{W}^{-1}(t_0, t_1)\, \mathbf{x}(t_1)$$

Substitution will verify that

$$\mathbf{x}(t_1) = \int_{t_0}^{t_1} \mathbf{\Phi}(t_1, \tau)\, \mathbf{B}(\tau)\, \mathbf{u}(\tau)\, d\tau$$

so that the system is totally controllable if $\mathbf{W}(t, \tau)$ is positive definite. Now suppose the system is totally controllable and show $\mathbf{W}(t, \tau)$ is positive definite. First note for any constant vector $\mathbf{k}$,

$$(\mathbf{k}, \mathbf{W}\mathbf{k}) = \int_\tau^t \mathbf{k}^\dagger \mathbf{\Phi}(\tau, \eta)\, \mathbf{B}(\eta)\, \mathbf{B}^\dagger(\eta)\, \mathbf{\Phi}^\dagger(\tau, \eta)\mathbf{k}\, d\eta$$

$$= \int_\tau^t \|\mathbf{B}^\dagger(\eta)\, \mathbf{\Phi}^\dagger(\tau, \eta)\mathbf{k}\|_2^2\, d\eta \geq 0$$

Therefore the problem is to show $\mathbf{W}$ is nonsingular if the system is totally controllable. Suppose $\mathbf{W}$ is singular, to obtain a contradiction. Then there exists a constant n-vector $\mathbf{z} \neq \mathbf{0}$ such that $(\mathbf{z}, \mathbf{W}\mathbf{z}) = 0$. Define a continuous, m-vector function of time $\mathbf{f}(t) = -\mathbf{B}^\dagger(t)\, \mathbf{\Phi}^\dagger(t_0, t)\mathbf{z}$. But

$$\int_{t_0}^{t_1} \|\mathbf{f}(\tau)\|_2^2\, d\tau = \int_{t_0}^{t_1} \mathbf{z}^\dagger \mathbf{\Phi}(t_0, t)\, \mathbf{B}(t)\, \mathbf{B}^\dagger(t)\, \mathbf{\Phi}^\dagger(t_0, t)\mathbf{z}\, dt$$

$$= (\mathbf{z}, \mathbf{W}(t_1, t_0)\mathbf{z}) = 0$$

Hence $\mathbf{f}(t) = \mathbf{0}$ for all t, so that $0 = \int_{t_0}^{t_1} \mathbf{f}^\dagger(t)\, \mathbf{u}(t)\, dt$ for any $\mathbf{u}(t)$. Substituting for $\mathbf{f}(t)$ gives

$$0 = -\int_{t_0}^{t_1} \mathbf{z}^\dagger \mathbf{\Phi}(t_0, t)\, \mathbf{B}(t)\, \mathbf{u}(t)\, dt \tag{6.18}$$

In particular, since the system is assumed totally controllable, take $\mathbf{u}(t)$ to be the control that transfers $\mathbf{0}$ to $\mathbf{\Phi}(t_1, t_0)\mathbf{z} \neq \mathbf{0}$. Then

$$\mathbf{z} = \int_{t_0}^{t_1} \mathbf{\Phi}(t_0, t)\, \mathbf{B}(t)\, \mathbf{u}(t)\, dt$$

Substituting this into equation (6.18) gives $0 = \mathbf{z}^\dagger\mathbf{z}$ which is impossible for any nonzero $\mathbf{z}$. Therefore no nonzero $\mathbf{z}$ exists for which $(\mathbf{z}, \mathbf{W}\mathbf{z}) = 0$, so that $\mathbf{W}$ must be positive definite.

Supplementary Problems

6.9. Consider the bilinear scalar system $d\xi/dt = u(t)\,\xi(t)$. It is linear in the initial state and in the control, but not both, so that it is not a linear system and the theorems of this chapter do not apply. The flow diagram is shown in Fig. 6-10. Is this system completely controllable according to Definition 6.1?

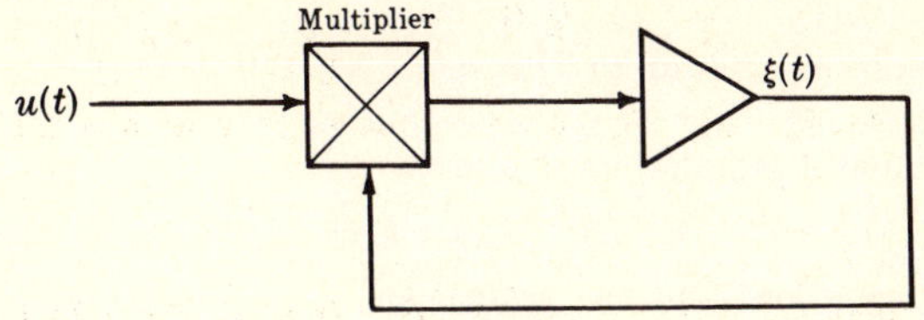

Fig. 6-10

6.10. Given the system

$$\frac{d\mathbf{x}}{dt} = \begin{pmatrix} 0 & 1 & 0 \\ 5 & 0 & 2 \\ -2 & 0 & -2 \end{pmatrix}\mathbf{x} + \begin{pmatrix} -1 \\ 1 \\ -1 \end{pmatrix}u, \qquad y = (-2 \;\; 1 \;\; 0)\mathbf{x}$$

Determine which states are observable and which are controllable, and check your work by deriving the transfer function.

6.11. Given the system

$$\frac{d\mathbf{x}}{dt} = -\frac{1}{4}\begin{pmatrix} 3 & 1 \\ 1 & 3 \end{pmatrix}\mathbf{x} + \begin{pmatrix} 1 \\ 1 \end{pmatrix}u, \qquad y = (1 \;\; 1)\mathbf{x}$$

Classify the states according to their observability and controllability, compute the **P** and **Q** matrices, and find the transfer function.

6.12. Six identical frictionless gears with inertia I are mounted on shafts as shown in Fig. 6-11, with a center crossbar keeping the outer two pairs diametrically opposite each other. A torque $u(t)$ is the input and a torque $y(t)$ is the output. Using the angular position of the two outer gearshafts as two of the elements in a state vector, show that the system is state uncontrollable but totally output controllable.

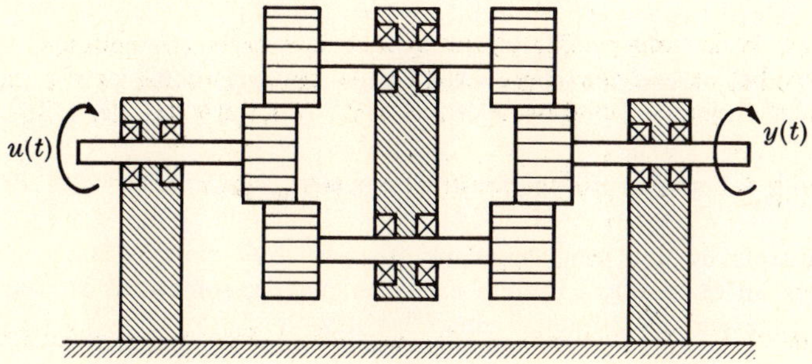

Fig. 6-11

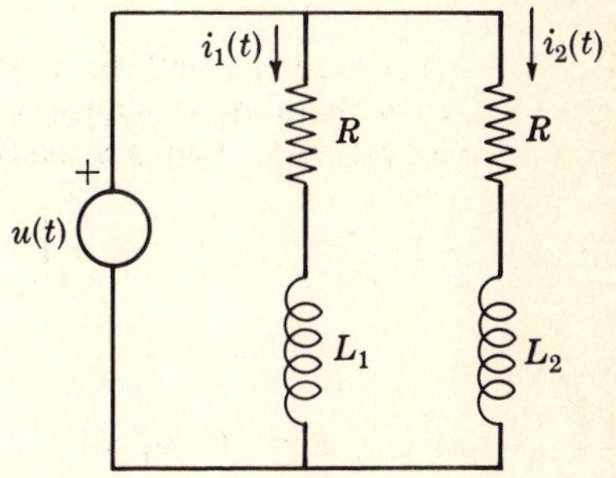

Fig. 6-12

6.13. Given the electronic circuit of Fig. 6-12 where $u(t)$ can be any voltage (function of time). Under what conditions on R, L_1 and L_2 can both $i_1(t_1)$ and $i_2(t_1)$ be arbitrarily prescribed for $t_1 > t_0$, given that $i_1(t_0)$ and $i_2(t_0)$ can be any numbers?

6.14. Consider the simplified model of a rocket vehicle

$$\frac{d}{dt}\begin{pmatrix} \theta \\ \dot\theta \\ z \end{pmatrix} = \begin{pmatrix} 0 & 1 & 0 \\ M_\alpha & 0 & -M_\alpha/v \\ Z_\alpha & 0 & -Z_\alpha/v \end{pmatrix}\begin{pmatrix} \theta \\ \dot\theta \\ z \end{pmatrix} + \begin{pmatrix} 0 \\ M_\delta \\ 0 \end{pmatrix}\delta$$

Under what conditions is the vehicle state controllable?

6.15. Find some other construction than equation (*6.2*) that will transfer a zero initial condition to an arbitrary $\mathbf{z}(t_1)$.

6.16. Prove Theorem 6.5.

6.17. Prove Theorem 6.6.

6.18. What are the conditions similar to Theorem 6.2 for which a two-input system is totally controllable?

6.19. Given the controllable sampled data system

$$\xi(n+2) + 3\xi(n+1) + 2\xi(n) = u(n+1) - u(n)$$

Write the state equation, find the transition matrix in closed form, and find the control that will force an arbitrary initial state to zero in the smallest number of steps. (This control depends upon these arbitrary initial conditions.)

6.20. Given the system with nondistinct eigenvalues

$$\frac{d\mathbf{x}}{dt} = \begin{pmatrix} -1 & 1 & -1 \\ 0 & 1 & -4 \\ 0 & 1 & -3 \end{pmatrix}\mathbf{x} + \begin{pmatrix} 0 \\ 2 \\ 1 \end{pmatrix}u, \qquad y = (0 \;\; 1 \;\; -1)\mathbf{x}$$

Classify the elements of the state vector $\mathbf{z}$ corresponding to the Jordan form into observable/not observable, controllable/not controllable.

6.21. Using the criterion $\mathbf{Q} = (\mathbf{b} \mid \mathbf{Ab} \mid \ldots \mid \mathbf{A}^{n-1}\mathbf{b})$, develop the result of Theorem 6.1.

6.22. Consider the discrete system $\quad \mathbf{x}(k+1) = \mathbf{A}\mathbf{x}(k) + \mathbf{b}u(k)$

where $\mathbf{x}$ is a 2-vector, u is a scalar, $\quad \mathbf{A} = \begin{pmatrix} 2 & 1 \\ 0 & 1 \end{pmatrix}, \quad \mathbf{b} = \begin{pmatrix} 1 \\ 1 \end{pmatrix}.$

(*a*) Is the system controllable? (*b*) If the initial condition is $\mathbf{x}(0) = \begin{pmatrix} 1 \\ 2 \end{pmatrix}$, find the control sequence $u(0), u(1)$ required to drive the state to the origin in two sample periods (i.e., $\mathbf{x}(2) = \mathbf{0}$).

6.23. Consider the discrete system $\quad \mathbf{x}(k+1) = \mathbf{A}\mathbf{x}(k), \quad y(k) = \mathbf{c}^\dagger\mathbf{x}(k)$

where $\mathbf{x}$ is a 2-vector, y is a scalar, $\quad \mathbf{A} = \begin{pmatrix} 2 & 1 \\ 0 & 1 \end{pmatrix}, \quad \mathbf{c}^\dagger = (1 \;\; 2).$

(*a*) Is the system observable? (*b*) Given the observation sequence $y(1) = 8$, $y(2) = 14$, find the initial state $\mathbf{x}(0)$.

6.24. Prove Theorem 6.8, constructing a bounded $\mathbf{u}(\tau)$ instead of equation (*6.7*). *Hint*: See Problem 6.8.

6.25. Given the multiple input–multiple output time-invariant system $d\mathbf{x}/dt = \mathbf{Ax} + \mathbf{Bu}$, $\mathbf{y} = \mathbf{Cx} + \mathbf{Du}$, where $\mathbf{y}$ is a k-vector and $\mathbf{u}$ is an m-vector. Find a criterion matrix somewhat similar to the $\mathbf{Q}$ matrix of Theorem 6.8 that assures complete output controllability.

6.26. Consider three linear time-invariant systems of the form

$$S_i: \quad d\mathbf{x}^{(i)}/dt = \mathbf{A}^{(i)}\mathbf{x}^{(i)} + \mathbf{B}^{(i)}\mathbf{u}^{(i)}, \qquad \mathbf{y}^{(i)} = \mathbf{C}^{(i)}\mathbf{x}^{(i)} \qquad i = 1, 2, 3$$

(*a*) Derive the transfer function matrix for the interconnected system of Fig. 6-13 in terms of $\mathbf{A}^{(i)}, \mathbf{B}^{(i)}$ and $\mathbf{C}^{(i)}$, $i = 1, 2, 3$.

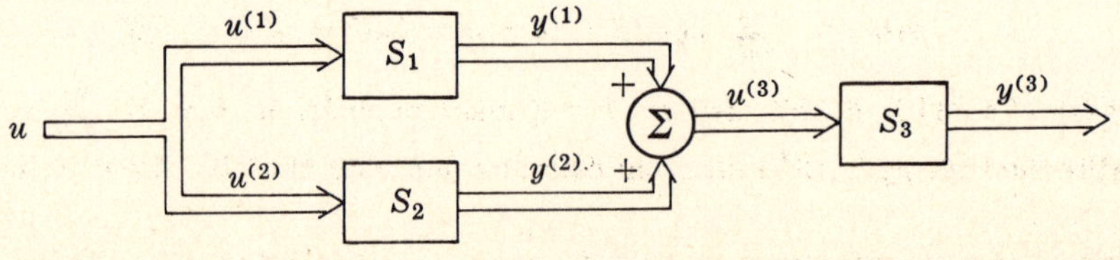

Fig. 6-13

(*b*) If the overall interconnected system in part (*a*) is observable, show that S_3 is observable. (Note that $\mathbf{u}^{(i)}$ and $\mathbf{y}^{(i)}$ are vectors.)

6.27. Given the time-varying system

$$\frac{d\mathbf{x}}{dt} = \begin{pmatrix} 1 & 0 \\ 0 & 2 \end{pmatrix}\mathbf{x} + \begin{pmatrix} t \\ 1 \end{pmatrix}u, \qquad y = (e^{-t} \;\; e^{-2t})\mathbf{x}$$

Is this system totally controllable and observable?

6.28. Prove Theorem 6.9 for the continuous-time case.

6.29. Prove a controllability theorem similar to Theorem 6.10 for the discrete time-varying case.

6.30. Similar to Problem 6.8, show that the time-varying system $d\mathbf{x}/dt = \mathbf{A}(t)\mathbf{x} + \mathbf{B}(t)\mathbf{u}$ is completely controllable for $t_1 > t$ if and only if the matrix $\mathbf{W}(t, \tau)$ is positive definite for every τ and some finite $t > \tau$. Also show this is equivalent to linear independence of the rows of $\mathbf{\Phi}(\tau, \eta)\,\mathbf{B}(\eta)$ for some finite $\eta > \tau$.

6.31. Prove that the linear time-varying system $d\mathbf{x}/dt = \mathbf{A}(t)\mathbf{x}$, $\mathbf{y} = \mathbf{C}(t)\mathbf{x}$ is totally observable if and only if $\mathbf{M}(t_1, t_0)$ is positive definite for all $t_1 > t_0$, where

$$\mathbf{M}(t_1, t_0) = \int_{t_0}^{t_1} \mathbf{\Phi}^\dagger(t, t_0)\,\mathbf{C}^\dagger(t)\,\mathbf{C}(t)\,\mathbf{\Phi}(t, t_0)\,dt$$

Answers to Supplementary Problems

6.9. No nonzero $\boldsymbol{\xi}(t_1)$ can be reached from $\boldsymbol{\xi}(t_0) = \mathbf{0}$, so the system is uncontrollable.

6.10. The states belonging to the eigenvalue 2 are unobservable and those belonging to -1 are uncontrollable. The transfer function is $3(s+3)^{-1}$, showing only the states belonging to -3 are both controllable and observable.

6.11. One state is observable and controllable, the other is neither observable nor controllable.

$$\mathbf{P} = \begin{pmatrix} 1 & 1 \\ -1 & -1 \end{pmatrix}, \qquad \mathbf{Q} = \begin{pmatrix} 1 & -1 \\ 1 & -1 \end{pmatrix} \qquad \text{and} \qquad h(s) = \frac{2}{s+1}$$

6.13. $L_1 \neq L_2$

6.14. $M_\delta \neq 0$ and $Z_\alpha \neq 0$

6.15. Many choices are possible, such as

$$u(t) = \sum_{k=1}^{n} \mu_k\{U[t - t_0 - (t_1 - t_0)(k-1)/n] - U[t - t_0 - (t_1 - t_0)k/n]\}$$

where $U(t - \tau)$ is a unit step at $t = \tau$; another choice is $u(t) = \sum_{k=1}^{n} \mu_k e^{-\lambda_k^* t}$. In both cases the expression for μ_k is different from equation (*6.4*) and must be shown to have an inverse.

6.18. Two Jordan blocks with the same eigenvalue can be controlled if $f_{11}f_{22} - f_{12}f_{21} \neq 0$, the f's being the coefficients of u_1 and u_2 in the last row of the Jordan blocks.

6.19. $$\mathbf{x}(n+1) = \begin{pmatrix} -3 & 1 \\ -2 & 0 \end{pmatrix}\mathbf{x}(n) + \begin{pmatrix} 1 \\ -1 \end{pmatrix}u(n); \quad y(n) = (1 \;\; 0)\,\mathbf{x}(n)$$

$$\mathbf{\Phi}(n,k) = (-1)^{n-k}\begin{pmatrix} -1 & 1 \\ -2 & 2 \end{pmatrix} + (-2)^{n-k}\begin{pmatrix} 2 & -1 \\ 2 & -1 \end{pmatrix}$$

$$\begin{pmatrix} u(0) \\ u(1) \end{pmatrix} = \frac{1}{6}\begin{pmatrix} 13 & -5 \\ 10 & -2 \end{pmatrix}\begin{pmatrix} x_1(0) \\ x_2(0) \end{pmatrix}$$

6.20. The flow diagram is shown in Fig. 6-14 where z_1 and z_2 are controllable, z_2 and z_3 are observable.

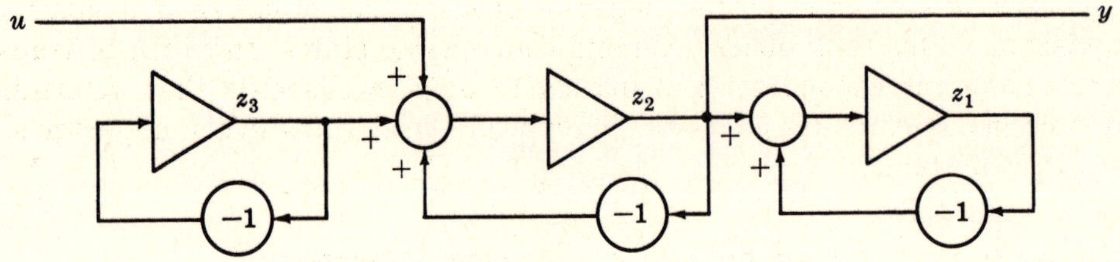

Fig. 6-14

for $\xi = 1$ and $\rho = 0$ in $\mathbf{T} = \begin{pmatrix} \xi & \rho & \kappa \\ 0 & 2\xi & 2\rho - \xi \\ 0 & \xi & \rho - \xi \end{pmatrix}$.

6.22. Yes; $u(0) = -4$, $u(1) = 2$.

6.23. Yes; $\mathbf{x}(0) = \begin{pmatrix} 1 \\ 2 \end{pmatrix}$.

6.25. rank $\mathbf{R} = k$, where $\mathbf{R} = (\mathbf{CB} \mid \mathbf{CAB} \mid \ldots \mid \mathbf{CA}^{n-1}\mathbf{B} \mid \mathbf{D})$

6.26. $\mathbf{H}(s) = \mathbf{C}^{(3)}(\mathbf{I}s - \mathbf{A}^{(3)})^{-1}\mathbf{B}^{(3)}[\mathbf{C}^{(1)}(\mathbf{I}s - \mathbf{A}^{(1)})^{-1}\mathbf{B}^{(1)} + \mathbf{C}^{(2)}(\mathbf{I}s - \mathbf{A}^{(2)})^{-1}\mathbf{B}^{(2)}]$

6.27. It is controllable but not observable.

Chapter 7

Canonical Forms of the State Equation

7.1 INTRODUCTION TO CANONICAL FORMS

The general state equation $d\mathbf{x}/dt = \mathbf{A}(t)\mathbf{x} + \mathbf{B}(t)\mathbf{u}$ appears to have all n^2 elements of the $\mathbf{A}(t)$ matrix determine the time behavior of $\mathbf{x}(t)$. The object of this chapter is to reduce the number of states to m observable and controllable states, and then to transform the m^2 elements of the corresponding $\mathbf{A}(t)$ matrix to only m elements that determine the input-output time behavior of the system. First we look at time-invariant systems, and then at time-varying systems.

7.2 JORDAN FORM FOR TIME-INVARIANT SYSTEMS

Section 2.4 showed how equation (*2.21*) in Jordan form can be found from the transfer function of a time-invariant system. For single-input systems, to go directly from the form $d\mathbf{x}/dt = \mathbf{A}\mathbf{x} + \mathbf{b}u$, let $\mathbf{x} = \mathbf{T}\mathbf{z}$ so that $d\mathbf{z}/dt = \mathbf{J}\mathbf{z} + \mathbf{T}^{-1}\mathbf{b}u$ where $\mathbf{T}^{-1}\mathbf{A}\mathbf{T} = \mathbf{J}$. The matrix $\mathbf{T}$ is arbitrary to within n constants so that $\mathbf{T} = \mathbf{T}_0\mathbf{K}$ as defined in Problem 4.41, page 97.

For distinct eigenvalues, $d\mathbf{z}/dt = \mathbf{\Lambda}\mathbf{z} + \mathbf{K}^{-1}\mathbf{T}_0^{-1}\mathbf{b}u$, where $\mathbf{K}$ is a diagonal matrix with elements k_{ii} on the diagonal. Defining $\mathbf{g} = \mathbf{T}_0^{-1}\mathbf{b}$, the equation for each state is $dz_i/dt = \lambda_i z_i + (g_i u/k_{ii})$. If $g_i = 0$, the state z_i is uncontrollable (Theorem 6.1) and does not enter into the transfer function (Theorem 6.7). For controllable states, choose $k_{ii} = g_i$. Then the canonical form of equation (*2.16*) is attained.

For the case of nondistinct eigenvalues, look at the $l \times l$ system of one Jordan block with one input, $d\mathbf{z}/dt = \mathbf{L}\mathbf{z} + \mathbf{T}^{-1}\mathbf{b}u$. If the system is controllable, it is desired that $\mathbf{T}^{-1}\mathbf{b} = \mathbf{e}_l$, as in equation (*2.21*). Then using $\mathbf{T} = \mathbf{T}_0\mathbf{K}$, we require $\mathbf{T}_0^{-1}\mathbf{b} = \mathbf{K}\mathbf{e}_l = (\alpha_l \;\; \alpha_{l-1} \;\; \ldots \;\; \alpha_1)^T$, where the α_i are the l arbitrary constants in the $\mathbf{T}$ matrix as given in Problem 4.41. In this manner the canonical form of equation (*2.21*) can be obtained.

Therefore by transformation to Jordan canonical form, the uncontrollable and unobservable states can be found and perhaps omitted from further input-output considerations. Also, the n^2 elements in the $\mathbf{A}$ matrix are transformed to the n eigenvalues that characterize the time behavior of the system.

7.3 REAL JORDAN FORM

Sometimes it is easier to program a computer if all the variables are real. A slight drawback of the Jordan form is that the canonical states $\mathbf{z}(t)$ are complex if $\mathbf{A}$ has any complex eigenvalues. This drawback is easily overcome by a change of variables. We keep the same z_i as in Section 7.2 when λ_i is real, but when λ_i is complex we use the following procedure. Since $\mathbf{A}$ is a real matrix, if λ is an eigenvalue then its complex conjugate λ^* is also an eigenvalue and if $\mathbf{t}$ is an eigenvector then its complex conjugate $\mathbf{t}^*$ is also. Without loss of generality we can look at two Jordan blocks for the case of complex eigenvalues.

$$\frac{d}{dt}\begin{pmatrix} \mathbf{z} \\ \mathbf{z}^* \end{pmatrix} = \begin{pmatrix} \mathbf{L} & \mathbf{0} \\ \mathbf{0} & \mathbf{L}^* \end{pmatrix}\begin{pmatrix} \mathbf{z} \\ \mathbf{z}^* \end{pmatrix}$$

If Re means "real part of" and Im means "imaginary part of", this is

$$\frac{d}{dt}\begin{pmatrix}\text{Re}\,\mathbf{z} + j\,\text{Im}\,\mathbf{z} \\ \text{Re}\,\mathbf{z} - j\,\text{Im}\,\mathbf{z}\end{pmatrix} = \begin{pmatrix}\text{Re}\,\mathbf{L} + j\,\text{Im}\,\mathbf{L} & \mathbf{0} \\ \mathbf{0} & \text{Re}\,\mathbf{L} - j\,\text{Im}\,\mathbf{L}\end{pmatrix}\begin{pmatrix}\text{Re}\,\mathbf{z} + j\,\text{Im}\,\mathbf{z} \\ \text{Re}\,\mathbf{z} - j\,\text{Im}\,\mathbf{z}\end{pmatrix}$$

$$= \begin{pmatrix}\text{Re}\,\mathbf{L}\,\text{Re}\,\mathbf{z} - \text{Im}\,\mathbf{L}\,\text{Im}\,\mathbf{z} + j\,\text{Re}\,\mathbf{L}\,\text{Im}\,\mathbf{z} + j\,\text{Im}\,\mathbf{L}\,\text{Re}\,\mathbf{z} \\ \text{Re}\,\mathbf{L}\,\text{Re}\,\mathbf{z} - \text{Im}\,\mathbf{L}\,\text{Im}\,\mathbf{z} - j\,\text{Re}\,\mathbf{L}\,\text{Im}\,\mathbf{z} - j\,\text{Im}\,\mathbf{L}\,\text{Re}\,\mathbf{z}\end{pmatrix}$$

By equating real and imaginary parts, the system can be rewritten in the "real" Jordan form as

$$\frac{d}{dt}\begin{pmatrix}\text{Re}\,\mathbf{z} \\ \text{Im}\,\mathbf{z}\end{pmatrix} = \begin{pmatrix}\text{Re}\,\mathbf{L} & -\text{Im}\,\mathbf{L} \\ \text{Im}\,\mathbf{L} & \text{Re}\,\mathbf{L}\end{pmatrix}\begin{pmatrix}\text{Re}\,\mathbf{z} \\ \text{Im}\,\mathbf{z}\end{pmatrix}$$

7.4 CONTROLLABLE AND OBSERVABLE FORMS FOR TIME-VARYING SYSTEMS

We can easily transform a linear time-invariant system into a controllable or observable subsystem by transformation to Jordan form. However, this cannot be done for time-varying systems because they cannot be transformed to Jordan form, in general. In this section we shall discuss a method of transformation to controllable and/or observable subsystems without solution of the transition matrix. Of course this method is applicable to time-invariant systems as a subset of time-varying systems.

We consider the transformation of the time-varying system

$$d\mathbf{x}/dt = \mathbf{A}^x(t)\mathbf{x} + \mathbf{B}^x(t)\mathbf{u}, \quad \mathbf{y} = \mathbf{C}^x(t)\mathbf{x} \tag{7.1}$$

into controllable and observable subsystems. The procedure for transformation can be extended to the case $\mathbf{y} = \mathbf{C}^x(t)\mathbf{x} + \mathbf{D}^x(t)\mathbf{u}$, but for simplicity we take $\mathbf{D}^x(t) = \mathbf{0}$. We adopt the notation of placing a superscript on the matrices **A**, **B** and **C** to refer to the state variable because we shall make many transformations of the state variable.

In this chapter it will always be assumed that $\mathbf{A}(t)$, $\mathbf{B}(t)$ and $\mathbf{C}(t)$ are differentiable $n-2$, $n-1$ and $n-1$ times, respectively. The transformations found in the following sections lead to the first and second ("phase-variable") canonical forms (*2.6*) and (*2.9*) when applied to time-invariant systems as a special case.

Before proceeding, we need two preliminary theorems.

Theorem 7.1: If the system (*7.1*) has a controllability matrix $\mathbf{Q}^x(t)$, and an *equivalence transformation* $\mathbf{x}(t) = \mathbf{T}(t)\,\mathbf{z}(t)$ is made, where $\mathbf{T}(t)$ is nonsingular and differentiable, then the controllability matrix of the transformed system $\mathbf{Q}^z(t) = \mathbf{T}^{-1}(t)\,\mathbf{Q}^x(t)$ and $\text{rank}\,\mathbf{Q}^z(t) = \text{rank}\,\mathbf{Q}^x(t)$.

Proof: The transformed system is $d\mathbf{z}/dt = \mathbf{A}^z(t)\mathbf{z} + \mathbf{B}^z(t)\mathbf{u}$, where $\mathbf{A}^z = \mathbf{T}^{-1}(\mathbf{A}^x\mathbf{T} - d\mathbf{T}/dt)$ and $\mathbf{B}^z = \mathbf{T}^{-1}\mathbf{B}^x$. Since $\mathbf{Q}^x = (\mathbf{Q}_1^x \mid \mathbf{Q}_2^x \mid \ldots \mid \mathbf{Q}_n^x)$ and $\mathbf{Q}^z$ is similarly partitioned, we need to show $\mathbf{Q}_k^z = \mathbf{T}^{-1}\mathbf{Q}_k^x$ for $k = 1, 2, \ldots, n$ using induction. First $\mathbf{Q}_1^z = \mathbf{B}^z = \mathbf{T}^{-1}\mathbf{B}^x = \mathbf{T}^{-1}\mathbf{Q}_1^x$. Then assuming $\mathbf{Q}_{k-1}^z = \mathbf{T}^{-1}\mathbf{Q}_{k-1}^x$,

$$\begin{aligned}\mathbf{Q}_k^z &= -\mathbf{A}^z\mathbf{Q}_{k-1}^z + d\mathbf{Q}_{k-1}^z/dt \\ &= -\mathbf{T}^{-1}(\mathbf{A}^x\mathbf{T} - d\mathbf{T}/dt)(\mathbf{T}^{-1}\mathbf{Q}_{k-1}^x) + d\mathbf{T}^{-1}/dt\,\mathbf{Q}_{k-1}^x + \mathbf{T}^{-1}d\mathbf{Q}_{k-1}^x/dt \\ &= \mathbf{T}^{-1}(-\mathbf{A}^x\mathbf{Q}_{k-1}^x + d\mathbf{Q}_{k-1}^x/dt) = \mathbf{T}^{-1}\mathbf{Q}_k^x\end{aligned}$$

for $k = 2, 3, \ldots, n$. Now $\mathbf{Q}^z(t) = \mathbf{T}^{-1}(t)\,\mathbf{Q}^x(t)$ and since $\mathbf{T}(t)$ is nonsingular for any t, $\text{rank}\,\mathbf{Q}^z(t) = \text{rank}\,\mathbf{Q}^x(t)$ for all t.

It is reassuring to know that the controllability of a system cannot be altered merely by a change of state variable. As we should expect, the same holds for observability.

Theorem 7.2: If the system (*7.1*) has an observability matrix $\mathbf{P}^x(t)$, then an equivalence transformation $\mathbf{x}(t) = \mathbf{T}(t)\,\mathbf{z}(t)$ gives $\mathbf{P}^z(t) = \mathbf{P}^x(t)\,\mathbf{T}(t)$ and $\operatorname{rank}\mathbf{P}^z(t) = \operatorname{rank}\mathbf{P}^x(t)$.

The proof is similar to that of Theorem 7.1.

Use of Theorem 7.1 permits construction of a $\mathbf{T}^c(t)$ that separates (*7.1*) into its controllable and uncontrollable states. Using the equivalence transformation $\mathbf{x}(t) = \mathbf{T}^c(t)\,\mathbf{z}(t)$, (*7.1*) becomes

$$\frac{d}{dt}\begin{pmatrix}\mathbf{z}_1\\ \mathbf{z}_2\end{pmatrix} = \begin{pmatrix}\mathbf{A}^z_{11}(t) & \mathbf{A}^z_{12}(t)\\ \mathbf{0} & \mathbf{A}^z_{22}(t)\end{pmatrix}\begin{pmatrix}\mathbf{z}_1\\ \mathbf{z}_2\end{pmatrix} + \begin{pmatrix}\mathbf{B}^z_1(t)\\ \mathbf{0}\end{pmatrix}\mathbf{u} \tag{7.2}$$

where the subsystem

$$d\mathbf{z}_1/dt = \mathbf{A}^z_{11}(t)\mathbf{z}_1 + \mathbf{B}^z_1(t)\mathbf{u} \tag{7.3}$$

is of order $n_1 \leq n$ and has a controllability matrix $\mathbf{Q}^{z_1}(t) = (\mathbf{Q}^{z_1}_1 \mid \mathbf{Q}^{z_1}_2 \mid \ldots \mid \mathbf{Q}^{z_1}_{n_1})$ with $\operatorname{rank} n_1$(e.d.). This shows $\mathbf{z}_1$ is controllable and $\mathbf{z}_2$ is uncontrollable in (*7.2*).

The main problem is to keep $\mathbf{T}^c(t)$ differentiable and nonsingular everywhere, i.e. for all values of t. Also, we will find $\mathbf{Q}^z$ such that it has $n - n_1$ zero rows.

Theorem 7.3: The system (*7.1*) is reducible to (*7.2*) by an equivalence transformation if and only if $\mathbf{Q}^x(t)$ has $\operatorname{rank} n_1$(e.d.) and can be factored as $\mathbf{Q}^x = \mathbf{V}_1(\mathbf{S} \mid \mathbf{R})$ where $\mathbf{V}_1$ is an $n \times n_1$ differentiable matrix with $\operatorname{rank} n_1$ everywhere, $\mathbf{S}$ is an $n_1 \times mn_1$ matrix with rank n_1(e.d.), and $\mathbf{R}$ is any $n_1 \times m(n - n_1)$ matrix.

Note here we do not say how the factorization $\mathbf{Q}^x = \mathbf{V}_1(\mathbf{S} \mid \mathbf{R})$ is obtained.

Proof: First assume (*7.1*) can be transformed by $\mathbf{x}(t) = \mathbf{T}^c(t)\,\mathbf{z}(t)$ to the form of (*7.2*). Using induction, $\mathbf{Q}^z_1 = \begin{pmatrix}\mathbf{B}^z_1\\ \mathbf{0}\end{pmatrix} = \begin{pmatrix}\mathbf{Q}^{z_1}_1\\ \mathbf{0}\end{pmatrix}$; and if $\mathbf{Q}^z_i = \begin{pmatrix}\mathbf{Q}^{z_1}_i\\ \mathbf{0}\end{pmatrix}$ then for $i = 1, 2, \ldots,$

$$\mathbf{Q}^z_{i+1} = -\begin{pmatrix}\mathbf{A}^z_{11} & \mathbf{A}^z_{12}\\ \mathbf{0} & \mathbf{A}^z_{22}\end{pmatrix}\begin{pmatrix}\mathbf{Q}^{z_1}_i\\ \mathbf{0}\end{pmatrix} + \frac{d}{dt}\begin{pmatrix}\mathbf{Q}^{z_1}_i\\ \mathbf{0}\end{pmatrix}$$

$$= \begin{pmatrix}-\mathbf{A}^z_{11}\mathbf{Q}^{z_1}_i + d\mathbf{Q}^{z_1}_i/dt\\ \mathbf{0}\end{pmatrix} = \begin{pmatrix}\mathbf{Q}^{z_1}_{i+1}\\ \mathbf{0}\end{pmatrix} \tag{7.4}$$

Therefore

$$\mathbf{Q}^z = \begin{pmatrix}\mathbf{Q}^{z_1}_1 \ldots \mathbf{Q}^{z_1}_{n_1} & \mathbf{Q}^{z_1}_{n_1+1} \ldots \mathbf{Q}^{z_1}_n\\ \mathbf{0} \ \ldots \mathbf{0} & \mathbf{0} \ \ldots \mathbf{0}\end{pmatrix} = \begin{pmatrix}\mathbf{Q}^{z_1} & \mathbf{F}\\ \mathbf{0} & \mathbf{0}\end{pmatrix}$$

where $\mathbf{F}(t)$ is the $n_1 \times m(n - n_1)$ matrix manufactured by the iteration process (*7.4*) for $i = n_1, n_1+1, \ldots, n-1$. Since $\mathbf{Q}^{z_1}$ has rank n_1(e.d.), $\mathbf{Q}^z$ must also. Use of Theorem 7.1 and the nonsingularity of $\mathbf{T}^c$ shows $\mathbf{Q}^x$ has $\operatorname{rank} n_1$(e.d.). Furthermore, let $\mathbf{T}^c(t) = (\mathbf{T}_1(t) \mid \mathbf{T}_2(t))$, so that

$$\mathbf{Q}^x(t) = \mathbf{T}^c(t)\,\mathbf{Q}^z(t) = (\mathbf{T}_1(t)\ \ \mathbf{T}_2(t))\begin{pmatrix}\mathbf{Q}^{z_1} & \mathbf{F}\\ \mathbf{0} & \mathbf{0}\end{pmatrix}$$

$$= (\mathbf{T}_1\mathbf{Q}^{z_1}\ \ \mathbf{T}_1\mathbf{F}) = \mathbf{T}_1(\mathbf{Q}^{z_1}\ \ \mathbf{F})$$

Since $\mathbf{T}_1(t)$ is an $n \times n_1$ differentiable matrix with $\operatorname{rank} n_1$ everywhere and $\mathbf{Q}^{z_1}$ is an $n_1 \times mn_1$ matrix with rank n_1(e.d.), the *only if* part of the theorem has been proven.

For the proof in the other direction, since $\mathbf{Q}^x$ factors, then

$$\mathbf{Q}^x(t) = (\mathbf{V}_1(t)\ \mathbf{V}_2(t))\begin{pmatrix}\mathbf{S}(t) & \mathbf{R}(t)\\ \mathbf{0} & \mathbf{0}\end{pmatrix}$$

where $\mathbf{V}_2(t)$ is any set of $n-n_1$ differentiable columns making $\mathbf{V}(t) = (\mathbf{V}_1\ \mathbf{V}_2)$ nonsingular. But what is the system corresponding to the controllability matrix on the right? From Theorem 6.10, $\begin{pmatrix}\mathbf{S}_1\\ \mathbf{0}\end{pmatrix} = \mathbf{B}^z$. Also,

$$\begin{pmatrix}\mathbf{S}_{i+1}\\ \mathbf{0}\end{pmatrix} = -\begin{pmatrix}\mathbf{A}^z_{11} & \mathbf{A}^z_{12}\\ \mathbf{A}^z_{21} & \mathbf{A}^z_{22}\end{pmatrix}\begin{pmatrix}\mathbf{S}_i\\ \mathbf{0}\end{pmatrix} + \frac{d}{dt}\begin{pmatrix}\mathbf{S}_i\\ \mathbf{0}\end{pmatrix} \qquad i = 1, 2, \ldots, n_1 - 1$$

and

$$\begin{pmatrix}\mathbf{R}_1\\ \mathbf{0}\end{pmatrix} = -\begin{pmatrix}\mathbf{A}^z_{11} & \mathbf{A}^z_{12}\\ \mathbf{A}^z_{21} & \mathbf{A}^z_{22}\end{pmatrix}\begin{pmatrix}\mathbf{S}_{n_1}\\ \mathbf{0}\end{pmatrix} + \frac{d}{dt}\begin{pmatrix}\mathbf{S}_{n_1}\\ \mathbf{0}\end{pmatrix}$$

Therefore $\mathbf{0} = -\mathbf{A}^z_{21}(t)\mathbf{S}(t)$, and since $\mathbf{S}(t)$ has rank n_1(e.d.) and $\mathbf{A}^z_{21}(t)$ is continuous, by Problem 3.11, $\mathbf{A}^z_{21}(t) = \mathbf{0}$. Therefore the transformation $\mathbf{V}(t) = (\mathbf{V}_1(t)\ \mathbf{V}_2(t))$ is the required equivalence transformation.

The dual relationship for observability is $\mathbf{x}(t) = \mathbf{T}^0(t)\,\mathbf{z}(t)$ that transforms *(7.1)* into

$$\frac{d}{dt}\begin{pmatrix}\mathbf{z}_3\\ \mathbf{z}_4\end{pmatrix} = \begin{pmatrix}\mathbf{A}^z_{33}(t) & \mathbf{0}\\ \mathbf{A}^z_{34}(t) & \mathbf{A}^z_{44}(t)\end{pmatrix}\begin{pmatrix}\mathbf{z}_3\\ \mathbf{z}_4\end{pmatrix} \qquad \mathbf{y} = (\mathbf{C}^z_3(t)\ \mathbf{0})\begin{pmatrix}\mathbf{z}_3\\ \mathbf{z}_4\end{pmatrix} \tag{7.5}$$

where the subsystem

$$d\mathbf{z}_3/dt = \mathbf{A}^z_{33}(t)\mathbf{z}_3 \qquad \mathbf{y} = \mathbf{C}^z_3(t)\mathbf{z}_3 \tag{7.6}$$

is of order $n_3 \leq n$ and has an observability matrix $\mathbf{P}^{z_3}(t)$ with rank n_3(e.d.). Then $\mathbf{P}^z(t)$ has $n - n_3$ zero columns.

Theorem 7.4: The system *(7.1)* is reducible to *(7.5)* by an equivalence transformation if and only if $\mathbf{P}^x$ has rank n_3(e.d.) and factors as $\mathbf{P}^x = \begin{pmatrix}\mathbf{S}\\ \hline \mathbf{V}\end{pmatrix}\mathbf{R}_3$ where $\mathbf{R}_3$ is an $n_3 \times n$ differentiable matrix with rank n_3 everywhere and $\mathbf{S}$ is a $kn_3 \times n_3$ matrix with rank n_3(e.d.).

The proof is similar to that of Theorem 7.3. Here $\mathbf{T}^0 = \begin{pmatrix}\mathbf{R}_3\\ \mathbf{R}_4\end{pmatrix}^{-1}$ where $\mathbf{R}_4$ is any set of $n - n_3$ differentiable rows making $\mathbf{T}^0$ nonsingular and $\mathbf{S}$ is the observability matrix of the totally observable subsystem.

We can extend this procedure to find states $\mathbf{w}_1, \mathbf{w}_2, \mathbf{w}_3$ and $\mathbf{w}_4$ that are controllable and observable, uncontrollable and observable, controllable and unobservable, and uncontrollable and unobservable respectively. The system *(7.1)* is transformed into

$$\frac{d}{dt}\begin{pmatrix}\mathbf{w}_1\\ \mathbf{w}_2\\ \mathbf{w}_3\\ \mathbf{w}_4\end{pmatrix} = \begin{pmatrix}\mathbf{A}^w_{11} & \mathbf{A}^w_{12} & \mathbf{0} & \mathbf{0}\\ \mathbf{0} & \mathbf{A}^w_{22} & \mathbf{0} & \mathbf{0}\\ \mathbf{A}^w_{31} & \mathbf{A}^w_{32} & \mathbf{A}^w_{33} & \mathbf{A}^w_{34}\\ \mathbf{0} & \mathbf{A}^w_{42} & \mathbf{0} & \mathbf{A}^w_{44}\end{pmatrix}\begin{pmatrix}\mathbf{w}_1\\ \mathbf{w}_2\\ \mathbf{w}_3\\ \mathbf{w}_4\end{pmatrix} + \begin{pmatrix}\mathbf{B}^w_1\\ \mathbf{0}\\ \mathbf{B}^w_3\\ \mathbf{0}\end{pmatrix}\mathbf{u}$$

$$\mathbf{y} = (\mathbf{C}^w_1\ \mathbf{C}^w_2\ \mathbf{0}\ \mathbf{0})\mathbf{w} \tag{7.7}$$

in which $\mathbf{w}_i$ is an n_i-vector for $i = 1, 2, 3, 4$ and where the subsystem

$$\frac{d}{dt}\begin{pmatrix}\mathbf{w}_1\\ \mathbf{w}_3\end{pmatrix} = \begin{pmatrix}\mathbf{A}_{11}^w & \mathbf{0}\\ \mathbf{A}_{31}^w & \mathbf{A}_{33}^w\end{pmatrix}\begin{pmatrix}\mathbf{w}_1\\ \mathbf{w}_3\end{pmatrix} + \begin{pmatrix}\mathbf{B}_1^w\\ \mathbf{B}_3^w\end{pmatrix}\mathbf{u}$$

has a controllability matrix $\begin{pmatrix}\mathbf{Q}_{11} & \mathbf{Q}_{13}\\ \mathbf{Q}_{31} & \mathbf{Q}_{33}\end{pmatrix}$ of rank $n_1+n_3 \leq n$(e.d.) and where the subsystem

$$\frac{d}{dt}\begin{pmatrix}\mathbf{w}_1\\ \mathbf{w}_2\end{pmatrix} = \begin{pmatrix}\mathbf{A}_{11}^w & \mathbf{A}_{12}^w\\ \mathbf{0} & \mathbf{A}_{22}^w\end{pmatrix}\begin{pmatrix}\mathbf{w}_1\\ \mathbf{w}_2\end{pmatrix} \qquad \mathbf{y} = (\mathbf{C}_1^w \ \mathbf{C}_2^w)\begin{pmatrix}\mathbf{w}_1\\ \mathbf{w}_2\end{pmatrix} \tag{7.8}$$

has an observability matrix $\begin{pmatrix}\mathbf{P}_{11} & \mathbf{P}_{12}\\ \mathbf{P}_{21} & \mathbf{P}_{22}\end{pmatrix}$ of rank $n_1+n_2 \leq n$(e.d.). Hence these subsystems are totally controllable and totally observable, respectively. Clearly if such a system as (*7.7*) can be found, the states $\mathbf{w}_1, \mathbf{w}_2, \mathbf{w}_3, \mathbf{w}_4$ will be as desired because the flow diagram of (*7.7*) shows $\mathbf{w}_2$ and $\mathbf{w}_4$ disconnected from the control and $\mathbf{w}_3$ and $\mathbf{w}_4$ disconnected from the output.

Theorem 7.5: The system (*7.1*) is reducible to (*7.7*) by an equivalence transformation if and only if $\mathbf{P}^x$ has rank n_1+n_2(e.d.) and factors as $\mathbf{P}^x = \mathbf{R}^1\mathbf{U}_1 + \mathbf{R}^2\mathbf{U}_2$ and $\mathbf{Q}^x$ has rank n_1+n_3 (e.d) and factors as $\mathbf{Q}^x = \mathbf{V}_1\mathbf{S}^1 + \mathbf{V}_3\mathbf{S}^3$.

Here $\mathbf{R}^i(t)$ is a $kn \times n_i$ matrix with rank n_i(e.d.), $\mathbf{S}^i(t)$ is an $n_i \times mn$ matrix with rank n_i(e.d.), $\mathbf{U}_i(t)$ is an $n_i \times n$ differentiable matrix with rank n_i everywhere, $\mathbf{V}_i(t)$ is an $n \times n_i$ differentiable matrix with rank n_i everywhere, and $\mathbf{U}_i(t)\,\mathbf{V}_j(t) = \delta_{ij}\mathbf{I}_{n_i}$. Furthermore, the rank of $\mathbf{R}^i$ and $\mathbf{S}^i$ must be such that the controllability and observability matrices of (*7.8*) have the correct rank.

Proof: First assume (*7.1*) can be transformed to (*7.7*). By reasoning similar to (*7.4*),

$$\mathbf{Q}^z = \begin{pmatrix}\mathbf{Q}_{11} & \mathbf{Q}_{13} & \mathbf{F}_{12} & \mathbf{F}_{14}\\ \mathbf{0} & \mathbf{0} & \mathbf{0} & \mathbf{0}\\ \mathbf{Q}_{31} & \mathbf{Q}_{33} & \mathbf{F}_{32} & \mathbf{F}_{34}\\ \mathbf{0} & \mathbf{0} & \mathbf{0} & \mathbf{0}\end{pmatrix} \qquad \mathbf{P}^z = \begin{pmatrix}\mathbf{P}_{11} & \mathbf{P}_{12} & \mathbf{0} & \mathbf{0}\\ \mathbf{P}_{21} & \mathbf{P}_{22} & \mathbf{0} & \mathbf{0}\\ \mathbf{G}_{31} & \mathbf{G}_{32} & \mathbf{0} & \mathbf{0}\\ \mathbf{G}_{41} & \mathbf{G}_{42} & \mathbf{0} & \mathbf{0}\end{pmatrix}$$

so that $\mathbf{Q}^x$ and $\mathbf{P}^x$ have rank n_1+n_3 (e.d.) and n_1+n_2 (e.d.), respectively. Let

$$\mathbf{T}(t) = (\mathbf{V}_1 \ \mathbf{V}_2 \ \mathbf{V}_3 \ \mathbf{V}_4) = \begin{pmatrix}\mathbf{U}_1\\ \mathbf{U}_2\\ \mathbf{U}_3\\ \mathbf{U}_4\end{pmatrix}^{-1}$$

Then

$$\mathbf{Q}^x = \mathbf{V}_1(\mathbf{Q}_{11} \ \mathbf{Q}_{13} \ \mathbf{F}_{12} \ \mathbf{F}_{14}) + \mathbf{V}_3(\mathbf{Q}_{31} \ \mathbf{Q}_{33} \ \mathbf{F}_{32} \ \mathbf{F}_{34})$$

and

$$\mathbf{P}^x = (\mathbf{P}_{11}^T \ \mathbf{P}_{12}^T \ \mathbf{G}_{13}^T \ \mathbf{G}_{14}^T)^T\mathbf{U}_1 + (\mathbf{P}_{21}^T \ \mathbf{P}_{22}^T \ \mathbf{G}_{23}^T \ \mathbf{G}_{24}^T)^T\mathbf{U}_2$$

which is the required form.

Now suppose $\mathbf{P}^x$ and $\mathbf{Q}^x$ factor in the required manner. Then, by reasoning similar to the proof of Theorem 7.3, $\mathbf{R}^i = (\mathbf{P}_{1i}^T \ \mathbf{P}_{2i}^T \ \mathbf{G}_{3i}^T \ \mathbf{G}_{4i}^T)^T$ for $i = 1,2$ and $\mathbf{S}^i = (\mathbf{Q}_{i1} \ \mathbf{Q}_{i3} \ \mathbf{F}_{i2} \ \mathbf{F}_{i4})$ for $i = 1,3$, and $\mathbf{A}_{13}^w, \mathbf{A}_{14}^w, \mathbf{A}_{21}^w, \mathbf{A}_{23}^w, \mathbf{A}_{24}^w, \mathbf{A}_{41}^w$ and $\mathbf{A}_{43}^w$ have all zero elements.

Theorem 7.5 leads to the following construction procedure:

1. Factor $\mathbf{P}^x\mathbf{Q}^x = \mathbf{R}^1\mathbf{S}^1$ to obtain $\mathbf{R}^1$ and $\mathbf{S}^1$.
2. Solve $\mathbf{P}^x\mathbf{V}_1 = \mathbf{R}^1$ and $\mathbf{U}_1\mathbf{Q}^x = \mathbf{S}^1$ for $\mathbf{V}_1$ and $\mathbf{U}_1$.
3. Check that $\mathbf{U}_1\mathbf{V}_1 = \mathbf{I}_{n_1}$.

4. Factor $\mathbf{P}^x - \mathbf{R}^1\mathbf{U}_1 = \mathbf{R}^2\mathbf{U}_2$ and $\mathbf{Q}^x - \mathbf{V}_1\mathbf{S}^1 = \mathbf{V}_3\mathbf{S}^3$ to obtain $\mathbf{R}^2, \mathbf{U}_2, \mathbf{V}_3$ and $\mathbf{S}^3$.
5. Find the reciprocal basis vectors of $\mathbf{U}_2$ to form $\mathbf{V}_2$.
6. Find $\mathbf{V}_4$ as any set of n_4 differentiable columns making $\mathbf{T}(t)$ nonsingular.
7. Check that $\mathbf{U}_i\mathbf{V}_j = \delta_{ij}\mathbf{I}_{n_i}$.
8. Form $\mathbf{T}(t) = (\mathbf{V}_1\ \mathbf{V}_2\ \mathbf{V}_3\ \mathbf{V}_4)$.

Unfortunately, factorization and finding sets of differentiable columns to make $\mathbf{T}(t)$ nonsingular is not easy in general.

Example 7.1.

Given $\mathbf{P}^x = \begin{pmatrix} \sin t & 1 \\ \sin t & 1 \end{pmatrix}$. Obviously rank $\mathbf{P}^x = 1$ and it is factored by inspection, but suppose we try to mechanize the procedure for the general case by attempting elementary column operations analogous to Example 3.13, page 44. Then

$$\begin{pmatrix} \sin t & 1 \\ \sin t & 1 \end{pmatrix}\begin{pmatrix} (\sin t)^{-1} & -(\sin t)^{-2} \\ 0 & (\sin t)^{-1} \end{pmatrix} = \begin{pmatrix} 1 & 0 \\ 1 & 0 \end{pmatrix}$$

The matrix on the right is perfect for $\mathbf{P}^z$, but the transformation matrix is not differentiable at $t = i\pi$ for $i = \ldots -1, 0, 1, \ldots$

However, if $\alpha(t)$ and $\beta(t)$ are *analytic* functions with no common zeros, let

$$\mathbf{E}(t) = \begin{pmatrix} \alpha(t) & -\beta(t) \\ \beta(t) & \alpha(t) \end{pmatrix}$$

Then $(\alpha(t)\ \beta(t))\mathbf{E}(t) = (\alpha^2(t) + \beta^2(t)\ \ 0)$ and $\mathbf{E}(t)$ is always nonsingular and differentiable. This gives us a means of attempting a factorization even if not all the elements are analytic.

If $\alpha(t)$ and $\beta(t)$ are analytic but have common zeros $\zeta_1, \zeta_2, \ldots, \zeta_k, \ldots$, the matrix $\mathbf{E}(t)$ can be fixed up as

$$\mathbf{E}(t) = \begin{pmatrix} \alpha(t) & -\beta(t) \\ \beta(t) & \alpha(t) \end{pmatrix} \prod_{k=1}^{\infty} (1 - t/\zeta_k)^{-\rho_k}\gamma_k(\zeta_k)$$

where ρ_k is the order of their common zero ζ_k and $\gamma_k(\zeta_k)$ is a convergence factor.

Example 7.2.

Let $\alpha(t) = t^2$ and $\beta(t) = t^3$. Their common zero is $\zeta_1 = 0$ with order $2 = \rho_1$. Choose $\gamma_1(\zeta_1) = \zeta_1^2$. Then

$$\mathbf{E}(t) = \begin{pmatrix} t^2 & -t^3 \\ t^3 & t^2 \end{pmatrix} t^{-2} = \begin{pmatrix} 1 & -t \\ t & 1 \end{pmatrix}$$

Note $\mathbf{E}(0)$ is nonsingular.

Using repeated applications of this form of elementary row or column operation, it is obvious that we can find $\mathbf{T}(t) = \mathbf{E}_n(t)\mathbf{E}_{n-1}(t) \cdots \mathbf{E}_1(t)$ such that $\mathbf{P}^x\mathbf{T} = (\mathbf{P}^{z_1} \mid \mathbf{0})$, and similarly for $\mathbf{Q}^x$ if $\mathbf{P}^x$ or $\mathbf{Q}^x$ is analytic and of fixed rank $r \leq n$. Also, denoting $\mathbf{T}^{-1} = \begin{pmatrix} \mathbf{U}_1 \\ \mathbf{U}_2 \end{pmatrix}$, then $\mathbf{P}^x$ factors as $\mathbf{P}^{z_1}\mathbf{U}_1$.

The many difficulties encountered if $\mathbf{P}^x$ or $\mathbf{Q}^x$ changes rank should be noted. Design of filters or controllers in this case can become quite complicated.

7.5 CANONICAL FORMS FOR TIME-VARYING SYSTEMS

As discussed in Section 5.3, no form analogous to the Jordan form exists for a general time-varying linear system. Therefore we shall study transformations to forms analogous to the first and second canonical forms of Section 2.3.

Consider first the transformation of a single-input, time-varying system

$$d\mathbf{x}/dt = \mathbf{A}(t)\mathbf{x} + \mathbf{b}(t)u \tag{7.9}$$

to a form analogous to (2.6),

$$\frac{d\mathbf{z}}{dt} = \begin{pmatrix} -\alpha_1(t) & 1 & 0 & \dots & 0 \\ -\alpha_2(t) & 0 & 1 & \dots & 0 \\ \dots & \dots & \dots & \dots & \dots \\ -\alpha_{n-1}(t) & 0 & 0 & \dots & 1 \\ -\alpha_n(t) & 0 & 0 & \dots & 0 \end{pmatrix} \mathbf{z} + \begin{pmatrix} 0 \\ 0 \\ \vdots \\ 0 \\ 1 \end{pmatrix} u \tag{7.10}$$

Theorem 7.6: The system (7.9) can be transformed to (7.10) by an equivalence transformation if and only if $\mathbf{Q}^x(t)$, the controllability matrix of (7.9), is differentiable and has rank n everywhere.

Note this implies (7.9) must be more than totally controllable to be put in the form (7.10) in that rank $\mathbf{Q} = n$ everywhere, not n(e.d.). However, using the methods of the previous section, the totally controllable states can be found to form a subsystem than can be put in canonical form.

Proof: This is proven by showing the needed equivalence transformation $\mathbf{T}(t) = \mathbf{Q}^x(t)\mathbf{K}$, where $\mathbf{K}$ is a nonsingular constant matrix. First let $\mathbf{x} = \mathbf{Q}^x(t)\mathbf{w}$, where the $n \times n$ matrix $\mathbf{Q}^x(t)$ is partitioned as $(\mathbf{q}_1^x \mid \dots \mid \mathbf{q}_n^x)$. Then

$$\frac{d\mathbf{w}}{dt} = \begin{pmatrix} 0 & 0 & \dots & 0 & (-1)^n\alpha_n \\ -1 & 0 & \dots & 0 & (-1)^{n-1}\alpha_{n-1} \\ 0 & -1 & \dots & 0 & (-1)^{n-2}\alpha_{n-2} \\ \dots & \dots & \dots & \dots & \dots \\ 0 & 0 & \dots & -1 & -\alpha_1 \end{pmatrix} \mathbf{w} + \begin{pmatrix} 1 \\ 0 \\ 0 \\ \vdots \\ 0 \end{pmatrix} u \tag{7.11}$$

This is true because $\mathbf{b}^x = \mathbf{Q}^x\mathbf{b}^w = \mathbf{q}_1^x$ and $\mathbf{Q}^{x-1}(\mathbf{A}^x\mathbf{Q}^x - d\mathbf{Q}^x/dt) = \mathbf{A}^w$ so that $-\mathbf{A}^x\mathbf{q}_i^x + d\mathbf{q}_i^x/dt = \mathbf{q}_{i+1}^x$ for $i = 1, 2, \dots, n-1$. Also $\mathbf{A}^w\mathbf{e}_n = \mathbf{Q}^{x-1}(\mathbf{A}^x\mathbf{q}_n^x - d\mathbf{q}_n^x/dt)$. Setting $\mathbf{w} = \mathbf{K}\mathbf{z}$, where

$$\mathbf{K} = \begin{pmatrix} 0 & \dots & 0 & 1 \\ 0 & \dots & -1 & 0 \\ \dots & \dots & \dots & \dots \\ (-1)^{n-1} & \dots & 0 & 0 \end{pmatrix}$$

will give the desired form (7.10).

If the system can be transformed to (7.10) by an equivalence transformation, from Theorem 7.1 $\mathbf{Q}^x = \mathbf{T}\mathbf{Q}^z$. Using Theorem 6.10 on (7.10), $\mathbf{Q}^z = \mathbf{K}^{-1}$ which has rank n everywhere and $\mathbf{T}$ by definition has rank n everywhere, so $\mathbf{Q}^x$ must have rank n everywhere.

Now we consider the transformation to the second canonical form of Section 2.3, which is also known as phase-variable canonical form.

$$\frac{d\mathbf{z}}{dt} = \begin{pmatrix} 0 & 1 & 0 & \dots & 0 \\ 0 & 0 & 1 & \dots & 0 \\ \dots & \dots & \dots & \dots & \dots \\ 0 & 0 & 0 & \dots & 1 \\ a_1(t) & a_2(t) & a_3(t) & \dots & a_n(t) \end{pmatrix} \mathbf{z} + \begin{pmatrix} 0 \\ 0 \\ \vdots \\ 0 \\ 1 \end{pmatrix} u \tag{7.12}$$

The controllability matrix $\mathbf{Q}^z$ of this system is

$$\mathbf{Q}^z = \begin{pmatrix} 0 & 0 & \dots & 0 & (-1)^{n-1} \\ 0 & 0 & \dots & (-1)^{n-2} & q_{n-1,n-1} \\ \dots & \dots & \dots & \dots & \dots \\ 0 & -1 & \dots & q_{n-2,2} & q_{n-1,2} \\ 1 & q_{11} & \dots & q_{n-2,1} & q_{n-1,1} \end{pmatrix} \quad \text{and} \quad \mathbf{q}^z_{n+1} = \begin{pmatrix} q_{n,n} \\ q_{n,n-1} \\ \vdots \\ q_{n,2} \\ q_{n,1} \end{pmatrix}$$

where $q_{ii} = (-1)^i a_n$ for $1 \le i \le n$; and for $1 \le k < i \le n$,

$$q_{ik} = (-1)^i a_{n-i+k} + (-1)^k \sum_{j=0}^{i-k-1} a_{n-j} q_{i-k,j+1} + \sum_{j=1}^{k} (-1)^{j+1} \frac{dq_{i-j,k-j+1}}{dt}$$

Theorem 7.7: The system *(7.9)* can be transformed to *(7.12)* by an equivalence transformation if and only if $\mathbf{Q}^x(t)$, the controllability matrix of *(7.9)*, is differentiable and has rank n everywhere.

Proof: The transformation matrix $\mathbf{T} = \mathbf{Q}^x\mathbf{Q}^{z-1}$, and $\mathbf{Q}^z$ has rank n everywhere. Therefore the proof proceeds similarly to that of Theorem 7.6, transforming to *(7.11)* and then setting $\mathbf{z} = \mathbf{Q}^z\mathbf{w}$. To determine the $a_i(t)$ from the $\alpha_i(t)$ obtained from *(7.11)*, note $-\mathbf{A}^z\mathbf{Q}^z + d\mathbf{Q}^z/dt = -\mathbf{Q}^z\mathbf{A}^w$ when written out in its columns gives

$$(\mathbf{q}^z_2 \mid \mathbf{q}^z_3 \mid \dots \mid \mathbf{q}^z_n \mid \mathbf{q}^z_{n+1}) = (\mathbf{q}^z_2 \mid \mathbf{q}^z_3 \mid \dots \mid \mathbf{q}^z_n \mid -\mathbf{Q}^z\mathbf{A}^w\mathbf{e}_n)$$

Therefore $\mathbf{q}^z_{n+1} = -\mathbf{Q}^z\mathbf{A}^w\mathbf{e}_n$, which gives a set of relations that can be solved recursively for the $a_i(t)$ in terms of the $\alpha_i(t)$ of *(7.11)*.

Example 7.3.

Suppose we have a second-order system. Then

$$\mathbf{q}^z_{n+1} = \begin{pmatrix} q_{22} \\ q_{21} \end{pmatrix} = \begin{pmatrix} a_2 \\ a_1 + a_2^2 - da_2/dt \end{pmatrix}$$

and

$$-\mathbf{Q}^z\mathbf{A}^w\mathbf{e}_n = -\begin{pmatrix} 0 & -1 \\ 1 & -a_2 \end{pmatrix}\begin{pmatrix} \alpha_2 \\ -\alpha_1 \end{pmatrix} = \begin{pmatrix} -\alpha_1 \\ -\alpha_2 - \alpha_1 a_2 \end{pmatrix}$$

By equating the two expressions it is possible to find, recursively, $a_2 = -\alpha_1$ and $a_1 = -\alpha_2 - d\alpha_1/dt$.

It appears that the conditions of Theorems 7.6 and 7.7 can be relaxed if the $\mathbf{b}^z(t)$ of equation *(7.10)* or *(7.12)* is left a general vector function of time instead of $\mathbf{e}_n$. No results are available at present for this case.

Note that if we are given *(7.9)*, defining $y = z_1$ in *(7.12)* permits us to find a corresponding scalar nth order equation $d^ny/dt^n = a_1y + a_2dy/dt + \cdots + a_nd^{n-1}y/dt^{n-1} + u$.

For the case where $\mathbf{u}$ is a vector instead of a scalar in *(7.9)* a possible approach is to set all elements in $\mathbf{u}$ except one equal to zero, and if the resulting $\mathbf{Q}^x$ has rank n everywhere then the methods developed previously are applicable. If this is not possible or desirable, the form

$$\frac{d\mathbf{w}}{dt} = \begin{pmatrix} \mathbf{A}^w_{11} & \mathbf{A}^w_{12} & \dots & \mathbf{A}^w_{1l} \\ \mathbf{A}^w_{21} & \mathbf{A}^w_{22} & \dots & \mathbf{A}^w_{2l} \\ \dots & \dots & \dots & \dots \\ \mathbf{A}^w_{l1} & \mathbf{A}^w_{l2} & \dots & \mathbf{A}^w_{ll} \end{pmatrix}\mathbf{w} + \left(\begin{matrix} \mathbf{b}^w_1 & 0 & \dots & 0 \\ 0 & \mathbf{b}^w_2 & \dots & 0 \\ \dots & \dots & \dots & \dots \\ 0 & 0 & \dots & \mathbf{b}^w_l \end{matrix} \;\middle|\; \mathbf{f}_{l+1} \;\; \dots \;\; \mathbf{f}_m \right)\mathbf{u} \qquad (7.13)$$

may be obtained, where the $\mathbf{f}_i$ are in general nonzero n-vector functions of time, and

$$d\mathbf{w}_i/dt = \mathbf{A}_{ii}^{w}\mathbf{w}_i + \mathbf{b}_i^{w}u_i$$

is of the form (*7.11*); and for $i \neq j$,

$$\mathbf{A}_{ij}^{w} = \begin{pmatrix} 0 & 0 & \ldots & \alpha_n^{ij} \\ 0 & 0 & \ldots & \alpha_{n-1}^{ij} \\ \cdots & \cdots & \cdots & \cdots \\ 0 & 0 & \ldots & \alpha_1^{ij} \end{pmatrix}$$

This form is obtained by calculating m $\mathbf{Q}^x$ matrices for each of the m columns of the $\mathbf{B}$ matrix, i.e. treating the systems as m single-input systems. Then from any l of these single-input $\mathbf{Q}^x$ matrices, choose the first n_1 columns of the first $\mathbf{Q}^x$ matrix, the first n_2 columns of the second $\mathbf{Q}^x$ matrix, ..., and the first n_l columns of the lth $\mathbf{Q}^x$ matrix such that these columns are independent and differentiable and $n_1 + n_2 + \cdots + n_l = n$. In this order these columns form the $\mathbf{T}$ matrix. The proof is similar to that of Theorem 7.6.

To transform from this form to a form analogous to the first canonical form (*7.10*), use a constant matrix similar to the $\mathbf{K}$ matrix used in the proof of Theorem 7.6. However, as yet a general theory of reduction of time-varying linear systems has not been developed.

Solved Problems

7.1. Transform

$$\mathbf{x}(m+1) = \begin{pmatrix} 2 & 1 & 1 \\ 0 & 3 & 1 \\ 0 & -1 & 1 \end{pmatrix}\mathbf{x}(m) + \begin{pmatrix} 2 \\ -2 \\ 1 \end{pmatrix}u(m)$$

to Jordan form.

From Example 4.8, page 74, and Problem 4.41, page 97,

$$\mathbf{J} = \begin{pmatrix} 2 & 1 & 0 \\ 0 & 2 & 0 \\ 0 & 0 & 2 \end{pmatrix}, \quad \mathbf{T}_0 = \begin{pmatrix} 1 & 0 & 0 \\ 1 & 1 & 1 \\ -1 & 0 & -1 \end{pmatrix}, \quad \mathbf{K} = \begin{pmatrix} \alpha_1 & \alpha_2 & 0 \\ 0 & \alpha_1 & 0 \\ 0 & 0 & \beta_1 \end{pmatrix}$$

Then

$$\mathbf{T}_0^{-1}\mathbf{b} = \begin{pmatrix} 1 & 0 & 0 \\ 0 & 1 & 1 \\ -1 & 0 & -1 \end{pmatrix}\begin{pmatrix} 2 \\ -2 \\ 1 \end{pmatrix} = \begin{pmatrix} \alpha_1 & \alpha_2 & 0 \\ 0 & \alpha_1 & 0 \\ 0 & 0 & \beta_1 \end{pmatrix}\begin{pmatrix} 0 \\ 1 \\ 1 \end{pmatrix}$$

from which $\alpha_2 = 2$, $\alpha_1 = -1$, $\beta_1 = -3$ are obtained. The proper transformation is then

$$\mathbf{T} = \mathbf{T}_0\mathbf{K} = \begin{pmatrix} 1 & 0 & 0 \\ 1 & 1 & 1 \\ -1 & 0 & -1 \end{pmatrix}\begin{pmatrix} -1 & 2 & 0 \\ 0 & -1 & 0 \\ 0 & 0 & -3 \end{pmatrix} = \begin{pmatrix} -1 & 2 & 0 \\ -1 & 1 & -3 \\ 1 & -2 & 3 \end{pmatrix}$$

Substitution of $\mathbf{x} = \mathbf{T}\mathbf{z}$ in the system equation then gives the canonical form

$$\mathbf{z}(m+1) = \begin{pmatrix} 2 & 1 & 0 \\ 0 & 2 & 0 \\ 0 & 0 & 2 \end{pmatrix}\mathbf{z}(m) + \begin{pmatrix} 0 \\ 1 \\ 1 \end{pmatrix}u(m)$$

7.2. Transform

$$d\mathbf{x}/dt = \begin{pmatrix} 1 & 1 & 2 \\ -1 & 1 & -2 \\ 0 & 0 & 1 \end{pmatrix}\mathbf{x} + \begin{pmatrix} 0 \\ 0 \\ 1 \end{pmatrix}u$$

to real Jordan form.

From Problem 4.8, page 93, and Problem 4.41, page 97,

$$\mathbf{x} = \begin{pmatrix} 1 & j & -j \\ 1 & 1 & 1 \\ -0.5 & 0 & 0 \end{pmatrix}\begin{pmatrix} \alpha & 0 & 0 \\ 0 & \beta & 0 \\ 0 & 0 & \beta^* \end{pmatrix}\mathbf{z}$$

Substitution into the system equation gives

$$d\mathbf{z}/dt = \begin{pmatrix} 1 & 0 & 0 \\ 0 & 1-j & 0 \\ 0 & 0 & 1+j \end{pmatrix}\mathbf{z} + \begin{pmatrix} -2/\alpha \\ (1-j)/\beta \\ (1+j)/\beta^* \end{pmatrix}u$$

Then $\alpha = -2$ and $\beta = 1 - j$. Putting this in real form gives

$$\frac{d}{dt}\begin{pmatrix} z_1 \\ \operatorname{Re} z_2 \\ \operatorname{Im} z_2 \end{pmatrix} = \begin{pmatrix} 1 & 0 & 0 \\ 0 & 1 & 1 \\ 0 & -1 & 1 \end{pmatrix}\begin{pmatrix} z_1 \\ \operatorname{Re} z_2 \\ \operatorname{Im} z_2 \end{pmatrix} + \begin{pmatrix} 1 \\ 1 \\ 0 \end{pmatrix}u$$

7.3. Using the methods of Section 7.4, reduce

$$d\mathbf{x}/dt = \begin{pmatrix} 1 & 0 & 0 \\ 2 & 2 & 3 \\ -2 & 0 & -1 \end{pmatrix}\mathbf{x} + \begin{pmatrix} 1 \\ 2 \\ -2 \end{pmatrix}u, \qquad y = (1\ 1\ 2)\mathbf{x}$$

(i) to a controllable system, (ii) to an observable system, (iii) to a controllable and observable system.

(i) Form the controllability matrix $\mathbf{Q} = \begin{pmatrix} 1 & -1 & 1 \\ 2 & 0 & 2 \\ -2 & 0 & -2 \end{pmatrix}$

Observe that $\mathbf{Q} = (\mathbf{b} \mid -\mathbf{Ab} \mid \mathbf{A}^2\mathbf{b})$ in accord with using the form of Theorem 6.10 in the time-invariant case, and not $\mathbf{Q} = (\mathbf{b} \mid \mathbf{Ab} \mid \mathbf{A}^2\mathbf{b})$ which is the form of Theorem 6.8. Performing elementary row operations to make the bottom row of $\mathbf{Q}^z = \mathbf{0}$ gives

$$\mathbf{Q} = \begin{pmatrix} 1 & 0 & 0 \\ -2 & 1 & 0 \\ 0 & 1 & 1 \end{pmatrix}^{-1}\begin{pmatrix} 1 & -1 & 1 \\ 0 & 2 & 0 \\ 0 & 0 & 0 \end{pmatrix}$$

Then the required transformation is $\mathbf{x} = \mathbf{Tz}$ where

$$\mathbf{T}^{-1} = \begin{pmatrix} 1 & 0 & 0 \\ -2 & 1 & 0 \\ 0 & 1 & 1 \end{pmatrix} \quad \text{and} \quad \mathbf{T} = \begin{pmatrix} 1 & 0 & 0 \\ 2 & 1 & 0 \\ -2 & -1 & 1 \end{pmatrix}$$

Using this change of variables, $d\mathbf{z}/dt = \mathbf{T}^{-1}(\mathbf{AT} - d\mathbf{T}/dt)\mathbf{z} + \mathbf{T}^{-1}\mathbf{b}u$.

$$d\mathbf{z}/dt = \begin{pmatrix} 1 & 0 & 0 \\ -2 & -1 & 3 \\ 0 & 0 & 2 \end{pmatrix}\mathbf{z} + \begin{pmatrix} 1 \\ 0 \\ 0 \end{pmatrix}u \qquad y = (-1 \;\; -1 \;\; 2)\mathbf{z}$$

(ii) Forming the observability matrix gives

$$\mathbf{P}^x = \begin{pmatrix} 1 & 1 & 2 \\ -1 & 2 & 1 \\ 1 & 4 & 5 \end{pmatrix} = \begin{pmatrix} 1 & 2 & 0 \\ -1 & 1 & 0 \\ 1 & 5 & 0 \end{pmatrix}\begin{pmatrix} 1 & -1 & 0 \\ 0 & 1 & 1 \\ 0 & 0 & 1 \end{pmatrix} = \mathbf{P}^z(\mathbf{T}^0)^{-1}$$

where the factorization is made using elementary column transformations. Using the transformation $\mathbf{x} = \mathbf{T}^0\mathbf{z}$,

$$d\mathbf{z}/dt = \begin{pmatrix} -1 & -3 & 0 \\ 0 & 2 & 0 \\ -2 & -2 & 1 \end{pmatrix}\mathbf{z} + \begin{pmatrix} -1 \\ 0 \\ -2 \end{pmatrix}u \qquad y = (1 \;\; 2 \;\; 0)\mathbf{z}$$

(iii) Using the construction procedure,

$$\mathbf{P}^x\mathbf{Q}^x = \begin{pmatrix} -1 & -1 & -1 \\ 1 & 1 & 1 \\ -1 & -1 & -1 \end{pmatrix} = \begin{pmatrix} 1 & 0 & 0 \\ -1 & 0 & 0 \\ 1 & 0 & 0 \end{pmatrix}\begin{pmatrix} -1 & -1 & -1 \\ 0 & 0 & 0 \\ 0 & 0 & 0 \end{pmatrix}$$

$$= \begin{pmatrix} 1 \\ -1 \\ 1 \end{pmatrix}(-1 \;\; -1 \;\; -1) = \mathbf{R}^1\mathbf{S}^1$$

Then

$$\mathbf{P}^x\mathbf{V}_1 = \begin{pmatrix} 1 & 1 & 2 \\ -1 & 2 & 1 \\ 1 & 4 & 5 \end{pmatrix}\begin{pmatrix} v_{11} \\ v_{21} \\ v_{31} \end{pmatrix} = \begin{pmatrix} 1 \\ -1 \\ 1 \end{pmatrix} = \mathbf{R}^1 \quad \text{gives} \quad \mathbf{V}_1 = \begin{pmatrix} 1 - v_{31} \\ -v_{31} \\ v_{31} \end{pmatrix}$$

and

$$\mathbf{U}_1\mathbf{Q}^x = (u_{11} \;\; u_{12} \;\; u_{13})\begin{pmatrix} 1 & -1 & 1 \\ 2 & 0 & 2 \\ -2 & 0 & -2 \end{pmatrix} = (-1 \;\; -1 \;\; -1) = \mathbf{S}^1 \quad \text{gives} \quad \mathbf{U}_1 = (1 \;\; u_{13} - 1 \;\; u_{13})$$

Note $\mathbf{U}_1\mathbf{V}_1 = 1$ for any values of u_{13} and v_{31}. The usual procedure would be to pick a value for each, but for instructional purposes we retain u_{13} and v_{31} arbitrary. Then

$$\mathbf{P}^x - \mathbf{R}^1\mathbf{U}_1 = \begin{pmatrix} 0 & 2 - u_{13} & 2 - u_{13} \\ 0 & 1 + u_{13} & 1 + u_{13} \\ 0 & 5 - u_{13} & 5 - u_{13} \end{pmatrix} = \begin{pmatrix} 2 - u_{13} \\ 1 + u_{13} \\ 5 - u_{13} \end{pmatrix}(0 \;\; 1 \;\; 1) = \mathbf{R}^2\mathbf{U}_2$$

and

$$\mathbf{Q}^x - \mathbf{V}_1\mathbf{S}^1 = \begin{pmatrix} 2 - v_{31} & -v_{31} & 2 - v_{31} \\ 2 - v_{31} & -v_{31} & 2 - v_{31} \\ -2 + v_{31} & v_{31} & -2 + v_{31} \end{pmatrix} = \begin{pmatrix} 1 \\ 1 \\ -1 \end{pmatrix}(2 - v_{31} \;\; -v_{31} \;\; 2 - v_{31}) = \mathbf{V}_3\mathbf{S}^3$$

Choosing $v_{31} = 0$ and $u_{13} = 1$ gives

$$\mathbf{T} = \begin{pmatrix} 1 & v_{12} & 1 \\ 0 & v_{22} & 1 \\ 0 & v_{32} & -1 \end{pmatrix} = \begin{pmatrix} 1 & 0 & 1 \\ 0 & 1 & 1 \\ u_{31} & u_{32} & u_{33} \end{pmatrix}^{-1}$$

Using $\mathbf{U}_i\mathbf{V}_j = \delta_{ij}$ we can determine

$$\mathbf{T} = \begin{pmatrix} 1 & 0 & 1 \\ 0 & 1 & 1 \\ 0 & 0 & -1 \end{pmatrix} = \begin{pmatrix} 1 & 0 & 1 \\ 0 & 1 & 1 \\ 0 & 0 & -1 \end{pmatrix}^{-1}$$

The reduced system is then

$$\frac{d}{dt}\begin{pmatrix} z_1 \\ z_2 \\ z_3 \end{pmatrix} = \begin{pmatrix} -1 & 0 & 0 \\ 0 & 2 & 0 \\ 2 & 0 & 3 \end{pmatrix}\begin{pmatrix} z_1 \\ z_2 \\ z_3 \end{pmatrix} + \begin{pmatrix} -1 \\ 0 \\ 2 \end{pmatrix} u \qquad y = (1 \ 1 \ 0)\mathbf{z}$$

where z_1 is controllable and observable, z_2 is observable and uncontrollable, and z_3 is controllable and unobservable. Setting $v_{31} = u_{13} = 1$ leads to the Jordan form, and setting $v_{31} = u_{13} = 0$ leads to the system found in (ii), which coincidentally happened to be in the correct controllability form in addition to the observable form.

7.4. Reduce the following system to controllable-observable form.

$$d\mathbf{x}/dt = \begin{pmatrix} -2 & 3+t & -t & 1-3t+t^2 \\ 0 & -1 & 0 & -3 \\ 0 & 0 & 2 & t^2-2t-1 \\ 0 & 0 & 0 & t \end{pmatrix}\mathbf{x} + \begin{pmatrix} t \\ 2 \\ 0 \\ 0 \end{pmatrix} u \qquad y = (0 \ t^2 \ 0 \ -1)\mathbf{x}$$

Using the construction procedure following Theorem 7.5,

$$\mathbf{Q}^x = \begin{pmatrix} t & -5 & -16-2t & -6t-40 \\ 2 & 2 & 2 & 2 \\ 0 & 0 & 0 & 0 \\ 0 & 0 & 0 & 0 \end{pmatrix}$$

$$\mathbf{P}^x = \begin{pmatrix} 0 & t^2 & 0 & -1 \\ 0 & -t^2+2t & 0 & -3t^2-t \\ 0 & t^2-4t+2 & 0 & -3t^3+2t^2-12t-1 \\ 0 & -t^2+6t-6 & 0 & -3t^4+2t^3-24t^2+15t-18 \end{pmatrix}$$

Then from $\mathbf{P}^x\mathbf{Q}^x = \mathbf{R}^1\mathbf{S}^1$,

$$\mathbf{R}^1 = \begin{pmatrix} t^2 \\ -t^2+2t \\ t^2-4t+2 \\ -t^2+6t-6 \end{pmatrix} \qquad \mathbf{S}^1 = (2 \ 2 \ 2 \ 2)$$

Solving $\mathbf{P}^x\mathbf{V}_1 = \mathbf{R}^1$ and $\mathbf{U}_1\mathbf{Q}^x = \mathbf{S}^1$ gives

$$\mathbf{V}_1 = \begin{pmatrix} 0 \\ 1 \\ 0 \\ 0 \end{pmatrix} \qquad \mathbf{U}_1 = (0 \ 1 \ 0 \ 0)$$

Factoring $\mathbf{P}^x - \mathbf{R}^1\mathbf{U}_1$ and $\mathbf{Q}^x - \mathbf{V}_1\mathbf{S}^1$ gives

$$\mathbf{U}_2 = (0 \ 0 \ 0 \ 1) \qquad \mathbf{V}_3 = \begin{pmatrix} 1 \\ 0 \\ 0 \\ 0 \end{pmatrix} \qquad \mathbf{R}^2 = \begin{pmatrix} -1 \\ -3t^2-t \\ -3t^3+2t^2-12t-1 \\ -3t^4+2t^3-24t^2+15t-18 \end{pmatrix}$$

$$\mathbf{S}^3 = (t \ -5 \ -16-2t \ -6t-40)$$

Then $\mathbf{V}_2^T = (0 \ 0 \ 0 \ 1)$ and $\mathbf{V}_4^T = (0 \ 0 \ 1 \ 0)$ and $\mathbf{T} = (\mathbf{V}_1 \mid \mathbf{V}_2 \mid \mathbf{V}_3 \mid \mathbf{V}_4)$. It is interesting to note that the equivalence transformation

$$\mathbf{T} = \begin{pmatrix} t & 0 & 1 & 0 \\ 1 & 0 & 0 & 0 \\ 0 & t & 0 & 1 \\ 0 & 1 & 0 & 0 \end{pmatrix}$$

will also put the system in the form (7.7).

7.5. Reduce the following system to controllable form:

$$d\mathbf{x}/dt = \begin{pmatrix} \dfrac{1}{t} & -t \\ -\dfrac{\sin t}{t} & -\cos t \end{pmatrix}\mathbf{x} + \begin{pmatrix} t \\ \cos t \end{pmatrix} u$$

First we calculate $\mathbf{Q}^x$ and use elementary row operations to obtain

$$\mathbf{E}(t)\mathbf{Q}^x = \begin{pmatrix} t & \cos t \\ -\cos t & t \end{pmatrix}\begin{pmatrix} t & t\cos t \\ \cos t & \cos^2 t \end{pmatrix} = \begin{pmatrix} t^2 + \cos^2 t & t^2\cos t + \cos^3 t \\ 0 & 0 \end{pmatrix}$$

The required transformation is

$$\mathbf{T} = \mathbf{E}^{-1}(t) = \frac{1}{t^2 + \cos^2 t}\begin{pmatrix} t & -\cos t \\ \cos t & t \end{pmatrix}$$

7.6. Put the time-invariant system

$$\mathbf{x}(m+1) = \begin{pmatrix} 1 & 0 & 0 \\ 2 & 2 & 3 \\ -2 & 0 & -1 \end{pmatrix}\mathbf{x}(m) + \begin{pmatrix} 1 \\ 2 \\ 2 \end{pmatrix} u(m)$$

(i) into first canonical form (*7.10*) and (ii) into phase-variable canonical form (*7.12*).

Note this is of the same form as the system of Problem 7.3, except that this is a discrete-time system. Since the controllability matrix has the same form in the time-invariant case, the procedures developed there can be used directly. (See Problem 7.19 for time-varying discrete-time systems.)

From part (i) of Problem 7.3, the system can be put in the form

$$\mathbf{z}(m+1) = \begin{pmatrix} 1 & 0 & 0 \\ -2 & -1 & 3 \\ 0 & 0 & 2 \end{pmatrix}\mathbf{z}(m) + \begin{pmatrix} 1 \\ 0 \\ 0 \end{pmatrix} u(m)$$

Therefore the best we can do is put the controllable subsystem

$$\mathbf{z}_1(m+1) = \begin{pmatrix} 1 & 0 \\ -2 & -1 \end{pmatrix}\mathbf{z}_1(m) + \begin{pmatrix} 1 \\ 0 \end{pmatrix} u(m)$$

into the desired form. The required transformation is

$$\mathbf{T} = \mathbf{Q}^{z_1}\mathbf{K} = \begin{pmatrix} 1 & -1 \\ 0 & 2 \end{pmatrix}\begin{pmatrix} 0 & 1 \\ -1 & 0 \end{pmatrix} = \begin{pmatrix} 1 & 1 \\ -2 & 0 \end{pmatrix}$$

to the first canonical form

$$\mathbf{z}_c(m+1) = \begin{pmatrix} 0 & 1 \\ 1 & 0 \end{pmatrix}\mathbf{z}_c(m) + \begin{pmatrix} 0 \\ 1 \end{pmatrix} u(m)$$

To obtain the phase-variable canonical form,

$$\mathbf{T} = \mathbf{Q}^{z_1}\mathbf{Q}^{z_p\,-1} = \begin{pmatrix} 1 & -1 \\ 0 & 2 \end{pmatrix}\begin{pmatrix} 0 & -1 \\ 1 & q_{11} \end{pmatrix}^{-1}$$

where $q_{11} = -a_2 = \alpha_1$, from Example 7.3. From $\mathbf{A}^w\mathbf{e}_n = \mathbf{Q}^{z_1 -1}(\mathbf{A}^{z_1}\mathbf{q}_n^{z_1} - d\mathbf{q}_n^{z_1}/dt)$ we obtain

$$\begin{pmatrix} \alpha_2 \\ -\alpha_1 \end{pmatrix} = \begin{pmatrix} 1 & -1 \\ 0 & 2 \end{pmatrix}^{-1} \begin{pmatrix} 1 & 0 \\ -2 & -1 \end{pmatrix} \begin{pmatrix} -1 \\ 2 \end{pmatrix} = \begin{pmatrix} -1 \\ 0 \end{pmatrix}$$

Then $\mathbf{T} = \begin{pmatrix} 1 & 1 \\ -2 & 0 \end{pmatrix}$ from which we obtain the phase-variable canonical form

$$\mathbf{z}_p(m+1) = \begin{pmatrix} 0 & 1 \\ 1 & 0 \end{pmatrix} \mathbf{z}_p(m) + \begin{pmatrix} 0 \\ 1 \end{pmatrix} u(m)$$

By chance, this also happens to be in the form of the first canonical form.

7.7. Put the time-varying system

$$d\mathbf{x}/dt = \begin{pmatrix} t & \sin t \\ 1 & -1 \end{pmatrix} \mathbf{x} + \begin{pmatrix} 2 \\ -1 \end{pmatrix} u \qquad \text{for } t \geq 0$$

(i) into first canonical form, (ii) into second canonical form, and (iii) find a scalar second-order equation with the given state space representation.

To obtain the first canonical form,

$$\mathbf{T} = \mathbf{Q}^x\mathbf{K} = \begin{pmatrix} 2 & -2t + \sin t \\ -1 & -3 \end{pmatrix} \begin{pmatrix} 0 & 1 \\ -1 & 0 \end{pmatrix} = \begin{pmatrix} 2t - \sin t & 2 \\ 3 & -1 \end{pmatrix}$$

from which

$$d\mathbf{z}_c/dt = \begin{pmatrix} -\alpha_1(t) & 1 \\ -\alpha_2(t) & 0 \end{pmatrix} \mathbf{z}_c + \begin{pmatrix} 0 \\ 1 \end{pmatrix} u$$

where

$$\begin{pmatrix} \alpha_1(t) \\ \alpha_2(t) \end{pmatrix} = \frac{1}{6 + 2t - \sin t} \begin{pmatrix} 8 - 4t - 2t^2 + (t-1)\sin t - \cos t \\ 6 - 6t - 2t^2 - (t+6)\sin t - 3\cos t + \sin^2 t \end{pmatrix}$$

To obtain the phase-variable form,

$$\mathbf{T} = \mathbf{Q}^x\mathbf{Q}^{z-1} = \begin{pmatrix} 2 & -2t + \sin t \\ -1 & -3 \end{pmatrix} \begin{pmatrix} 0 & -1 \\ 1 & -\alpha_1(t) \end{pmatrix}^{-1} = \begin{pmatrix} 2t - 2\alpha_1 - \sin t & 2 \\ \alpha_1 + 3 & -1 \end{pmatrix}$$

from which

$$d\mathbf{z}_p/dt = \begin{pmatrix} 0 & 1 \\ \alpha_2 + d\alpha_1/dt & \alpha_1 \end{pmatrix} \mathbf{z}_p + \begin{pmatrix} 0 \\ 1 \end{pmatrix} u$$

The second-order scalar equation corresponding to this is

$$d^2y/dt^2 - \alpha_1\, dy/dt - (\alpha_2 + d\alpha_1/dt)y = u$$

where $y = z_{p1}$.

7.8. Transform to canonical form the multiple-input system

$$d\mathbf{x}/dt = \begin{pmatrix} 1 & 2 & -1 \\ 0 & 1 & 0 \\ 1 & -4 & 3 \end{pmatrix} \mathbf{x} + \begin{pmatrix} 0 & 1 \\ 0 & 1 \\ 1 & 1 \end{pmatrix} \mathbf{u}$$

To obtain the form (*7.13*), we calculate the two $\mathbf{Q}^x$ matrices resulting from each of the columns of the $\mathbf{B}^x$ matrix separately:

$$\mathbf{Q}^x\ (u_1 \text{ only}) \;=\; \begin{pmatrix} 0 & 1 & -4 \\ 0 & 0 & 0 \\ 1 & -3 & 8 \end{pmatrix} \qquad \mathbf{Q}^x\ (u_2 \text{ only}) \;=\; \begin{pmatrix} 1 & -2 & 4 \\ 1 & -1 & 1 \\ 1 & 0 & -2 \end{pmatrix}$$

Note both of these matrices are singular. However, choosing $\mathbf{T}$ as the first two columns of $\mathbf{Q}^x\ (u_1 \text{ only})$ and the first column of $\mathbf{Q}^x\ (u_2 \text{ only})$ gives

$$\mathbf{T} \;=\; \begin{pmatrix} 0 & 1 & 1 \\ 0 & 0 & 1 \\ 1 & -3 & 1 \end{pmatrix} \qquad \mathbf{T}^{-1} \;=\; \begin{pmatrix} 3 & -4 & 1 \\ 1 & -1 & 0 \\ 0 & 1 & 0 \end{pmatrix}$$

so that

$$d\mathbf{w}/dt \;=\; \begin{pmatrix} 0 & 4 & 2 \\ -1 & 4 & 1 \\ 0 & 0 & 1 \end{pmatrix}\mathbf{w} + \begin{pmatrix} 1 & 0 \\ 0 & 0 \\ 0 & 1 \end{pmatrix}\mathbf{u}$$

Also, $\mathbf{T}$ could have been chosen as the first column of $\mathbf{Q}^x\ (u_1 \text{ only})$ and the first two columns of $\mathbf{Q}^x\ (u_2 \text{ only})$.

To transform to a form analogous to the first canonical form, let $\mathbf{w} = \mathbf{Kz}$ where

$$\mathbf{K} \;=\; \begin{pmatrix} 0 & 0 & 1 \\ 0 & -1 & 0 \\ 1 & 0 & 0 \end{pmatrix}$$

Then

$$d\mathbf{z}/dt \;=\; \begin{pmatrix} 1 & 0 & 0 \\ -1 & 4 & 1 \\ 2 & -4 & 0 \end{pmatrix}\mathbf{z} + \begin{pmatrix} 0 & 1 \\ 0 & 0 \\ 1 & 0 \end{pmatrix}u$$

Supplementary Problems

7.9. Transform $d\mathbf{x}/dt \;=\; \begin{pmatrix} 1 & 2 & -1 \\ 0 & 1 & 0 \\ 1 & -4 & 3 \end{pmatrix}\mathbf{x} + \begin{pmatrix} -1 \\ 0 \\ 0 \end{pmatrix}u$ to Jordan form.

7.10. Transform $d\mathbf{x}/dt \;=\; \begin{pmatrix} -1 & 1 \\ -2 & 1 \end{pmatrix}\mathbf{x} + \begin{pmatrix} 4 \\ -4 \end{pmatrix}u$ to real Jordan form.

7.11. Using the methods of Section 7.4, reduce

$$d\mathbf{x}/dt \;=\; \begin{pmatrix} 1 & 2 & -1 \\ 0 & 1 & 0 \\ 1 & -4 & 3 \end{pmatrix}\mathbf{x} + \begin{pmatrix} 0 \\ 0 \\ 1 \end{pmatrix}u \qquad y \;=\; (1\ \ -1\ \ 1)\mathbf{x}$$

(i) to a controllable system, (ii) to an observable system, (iii) to an observable and controllable system.

7.12. Prove Theorem 7.2, page 149.

7.13. Prove Theorem 7.4, page 150.

7.14. Show that the factorization requirement on Theorem 7.3 can be dropped if $\mathbf{T}(t)$ can be nonsingular and differentiable for times t everywhere dense in $[t_0, t_1]$.

7.15. Consider the system $\frac{d}{dt}\begin{pmatrix} x_1 \\ x_2 \end{pmatrix} = \begin{pmatrix} f_1(t) \\ f_2(t) \end{pmatrix} u$ where $f_1(t) = 1 - \cos t$ for $0 \leq t \leq 2$ and zero elsewhere, and $f_2(t) = 0$ for $0 \leq t \leq 2$ and $1 - \cos t$ elsewhere. Can this system be put in the form of equation *(7.2)*?

7.16. Find the observable states of the system

$$d\mathbf{x}/dt = \begin{pmatrix} 4 & 6 & 2 & 8 \\ 6 & 2 & 8 & 4 \\ 2 & 8 & 4 & 6 \\ 8 & 4 & 6 & 2 \end{pmatrix}\mathbf{x} \qquad y = (1 \; 1 \; 1 \; 1)\mathbf{x}$$

7.17. Check that the transformation of Problem 7.5, page 159, puts the system in the form of equation *(7.2)*, page 149, by calculating $\mathbf{A}^z$ and $\mathbf{B}^z$.

7.18. Develop Theorem 7.3 for time-varying discrete-time systems.

7.19. Develop the transformation to a form similar to equation *(7.11)* for time-varying discrete-time systems.

7.20. Reduce the system $$d\mathbf{x}/dt = \begin{pmatrix} t-1 & 0 & -t+2 \\ -t-2 & 1 & t+2 \\ t & 0 & -t+1 \end{pmatrix}\mathbf{x} + \begin{pmatrix} 1 \\ 1 \\ 0 \end{pmatrix} u$$

to a system of the form of equation *(7.2)*.

7.21. Given the time-invariant system $d\mathbf{x}/dt = \mathbf{A}\mathbf{x} + \mathbf{e}_n u$ where the system is in phase-variable canonical form as given by equation *(7.12)*. Let $\mathbf{z} = \mathbf{T}\mathbf{x}$ where $\mathbf{z}$ is in the Jordan canonical form $d\mathbf{z}/dt = \mathbf{\Lambda}\mathbf{z} + \mathbf{b}u$ and $\mathbf{\Lambda}$ is a diagonal matrix. Show that $\mathbf{T}$ is the Vandermonde matrix of eigenvalues.

7.22. Verify the relationship for the q_{ik} in terms of the a_k following equation *(7.12)* for a third-order system.

7.23. Solve for the a_i in terms of the α_i for $i = 1, 2, 3$ (third-order system) in a manner analogous to Example 7.3, page 154.

7.24. Transform the system $$\frac{d\mathbf{x}}{dt} = \begin{pmatrix} -11 & 4 & 6 \\ -1/2 & 1 & 0 \\ -27/2 & 6 & 7 \end{pmatrix}\mathbf{x} + \begin{pmatrix} 6 \\ 1 \\ 8 \end{pmatrix} u$$

to phase-variable canonical form.

7.25. Transform the system $$d\mathbf{x}/dt = \begin{pmatrix} -1 & 1 & 0 \\ 0 & -1 & -1 \\ -e^{-t} & e^{-t} & -2 \end{pmatrix}\mathbf{x} + \begin{pmatrix} e^{-t} \\ 0 \\ 0 \end{pmatrix} u$$

to phase-variable canonical form.

7.26. Using the results of Section 7.5, find the transformation $\mathbf{x} = \mathbf{T}\mathbf{w}$ that puts the system

$$d\mathbf{x}/dt = \begin{pmatrix} 1 & -1 & 2 \\ 0 & -2 & 1 \\ -1 & 3 & 0 \end{pmatrix}\mathbf{x} + \begin{pmatrix} 1 & 3 \\ 0 & 1 \\ 0 & 2 \end{pmatrix}\mathbf{u}$$

into the form $$d\mathbf{w}/dt = \begin{pmatrix} 0 & -3 & 6 \\ -1 & -1 & 0 \\ 0 & 1 & 0 \end{pmatrix}\mathbf{w} + \begin{pmatrix} 1 & 0 \\ 0 & 0 \\ 0 & 1 \end{pmatrix}\mathbf{u}$$

7.27. Obtain explicit formulas to go to phase-variable canonical form directly in the case of time-invariant systems.

7.28. Use the duality principle to find a transformation that puts the system $d\mathbf{x}/dt = \mathbf{A}(t)\mathbf{x}$ and $\mathbf{y} = \mathbf{C}(t)\mathbf{x}$ into the form

$$\frac{d\mathbf{z}}{dt} = \begin{pmatrix} 0 & 0 & \ldots & 0 & \alpha_1(t) \\ 1 & 0 & \ldots & 0 & \alpha_2(t) \\ 0 & 1 & \ldots & 0 & \alpha_3(t) \\ \cdots & \cdots & \cdots & \cdots & \cdots \\ 0 & 0 & \ldots & 1 & \alpha_n(t) \end{pmatrix} \mathbf{z} \qquad \mathbf{y} = (0 \;\; 0 \;\; \ldots \;\; 0 \;\; 1)\mathbf{z}$$

7.29. Prove that $\|\mathbf{T}\| < \infty$ for the transformation to phase-variable canonical form.

Answers to Supplementary Problems

7.9. $\mathbf{T} = \begin{pmatrix} 1 & -1 & 0 \\ 0 & 0 & \beta \\ -1 & 0 & 2\beta \end{pmatrix}$ where β is any number $\neq 0$.

7.10. $\mathbf{T} = \begin{pmatrix} 4 & -8 \\ -4 & -12 \end{pmatrix}$

7.11. There is one controllable and observable state, one controllable and unobservable state, and one uncontrollable and observable state.

7.15. No. $\mathbf{Q}^x = \mathbf{V}_1(1 \;\; 0)$ but $\mathbf{V}_1$ does not have rank one everywhere.

7.16. The transformation $\mathbf{T} = \left(\begin{array}{c|c|c|c} 1 & & & \\ 20 & \mathbf{r}_{41} & \mathbf{r}_{42} & \mathbf{r}_{43} \\ 400 & & & \\ 8000 & & & \end{array}\right)$

puts the system into the form of equation *(7.6)*, for any $\mathbf{r}_{4i}$ that make $\mathbf{T}$ nonsingular. Also, Jordan form can be used but is more difficult algebraically.

7.23. $-a_3 = \alpha_1, \;\; -a_2 = \alpha_2 + 2\dot{\alpha}_1, \;\; -a_1 = \alpha_3 + \dot{\alpha}_2 + \ddot{\alpha}_1$

7.24. $\mathbf{T}^{-1} = \begin{pmatrix} -3/2 & 1 & 1 \\ 5/2 & 1 & -2 \\ -1 & -1 & 1 \end{pmatrix}$

7.25. $\mathbf{T}^{-1} = \begin{pmatrix} 0 & e^{2t} & 0 \\ 0 & e^{2t} & -e^{2t} \\ e^t & e^{2t} - e^t & -e^{2t} \end{pmatrix} \qquad \mathbf{A} = \begin{pmatrix} 0 & 1 & 0 \\ 0 & 0 & 1 \\ 2e^{-t} & -e^{-t} & 1 \end{pmatrix}$

7.26. $\mathbf{T} = \begin{pmatrix} 1 & -1 & 3 \\ 0 & 0 & 1 \\ 0 & 1 & 2 \end{pmatrix}$

7.28. This form is obtained using the same transformation that puts the system $d\mathbf{w}/dt = \mathbf{A}^\dagger(t)\mathbf{w} + \mathbf{C}^\dagger(t)\mathbf{u}$ into phase-variable canonical form.

7.29. The elements of $\mathbf{Q}^{z-1}$ are a linear combination of the elements of $\mathbf{Q}^z$, which are always finite as determined by the recursion relation.

Chapter 8

Relations with Classical Techniques

8.1 INTRODUCTION

Classical techniques such as block diagrams, root locus, error constants, etc., have been used for many years in the analysis and design of time-invariant single input–single output systems. Since this type of system is a subclass of the systems that can be analyzed by state space methods, we should expect that these classical techniques can be formulated within the framework already developed in this book. This formulation is the purpose of the chapter.

8.2 MATRIX FLOW DIAGRAMS

We have already studied flow diagrams in Chapter 2 as a graphical aid to obtaining the state equations. The flow diagrams studied in Chapter 2 used only four basic objects (summer, scalor, integrator, delayor) whose inputs and outputs were scalar functions of time. Here we consider vector inputs and outputs to these basic objects. In this chapter these basic objects will have the same symbols, and Definitions 2.1–2.4, pages 16–17, hold with the following exceptions. A summer has n m-vector inputs $\mathbf{u}_1(t), \mathbf{u}_2(t), \ldots, \mathbf{u}_n(t)$ and one output m-vector $\mathbf{y}(t) = \pm\mathbf{u}_1(t) \pm \mathbf{u}_2(t) \pm \cdots \pm \mathbf{u}_n(t)$. A scalor has one m-vector input $\mathbf{u}(t)$ and one output k-vector $\mathbf{y}(t) = \mathbf{A}(t)\mathbf{u}(t)$, where $\mathbf{A}(t)$ is a $k \times m$ matrix. An integrator has one m-vector input $\mathbf{u}(t)$ and one output m-vector $\mathbf{y}(t) = \mathbf{y}(t_0) + \int_{t_0}^{t} \mathbf{u}(\tau)\, d\tau$. To denote vector (instead of purely scalar) time function flow from one basic object to another, thick arrows will be used.

Example 8.1.

Consider the two input – one output system

$$\frac{d}{dt}\begin{pmatrix} x_1 \\ x_2 \end{pmatrix} = \begin{pmatrix} 0 & 1 \\ 0 & 0 \end{pmatrix}\begin{pmatrix} x_1 \\ x_2 \end{pmatrix} + \begin{pmatrix} 2 & 0 \\ 0 & 1 \end{pmatrix}\begin{pmatrix} u_1 \\ u_2 \end{pmatrix} \qquad y = (3 \quad 2)\mathbf{x}$$

This can be diagrammed as in Fig. 8-1.

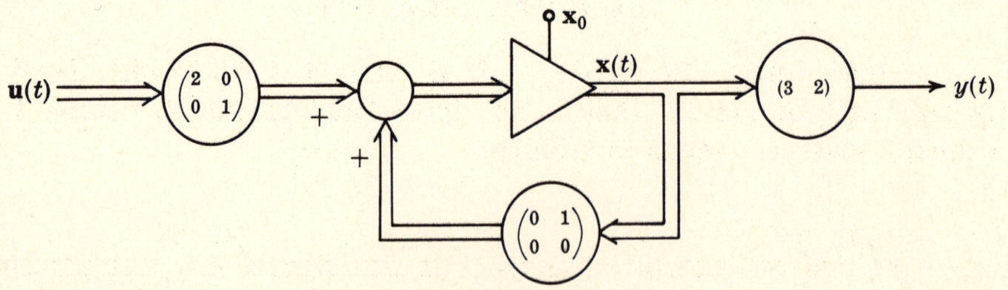

Fig. 8-1

Also, flow diagrams of transfer functions (block diagrams) can be drawn in a similar manner for time-invariant systems. We denote the Laplace transform of $x(t)$ as $\mathcal{L}\{x\}$, etc.

Example 8.2.

The block diagram of the system considered in Example 8.1 is shown in Fig. 8-2.

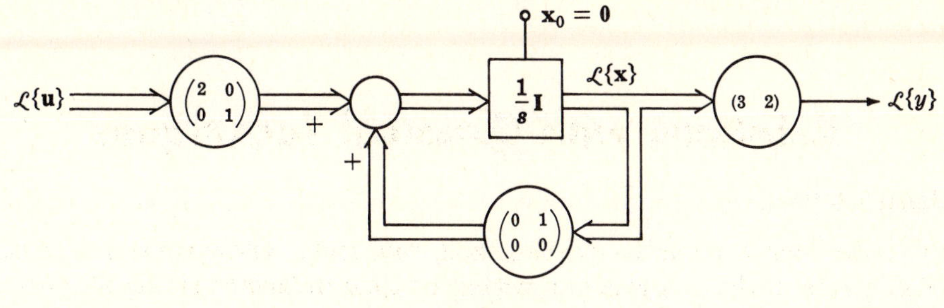

Fig. 8-2

Using equation (5.56) or proceeding analogously from block diagram manipulation, this can be reduced to the diagram of Fig. 8-3 where

$$\mathbf{H}(s) = (6/s \quad (2s+3)/s^2)$$

$$= (3 \quad 2)\left[s\begin{pmatrix}1 & 0\\0 & 1\end{pmatrix} - \begin{pmatrix}0 & 1\\0 & 0\end{pmatrix}\right]^{-1}\begin{pmatrix}2 & 0\\0 & 1\end{pmatrix}$$

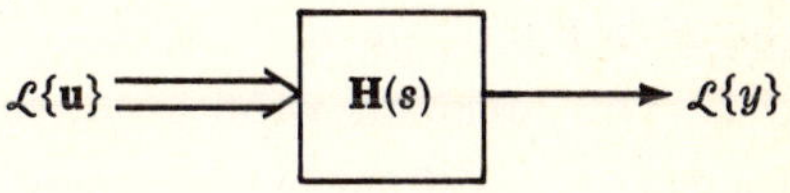

Fig. 8-3

Vector block diagram manipulations are similar to the scalar case, and are as useful to the system designer. Keeping the system representation in matrix form is often helpful, especially when analyzing multiple input–multiple output devices.

8.3 STEADY STATE ERRORS

Knowledge of the type of feedback system that will follow an input with zero steady state error is useful for designers. In this section we shall investigate steady state errors of systems in which only the output is fed back (unity feedback). The development can be extended to nonunity feedback systems, but involves comparing the plant output with a desired output which greatly complicates the notation (see Problem 8.22). Here we extend the classical steady state error theory for systems with scalar unity feedback to time-varying multiple input–multiple output systems. By steady state we mean the asymptotic behavior of a function for large t. The system considered is diagrammed in Fig. 8-4. The plant equation is $d\mathbf{x}/dt = \mathbf{A}(t)\mathbf{x} + \mathbf{B}(t)\mathbf{e}$, the output is $\mathbf{y} = \mathbf{C}(t)\mathbf{x}$ and the reference input is $\mathbf{d}(t) = \mathbf{y}(t) + \mathbf{e}(t)$, where $\mathbf{y}$, $\mathbf{d}$ and $\mathbf{e}$ are all m-vectors. For this system it will always be assumed that the zero output is asymptotically stable, i.e.

$$\lim_{t\to\infty} \mathbf{C}(t)\mathbf{\Phi}_{A-BC}(t,\tau) = \mathbf{0}$$

where $\partial\mathbf{\Phi}_{A-BC}(t,\tau)/\partial t = [\mathbf{A}(t) - \mathbf{B}(t)\mathbf{C}(t)]\mathbf{\Phi}_{A-BC}(t,\tau)$ and $\mathbf{\Phi}_{A-BC}(t,t) = \mathbf{I}$. Further, we shall be concerned only with inputs $\mathbf{d}(t)$ that do not drive $\|\mathbf{y}(t)\|$ to infinity before $t = \infty$, so that we obtain a steady state as t tends to infinity.

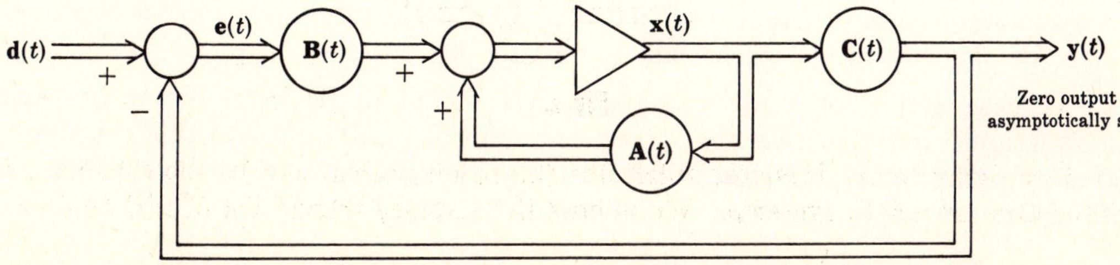

Fig. 8-4. Unity Feedback System with Asymptotically Stable Zero Output

Theorem 8.1: For the system of Fig. 8-4, $\lim_{t\to\infty} \mathbf{e}(t) = \mathbf{0}$ if and only if $\mathbf{d} = \mathbf{C}(t)\mathbf{w} + \mathbf{g}$ where $d\mathbf{w}/dt = \mathbf{A}(t)\mathbf{w} + \mathbf{B}(t)\mathbf{g}$ for all $t \geq t_0$ in which $\mathbf{g}(t)$ is any function such that $\lim_{t\to\infty} \mathbf{g}(t) = \mathbf{0}$ and $\mathbf{A}, \mathbf{B}, \mathbf{C}$ are unique up to a transformation on $\mathbf{w}$.

Proof: Consider two arbitrary functions $\mathbf{f}(t)$ and $\mathbf{h}(t)$ whose limits may not exist as t tends to ∞. If $\lim_{t\to\infty} [\mathbf{f}(t) - \mathbf{h}(t)] = \mathbf{0}$, then $\mathbf{f}(t) - \mathbf{h}(t) = \mathbf{r}(t)$ for all t, where $\mathbf{r}(t)$ is an arbitrary function such that $\lim_{t\to\infty} \mathbf{r}(t) = \mathbf{0}$. From this, if $\mathbf{0} = \lim_{t\to\infty} \mathbf{e}(t) = \lim_{t\to\infty} [\mathbf{d}(t) - \mathbf{y}(t)]$, then for all t,

$$\mathbf{d}(t) = \mathbf{y}(t) + \mathbf{r}(t) = \mathbf{C}(t)\,\mathbf{\Phi}_{A-BC}(t, t_0)\,\mathbf{x}(t_0) + \int_{t_0}^{t} \mathbf{C}(t)\,\mathbf{\Phi}_{A-BC}(t, \tau)\,\mathbf{B}(\tau)\,\mathbf{d}(\tau)\,d\tau + \mathbf{r}(t)$$

$$= \int_{t_0}^{t} \mathbf{C}(t)\,\mathbf{\Phi}_{A-BC}(t, \tau)\,\mathbf{B}(\tau)\,\mathbf{d}(\tau)\,d\tau + \mathbf{g}(t) + \mathbf{C}(t)\,\mathbf{\Phi}_{A-BC}(t, t_0)\,\mathbf{w}(t_0)$$

where the change of variables $\mathbf{g}(t) = \mathbf{r}(t) + \mathbf{C}(t)\mathbf{\Phi}_{A-BC}(t, t_0)[\mathbf{x}(t_0) - \mathbf{w}(t_0)]$ is one-to-one for arbitrary constant $\mathbf{w}(t_0)$ because $\lim_{t\to\infty} \mathbf{C}(t)\mathbf{\Phi}_{A-BC}(t, t_0) = \mathbf{0}$. This Volterra integral equation for $\mathbf{d}(t)$ is equivalent to the differential equations $d\mathbf{w}/dt = [\mathbf{A}(t) - \mathbf{B}(t)\,\mathbf{C}(t)]\mathbf{w} + \mathbf{B}(t)\mathbf{d}$ and $\mathbf{d} = \mathbf{C}(t)\mathbf{w} + \mathbf{g}$. Substituting the latter equation into the former gives the set of equations that generate any $\mathbf{d}(t)$ such that $\lim_{t\to\infty} \mathbf{e}(t) = \mathbf{0}$.

Conversely, from Fig. 8-4, $d\mathbf{x}/dt = \mathbf{A}(t)\mathbf{x} + \mathbf{B}(t)\mathbf{e} = [\mathbf{A}(t) - \mathbf{B}(t)\,\mathbf{C}(t)]\mathbf{x} + \mathbf{B}(t)\mathbf{d}$. Assuming $\mathbf{d} = \mathbf{C}(t)\mathbf{w} + \mathbf{g}$ and subtracting $d\mathbf{w}/dt = \mathbf{A}(t)\mathbf{w} + \mathbf{B}(t)\mathbf{g}$ gives

$$d(\mathbf{x} - \mathbf{w})/dt = [\mathbf{A}(t) - \mathbf{B}(t)\,\mathbf{C}(t)](\mathbf{x} - \mathbf{w})$$

Then
$$\lim_{t\to\infty} \mathbf{e} = \lim_{t\to\infty} (\mathbf{d} - \mathbf{y}) = \lim_{t\to\infty} [\mathbf{g} - \mathbf{C}(t)(\mathbf{x} - \mathbf{w})]$$

$$= \lim_{t\to\infty} \mathbf{g} - [\lim_{t\to\infty} \mathbf{C}(t)\,\mathbf{\Phi}_{A-BC}(t, t_0)][\mathbf{x}(t_0) - \mathbf{w}(t_0)] = \mathbf{0}$$

From the last part of the proof we see that $\mathbf{e}(t) = \mathbf{g}(t) - \mathbf{C}(t)\,\mathbf{\Phi}_{A-BC}(t, t_0)[\mathbf{x}(t_0) - \mathbf{w}(t_0)]$ regardless of what the function $\mathbf{g}(t)$ is. Therefore, the system $d\mathbf{w}/dt = \mathbf{A}(t)\mathbf{w} + \mathbf{B}(t)\mathbf{g}$ with $\mathbf{d} = \mathbf{C}(t)\mathbf{w} + \mathbf{g}$ and the system $d\mathbf{x}/dt = [\mathbf{A}(t) - \mathbf{B}(t)\,\mathbf{C}(t)]\mathbf{x} + \mathbf{B}(t)\mathbf{d}$ with $\mathbf{e} = \mathbf{d} - \mathbf{C}(t)\mathbf{x}$ are inverse systems. Another way to see this is that in the time-invariant case we have the transfer function matrix of the open loop system $\mathbf{H}(s) = \mathbf{C}(s\mathbf{I} - \mathbf{A})^{-1}\mathbf{B}$ relating $\mathbf{e}$ to $\mathbf{y}$. Then for zero initial conditions, $\mathcal{L}\{\mathbf{d}\} = [\mathbf{H}(s) + \mathbf{I}]\mathcal{L}\{\mathbf{g}\}$ and $\mathcal{L}\{\mathbf{e}\} = [\mathbf{H}(s) + \mathbf{I}]^{-1}\mathcal{L}\{\mathbf{d}\}$ so that $\mathcal{L}\{\mathbf{g}\} = \mathcal{L}\{\mathbf{e}\}$. Consequently the case where $\mathbf{g}(t)$ is a constant vector forms a sort of boundary between functions that grow with time and those that decay. Of course this neglects those functions (like $\sin t$) that oscillate, for which we can also use Theorem 8.1.

Furthermore, the effect of nonzero initial conditions $\mathbf{w}(t_0)$ can be incorporated into $\mathbf{g}(t)$. Since we are interested in only the output characteristics of the plant, we need concern ourselves only with observable states. Also, because uncontrollable but observable states of the plant must tend to zero by the assumed asymptotic stability of the closed loop system, we need concern ourselves only with states that are both observable and controllable. Use of equation *(6.9)* shows that the response due to any $w_i(t_0)$ is identical to the response due to an input made up of delta functions and derivatives of delta functions. These are certainly included in the class of all $\mathbf{g}(t)$ such that $\lim_{t\to\infty} \mathbf{g}(t) = \mathbf{0}$.

Since the case $\mathbf{g}(t) = \text{constant}$ forms a sort of boundary between increasing and decreasing functions, and since we can incorporate initial conditions into this class, we may take $\mathbf{g}(t)$ as the unit vectors to give an indication of the kind of input the system can follow with zero error. In other words, consider inputs

$$\mathbf{d}(t) = \mathbf{C}(t)\int_{t_0}^{t} \mathbf{\Phi}_A(t, \tau)\,\mathbf{B}(\tau)\,\mathbf{g}(\tau)\,d\tau = \mathbf{C}(t)\int_{t_0}^{t} \mathbf{\Phi}_A(t, \tau)\,\mathbf{B}(\tau)\,\mathbf{e}_i\,d\tau \qquad \text{for } i = 1, 2, \ldots, m$$

which can be combined into the matrix function

$$\mathbf{C}(t)\int_{t_0}^{t} \mathbf{\Phi}_A(t,\tau)\,\mathbf{B}(\tau)(\mathbf{e}_1\,|\,\mathbf{e}_2\,|\,\ldots\,|\,\mathbf{e}_m)d\tau \quad = \quad \mathbf{C}(t)\int_{t_0}^{t} \mathbf{\Phi}_A(t,\tau)\,\mathbf{B}(\tau)\,d\tau$$

Inputs of this form give unity error, and probably inputs that go to infinity any little bit slower than this will give zero error.

Example 8.3.

Consider the system of Example 8.1 in which $e(t) = u_2(t)$ and there is no input $u_1(t)$. The zero input, unity feedback system is then $\dfrac{d\mathbf{x}}{dt} = \left[\begin{pmatrix} 0 & 1 \\ 0 & 0 \end{pmatrix} - \begin{pmatrix} 0 \\ 1 \end{pmatrix}(3 \;\; 2)\right]\mathbf{x}$ whose output

$$y \quad = \quad (3 \;\; 2)\mathbf{x} \quad = \quad e^{-t}\left\{[3x_1(0) + 2x_2(0)]\cos\sqrt{2}\,t + \frac{1}{\sqrt{2}}[x_1(0) + x_2(0)]\sin\sqrt{2}\,t\right\}$$

tends to zero asymptotically. Consequently Theorem 8.1 applies to the unity feedback system, so that the equations

$$\frac{d}{dt}\begin{pmatrix} w_1 \\ w_2 \end{pmatrix} \quad = \quad \begin{pmatrix} 0 & 1 \\ 0 & 0 \end{pmatrix}\begin{pmatrix} w_1 \\ w_2 \end{pmatrix} + \begin{pmatrix} 0 \\ 1 \end{pmatrix} g \qquad\qquad d \quad = \quad (3 \;\; 2)\begin{pmatrix} w_1 \\ w_2 \end{pmatrix} + g$$

where $\lim_{t\to\infty} g(t) = 0$, generate the class of inputs $d(t)$ that the system can follow with zero error. Solving this system of equations gives

$$d(t) \quad = \quad 3w_1(0) + 2w_2(0) + 3tw_2(0) + \int_0^t [3(t-\tau) + 2]g(\tau)\,d\tau + g(t)$$

For $g(t) = 0$, we see that the system can follow arbitrary steps and ramps with zero error, which is in agreement with the classical conclusion that the system is of type 2. Also, evaluating

$$\mathbf{C}(t)\int_0^t \mathbf{\Phi}_A(t,\tau)\,\mathbf{B}(\tau)\,d\tau \quad = \quad \int_0^t [3(t-\tau) + 2]\,d\tau \quad = \quad 1.5t^2 + 2t$$

shows the system will follow t^2 with constant error and will probably follow with zero error any function $t^{2-\epsilon}$ for any $\epsilon > 0$. This is in fact the case, as can be found by taking $\mathbf{g}(t) = t^{-\epsilon}$.

Now if we consider the system of Example 8.1 in which $e(t) = u_1(t)$ and there is no input $u_2(t)$, then the closed loop system is $\dfrac{d\mathbf{x}}{dt} = \left[\begin{pmatrix} 0 & 1 \\ 0 & 0 \end{pmatrix} - \begin{pmatrix} 2 \\ 0 \end{pmatrix}(3 \;\; 2)\right]\mathbf{x}$. The output of this system is $y = 0.5x_2(0) + [3x_1(0) + 1.5x_2(0)]e^{-6t}$ which does not tend to zero asymptotically so that Theorem 8.1 cannot be used.

Definition 8.1: The system of Fig. 8-4 is called a *type-l system* $(l = 1, 2, \ldots)$ when $\lim_{t\to\infty} \mathbf{e}(t) = \mathbf{0}$ for the inputs $\mathbf{d}_i = (t-t_0)^{l-1}U(t-t_0)\mathbf{e}_i$ for all $i = 1, 2, \ldots, m$.

In the definition, $U(t-t_0)$ is the unit step function starting at $t = t_0$ and $\mathbf{e}_i$ is the ith unit vector. All systems that do not satisfy Definition 8.1 will be called type-0 systems.

Use of Theorem 8.1 involves calculation of the transition matrix and integration of the superposition integral. For classical scalar type-l systems the utility of Definition 8.1 is that the designer can simply observe the power of s in the denominator of the plant transfer function and know exactly what kind of input the closed loop system will follow. The following theorem is the extension of this, but is applicable only to time-invariant systems with the plant transfer function matrix $\mathbf{H}(s) = \mathbf{C}(s\mathbf{I} - \mathbf{A})^{-1}\mathbf{B}$.

Theorem 8.2: The time-invariant system of Fig. 8-4 is of type $l \geq 1$ if and only if $\mathbf{H}(s) = s^{-l}\mathbf{R}(s) + \mathbf{P}(s)$ where $\mathbf{R}(s)$ and $\mathbf{P}(s)$ are any matrices such that $\lim_{s\to 0} s\mathbf{R}^{-1}(s) = \mathbf{0}$ and $\|\lim_{s\to 0} s^{l-1}\mathbf{P}(s)\| < \infty$.

Proof: From Theorem 8.1, the system is of type l if and only if $\mathcal{L}\{(\mathbf{d}_1 | \mathbf{d}_2 | \ldots | \mathbf{d}_m)\} = (l-1)!\,s^{-l}\mathbf{I} = [\mathbf{H}(s)+\mathbf{I}]\mathbf{G}(s)$ where $\mathcal{L}\{\mathbf{g}_i\}$, the columns of $\mathbf{G}(s)$, are the Laplace transforms of any functions $\mathbf{g}_i(t)$ such that $\mathbf{0} = \lim_{t\to\infty} \mathbf{g}_i(t) = \lim_{s\to 0} s\mathcal{L}\{\mathbf{g}_i\}$ where $s\mathcal{L}\{\mathbf{g}_i\}$ is analytic for $\operatorname{Re} s \geqq 0$.

First, assume $\mathbf{H}(s) = s^{-l}\mathbf{R}(s) + \mathbf{P}(s)$ where $\lim_{s\to 0} s\mathbf{R}^{-1}(s) = \mathbf{0}$ so that $\mathbf{R}^{-1}(s)$ exists in a neighborhood of $s = 0$. Choose

$$\mathbf{G}(s) = (l-1)!\,s^{-l}[\mathbf{H}(s)+\mathbf{I}]^{-1} = (l-1)!\,[\mathbf{R}(s)+s^l\mathbf{P}(s)+s^l\mathbf{I}]^{-1}$$

Since $[\mathbf{H}(s)+\mathbf{I}]^{-1}$ is the asymptotically stable closed loop transfer function matrix, it is analytic for $\operatorname{Re} s \geqq 0$. Then $s\mathbf{G}(s)$ has at most a pole of order $l-1$ at $s=0$ in the region $\operatorname{Re} s \geqq 0$. In some neighborhood of $s=0$ where $\mathbf{R}^{-1}(s)$ exists we can expand

$$s\mathbf{G}(s) = (l-1)!\,s[\mathbf{R}(s)+s^l\mathbf{P}(s)+s^l\mathbf{I}]^{-1} = (l-1)!\,s\mathbf{R}^{-1}(s)[\mathbf{I}+\mathbf{Z}(s)+\mathbf{Z}^2(s)+\cdots]$$

where $\mathbf{Z}(s) = s\mathbf{R}^{-1}(s)[s^{l-1}\mathbf{P}(s)+s^{l-1}\mathbf{I}]$. Since $\lim_{s\to 0}\mathbf{Z}(s) = \mathbf{0}$, this expansion is valid for small $|s|$, and $\lim_{s\to 0} s\mathbf{G}(s) = \mathbf{0}$. Consequently $s\mathbf{G}(s)$ has no pole at $s = 0$ and must be analytic in $\operatorname{Re} s \geqq 0$ which satisfies Theorem 8.1.

Conversely, assume $\lim_{s\to 0} s\mathbf{G}(s) = \mathbf{0}$ where $s\mathbf{G}(s)$ is analytic in $\operatorname{Re} s \geqq 0$. Write $\mathbf{H}(s) = s^{-l}\mathbf{R}(s)+\mathbf{P}(s)$ where $\mathbf{P}(s)$ is any matrix such that $\|\lim_{s\to 0} s^{l-1}\mathbf{P}(s)\| < \infty$ and $\mathbf{R}(s) = s^l[\mathbf{H}(s)-\mathbf{P}(s)]$ is still arbitrary. Then

$$(l-1)!\,s^{-l}\mathbf{I} = [s^{-l}\mathbf{R}(s)+\mathbf{P}(s)+\mathbf{I}]\mathbf{G}(s)$$

can be solved for $s\mathbf{R}^{-1}(s)$ as

$$(l-1)!\,s\mathbf{R}^{-1}(s) = s\mathbf{G}(s)(\mathbf{I}+\mathbf{W}(s))^{-1} = s\mathbf{G}(s)[\mathbf{I}+\mathbf{W}(s)+\mathbf{W}^2(s)+\cdots]$$

where $(l-1)!\,\mathbf{W}(s) = [s^{l-1}\mathbf{P}(s)+s^{l-1}\mathbf{I}]s\mathbf{G}(s)$. This expansion is valid for $\|s\mathbf{G}(s)\|$ small enough, so that $\mathbf{R}^{-1}(s)$ exists in some neighborhood of $s=0$. Taking limits then gives $\lim_{s\to 0} s\mathbf{R}^{-1}(s) = \mathbf{0}$.

We should be careful in the application of Theorem 8.2, however, in light of Theorem 8.1. The classification of systems into type l is not as clear cut as it appears. A system with $H(s) = (s+\epsilon)^{-1}$ can follow inputs of the form $e^{-\epsilon t}$. As ϵ tends to zero this tends to a step function, so that we need only take ϵ^{-1} on the order of the time of operation of the system.

Unfortunately, for time-varying systems there is no guarantee that if a system is of type N, then it is of type $N-k$ for all $k \geqq 0$. However, this is true for time-invariant systems. (See Problem 8.24.).

Example 8.4.

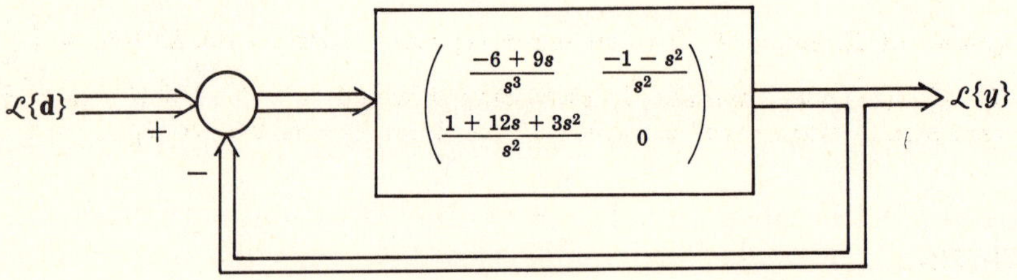

Fig. 8-5

The system shown in Fig. 8-5 has a plant transfer function matrix $\mathbf{H}(s)$ that can be written in the form

$$\mathbf{H}(s) = \frac{1}{s^2}\begin{pmatrix} -6s^{-1}+9 & -1 \\ 1 & 0 \end{pmatrix} + \begin{pmatrix} 0 & -1 \\ 12s^{-1}+3 & 0 \end{pmatrix} = s^{-2}\mathbf{R}(s) + \mathbf{P}(s)$$

in which

$$\left\| \lim_{s\to 0} s\mathbf{P}(s) \right\| = \left\| \lim_{s\to 0} \begin{pmatrix} 0 & -s \\ 12+3s & 0 \end{pmatrix} \right\| < \infty$$

and where

$$\lim_{s\to 0} s\mathbf{R}^{-1}(s) \quad = \quad \lim_{s\to 0} s\begin{pmatrix} 0 & 1 \\ -1 & 9-6s^{-1} \end{pmatrix} \quad = \quad \begin{pmatrix} 0 & 0 \\ 0 & -6 \end{pmatrix} \quad \neq \quad \mathbf{0}$$

Since $\lim_{s\to 0} s\mathbf{R}^{-1}(s)$ has a nonzero element, the system is not of type 2 as appears to be the case upon first inspection. Rewrite $\mathbf{H}(s)$ in the form (where $\mathbf{R}(s)$ and $\mathbf{P}(s)$ are different)

$$\mathbf{H}(s) \quad = \quad \frac{1}{s}\begin{pmatrix} -6s^{-2}+9s^{-1} & -s^{-1} \\ s^{-1}+12 & 0 \end{pmatrix} + \begin{pmatrix} 0 & -1 \\ 3 & 0 \end{pmatrix} \quad = \quad s^{-1}\mathbf{R}(s) + \mathbf{P}(s)$$

Again, $\|\lim_{s\to 0} \mathbf{P}(s)\| < \infty$ but now

$$\lim_{s\to 0} s\mathbf{R}^{-1}(s) \quad = \quad \lim_{s\to 0} s\begin{pmatrix} 0 & \dfrac{s}{1+12s} \\ -s & \dfrac{9s-6}{1+12s} \end{pmatrix} \quad = \quad \begin{pmatrix} 0 & 0 \\ 0 & 0 \end{pmatrix}$$

Since the closed loop system has poles at $-1, -1, -0.5$, and -0.5, the zero output is asymptotically stable. Therefore the system is of type 1.

To find the error constant matrix of a type-l system, we use block diagram manipulations on Fig. 8-4 to get $\mathcal{L}\{\mathbf{e}\} = [\mathbf{I}+\mathbf{H}(s)]^{-1}\mathcal{L}\{\mathbf{d}\}$. If it exists, then

$$\lim_{t\to\infty} \mathbf{e}(t) \quad = \quad \lim_{s\to 0} s[\mathbf{I}+s^{-l}\mathbf{R}(s)+\mathbf{P}(s)]^{-1}\mathcal{L}\{\mathbf{d}\}$$

$$= \quad \lim_{s\to 0} s^{l+1}[s^l\mathbf{I}+\mathbf{R}(s)+s^l\mathbf{P}(s)]^{-1}\mathcal{L}\{\mathbf{d}\} \quad = \quad \lim_{s\to 0} s^{l+1}\mathbf{R}^{-1}(s)\mathcal{L}\{\mathbf{d}\}$$

for any $l > 0$. Then an error constant matrix table can be formed for time-invariant systems of Fig. 8-4.

Steady State Error Constant Matrices

System Type	Step Input	Ramp Input	Parabolic Input
0	$\lim_{s\to 0} [\mathbf{I}+\mathbf{H}(s)]^{-1}$	*	*
1	$\mathbf{0}$	$\lim_{s\to 0} \mathbf{R}^{-1}(s)$	*
2	$\mathbf{0}$	$\mathbf{0}$	$\lim_{s\to 0} \mathbf{R}^{-1}(s)$

In the table * means the system cannot follow all such inputs.

Example 8.5.

The type-1 system of Example 8.4 has an error constant matrix $\lim_{s\to 0} \mathbf{R}^{-1}(s) = \begin{pmatrix} 0 & 0 \\ 0 & -6 \end{pmatrix}$. Thus if the input were $(t-t_0)U(t-t_0)\mathbf{e}_2$, the steady state output would be $[(t-t_0)U(t-t_0)+6]\mathbf{e}_2$. The system can follow with zero steady state error an input of the form $(t-t_0)U(t-t_0)\mathbf{e}_1$.

8.4 ROOT LOCUS

Because root locus is useful mainly for time-invariant systems, we shall consider only time-invariant systems in this section. Both single and multiple inputs and outputs can be considered using vector notation, i.e. we consider

$$d\mathbf{x}/dt \; = \; \mathbf{A}\mathbf{x}+\mathbf{B}\mathbf{u} \qquad \mathbf{y} \; = \; \mathbf{C}\mathbf{x}+\mathbf{D}\mathbf{u} \tag{8.1}$$

Then the transfer function from $\mathbf{y}$ to $\mathbf{u}$ is $\mathbf{H}(s) = \mathbf{C}(s\mathbf{I}-\mathbf{A})^{-1}\mathbf{B}+\mathbf{D}$, with poles determined by $\det(s\mathbf{I}-\mathbf{A}) = 0$. Note for the multiple input and output case these are the poles of the whole system. The eigenvalues of $\mathbf{A}$ determine the time behavior of all the outputs.

We shall consider the case where equations (*8.1*) represent the closed loop system. Suppose that the characteristic equation $\det(s\mathbf{I} - \mathbf{A}) = 0$ is linear in some parameter κ so that it can be written as

$$s^n + (\theta_1 + \psi_1\kappa)s^{n-1} + (\theta_2 + \psi_2\kappa)s^{n-2} + \cdots + (\theta_{n-1} + \psi_{n-1}\kappa)s + \theta_n + \psi_n\kappa = 0$$

This can be rearranged to standard root locus form under κ variation,

$$-1 = \kappa\frac{\psi_1 s^{n-1} + \psi_2 s^{n-2} + \cdots + \psi_{n-1}s + \psi_n}{s^n + \theta_1 s^{n-1} + \cdots + \theta_{n-1}s + \theta_n}$$

The roots of the characteristic equation can be found as κ varies using standard root locus techniques. The assumed form of the characteristic equation results from both loop gain variation and parameter variation of the open loop system.

Example 8.6.

Given the system of Fig. 8-6 with variable feedback gain κ.

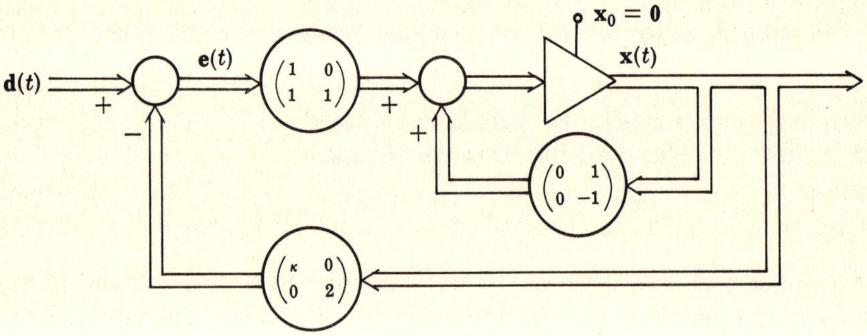

Fig. 8-6

The closed loop system can be written as

$$\frac{d\mathbf{x}}{dt} = \begin{pmatrix} -\kappa & 1 \\ -\kappa & -3 \end{pmatrix}\mathbf{x} + \begin{pmatrix} 1 & 0 \\ 1 & 1 \end{pmatrix}\mathbf{d}$$

The characteristic equation is $s^2 + (3+\kappa)s + 4\kappa = 0$. Putting it into standard root locus form leads to the root locus shown in Fig. 8-7.

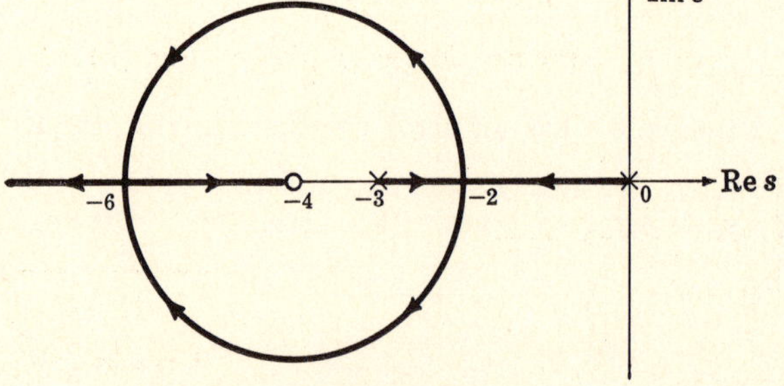

Fig. 8-7

Example 8.7.

The feedback system of Fig. 8-8 has an unknown parameter α. Find the effect of variations in α upon the closed loop roots.

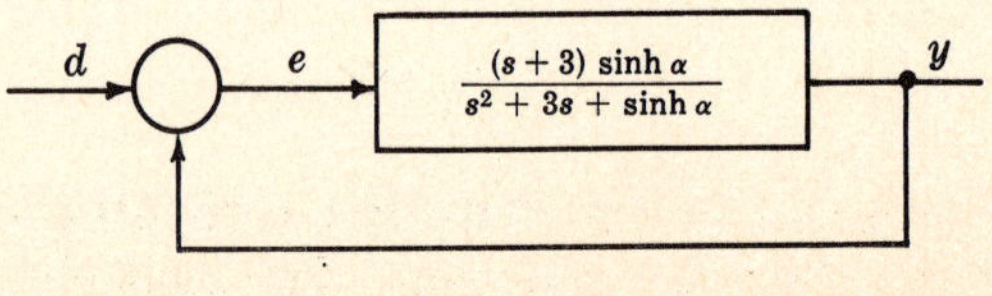

Fig. 8-8

Let $\sinh\alpha = \kappa$. The usual procedure is to set the open loop transfer function equal to -1 to find the closed loop poles of a unity feedback system.

$$-1 \quad = \quad \frac{(s+3)\kappa}{s^2+3s+\kappa}$$

This can be rearranged to form the characteristic equation of the closed loop system, $s^2+3s+\kappa+(s+3)\kappa = 0$. Further rearrangement gives the standard root locus form under κ variation.

$$-1 \quad = \quad \kappa\frac{s+4}{s(s+3)}$$

This happens to give the same root locus as in the previous example for $\sinh\alpha \geqq 0$.

8.5 NYQUIST DIAGRAMS

First we consider the time-invariant single input–single output system whose block diagram is shown in Fig. 8-9. The standard Nyquist procedure is to plot $G(s)H(s)$ where s varies along the Nyquist path enclosing the right half s-plane. To do this, we need polar plots of $G(j\omega)H(j\omega)$ where ω varies from $-\infty$ to $+\infty$.

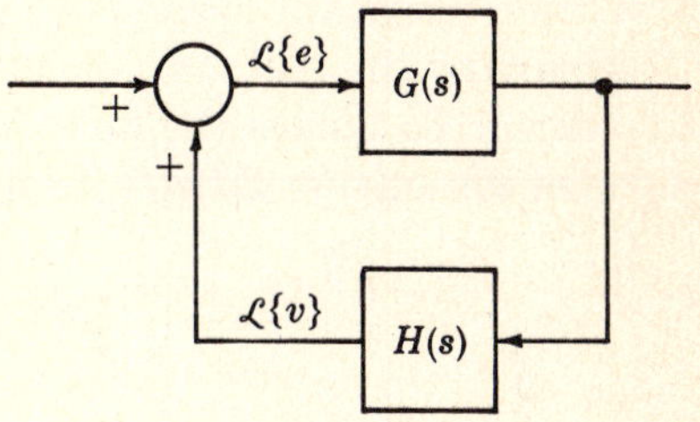

Fig. 8-9

Using standard procedures, we break the closed loop between e and v. Then setting this up in state space form gives

$$d\mathbf{x}/dt = \mathbf{A}\mathbf{x} + \mathbf{b}e \qquad v = \mathbf{c}^\dagger\mathbf{x} + de \tag{8.2}$$

Then $G(j\omega)H(j\omega) = \mathbf{c}^\dagger(j\omega\mathbf{I}-\mathbf{A})^{-1}\mathbf{b}+d$. Usually a choice of state variable $\mathbf{x}$ can be found such that the gain or parameter variation κ of interest can be incorporated into the $\mathbf{c}$ vector only. Digital computer computation of $(j\omega\mathbf{I}-\mathbf{A})^{-1}\mathbf{b}$ as ω varies can be most easily done by iterative techniques, such as Gauss-Seidel. Each succeeding evaluation of $(j\omega_{i+1}\mathbf{I}-\mathbf{A})^{-1}\mathbf{b}$ can be started with the initial condition $(j\omega_i\mathbf{I}-\mathbf{A})^{-1}\mathbf{b}$, which usually gives fast convergence.

Example 8.8.

Given the system shown in Fig. 8-10. The state space form of this is, in phase-variable canonical form for the transfer function from e to v,

$$\frac{d\mathbf{x}}{dt} \quad = \quad \begin{pmatrix} 0 & 1 \\ 0 & -1 \end{pmatrix}\mathbf{x} + \begin{pmatrix} 0 \\ 1 \end{pmatrix}e \qquad\qquad v \quad = \quad (\kappa \quad 0)\mathbf{x}$$

Then $\mathbf{c}^\dagger(j\omega\mathbf{I}-\mathbf{A})^{-1}\mathbf{b} \;=\; \dfrac{\kappa}{j\omega(j\omega+1)}$, giving the polar plot of Fig. 8-11.

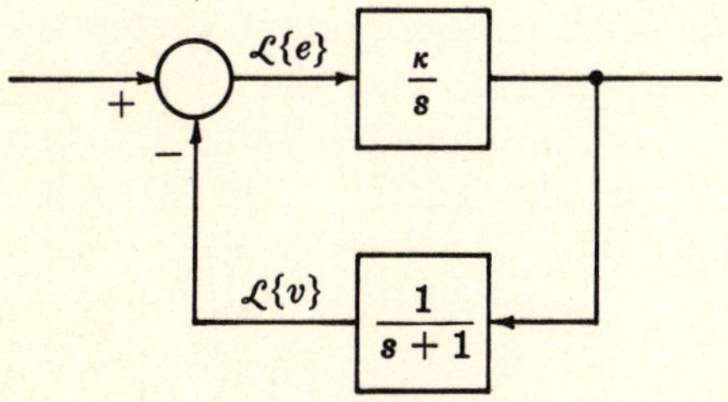

Fig. 8-10

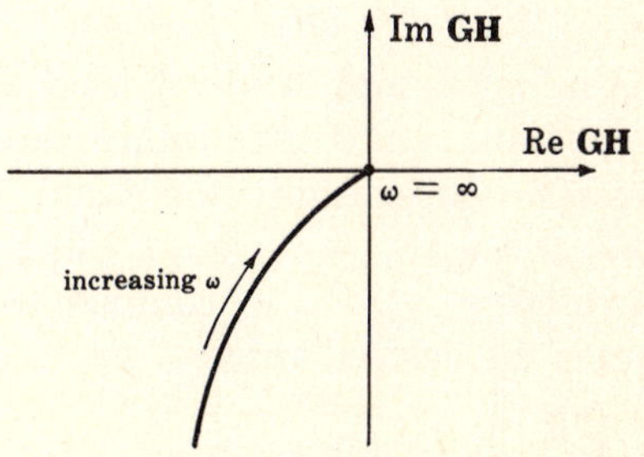

Fig. 8-11

About the only advantage of this over standard techniques is that it is easily mechanized for a computer program. For multiple-loop or multiple-input systems, matrix block diagram manipulations give such a computer routine even more flexibility.

Example 8.9.

Given the 2 input – 2 output system with block diagram shown in Fig. 8-12.

Then $d\mathbf{x}/dt = \mathbf{A}\mathbf{x} + \mathbf{b}_1 e_1 + \mathbf{b}_2 e_2$ and $v_1 = \mathbf{c}_1^\dagger \mathbf{x}$ and $v_2 = \mathbf{c}_2^\dagger \mathbf{x}$. The loop connecting v_1 and e_1 can be closed, so that $e_1 = v_1 = \mathbf{c}_1^\dagger \mathbf{x}$. Then

$$\mathcal{L}\{v_2\} \;=\; \mathbf{c}_2^\dagger (s\mathbf{I} - \mathbf{A} - \mathbf{b}_1 \mathbf{c}_1^\dagger)^{-1} \mathbf{b}_2 \mathcal{L}\{e_2\}$$

so that we can ask the computer to give us the polar plot of $\mathbf{c}_2^\dagger (j\omega\mathbf{I} - \mathbf{A} - \mathbf{b}_1\mathbf{c}_1^\dagger)^{-1}\mathbf{b}_2$.

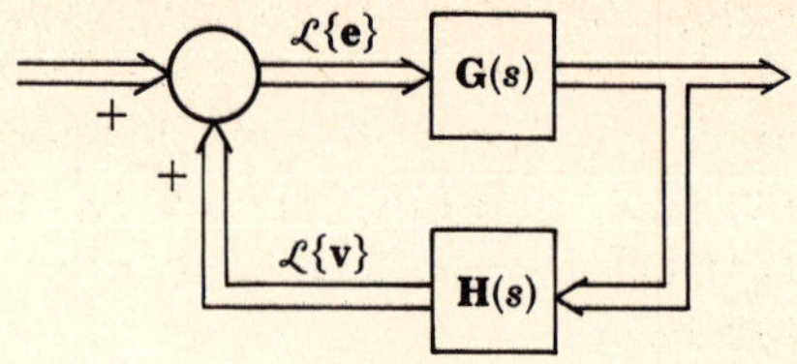

Fig. 8-12

8.6 STATE FEEDBACK POLE PLACEMENT

Here we discuss a way to feed back the state vector to shape the closed loop transition matrix to correspond to that of any desired nth-order scalar linear differential equation. For time-invariant systems in particular, the closed loop poles can be placed where desired. This is why the method is called pole placement, though applicable to general time-varying systems. To "place the poles," the totally controllable part of the state equation is transformed via $\mathbf{x}(t) = \mathbf{T}(t)\mathbf{z}(t)$ to phase-variable canonical form *(7.12)* as repeated here:

$$\frac{d\mathbf{z}}{dt} = \begin{pmatrix} 0 & 1 & 0 & \cdots & 0 \\ 0 & 0 & 1 & \cdots & 0 \\ \cdots & \cdots & \cdots & \cdots & \cdots \\ 0 & 0 & 0 & \cdots & 1 \\ a_1(t) & a_2(t) & a_3(t) & \cdots & a_n(t) \end{pmatrix} \mathbf{z} + \begin{pmatrix} 0 \\ 0 \\ \vdots \\ 0 \\ 1 \end{pmatrix} u \qquad (7.12)$$

Now the scalar control u is constructed by feeding back a linear combination of the $\mathbf{z}$ state variables as $u = \mathbf{k}^\dagger(t)\mathbf{z}$ where each $k_i(t) = -\alpha_i(t) - a_i(t)$.

$$u \;=\; [-\alpha_1(t) - a_1(t)]\mathbf{z}_1 + [-\alpha_2(t) - a_2(t)]\mathbf{z}_2 + \cdots + [-\alpha_n(t) - a_n(t)]\mathbf{z}_n$$

Each $\alpha_i(t)$ is a time function to be chosen. This gives the closed loop system

$$\frac{d\mathbf{z}}{dt} = \begin{pmatrix} 0 & 1 & 0 & \cdots & 0 \\ 0 & 0 & 1 & \cdots & 0 \\ \cdots & \cdots & \cdots & \cdots & \cdots \\ 0 & 0 & 0 & \cdots & 1 \\ -\alpha_1(t) & -\alpha_2(t) & -\alpha_3(t) & \cdots & -\alpha_n(t) \end{pmatrix} \mathbf{z}$$

Then $z_1(t)$ obeys $z_1^{(n)} + \alpha_n(t) z_1^{(n-1)} + \cdots + \alpha_2(t)\dot{z}_1 + \alpha_1(t) z_1 = 0$ and each $z_{i+1}(t)$ for $i = 1, 2, \ldots, n-1$ is the ith derivative of $z_1(t)$. Since the $\alpha_i(t)$ are to be chosen, the corresponding closed loop transition matrix $\Phi_z(t, t_0)$ can be shaped accordingly. Note, however, that $\mathbf{x}(t) = \mathbf{T}(t)\Phi_z(t, t_0)\mathbf{z}_0$ *so that shaping of the transition matrix* $\Phi_z(t, t_0)$ *must be done keeping in mind the effect of* $\mathbf{T}(t)$.

This minor complication disappears when dealing with time-invariant systems. Then $\mathbf{T}(t)$ is constant, and furthermore each $\alpha_i(t)$ is constant. In this case the time behavior of $\mathbf{x}(t)$ and $\mathbf{z}(t)$ is essentially the same, in that both $\mathbf{A}^x$ and $\mathbf{A}^z$ have the same characteristic equation $\lambda^n + \alpha_n \lambda^{n-1} + \cdots + \alpha_2 \lambda + \alpha_1 = 0$. For the closed loop system to have poles at the desired values $\gamma_1, \gamma_2, \ldots, \gamma_n$, comparison of coefficients of the λ's in $(\lambda - \gamma_1)(\lambda - \alpha_2)\cdots(\lambda - \alpha_n) = 0$ determines the desired value of each α_i.

Example 8.10.

Given the system

$$\frac{d\mathbf{x}}{dt} = \begin{pmatrix} 0 & 1 & 0 \\ 0 & 0 & 1 \\ 2 & -3 & t \end{pmatrix} \mathbf{x} + \begin{pmatrix} 0 \\ 0 \\ 1 \end{pmatrix} u$$

It is desired to have a time-invariant closed loop system with poles at 0, -1 and -2. Then the desired system will have a characteristic equation $\lambda^3 + 3\lambda^2 + 2\lambda = 0$. Therefore we choose $u = (-2 \;\; 1 \;\; -(t+3))\mathbf{x}$, so $\mathbf{k}^T = (-2 \;\; 1 \;\; -(t+3))$.

For multiple-input systems, the system is transformed to the form of equation (*7.13*), except that the subsystem $d\mathbf{w}_i/dt = \mathbf{A}_{ii}^w \mathbf{w}_i + \mathbf{b}_i^w u_i$ must be in phase-variable canonical form (*7.12*) and for $i \neq j$, the $\mathbf{A}_{ij}^w(t)$ must be all zeros except for the bottom row. Procedures similar to those used in Chapter 7 can usually attain this form, although general conditions are not presently available. If this form can be attained, each control is chosen as $u_i = \mathbf{k}_i^\dagger(t)\mathbf{w}_i - \mathbf{e}_n^T \mathbf{A}_{ij}^w \mathbf{w}_j$ for $j \neq i$ to "place the poles" of $\mathbf{A}_{ii}(t)$ and to subtract off the coupling terms.

Why bother to transform to canonical form when trial and error can determine $\mathbf{k}$?

Example 8.11.

Place the poles of the system of Problem 7.8 at p_1, p_2 and p_3. We calculate

$$\det\left[\begin{pmatrix} \lambda-1 & -2 & 1 \\ 0 & \lambda-1 & 0 \\ -1 & 4 & \lambda-3 \end{pmatrix} - \begin{pmatrix} 0 & 1 \\ 0 & 1 \\ 1 & 1 \end{pmatrix}\begin{pmatrix} k_{11} & k_{12} & k_{13} \\ k_{21} & k_{22} & k_{23} \end{pmatrix}\right]$$

This is

$$\lambda^3 - (k_{13}+k_{21}+k_{22}+k_{23}+5)\lambda^2 + [k_{11}+2k_{13}+3k_{21}+4k_{22}+5k_{23}+k_{13}(k_{21}+k_{22})-k_{23}(k_{11}+k_{12})+8]\lambda$$
$$- k_{11} - k_{13} - 2k_{21} - 4k_{22} - 6k_{23} - k_{11}(k_{22}+k_{23}) + k_{12}(k_{21}+k_{23}) + k_{13}(k_{21}-k_{22}) - 4$$

It would take much trial and error to choose the k's to match

$$(\lambda-p_1)(\lambda-p_2)(\lambda-p_3) = \lambda^3 - (p_1+p_2+p_3)\lambda^2 + (p_1p_2+p_2p_3+p_1p_3)\lambda - p_1p_2p_3$$

Trial and error is usually no good, because the algebra is nonlinear and increases greatly with the order of the system. Also, Theorem 7.7 tells us when it is possible to "place the poles", namely when $\mathbf{Q}(t)$ has rank n everywhere. Transformation to canonical form seems the best method, as it can be programmed on a computer.

State feedback pole placement has a number of possible defects: (1) The solution appears after transformation to canonical form, with no opportunity for obtaining an engineering feeling for the system. (2) The compensation is in the feedback loop, and experience has shown that cascade compensation is usually better. (3) All the state variables must be available for measurement. (4) The closed loop system may be quite sensitive to small variation in plant parameters. Despite these defects state feedback pole placement may lead to a very good system. Furthermore, it can be used for very high-order and/or time-varying systems for which any compensation may be quite difficult to find. Perhaps the best approach is to try it and then test the system, especially for sensitivity.

Example 8.12.

Suppose that the system of Example 8.10 had $t-\epsilon$ instead of t in the lower right hand corner of the $\mathbf{A}(t)$ matrix, where ϵ is a small positive constant. Then the closed loop system has a characteristic equation $\lambda^3 + 3\lambda^2 + 2\lambda - \epsilon = 0$, which has an unstable root. Therefore this system is extremely sensitive.

8.7 OBSERVER SYSTEMS

Often we need to know the state of a system, and we can measure only the output of the system. There are many practical situations in which knowledge of the state vector is required, but only a linear combination of its elements is known. Knowledge of the state, not the output, determines the future output if the future input is known. Conversely knowledge of the present state and its derivative can be used in conjunction with the state equation to determine the present input. Furthermore, if the state can be reconstructed from the output, state feedback pole placement could be used in a system in which only the output is available for measurement.

In a noise-free environment, n observable states can be reconstructed by differentiating a single output $n-1$ times (see Section 10.6). In a noisy environment, the optimal reconstruction of the state from the output of a linear system is given by the Kalman-Bucy filter. A discussion of this is beyond the scope of this book. In this section, we discuss an observer system that can be used in a noisy environment because it does not contain differentiators. However, in general it does not reconstruct the state in an optimal manner.

To reconstruct all the states at all times, we assume the physical system to be observed is totally observable. For simplicity, at first only single-output systems will be considered. We wish to estimate the state of $d\mathbf{x}/dt = \mathbf{A}(t)\mathbf{x} + \mathbf{B}(t)\mathbf{u}$, where the output $y = \mathbf{c}^\dagger(t)\mathbf{x}$. The state, as usual, is denoted $\mathbf{x}(t)$, and here we denote the estimate of the state as $\hat{\mathbf{x}}(t)$.

First, consider an observer system of dimension n. The observer system is constructed as

$$d\hat{\mathbf{x}}/dt \;=\; \mathbf{A}(t)\hat{\mathbf{x}} + \mathbf{k}(t)[\mathbf{c}^\dagger(t)\hat{\mathbf{x}} - y] + \mathbf{B}(t)\mathbf{u} \tag{8.3}$$

where $\mathbf{k}(t)$ is an n-vector to be chosen. Then the observer system can be incorporated into the flow diagram as shown in Fig. 8-13.

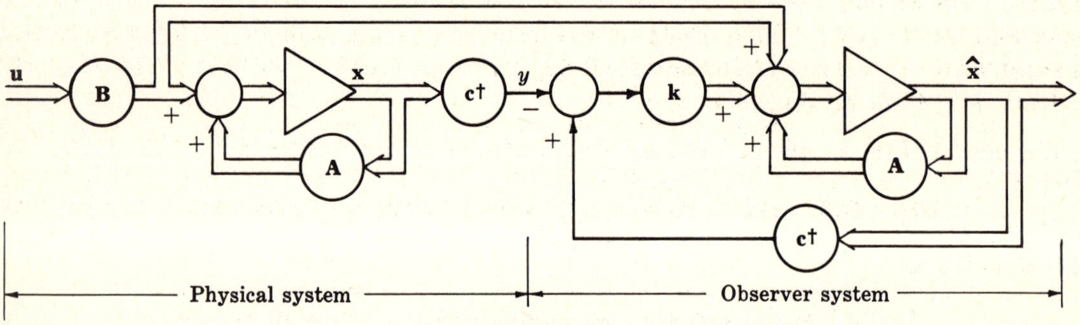

Fig. 8-13

Since the initial state $\mathbf{x}(t_0)$, where t_0 is the time the observer system is started, is not known, we choose $\hat{\mathbf{x}}(t_0) = \mathbf{0}$. Then we can investigate the conditions under which $\mathbf{x}(t)$ tends to $\hat{\mathbf{x}}(t)$. Define the error $\mathbf{e}(t) = \mathbf{x}(t) - \hat{\mathbf{x}}(t)$. Then

$$d\mathbf{e}/dt \;=\; d\mathbf{x}/dt - d\hat{\mathbf{x}}/dt \;=\; [\mathbf{A}(t) + \mathbf{k}(t)\mathbf{c}^\dagger(t)]\mathbf{e} \tag{8.4}$$

Similar to the method of the previous Section 8.6, $\mathbf{k}(t)$ can be chosen to "place the poles" of the error equation (8.4). By duality, the closed loop transition matrix $\boldsymbol{\Psi}(t, t_0)$ of the adjoint equation $d\mathbf{p}/dt = -\mathbf{A}^\dagger(t)\mathbf{p} - \mathbf{c}(t)v$ is shaped using $v = \mathbf{k}^\dagger(t)\mathbf{p}$. Then the transition matrix $\boldsymbol{\Phi}(t, t_0)$ of equation (8.4) is found as $\boldsymbol{\Phi}(t, t_0) = \boldsymbol{\Psi}^\dagger(t_0, t)$, from equation (5.64). For time-invariant systems, it is simpler to consider $d\mathbf{w}/dt = \mathbf{A}^\dagger\mathbf{w} + \mathbf{c}v$ rather than the adjoint. This is because the matrix $\mathbf{A}^\dagger + \mathbf{c}\mathbf{k}^\dagger$ and the matrix $\mathbf{A} + \mathbf{k}\mathbf{c}^\dagger$ have the same eigenvalues. This is easily proved by noting that if λ is an eigenvalue of $\mathbf{A}^\dagger + \mathbf{c}\mathbf{k}^\dagger$, its complex conjugate λ^* is also. Then λ^* satisfies the characteristic equation $\det(\lambda^*\mathbf{I} - \mathbf{A}^\dagger - \mathbf{c}\mathbf{k}^\dagger) = 0$. Taking the complex conjugate of this equation and realizing the determinant is invariant under matrix transposition completes the proof. Hence the poles of equations (8.3) and (8.4) can be placed where desired. Consequently the error $\mathbf{e}(t)$ can be made to decay as quickly as desired, and the state of the observer system tends to the state of the physical system. However, as is indicated in Problem 8.3, we do not want to make the error tend to zero too quickly in a practical system.

Example 8.13.

Given the physical system

$$\frac{d\mathbf{x}}{dt} \;=\; \begin{pmatrix} -2 & 1 \\ 2 & -2 \end{pmatrix}\mathbf{x} + \begin{pmatrix} -1 \\ 0 \end{pmatrix} u \qquad\qquad y \;=\; (1 \;\; 1)\mathbf{x}$$

Construct an observer system such that the error decays with poles at -2 and -3.

First we transform the hypothetical system

$$\frac{d\mathbf{w}}{dt} \;=\; \begin{pmatrix} -2 & 2 \\ 1 & -2 \end{pmatrix}\mathbf{w} + \begin{pmatrix} 1 \\ 1 \end{pmatrix} v$$

to the phase variable canonical form

$$\frac{d\mathbf{z}}{dt} = \begin{pmatrix} 0 & 1 \\ -2 & -4 \end{pmatrix}\mathbf{z} + \begin{pmatrix} 0 \\ 1 \end{pmatrix} v$$

where $\mathbf{w} = \begin{pmatrix} 4 & 1 \\ 3 & 1 \end{pmatrix}\mathbf{z}$, obtained by Theorem 7.7. We desire the closed loop system to have the characteristic equation $0 = (\lambda+2)(\lambda+3) = \lambda^2 + 5\lambda + 6$. Therefore choose $v = (-4 \;\; -1)\mathbf{z} = (-1 \;\; 0)\mathbf{w}$. Then $\mathbf{k} = (-1 \;\; 0)^\dagger$ and the observer system is constructed as

$$\frac{d\hat{\mathbf{x}}}{dt} = \begin{pmatrix} -3 & 0 \\ 2 & -2 \end{pmatrix}\hat{\mathbf{x}} + \begin{pmatrix} 1 \\ 0 \end{pmatrix} y + \begin{pmatrix} -1 \\ 0 \end{pmatrix} u$$

Now we consider an observer system of dimension less than n. In the case of a single-output system we only need to estimate $n-1$ elements of the state vector because the known output and the $n-1$ estimated elements will usually give an estimate of the nth element of the state vector. In general for a system having k independent outputs we shall construct an observer system of dimension $n-k$.

We choose $\mathbf{P}(t)$ to be certain $n-k$ differentiable rows such that the $n \times n$ matrix $\begin{pmatrix} \mathbf{P}(t) \\ \hline \mathbf{C}(t) \end{pmatrix}^{-1} = (\mathbf{H}(t) \mid \mathbf{G}(t))$ exists at all times where $\mathbf{H}$ has $n-k$ columns. The estimate $\hat{\mathbf{x}}$ is constructed as

$$\hat{\mathbf{x}}(t) = \mathbf{H}(t)\mathbf{w} + \mathbf{G}(t)\mathbf{y} \quad \text{or, equivalently,} \quad \begin{pmatrix} \mathbf{P}(t) \\ \hline \mathbf{C}(t) \end{pmatrix}\hat{\mathbf{x}} = \begin{pmatrix} \mathbf{w} \\ \hline \mathbf{y} \end{pmatrix} \tag{8.5}$$

Analogous to equation (*8.3*), we require

$$\mathbf{P}\,d\hat{\mathbf{x}}/dt = \mathbf{P}[\mathbf{A}\hat{\mathbf{x}} + \mathbf{L}(\mathbf{C}\hat{\mathbf{x}} - \mathbf{y}) + \mathbf{B}\mathbf{u}]$$

where $\mathbf{L}(t)$ is an $n \times k$ matrix to be found. (It turns out we only need to find $\mathbf{PL}$, not $\mathbf{L}$.) This is equivalent to constructing the following system to generate $\mathbf{w}$, from equation (*8.5*),

$$d\mathbf{w}/dt = (d\mathbf{P}/dt)\hat{\mathbf{x}} + \mathbf{P}\,d\hat{\mathbf{x}}/dt = \mathbf{F}\mathbf{w} - \mathbf{PL}\mathbf{y} + \mathbf{PB}\mathbf{u} \tag{8.6}$$

where $\mathbf{F}$ is determined from $\mathbf{FP} = d\mathbf{P}/dt + \mathbf{PA} + \mathbf{PLC}$. Then $(\mathbf{F} \mid -\mathbf{PL})\begin{pmatrix} \mathbf{P} \\ \hline \mathbf{C} \end{pmatrix} = d\mathbf{P}/dt + \mathbf{PA}$ so that $\mathbf{F}$ and $\mathbf{PL}$ are determined from $(\mathbf{F} \mid -\mathbf{PL}) = (d\mathbf{P}/dt + \mathbf{PA})(\mathbf{H} \mid \mathbf{G})$. From (*8.5*) and (*8.6*) the error $\mathbf{e} = \mathbf{P}(\mathbf{x} - \hat{\mathbf{x}}) = \mathbf{Px} - \mathbf{w}$ obeys the equation $d\mathbf{e}/dt = \mathbf{Fe}$.

The flow diagram is then as shown in Fig. 8-14.

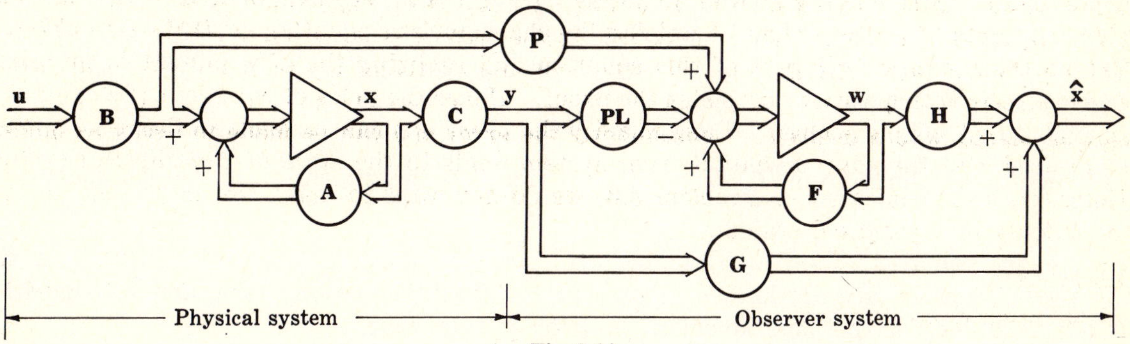

Fig. 8-14

Example 8.14.

Given the system of Example 8.13, construct a first-order observer system.

Since $\mathbf{C} = (1 \;\; 1)$, choose $\mathbf{P} = (p_1 \;\; p_2)$ with $p_1 \neq p_2$. Then

$$\begin{pmatrix} \mathbf{P} \\ \hline \mathbf{C} \end{pmatrix}^{-1} = \begin{pmatrix} p_1 & p_2 \\ 1 & 1 \end{pmatrix}^{-1} = \frac{1}{p_1 - p_2}\begin{pmatrix} 1 & -p_2 \\ -1 & p_1 \end{pmatrix} = (\mathbf{H} \mid \mathbf{G})$$

Therefore

$$\hat{\mathbf{x}} = \frac{1}{p_1 - p_2}\begin{pmatrix}1\\-1\end{pmatrix} w + \frac{1}{p_1 - p_2}\begin{pmatrix}0\\1\end{pmatrix} y$$

and

$$\mathbf{F} = (d\mathbf{P}/dt + \mathbf{PA})\mathbf{H} = \frac{(p_1 \ \ p_2)}{p_1 - p_2}\begin{pmatrix}-2 & 1\\ 2 & -2\end{pmatrix}\begin{pmatrix}1\\-1\end{pmatrix} = \frac{-3p_1 + 4p_2}{p_1 - p_2}$$

$$\mathbf{PL} = -(d\mathbf{P}/dt + \mathbf{PA})\mathbf{G} = \frac{(p_1 \ \ p_2)}{p_1 - p_2}\begin{pmatrix}-2 & 1\\ 2 & -2\end{pmatrix}\begin{pmatrix}-p_2\\ p_1\end{pmatrix} = \frac{p_1^2 - 2p_2^2}{p_1 - p_2}$$

so that $(p_1 - p_2)dw/dt = (-3p_1 + 4p_2)w - (p_1^2 - 2p_2^2)y - p_1(p_1 - p_2)u$ is the first-order observer. A bad choice of p_1/p_2 with $1 < p_1/p_2 < 4/3$ gives an unstable observer and makes the error blow up.

The question is, can we place the poles of $\mathbf{F}$ by proper selection of $\mathbf{P}$ in a manner similar to that of the n-dimensional observer? One method is to use trial and error, which is sometimes more rapid for low-order, time-invariant systems. However, to show that the poles of $\mathbf{F}$ can be placed arbitrarily, we use the transformation $\mathbf{x} = \mathbf{Tz}$ to obtain the canonical form

$$\frac{d}{dt}\begin{pmatrix}\mathbf{z}_1\\ \mathbf{z}_2\\ \vdots\\ \mathbf{z}_l\end{pmatrix} = \begin{pmatrix}\mathbf{A}_{11} & \mathbf{0} & \cdots & \mathbf{0}\\ \mathbf{A}_{21} & \mathbf{A}_{22} & \cdots & \mathbf{0}\\ \cdots & \cdots & \cdots & \cdots\\ \mathbf{A}_{l1} & \mathbf{A}_{l2} & \cdots & \mathbf{A}_{ll}\end{pmatrix}\begin{pmatrix}\mathbf{z}_1\\ \mathbf{z}_2\\ \vdots\\ \mathbf{z}_l\end{pmatrix} + \mathbf{T}^{-1}\mathbf{Bu}$$

$$\mathbf{y} = \begin{pmatrix}\mathbf{c}_1^\dagger & \mathbf{0} & \cdots & \mathbf{0}\\ \mathbf{0} & \mathbf{c}_2^\dagger & \cdots & \mathbf{0}\\ \cdots & \cdots & \cdots & \cdots\\ \mathbf{0} & \mathbf{0} & \cdots & \mathbf{c}_l^\dagger\end{pmatrix}\mathbf{z}$$

where the subsystem $d\mathbf{z}_i/dt = \mathbf{A}_{ii}\mathbf{z}_i + \mathbf{B}_i\mathbf{u}$ and $y_i = \mathbf{c}_i^\dagger\mathbf{z}_i$ is in the dual phase variable canonical form

$$\frac{d\mathbf{z}_i}{dt} = \begin{pmatrix}0 & 0 & \cdots & 0 & a_1(t)\\ 1 & 0 & \cdots & 0 & a_2(t)\\ \cdots & \cdots & \cdots & \cdots & \cdots\\ 0 & 0 & \cdots & 1 & a_{n_i}(t)\end{pmatrix}\mathbf{z}_i + \mathbf{B}_i\mathbf{u} \qquad i = 1, 2, \ldots, l$$

$$y_i = (0 \ \ldots \ 0 \ \ 1)\mathbf{z}_i = z_{n_i} \tag{8.7}$$

in which $\mathbf{B}_i$ is defined from $\mathbf{T}^{-1}\mathbf{B} = \begin{pmatrix}\mathbf{B}_1\\ \hline \vdots\\ \hline \mathbf{B}_l\end{pmatrix}$ and n_i is the dimension of the ith subsystem.

As per the remarks following Example 8.10, the conditions under which this form can always be obtained are not known at present for the time-varying case, and an algorithm is not available for the time-invariant multiple-output case.

However, assuming the subsystem *(8.7)* can be obtained, we construct the observer equation *(8.6)* for the subsystem *(8.7)* by the choice of $\mathbf{P}_i = (\mathbf{I} \mid \mathbf{k}_i)$ where $\mathbf{k}_i(t)$ is an $(n_i - 1)$-vector that will set the poles of the observer. We assume $\mathbf{k}_i(t)$ is differentiable. Then

$$\begin{pmatrix}\mathbf{P}_i\\ \hline \mathbf{c}_i\end{pmatrix} = \left(\begin{array}{c|c}\mathbf{I} & \mathbf{k}_i\\ \hline \mathbf{0} & 1\end{array}\right) \quad\text{and}\quad \begin{pmatrix}\mathbf{P}_i\\ \hline \mathbf{a}_i\end{pmatrix}^{-1} = \left(\begin{array}{c|c}\mathbf{I} & -\mathbf{k}_i\\ \hline \mathbf{0} & 1\end{array}\right) = (\mathbf{H}_i \mid \mathbf{G}_i) \tag{8.8}$$

We find $\mathbf{F}_i = (d\mathbf{P}_i/dt + \mathbf{P}_i\mathbf{A}_{ii})\mathbf{H}_i = [(\mathbf{0} \mid d\mathbf{k}_i/dt) + (\mathbf{I} \mid \mathbf{k}_i)\mathbf{A}_{ii}]\left(\frac{\mathbf{I}}{\mathbf{0}}\right)$ from which

$$\mathbf{F}_i = \begin{pmatrix} 0 & 0 & \dots & 0 & k_{i1}(t) \\ 1 & 0 & \dots & 0 & k_{i2}(t) \\ 0 & 1 & \dots & 0 & k_{i3}(t) \\ \dots & \dots & \dots & \dots & \dots \\ 0 & 0 & \dots & 1 & k_{i,n_i-1}(t) \end{pmatrix}$$

By matching coefficients of the "characteristic equation" with "desired pole positions," we make the error decay as quickly as desired.

Also, we find $\mathbf{P}_i\mathbf{L}_i$ as $\mathbf{P}_i\mathbf{L}_i = -(d\mathbf{P}_i/dt + \mathbf{P}_i\mathbf{A}_{ii})\mathbf{G}_i$.

Then $\hat{\mathbf{x}} = \mathbf{T}\hat{\mathbf{z}} = \mathbf{T}\begin{pmatrix} \hat{\mathbf{z}}_1 \\ \vdots \\ \hat{\mathbf{z}}_l \end{pmatrix}$ where $\hat{\mathbf{z}}_i = \mathbf{H}_i\mathbf{w}_i + \mathbf{G}_i y_i = \left(\frac{\mathbf{I}}{\mathbf{0}}\right)\mathbf{w}_i + \left(\frac{-\mathbf{k}_i}{1}\right)y_i$

Example 8.15.

Again, consider the system of Example 8.13.

$$\frac{d\mathbf{x}}{dt} = \begin{pmatrix} -2 & 1 \\ 2 & -2 \end{pmatrix}\mathbf{x} + \begin{pmatrix} -1 \\ 0 \end{pmatrix}u \qquad y = (1 \;\; 1)\mathbf{x}$$

To construct an observer system with a pole at -2, use the transformation $\mathbf{x} = \mathbf{T}\mathbf{z}$ where $(\mathbf{T}^\dagger)^{-1} = \begin{pmatrix} 4 & 1 \\ 3 & 1 \end{pmatrix}$. Then equations (*8.7*) are

$$\frac{d\mathbf{z}}{dt} = \begin{pmatrix} 0 & -2 \\ 1 & -4 \end{pmatrix}\mathbf{z} + \begin{pmatrix} -4 \\ -1 \end{pmatrix}u \qquad y = (0 \;\; 1)\mathbf{z} = z_2$$

The estimate $\hat{\mathbf{z}}$ according to equation (*8.8*) is now obtained as

$$\hat{\mathbf{z}} = \left(\frac{\mathbf{P}}{\mathbf{C}}\right)^{-1}\begin{pmatrix} w \\ y \end{pmatrix} = \left(\begin{array}{c|c} 1 & 2 \\ \hline 0 & 1 \end{array}\right)\begin{pmatrix} w \\ y \end{pmatrix}$$

where $k_1 = -2$ sets the pole of the observer at -2. Then $F = -2$ and $PL = -2$ so that the observer system is $dw/dt = -2w + 2y - 2u$. Therefore

$$\hat{\mathbf{x}} = \mathbf{T}\hat{\mathbf{z}} = \begin{pmatrix} 1 & -3 \\ -1 & 4 \end{pmatrix}\begin{pmatrix} 1 & 2 \\ 0 & 1 \end{pmatrix}\begin{pmatrix} w \\ y \end{pmatrix} = \begin{pmatrix} 1 & -1 \\ -1 & 2 \end{pmatrix}\begin{pmatrix} w \\ y \end{pmatrix}$$

and the error $\mathbf{P}\mathbf{T}^{-1}\mathbf{x} - w = 2x_1 + x_2 - w$ decays with a time constant of 1/2. This gives the block diagram of Fig. 8-15.

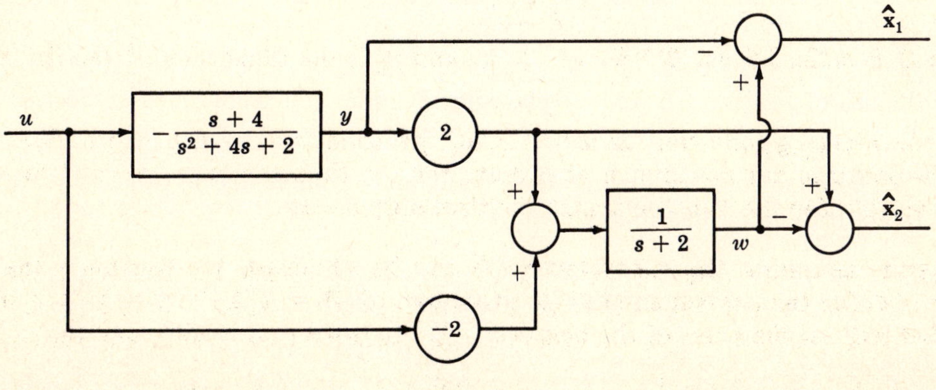

Fig. 8-15

8.8 ALGEBRAIC SEPARATION

In this section we use the observer system of Section 8.7 to generate a feedback control to place the closed loop poles where desired, as discussed in Section 8.6. Specifically, we consider the physical open loop system

$$d\mathbf{x}/dt = \mathbf{A}(t)\mathbf{x} + \mathbf{B}(t)\mathbf{u} + \mathbf{J}(t)\mathbf{d}$$
$$\mathbf{y} = \mathbf{C}(t)\mathbf{x} \tag{8.9}$$

with an observer system (see equation (8.6)),

$$d\mathbf{w}/dt = \mathbf{F}(t)\mathbf{w} - \mathbf{P}(t)\mathbf{L}(t)\mathbf{y} + \mathbf{P}(t)\mathbf{B}(t)\mathbf{u} + \mathbf{P}(t)\mathbf{J}(t)\mathbf{d}$$
$$\hat{\mathbf{x}} = \mathbf{H}(t)\mathbf{w} + \mathbf{G}(t)\mathbf{y} \tag{8.10}$$

and a feedback control $\mathbf{u}(t)$ that has been formed to place the poles of the closed loop system as

$$\mathbf{u} = \mathbf{W}(t)\hat{\mathbf{x}} \tag{8.11}$$

Then the closed loop system block diagram is as in Fig. 8-16.

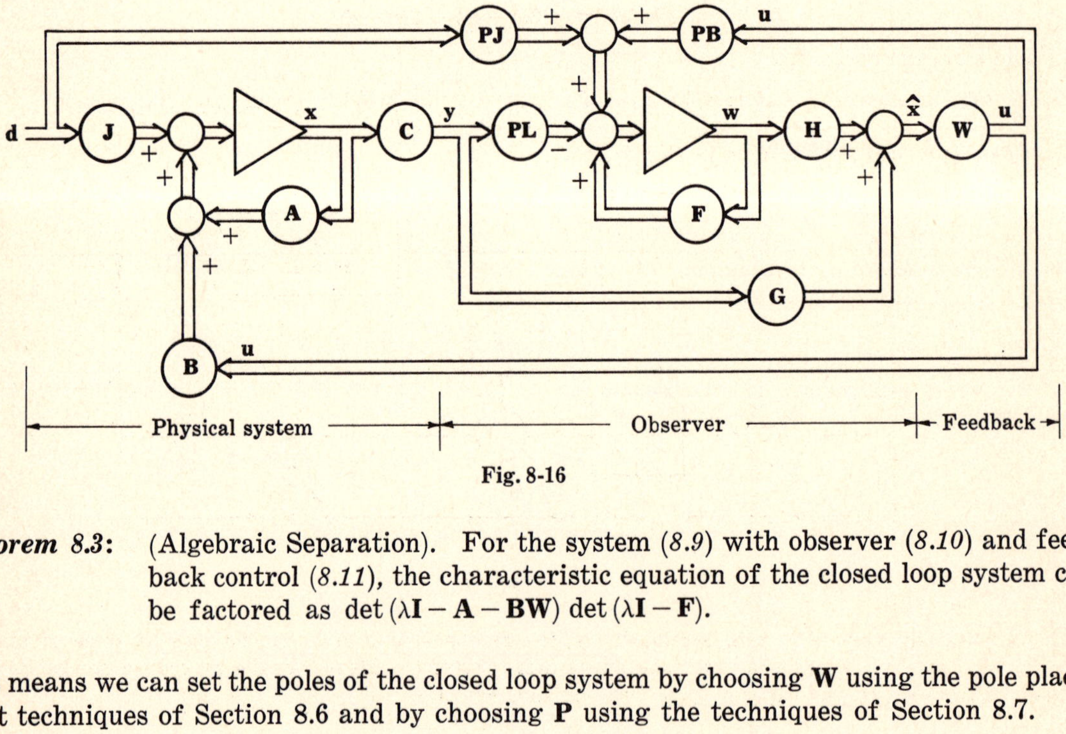

Fig. 8-16

Theorem 8.3: (Algebraic Separation). For the system (8.9) with observer (8.10) and feedback control (8.11), the characteristic equation of the closed loop system can be factored as $\det(\lambda\mathbf{I} - \mathbf{A} - \mathbf{BW})\det(\lambda\mathbf{I} - \mathbf{F})$.

This means we can set the poles of the closed loop system by choosing $\mathbf{W}$ using the pole placement techniques of Section 8.6 and by choosing $\mathbf{P}$ using the techniques of Section 8.7.

Proof: The equations governing the closed loop system are obtained by substituting equation (8.11) into equations (8.9) and (8.10):

$$\frac{d}{dt}\begin{pmatrix}\mathbf{x}\\ \mathbf{w}\end{pmatrix} = \begin{pmatrix}\mathbf{A}+\mathbf{BWGC} & \mathbf{BWH}\\ \mathbf{PBWGC}-\mathbf{PLC} & \mathbf{F}+\mathbf{PBWH}\end{pmatrix}\begin{pmatrix}\mathbf{x}\\ \mathbf{w}\end{pmatrix} + \begin{pmatrix}\mathbf{J}\\ \mathbf{PJ}\end{pmatrix}\mathbf{d}$$

Changing variables to $\mathbf{e} = \mathbf{Px} - \mathbf{w}$ and using $\mathbf{HP} + \mathbf{GC} = \mathbf{I}$ and $\mathbf{FP} = d\mathbf{P}/dt + \mathbf{PA} + \mathbf{PLC}$ gives

$$\frac{d}{dt}\begin{pmatrix}\mathbf{x}\\ \mathbf{e}\end{pmatrix} = \begin{pmatrix}\mathbf{A}+\mathbf{BW} & -\mathbf{BWH}\\ \mathbf{0} & \mathbf{F}\end{pmatrix}\begin{pmatrix}\mathbf{x}\\ \mathbf{e}\end{pmatrix} + \begin{pmatrix}\mathbf{J}\\ \mathbf{0}\end{pmatrix}\mathbf{d} \qquad \mathbf{y} = (\mathbf{C}\ \ \mathbf{0})\begin{pmatrix}\mathbf{x}\\ \mathbf{e}\end{pmatrix}$$

Note that the bottom equation $d\mathbf{e}/dt = \mathbf{Fe}$ generates an input $-\mathbf{WHe}$ to the closed loop of observer system $d\mathbf{x}/dt = (\mathbf{A}+\mathbf{BW})\mathbf{x}$. Use of Problem 3.5 then shows the characteristic equation factors as hypothesized. Furthermore, the observer dynamics are in general observable at the output (through coupling with $\mathbf{x}$) but are uncontrollable by $\mathbf{d}$ and hence cancel out of the closed loop transfer function.

Example 8.16.

For the system of Example 8.13, construct a one-dimensional observer system with a pole at -2 to generate a feedback that places both the system poles at -1.

We employ the algebraic separation theorem to separately consider the system pole placement and the observer pole placement. To place the pole of

$$\frac{d\mathbf{x}}{dt} = \begin{pmatrix} -2 & 1 \\ 2 & -2 \end{pmatrix}\mathbf{x} + \begin{pmatrix} -1 \\ 0 \end{pmatrix}u + \begin{pmatrix} -1 \\ 0 \end{pmatrix}d \qquad y = (1 \;\; 1)\mathbf{x}$$

using the techniques of Section 8.6 we would like

$$u = (-2 \;\; 3/2)\mathbf{x}$$

which gives closed loop poles at -1. However, we cannot use $\mathbf{x}$ to form u, but must use $\hat{\mathbf{x}}$ as found from the observer system with a pole at -2, which was constructed in Example 8.15.

$$dw/dt = -2w + 2y - 2(u+d)$$

We then form the control as

$$u = (-2 \;\; 3/2)\hat{\mathbf{x}} = (-2 \;\; 3/2)\begin{pmatrix} 1 & -1 \\ -1 & 2 \end{pmatrix}\begin{pmatrix} w \\ y \end{pmatrix} = -7w/2 + 5y$$

Thus the closed loop system is as in Fig. 8-17.

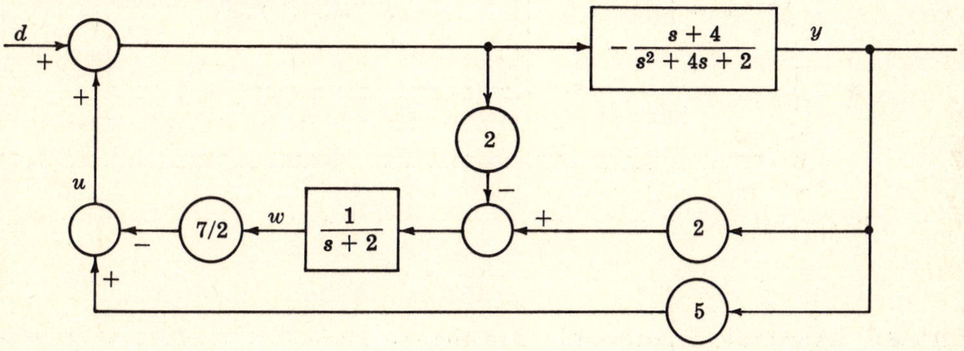

Fig. 8-17

Note that the control is still essentially in the feedback loop and that no reasons were given as to why plant poles at -1 and observer pole at -2 were selected. However, the procedure works for high-order, multiple input—multiple output, time-varying systems.

8.9 SENSITIVITY, NOISE REJECTION, AND NONLINEAR EFFECTS

Three major reasons for the use of feedback control, as opposed to open loop control, are (1) to reduce the sensitivity of the response to parameter variations, (2) to reduce the effect of noise disturbances, and (3) to make the response of nonlinear elements more linear. A proposed design of a feedback system should be evaluated for sensitivity, noise rejection, and the effect of nonlinearities. Certainly any system designed using the pole placement techniques of Sections 8.6, 8.7 and 8.8 must be evaluated in these respects because of the cookbook nature of pole placement.

In this section we consider these topics in a very cursory manner, mainly to show the relationship with controllability and observability. Consequently we consider only small percentage changes in parameter variations, small noise compared with the signal, and nonlinearities that are almost linear. Under these assumptions we will show how each effect produces an unwanted input into a linear system and then how to minimize this unwanted input.

First we consider the effect of parameter variations. Let the subscript N refer to the nominal values and the subscript a refer to actual values. Then the nominal system (the system with zero parameter variations) can be represented by

$$\begin{aligned} d\mathbf{x}_N/dt &= \mathbf{A}_N(t)\mathbf{x}_N + \mathbf{B}(t)\mathbf{u} \\ \mathbf{y}_N &= \mathbf{C}(t)\mathbf{x}_N + \mathbf{D}(t)\mathbf{u} \end{aligned} \tag{8.12}$$

These equations determine $\mathbf{x}_N(t)$ and $\mathbf{y}_N(t)$, so these quantities are assumed known. If some of the elements of $\mathbf{A}_N$ drift to some actual $\mathbf{A}_a$ (keeping $\mathbf{B}$, $\mathbf{C}$ and $\mathbf{D}$ fixed only for simplicity), then

$$\begin{aligned} d\mathbf{x}_a/dt &= \mathbf{A}_a(t)\mathbf{x}_a + \mathbf{B}(t)\mathbf{u} \\ \mathbf{y}_a &= \mathbf{C}(t)\mathbf{x}_a + \mathbf{D}(t)\mathbf{u} \end{aligned} \tag{8.13}$$

Then let $\delta\mathbf{x} = \mathbf{x}_a - \mathbf{x}_N$, $\delta\mathbf{A} = \mathbf{A}_a - \mathbf{A}_N$, $\delta\mathbf{y} = \mathbf{y}_a - \mathbf{y}_N$, subtract equations (*8.12*) from (*8.13*), and neglect the product of small quantities $\delta\mathbf{A}\,\delta\mathbf{x}$. *Warning*: That $\delta\mathbf{A}\,\delta\mathbf{x}$ is truly small at all times must be verified by simulation. If this is so, then

$$\begin{aligned} d(\delta\mathbf{x})/dt &= \mathbf{A}_N(t)\,\delta\mathbf{x} + \delta\mathbf{A}(t)\,\mathbf{x}_N \\ \delta\mathbf{y} &= \mathbf{C}(t)\,\delta\mathbf{x} \end{aligned} \tag{8.14}$$

In these equations $\mathbf{A}_N(t)$, $\mathbf{C}(t)$ and $\mathbf{x}_N(t)$ are known and $\delta\mathbf{A}(t)$, the variation of the parameters of the $\mathbf{A}(t)$ matrix, is the input to drive the unwanted signal $\delta\mathbf{x}$.

For the case of noise disturbances $\mathbf{d}(t)$, the nominal system remains equations (*8.12*) but the actual system is

$$\begin{aligned} d\mathbf{x}_a/dt &= \mathbf{A}_N(t)\mathbf{x}_a + \mathbf{B}(t)\mathbf{u} + \mathbf{J}(t)\mathbf{d} \\ \mathbf{y}_a &= \mathbf{C}(t)\mathbf{x}_a + \mathbf{D}(t)\mathbf{u} + \mathbf{K}(t)\mathbf{d} \end{aligned} \tag{8.15}$$

Then we subtract equations (*8.12*) from (*8.15*) to obtain

$$\begin{aligned} d(\delta\mathbf{x})/dt &= \mathbf{A}_N(t)\,\delta\mathbf{x} + \mathbf{J}(t)\mathbf{d} \\ \delta\mathbf{y} &= \mathbf{C}(t)\,\delta\mathbf{x} + \mathbf{K}(t)\mathbf{d} \end{aligned} \tag{8.16}$$

Here the noise $\mathbf{d}(t)$ drives the unwanted signal $\delta\mathbf{x}$.

Finally, to show how a nonlinearity produces an unwanted signal, consider a scalar input x_N into a nonlinearity as shown in Fig. 8-18. This can be redrawn into a linear system with a large output and a nonlinear system with a small output (Fig. 8-19).

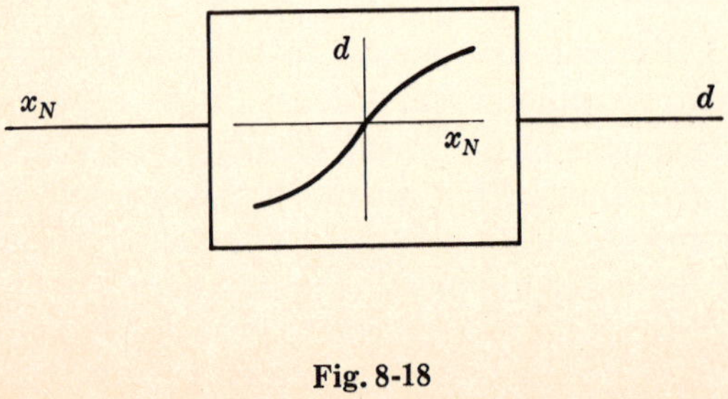

Fig. 8-18

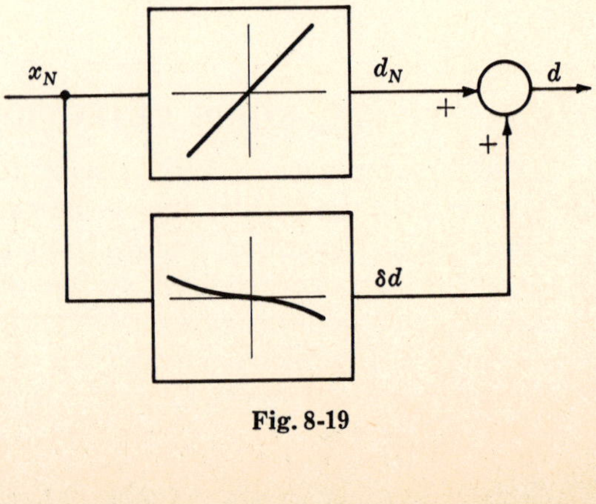

Fig. 8-19

Here the unwanted signal is δd which is generated by the nominal x_N. This can be incorporated into a block diagram containing linear elements, and the effect of the nonlinearity can be evaluated in a manner similar to that used in deriving equations (*8.16*).

$$d(\delta\mathbf{x})/dt = \mathbf{A}_N(t)\,\delta\mathbf{x} + \mathbf{j}(t)\,\delta d$$

$$\delta\mathbf{y} = \mathbf{C}(t)\,\delta\mathbf{x} + \mathbf{k}(t)\,\delta d \qquad (8.17)$$

Now observability and controllability theory can be applied to equations (*8.14*), (*8.16*) and (*8.17*). We conclude that, if possible, we will choose $\mathbf{C}(t)$, $\mathbf{A}_N(t)$ and the corresponding input matrices $\mathbf{B}(t), \mathbf{D}(t)$ or $\mathbf{J}(t), \mathbf{K}(t)$ such that the unwanted signal is unobservable with respect to the output $\delta\mathbf{y}(t)$, or at least the elements of the state vector associated with the dominant poles are uncontrollable with respect to the unwanted signal. If this is impossible, the system gain with respect to the unwanted signal should be made as low as possible.

Example 8.17.

Consider the system $d\mathbf{x}/dt = \begin{pmatrix} -2 & \alpha \\ 0 & -1 \end{pmatrix}\mathbf{x} + \begin{pmatrix} 1 \\ 1 \end{pmatrix}u$. The nominal value of the parameter α is zero and the nominal input $u(t)$ is a unit step function.

$$d\mathbf{x}_N/dt = \begin{pmatrix} -2 & 0 \\ 0 & -1 \end{pmatrix}\mathbf{x}_N + \begin{pmatrix} 1 \\ 1 \end{pmatrix}$$

If $\mathbf{x}_N(0) = \mathbf{0}$, then $\mathbf{x}_N(t) = \begin{pmatrix} 1 - e^{-2t} \\ 1 - e^{-t} \end{pmatrix}$. The effect of small variations in α can be evaluated from equation (*8.14*),

$$d(\delta\mathbf{x})/dt = \begin{pmatrix} -2 & 0 \\ 0 & -1 \end{pmatrix}\delta\mathbf{x} + \begin{pmatrix} 0 & \alpha \\ 0 & 0 \end{pmatrix}\begin{pmatrix} 1 - e^{-2t} \\ 1 - e^{-t} \end{pmatrix}$$

Simplifying,

$$d(\delta\mathbf{x})/dt = \begin{pmatrix} -2 & 0 \\ 0 & -1 \end{pmatrix}\delta\mathbf{x} + \begin{pmatrix} 1 \\ 0 \end{pmatrix}\alpha(1 - e^{-t})$$

We can eliminate the effects of α variation upon the output if $\mathbf{c}$ is chosen such that the output observability matrix $(\mathbf{c}^\dagger\mathbf{b} \;\; \mathbf{c}^\dagger\mathbf{A}\mathbf{b}) = \mathbf{0}$. This results in a choice $\mathbf{c}^\dagger = (0 \;\; \gamma)$ where γ is any number.

Furthermore, all the analysis and synthesis techniques developed in this chapter can be used to analyze and design systems to reduce sensitivity and nonlinear effects and to reject noise. This may be done by using the error constant matrix table, root locus, Nyquist, and/or pole placement techniques on the equations (*8.14*), (*8.16*) and (*8.17*).

Solved Problems

8.1. For the system of Fig. 8-4, let $\frac{d\mathbf{x}}{dt} = \begin{pmatrix} 0 & 1 \\ \frac{1}{t+1} & -\frac{t}{t+1} \end{pmatrix}\mathbf{x} + \begin{pmatrix} 0 \\ \frac{t+2}{t+1} \end{pmatrix} e$ for $t \geq 0$, with $y = (1 \;\; 1)\mathbf{x}$. Find the class of inputs that give a constant steady state error.

The closed loop system with zero input ($d(t) = 0$) is

$$\frac{d\mathbf{x}}{dt} = \left[\begin{pmatrix} 0 & 1 \\ \frac{1}{t+1} & -\frac{t}{t+1} \end{pmatrix} - \begin{pmatrix} 0 \\ \frac{t+2}{t+1} \end{pmatrix}(1 \;\; 1)\right]\mathbf{x} = \begin{pmatrix} 0 & 1 \\ -1 & -2 \end{pmatrix}\mathbf{x}$$

which has a double pole at -1. Therefore the zero output of the closed loop system is asymptotically stable and the remarks following Theorem 8.1 are valid. The class of all $d(t)$ that the system can follow with $\lim_{t\to\infty} e(t) = 0$ is given by

$$\frac{d\mathbf{w}}{dt} = \begin{pmatrix} 0 & 1 \\ \frac{1}{t+1} & -\frac{t}{t+1} \end{pmatrix}\mathbf{w} + \begin{pmatrix} 0 \\ \frac{t+2}{t+1} \end{pmatrix} g$$

with $d = (1 \;\; 1)\mathbf{w} + g$. The transition matrix for this system is

$$\mathbf{\Phi}(t,\tau) = \frac{1}{\tau+1}\begin{pmatrix} t + e^{\tau-t} & t(1-e^{\tau-t}) \\ 1 - e^{\tau-t} & 1 + te^{\tau-t} \end{pmatrix}$$

Then

$$d(t) = \frac{t+1}{t_0+1}[w_1(t_0) + w_2(t_0)] + (t+1)\int_{t_0}^{t} \frac{\tau+2}{(\tau+1)^2} g(\tau)\, d\tau + g(t)$$

(Notice this system is unobservable.) For constant error let $g(t) = \kappa$, an arbitrary constant. Then

$$\kappa(t+1)\int_{t_0}^{t} \frac{\tau+2}{(\tau+1)^2}\, d\tau = \kappa(t+1)[\ln(t+1) - \ln(t_0+1) + (t_0+1)^{-1} - (t+1)^{-1}]$$

and the system follows with error κ all functions that are asymptotic to this. Since the system is reasonably well behaved, we can assume that the system will follow all functions going to infinity slower than $\kappa(t+1)\ln(t+1)$.

8.2. Given the multiple-input system

$$\frac{d}{dt}\begin{pmatrix} z_1 \\ z_2 \\ z_3 \end{pmatrix} = \begin{pmatrix} -1 & 0 & 0 \\ 0 & -2 & 0 \\ 0 & 0 & -3 \end{pmatrix}\begin{pmatrix} z_1 \\ z_2 \\ z_3 \end{pmatrix} + \begin{pmatrix} \alpha & 1 \\ 1 & -1 \\ 1 & 0 \end{pmatrix}\begin{pmatrix} u_1 \\ u_2 \end{pmatrix}$$

Place the poles of the closed loop system at $-4, -5$ and -6.

Transform the system to phase variable canonical form using the results of Problem 7.21.

$$\mathbf{x} = \begin{pmatrix} 1 & 1 & 1 \\ -1 & -2 & -3 \\ 1 & 4 & 9 \end{pmatrix}\begin{pmatrix} \kappa_1 & 0 & 0 \\ 0 & \kappa_2 & 0 \\ 0 & 0 & \kappa_3 \end{pmatrix}\mathbf{z}$$

Then

$$\mathbf{z} = \begin{pmatrix} \kappa_1^{-1} & 0 & 0 \\ 0 & \kappa_2^{-1} & 0 \\ 0 & 0 & \kappa_3^{-1} \end{pmatrix}\begin{pmatrix} 3 & 5/2 & 1/2 \\ -3 & -4 & -1 \\ 1 & 3/2 & 1/2 \end{pmatrix}\mathbf{x}$$

and

$$\frac{d\mathbf{x}}{dt} = \begin{pmatrix} 0 & 1 & 0 \\ 0 & 0 & 1 \\ -6 & -11 & -6 \end{pmatrix}\mathbf{x} + \begin{pmatrix} 1 & 1 & 1 \\ -1 & -2 & -3 \\ 1 & 4 & 9 \end{pmatrix}\begin{pmatrix} \alpha\kappa_1 & \kappa_1 \\ \kappa_2 & -\kappa_2 \\ \kappa_3 & 0 \end{pmatrix}\begin{pmatrix} u_1 \\ u_2 \end{pmatrix}$$

To obtain phase variable canonical form for u_1, we set

$$\begin{pmatrix} 1 & 1 & 1 \\ -1 & -2 & -3 \\ 1 & 4 & 9 \end{pmatrix}\begin{pmatrix} \alpha\kappa_1 \\ \kappa_2 \\ \kappa_3 \end{pmatrix} = \begin{pmatrix} 0 \\ 0 \\ 1 \end{pmatrix}$$

which gives $\alpha\kappa_1 = 1/2$, $\kappa_2 = -1$, $\kappa_3 = 1/2$.

For $\alpha \neq 0$, we have the phase variable canonical system

$$\frac{d\mathbf{x}}{dt} = \begin{pmatrix} 0 & 1 & 0 \\ 0 & 0 & 1 \\ -6 & -11 & -6 \end{pmatrix}\mathbf{x} + \begin{pmatrix} 0 & 1 + 1/2\alpha \\ 0 & -2 - 1/2\alpha \\ 1 & 4 + 1/2\alpha \end{pmatrix}\begin{pmatrix} u_1 \\ u_2 \end{pmatrix}$$

To have the closed loop poles at -4, -5 and -6 we desire a charateristic polynomial $\lambda^3 + 15\lambda^2 + 74\lambda + 120$. Therefore we choose $u_1 = -114x_1 - 63x_2 - 9x_3$ and $u_2 = 0x_1 + 0x_2 + 0x_3$.

In the case $\alpha = 0$, the state z_1 is uncontrollable with respect to u_1 and, from Theorem 7.7, cannot be put into phase variable canonical form with respect to u_1 alone. Hence we must use u_2 to control z_1, and can assure a pole at -4 by choosing $u_2 = -3z_1$. Then we have the single-input system

$$\frac{d}{dt}\begin{pmatrix} z_1 \\ z_2 \\ z_3 \end{pmatrix} = \begin{pmatrix} -4 & 0 & 0 \\ -3 & -2 & 0 \\ 0 & 0 & -3 \end{pmatrix}\begin{pmatrix} z_1 \\ z_2 \\ z_3 \end{pmatrix} + \begin{pmatrix} 0 \\ 1 \\ 1 \end{pmatrix} u_1$$

whose controllable subsystem

$$\frac{d}{dt}\begin{pmatrix} z_2 \\ z_3 \end{pmatrix} = \begin{pmatrix} -2 & 0 \\ 0 & -3 \end{pmatrix}\begin{pmatrix} z_2 \\ z_3 \end{pmatrix} + \begin{pmatrix} 1 \\ 1 \end{pmatrix} u_1$$

can be transformed to phase variable canonical form, and $u_1 = -12z_2 + 6z_3$.

The above procedure can be generalized to give a means of obtaining multiple-input pole placement.

8.3. Given the system $d^2y/dt^2 = 0$. Construct the observer system such that it has a double pole at $-\gamma$. Then find the error $\mathbf{e} = \mathbf{x} - \hat{\mathbf{x}}$ as a function of γ, if the output really is $y(t) + \eta(t)$, where $\eta(t)$ is noise.

The observer system has the form

$$\frac{d\hat{\mathbf{x}}}{dt} = \begin{pmatrix} 0 & 1 \\ 0 & 0 \end{pmatrix}\hat{\mathbf{x}} + \begin{pmatrix} k_1 \\ k_2 \end{pmatrix}[(1 \;\; 0)\hat{\mathbf{x}} - y - \eta]$$

The characteristic equation of the closed loop system is $\lambda(\lambda - k_1) - k_2 = 0$. A double pole at $-\gamma$ has the characteristic equation $\lambda^2 + 2\gamma\lambda + \gamma^2 = 0$. Hence set $k_1 = -2\gamma$ and $k_2 = -\gamma^2$. Then the equation for the error is

$$\frac{d}{dt}\begin{pmatrix} e_1 \\ e_2 \end{pmatrix} = \begin{pmatrix} -2\gamma & 1 \\ \gamma^2 & 0 \end{pmatrix}\begin{pmatrix} e_1 \\ e_2 \end{pmatrix} + \begin{pmatrix} -2\gamma \\ \gamma^2 \end{pmatrix}\eta$$

Note the noise drives the error and prevents it from reaching zero.

The transfer function is found from

$$\mathcal{L}\left\{\begin{pmatrix} e_1 \\ e_2 \end{pmatrix}\right\} = \frac{\mathcal{L}\{\eta\}}{s^2 + 2\gamma s + \gamma^2}\begin{pmatrix} 2\gamma s + \gamma^2 \\ \gamma^2 s \end{pmatrix}$$

As $\gamma \to \infty$, then $e_1 \to \eta$ and $e_2 \to d\eta/dt$. If $\eta(t) = \eta_0 \cos \omega t$, then $\omega\eta_0$, the amplitude of $d\eta/dt$, may be large even though η_0 is small, because the noise may be of very high frequency. We conclude that it is not a good idea to set the observer system gains too high, so that the observer system can filter out some of the noise.

8.4. Given the discrete-time system

$$\mathbf{x}(n+1) = \begin{pmatrix} 0 & 1 \\ -2 & 3 \end{pmatrix} \mathbf{x}(n) + \begin{pmatrix} 0 \\ 1 \end{pmatrix} u(n), \qquad y(n) = (1 \;\; 0)\mathbf{x}(n)$$

Design a feedback controller given dominant closed loop poles in the z plane at $(1 \pm j)/2$.

We shall construct an observer to generate the estimate of the state from which we can construct the desired control. The desired closed loop characteristic equation is $\lambda^2 - \lambda + 1/2 = 0$. Hence we choose $u = 3\hat{x}_1/2 - 2\hat{x}_2$. To generate $\hat{x}_1$ and $\hat{x}_2$ we choose a first-order observer with a pole at 0.05 so that it will hardly affect the response due to the dominant poles and yet will be large enough to filter high-frequency noise. The transformation of variables

$$\mathbf{x} = \begin{pmatrix} 0 & 1 \\ 1 & 3 \end{pmatrix} \mathbf{z} \quad \text{gives} \quad \mathbf{z}(n+1) = \begin{pmatrix} 0 & -2 \\ 1 & 3 \end{pmatrix} \mathbf{z}(n) + \begin{pmatrix} 1 \\ 0 \end{pmatrix} u(n), \qquad y(n) = (0 \;\; 1)\mathbf{z}(n)$$

Use of equation *(8.8)* gives

$$\mathbf{P} = (1 \;\; -0.05) \quad \text{and} \quad \mathbf{PL} = -(1 \;\; -0.05)\begin{pmatrix} 0 & -2 \\ 1 & 3 \end{pmatrix}\begin{pmatrix} 0.05 \\ 1 \end{pmatrix} = 2.1525$$

Then the observer is

$$\hat{\mathbf{x}} = \begin{pmatrix} -3 & 1 \\ 1 & 0 \end{pmatrix} \hat{\mathbf{z}} \quad \text{where} \quad \hat{\mathbf{z}}(n) = \begin{pmatrix} 1 \\ 0 \end{pmatrix} w(n) + \begin{pmatrix} 0.05 \\ 1 \end{pmatrix} y(n)$$

and $w(n)$ is obtained from $w(n+1) = 0.05w(n) - 2.1525y(n) + u(n)$.

8.5. Find the sensitivity of the poles of the system

$$\frac{d\mathbf{x}}{dt} = \begin{pmatrix} -1+\alpha^2 & \alpha \\ 2 & -2 \end{pmatrix} \mathbf{x}$$

to changes in the parameter α where $|\alpha|$ is small.

We denote the actual system as $d\mathbf{x}_a/dt = \mathbf{A}_a\mathbf{x}_a$ and the nominal system as $d\mathbf{x}_N/dt = \mathbf{A}_N\mathbf{x}_N$, where $\mathbf{A}_N = \mathbf{A}_a$ when $\alpha = 0$. In general, $\mathbf{A}_N = \mathbf{A}_a - \delta\mathbf{A}$ where $\|\delta\mathbf{A}\|$ is small since $|\alpha|$ is small. We assume $\mathbf{A}_N$ has distinct eigenvalues λ_i^N so that we can always find a corresponding eigenvector $\mathbf{w}_i$ from $\mathbf{A}_N\mathbf{w}_i = \lambda_i^N\mathbf{w}_i$. Denote the eigenvectors of $\mathbf{A}_N^T$ as $\mathbf{v}_i^*$, so that $\mathbf{A}_N^T\mathbf{v}_i^* = \lambda_i^N\mathbf{v}_i^*$. Note the eigenvalues λ_i^N are the same for $\mathbf{A}_N$ and $\mathbf{A}_N^T$. Taking the transpose gives

$$\mathbf{v}_i^\dagger \mathbf{A}_N = \lambda_i^N \mathbf{v}_i^\dagger \tag{8.18}$$

which we shall need later. Next we let the actual eigenvalues $\lambda_i^a = \lambda_i^N + \delta\lambda_i$. Substituting this into the eigenvalue equation for the actual $\mathbf{A}_a$ gives

$$(\mathbf{A}_N + \delta\mathbf{A})(\mathbf{w}_i + \delta\mathbf{w}_i) = (\lambda_i^N + \delta\lambda_i)(\mathbf{w}_i + \delta\mathbf{w}_i) \tag{8.19}$$

Subtracting $\mathbf{A}_N\mathbf{w}_i = \lambda_i^N\mathbf{w}_i$ and multiplying by $\mathbf{v}_i^\dagger$ gives

$$\mathbf{v}_i^\dagger\,\delta\mathbf{A}\,\mathbf{w}_i + \mathbf{v}_i^\dagger\mathbf{A}_N\,\delta\mathbf{w}_i = \delta\lambda_i\,\mathbf{v}_i^\dagger\mathbf{w}_i + \lambda_i^N\mathbf{v}_i^\dagger\,\delta\mathbf{w}_i + \mathbf{v}_i^\dagger(\delta\lambda_i\mathbf{I} - \delta\mathbf{A})\,\delta\mathbf{w}_i$$

Neglecting the last quantity on the right since it is of second order and using equation *(8.18)* then leads to

$$\delta\lambda_i = \frac{\mathbf{v}_i^\dagger\,\delta\mathbf{A}\,\mathbf{w}_i}{\mathbf{v}_i^\dagger\mathbf{w}_i} \tag{8.20}$$

Therefore for the particular system in question,

$$\mathbf{A}_a = \begin{pmatrix} -1+\alpha^2 & \alpha \\ 2 & -2 \end{pmatrix} = \begin{pmatrix} -1 & 0 \\ 2 & -2 \end{pmatrix} + \begin{pmatrix} \alpha^2 & \alpha \\ 0 & 0 \end{pmatrix} = \mathbf{A}_N + \delta\mathbf{A}$$

Then for $\mathbf{A}_N$ we find

$$\lambda_1^N = -1 \qquad \mathbf{w}_1 = \begin{pmatrix}1\\2\end{pmatrix} \qquad \mathbf{v}_1 = \begin{pmatrix}1\\0\end{pmatrix}$$

$$\lambda_2^N = -2 \qquad \mathbf{w}_2 = \begin{pmatrix}0\\1\end{pmatrix} \qquad \mathbf{v}_2 = \begin{pmatrix}-2\\1\end{pmatrix}$$

Using equation *(8.20)*,

$$\lambda_1^a \approx -1 + (1\ \ 0)\begin{pmatrix}\alpha^2 & \alpha\\0 & 0\end{pmatrix}\begin{pmatrix}1\\2\end{pmatrix} = -1 + \alpha^2 + 2\alpha$$

$$\lambda_2^a \approx -2 + (-2\ \ 1)\begin{pmatrix}\alpha^2 & \alpha\\0 & 0\end{pmatrix}\begin{pmatrix}0\\1\end{pmatrix} = -2 - 2\alpha$$

For larger values of α note we can use root locus under parameter variation to obtain exact values. However, the root locus is difficult computationally for very high-order systems, whereas the procedure just described has been applied to a 51st-order system.

8.6. Given the scalar nonlinear system of Fig. 8-20 with input $\alpha \sin t$. Should $K > 1$ be increased or decreased to minimize the effect of the nonlinearity?

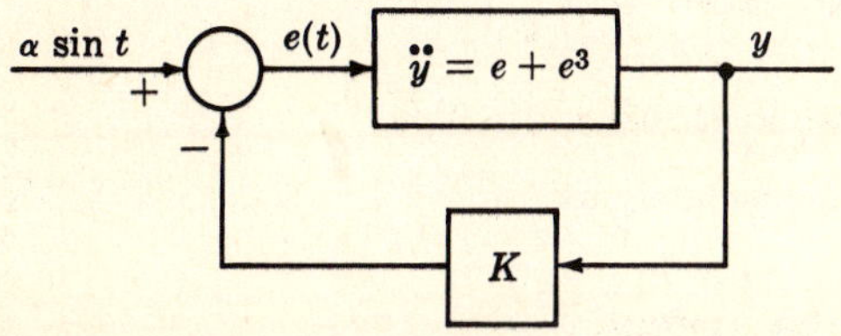

Fig. 8-20

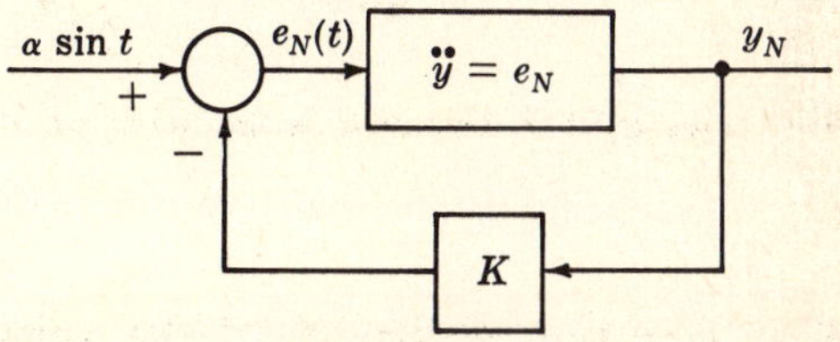

Fig. 8-21

The nominal linear system is shown in Fig. 8-21. The steady state value of $e_N = (\alpha \sin t)/(K-1)$. We approximate this as the input to the unwanted signal $d^2(\delta y)/dt^2 = e_N^3$, which gives the steady state value of $\delta y = \alpha^3(27 \sin t - \sin 3t)/[36(K-1)^3]$. This approximation that $d^2(\delta y)/dt^2 = e_N^3$ instead of e_a^3 must be verified by simulation. It turns out this a good approximation for $|\alpha| \ll 1$, and we can conclude that for $|\alpha| \ll 1$ we increase K to make $\delta y/y$ become smaller.

8.7. A simplified, normalized representation of the control system of a solid-core nuclear rocket engine is shown in Fig. 8-22 where $\delta n, \delta T, \delta P, \delta\rho_{cr}$ and δV are the changes from nominal in neutron density, core temperature, core hydrogen pressure, control rod setting, and turbine power control valve setting, respectively. Also, $G_1(s), G_2(s)$ and $G_3(s)$ are scalar transfer functions and in the compensation $\kappa_1, \kappa_2, \kappa_3$ and κ_4 are scalar constants so that the control is proportional plus integral. Find a simple means of improving response.

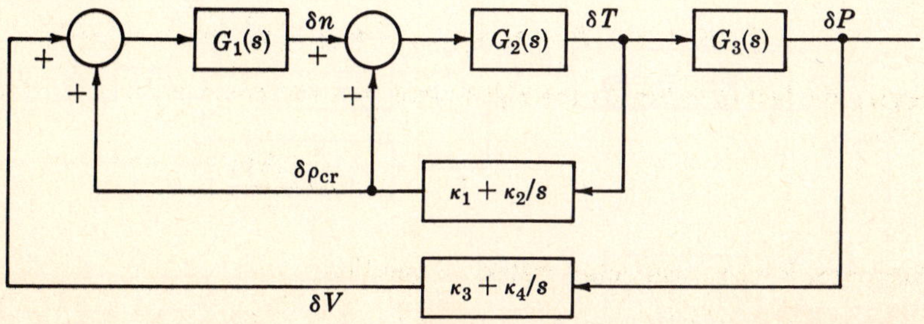

Fig. 8-22

The main point is to realize this "multiple loop" system is really a multiple input–multiple output system. The control system shown has constrained $\delta\rho_{cr}$ to be an inner loop. A means of improving response is to rewrite the system in matrix form as shown in Fig. 8-23. This opens up the possibility of "cross coupling" the feedback, such as having $\delta\rho_{cr}$ depend on δP as well as δT. Furthermore it is evident that the system is of type 1 and can follow a step input with zero steady state error.

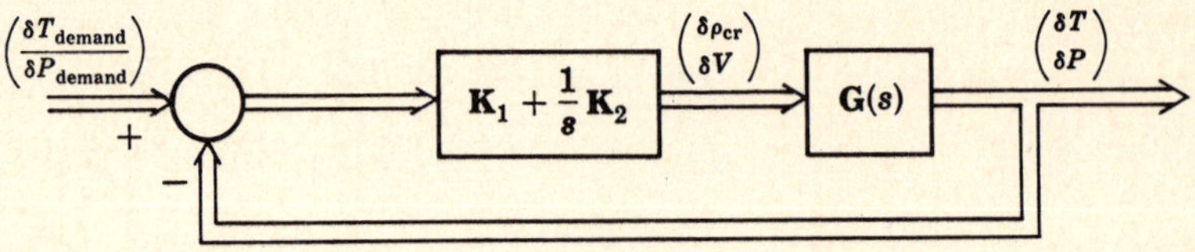

Fig. 8-23

8.8. Compensate the system $(s+p_1)^{-1}(s+p_2)^{-1}$ to have poles at $-\pi_1$ and $-\pi_2$, with an observer pole at $-\pi_0$, by using the algebraic separation theorem, and discuss the effect of noise η at the output.

The state space representation of the plant is

$$\frac{d}{dt}\begin{pmatrix} x_1 \\ x_2 \end{pmatrix} = \begin{pmatrix} 0 & 1 \\ -p_1p_2 & -p_1-p_2 \end{pmatrix}\begin{pmatrix} x_1 \\ x_2 \end{pmatrix} + \begin{pmatrix} 0 \\ 1 \end{pmatrix} u + \begin{pmatrix} 0 \\ 1 \end{pmatrix} d$$

$$y = (1 \quad 0)\mathbf{x}$$

The feedback compensation can be found immediately as

$$u = (p_1p_2 - \pi_1\pi_2 \quad p_1+p_2-\pi_1-\pi_2)\hat{\mathbf{x}}$$

To construct the observer system, let $\mathbf{P} = (\alpha_1 \quad \alpha_2)$. Then

$$(\mathbf{F} \mid -\mathbf{PL})\left(\frac{\mathbf{P}}{\mathbf{C}}\right) = (-\pi_0 \mid -\mathbf{PL})\begin{pmatrix} \alpha_1 & \alpha_2 \\ 1 & 0 \end{pmatrix} = (\alpha_1 \quad \alpha_2)\begin{pmatrix} 1 & 1 \\ -p_1p_2 & -p_1-p_2 \end{pmatrix} = \mathbf{PA}$$

from which

$$\alpha_1 = (p_1+p_2-\pi_0)\alpha_2 \qquad \text{and} \qquad -\mathbf{PL} = [\pi_0(p_1+p_2-\pi_0) - p_1p_2]\alpha_2$$

Also $\mathbf{PB} = \mathbf{PJ} = \alpha_2$. Therefore the estimator dynamics are

$$\frac{d}{dt}\left(\frac{w}{\alpha_2}\right) = -\pi_0\left(\frac{w}{\alpha_2}\right) + [\pi_0(p_1+p_2-\pi_0) - p_1p_2]y + u + d$$

To construct the estimate,

$$\begin{pmatrix} \hat{x}_1 \\ \hat{x}_2 \end{pmatrix} = \begin{pmatrix} \mathbf{P} \\ \mathbf{C} \end{pmatrix}^{-1}\begin{pmatrix} w \\ y \end{pmatrix} = \begin{pmatrix} 0 & 1 \\ \alpha_2^{-1} & \pi_0-p_1-p_2 \end{pmatrix}\begin{pmatrix} w \\ y \end{pmatrix}$$

The flow diagram of the closed loop system is shown in Fig. 8-24.

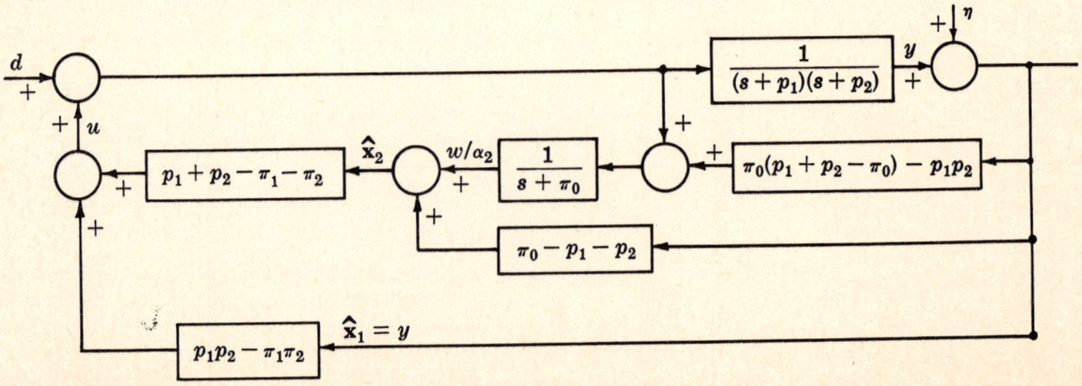

Fig. 8-24

Note the noise η is fed through a first-order system and gain elements to form u. If the noise level is high, it is better to use a second-order observer or a Kalman filter because then the noise goes through no gain elements directly to the control, but instead is processed through first- and second-order systems. If there is no noise whatsoever, flow diagram manipulations can be used to show the closed loop system is equivalent to one compensated by a lead network, i.e. the above flow diagram with $\eta = 0$ can be rearranged as in Fig. 8-25.

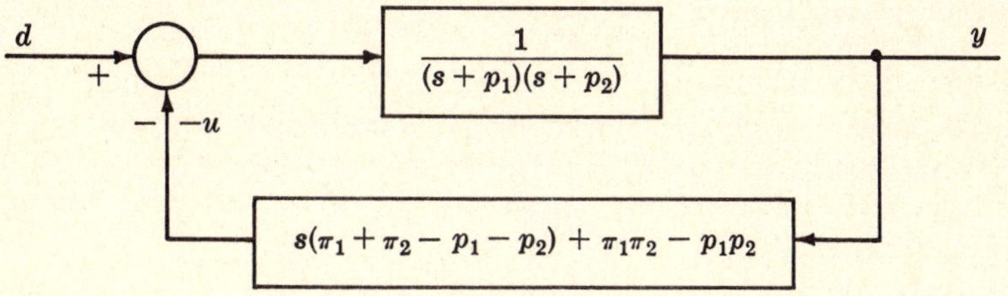

Fig. 8-25

Note the observer dynamics have cancelled out, and the closed loop system remains second-order. This corresponds to the conclusion the observer dynamics are uncontrollable by d. However any initial condition in the observer will produce an effect on y, as can be seen from the first flow diagram.

Supplementary Problems

8.9. Given the matrix block diagram of Fig. 8-26. Show that this reduces to Fig. 8-27 when the indicated inverse exists.

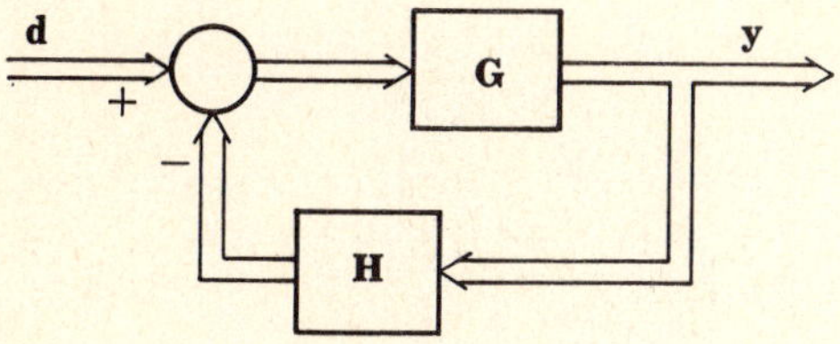

Fig. 8-26

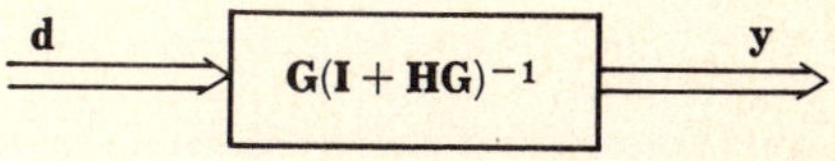

Fig. 8-27

8.10. Given the matrix block diagram of Fig. 8-28. Reduce the block diagram to obtain $\mathbf{H}_1$ isolated in a single feedback loop.

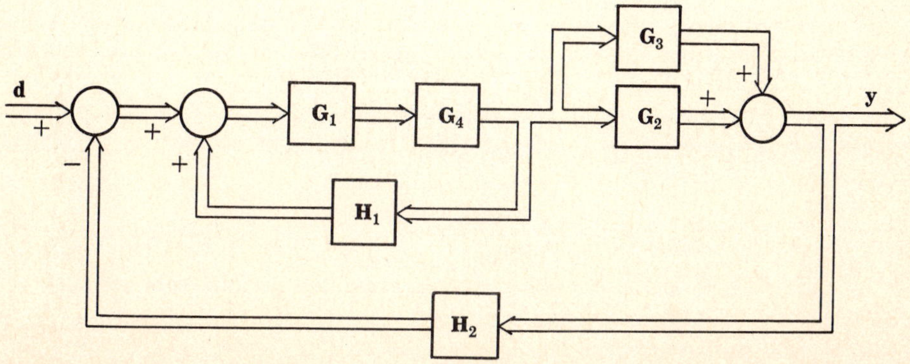

Fig. 8-28

8.11. Determine whether the scalar feedback system of Fig. 8-29 in which $y(t)$ is related to $e(t)$ as

$$\frac{d\mathbf{x}}{dt} = \begin{pmatrix} 0 & 1 \\ 0 & -4t^{-1} \end{pmatrix}\mathbf{x} + \begin{pmatrix} 0 \\ t^{-2} \end{pmatrix} e \qquad t > 0$$

$$y = (1 \quad 0)\mathbf{x}$$

(i) can follow a step input α with zero error, (ii) can follow a ramp input αt with zero error.

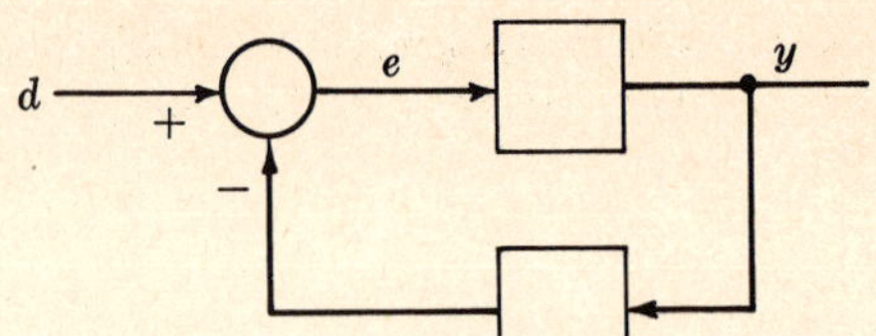

Fig. 8-29

8.12. Given the time-varying system $d^2y/dt^2 + \alpha(t)\,dy/dt + \beta(t)y = u$. Find a feedback control u such that the closed loop system behaves like $d^2z/dt^2 + \theta(t)\,dz/dt + \phi(t)z = 0$.

8.13. Given the system

$$\frac{d\mathbf{x}}{dt} = \begin{pmatrix} 1 & 1 & -2 \\ 0 & -1 & 1 \\ 1 & 0 & -1 \end{pmatrix}\mathbf{x} \qquad y = (0 \quad 1 \quad -1)\mathbf{x}$$

Construct a third-order observer system with poles at $0, -1$ and -1.

8.14. Given the system

$$\frac{d\mathbf{x}}{dt} = \begin{pmatrix} 1 & 1 & -2 \\ 0 & -1 & 1 \\ 1 & 0 & -1 \end{pmatrix}\mathbf{x} \qquad \mathbf{y} = \begin{pmatrix} 0 & 1 & -1 \\ 1 & 0 & 0 \end{pmatrix}\mathbf{x}$$

Construct a first-order observer system with a pole at -1.

8.15. Given the system

$$\frac{d\mathbf{x}}{dt} = \begin{pmatrix} -2 & 1 \\ 0 & -1 \end{pmatrix}\mathbf{x} + \begin{pmatrix} 0 \\ 1 \end{pmatrix} u \qquad y = (1 \quad 0)\mathbf{x}$$

Construct a first-order observer system with a pole at -3 and then find a feedback control $u = k_1\hat{x}_1 + k_2\hat{x}_2$ that places both the closed loop system poles at -2.

8.16. Given the system

$$\frac{d\mathbf{x}}{dt} = -\frac{1}{4}\begin{pmatrix} 3 & 1 \\ 1 & 3 \end{pmatrix}\mathbf{x} \qquad y = (1 \quad 1)\mathbf{x}$$

Construct a first-order observer system with a pole at -4.

8.17. Given the system $d\mathbf{x}/dt = \mathbf{A}(t)\mathbf{x} + \mathbf{B}(t)\mathbf{u}$ where $\mathbf{y} = \mathbf{C}(t)\mathbf{x} + \mathbf{D}(t)\mathbf{u}$. What is the form of the observer system when $\mathbf{D}(t) \neq \mathbf{0}$? What is the algebraic separation theorem when $\mathbf{D}(t) \neq \mathbf{0}$?

8.18. Given the system

$$\frac{d\mathbf{x}}{dt} = -\frac{1}{4}\begin{pmatrix} 3 & 1 \\ 1 & 3 + 4\alpha(t) \end{pmatrix}\mathbf{x} + \begin{pmatrix} 1 \\ 0 \end{pmatrix} u \qquad \text{and} \qquad y = (0 \quad 1)\mathbf{x}$$

where $\mathbf{x}(0) = \mathbf{0}$. As a measure of the sensitivity of the system, assuming $|\alpha(t)| \ll 1$, find $\delta y(t)$ as an integral involving $\alpha(t)$ and $u(t)$. *Hint*: It is easiest to use Laplace transforms.

8.19. Given the nominal system $d\mathbf{x}_N/dt = \mathbf{A}_N(t)\mathbf{x}_N + \mathbf{B}_N(t)\mathbf{u}_N$ and $\mathbf{y}_N = \mathbf{C}_N(t)\mathbf{x}_N + \mathbf{D}_N(t)\mathbf{u}_N$. What is the equation for $\delta\mathbf{x}$ corresponding to equation *(8.14)* when the parameters of the system become $\mathbf{A}_a(t)$, $\mathbf{B}_a(t)$, $\mathbf{C}_a(t)$, $\mathbf{D}_a(t)$ and $\mathbf{u}_a(t)$?

8.20. Given the system $d\mathbf{x}/dt = \begin{pmatrix} 1 + t + f(t) & u(t) \\ f(t) & t^2 \end{pmatrix}\mathbf{x}$. Choose $u(t)$ such that at least one state will be insensitive to small variations in $f(t)$, given the nominal solution $\mathbf{x}_N(t)$.

8.21. Given that the input $d(t)$ is generated by the scalar system $dw/dt = \alpha(t)w$ and $d(t) = [\gamma(t) + \rho(t)]w$, under what conditions on $\rho(t)$ can the system $dx/dt = \alpha(t)x + \beta(t)e$ with $y = \gamma(t)x$ follow any such $d(t)$ with zero error?

8.22. Given the general nonunity feedback time-invariant system of Fig. 8-30. Under what conditions can $\lim_{t\to\infty} \mathbf{e}(t) = \mathbf{0}$? Set $\mathbf{F} = \mathbf{I}$ and $\mathbf{H} = \mathbf{I}$ and derive Theorem 8.2 from these conditions.

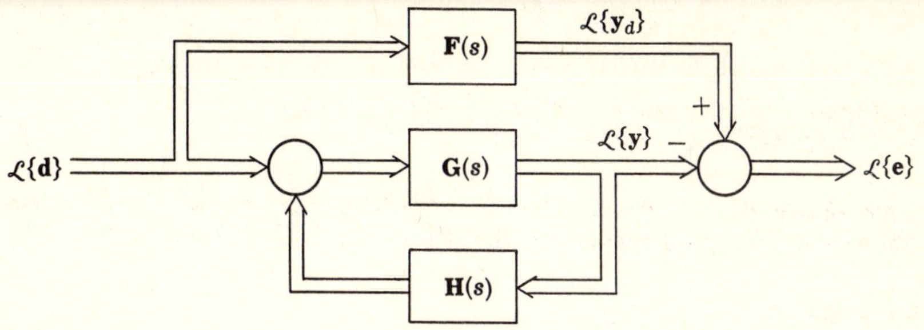

Fig. 8-30

8.23. In the proof of Theorem 8.1, why is

$$\mathbf{d}(t) = \int_{t_0}^{t} \mathbf{C}(t)\, \mathbf{\Phi}_{A-BC}(t,\tau)\, \mathbf{B}(\tau)\, \mathbf{d}(\tau)\, d\tau + \mathbf{g}(t) + \mathbf{C}(t)\, \mathbf{\Phi}_{A-BC}(t,t_0)\, \mathbf{w}(t_0)$$

equivalent to $d\mathbf{w}/dt = [\mathbf{A}(t) - \mathbf{B}(t)\,\mathbf{C}(t)]\mathbf{w} + \mathbf{B}(t)\mathbf{d}$ and $\mathbf{d} = \mathbf{C}(t)\mathbf{w} + \mathbf{g}$?

8.24. Show that if a time-invariant system is of type N, then it is of type $N-k$ for all integers k such that $0 \le k \le N$.

8.25. Given the system of Fig. 8-31 where the constant matrix **K** has been introduced as compensation. Show that

(*a*) The type number of the system cannot change if **K** is nonsingular.

(*b*) The system is of type zero if **K** is singular.

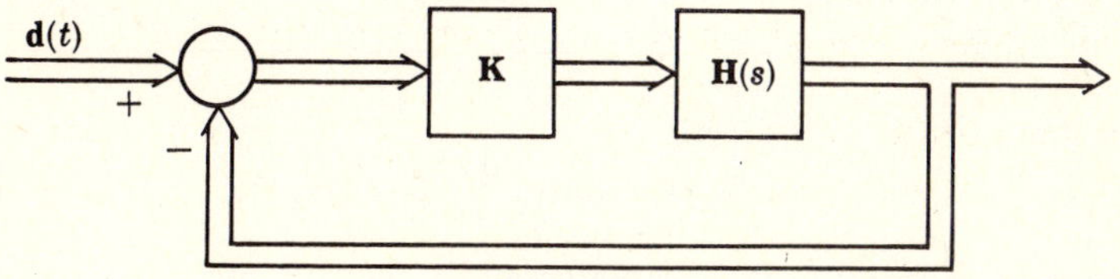

Fig. 8-31

8.26. Design a system that will follow $d(t) = \sin t$ with zero steady state error.

Answers to Supplementary Problems

8.10. This cannot be done unless the indicated inverse exists (Fig. 8-32). The matrices must be in the order given.

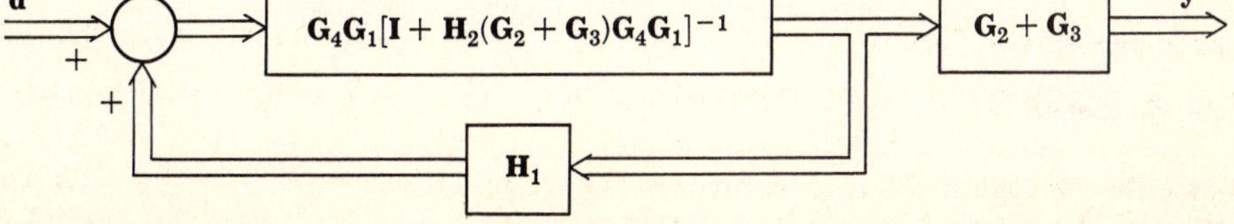

Fig. 8-32

8.11. The closed loop system output $y = \left(\frac{t_0}{t}\right)^2\left[\left(2\frac{t}{t_0} - 1\right)x_1(t_0) + (t - t_0)\,x_2(t_0)\right]$ tends to zero, so Theorem 8.1 applies. Then

$$d(t) = w_1(t_0) + \frac{t_0}{3}\left[1 - \left(\frac{t_0}{t}\right)^3\right]w_2(t_0) + \int_{t_0}^{t}\frac{1}{3\tau}\left[1 - \left(\frac{\tau}{t}\right)^3\right]g(\tau)\,d\tau \;\leqq\; \ln t \qquad \text{for } t > \text{some } t_1$$

so that the system can follow steps but not ramps with zero error.

8.12. For the state equation

$$\frac{d\mathbf{x}}{dt} = \begin{pmatrix} 0 & 1 \\ -\beta(t) & -\alpha(t) \end{pmatrix}\mathbf{x} + \begin{pmatrix} 0 \\ 1 \end{pmatrix}u$$

corresponding to the given input-output relation,

$$u = (\beta - \phi \quad \alpha - \theta)\mathbf{x}$$

which shows the basic idea behind "pole placement,"

$$d^2y/dt^2 + \alpha\, dy/dt + \beta y = (\alpha - \theta)\,dy/dt + (\beta - \phi)y$$

8.13. $$\frac{d\hat{\mathbf{x}}}{dt} = \begin{pmatrix} 1 & 1 & -2 \\ 0 & -1 & 1 \\ 1 & 0 & 1 \end{pmatrix}\hat{\mathbf{x}} + \begin{pmatrix} 1 \\ 0 \\ 1 \end{pmatrix}[(0 \;\; 1 \;\; -1)\hat{\mathbf{x}} - y]$$

8.14. $$dw/dt = -w + (0 \;\; 1)\mathbf{y}, \qquad \hat{\mathbf{x}} = \begin{pmatrix} 0 \\ 1 \\ 1 \end{pmatrix}w + \begin{pmatrix} 0 & 1 \\ 1 & 0 \\ 0 & 0 \end{pmatrix}\mathbf{y}$$

8.15.

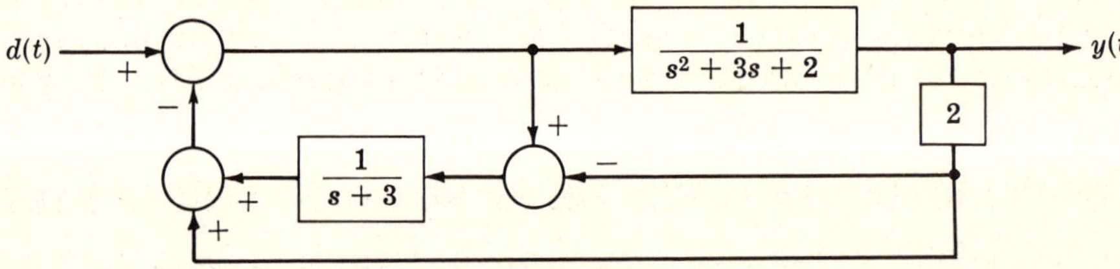

Fig. 8-33

8.16. This cannot be done because the system is unobservable.

8.17. Subtract $\mathbf{D}(t)\mathbf{u}$ from y entering the observer and this reduces to the given formulation.

8.18. $$\mathcal{L}\{\delta y(t)\} = -\frac{\mathcal{L}\{\alpha(t)\,x_N(t)\}}{(s+1)(s+\frac{1}{2})} \quad \text{where} \quad \mathcal{L}\{x_N(t)\} = -\frac{\mathcal{L}\{u(t)\}}{4(s+1)(s+\frac{1}{2})}$$

8.20. $$d(\delta\mathbf{x})/dt = \begin{pmatrix} 1 + t + f(t) & u(t) \\ f(t) & t^2 \end{pmatrix}\delta\mathbf{x} + \begin{pmatrix} x_{N1} \\ x_{N2} \end{pmatrix}\delta f$$

$$\mathbf{Q}(t) = \begin{pmatrix} x_{N1} & dx_{N1}/dt - (1 + t + f(t))x_{N1} - u(t)x_{N2} \\ x_{N2} & dx_{N2}/dt - f(t)x_{N1} - t^2x_{N2} \end{pmatrix}$$

Choose $u(t)$ such that $\det \mathbf{Q}(t) = 0$ for all t.

8.21. $\rho(t) = e^{-\int_{t_0}^{t}\alpha(\eta)\,d\eta}\left[\theta(t) + \int_{t_0}^{t}\gamma(t)e^{\int_{\tau}^{t}\alpha(\eta)\,d\eta}\beta(\tau)\,\theta(\tau)\,d\tau\right] + \gamma(t)\kappa$ where $\theta(t)$ is any function such that $\lim_{t\to\infty}\theta(t) = 0$ and κ is an arbitrary constant.

8.22. $\lim_{s\to 0}\,[\mathbf{F}(s) - \mathbf{G}(s)(\mathbf{I} + \mathbf{G}(s)\,\mathbf{H}(s))^{-1}]s\mathcal{L}\{\mathbf{d}\} = \mathbf{0}$ and is analytic for $s \geqq 0$.

8.24. Use Theorem 8.2.

8.25. Use Theorem 8.2.

8.26. One answer is $H(s) = s(s^2+1)^{-1}$.

Chapter 9

Stability of Linear Systems

9.1 INTRODUCTION

Historically, the concept of stability has been of great importance to the system designer. The concept of stability seems simple enough for linear time-invariant systems. However, we shall find that its extension to nonlinear and/or time-varying systems is quite complicated. For the unforced linear time-invariant system $d\mathbf{x}/dt = \mathbf{A}\mathbf{x}$, we are assured that the solution $\mathbf{x}(t) = \mathbf{T}e^{\mathbf{J}(t-t_0)}\mathbf{T}^{-1}\mathbf{x}_0$ does not blow up if all the eigenvalues of $\mathbf{A}$ are in the left half of the complex plane.

Other than transformation to Jordan form, there are many direct tests for stability. The Routh-Hurwitz criterion can be applied to the characteristic polynomial of $\mathbf{A}$ as a yes-or-no test for the existence of poles in the right half plane. More useful techniques are those of root locus, Nyquist, Bode, etc., which indicate the degree of stability in some sense. These are still the best techniques for the analysis of low-order time-invariant systems, and we have seen how to apply them to multiple input–multiple output systems in the previous chapter. Now we wish to extend the idea of stability to time-varying systems, and to do this we must first examine carefully what is meant by stability for this type of system.

9.2 DEFINITIONS OF STABILITY FOR ZERO-INPUT LINEAR SYSTEMS

A type of "stability" results if we can say the response is bounded.

Definition 9.1: For every $\mathbf{x}_0$ and every t_0, if there exists a constant κ depending on $\mathbf{x}_0$ and t_0 such that $\|\mathbf{x}(t)\| \leq \kappa$ for all $t \geq t_0$, then the response $\mathbf{x}(t)$ is *bounded.*

Even this simple definition has difficulties. The trouble is that we must specify the response to what. The trajectory of $\mathbf{x}(t) = \boldsymbol{\phi}(t; \mathbf{u}(\tau), \mathbf{x}(t_0), t_0)$ depends implicitly on three quantities: $\mathbf{u}(\tau)$, $\mathbf{x}(t_0)$ and t_0. By considering only $\mathbf{u}(\tau) = \mathbf{0}$ in this section, we have eliminated one difficulty for a while. But can the boundedness of the response depend on $\mathbf{x}(t_0)$?

Example 9.1.

Consider the scalar zero-input nonlinear equation $dx/dt = -x + x^2$ with initial condition $x(0) = x_0$. The solution to the linearized equation $dx/dt = -x$ is obviously bounded for all x_0. However, the solution to the nonlinear equation is

$$x(t) = \frac{x_0}{x_0 + e^t(1 - x_0)}$$

For all negative values of x_0, this is well behaved. For values of $x_0 > 1$, the denominator vanishes at a time $t_1 = \ln x_0 - \ln(x_0 - 1)$, so that $\lim_{t \to t_1} x(t) = \infty$.

It can be concluded that boundedness depends upon the initial conditions for nonlinear equations in general.

Theorem 9.1: The boundedness of the response $\mathbf{x}(t)$ of the linear system $d\mathbf{x}/dt = \mathbf{A}(t)\mathbf{x}$ is independent of the initial condition $\mathbf{x}_0$.

Proof: $\|\mathbf{x}(t)\| = \|\boldsymbol{\phi}(t; \mathbf{0}, \mathbf{x}_0, t_0)\| = \|\boldsymbol{\Phi}(t, t_0)\mathbf{x}_0\| \leq \|\boldsymbol{\Phi}(t, t_0)\| \, \|\mathbf{x}_0\|$

Since $\|\mathbf{x}_0\|$ is a constant, if $\|\mathbf{x}(t)\|$ becomes unbounded as $t \to \infty$, it is solely due to $\boldsymbol{\Phi}(t, t_0)$.

Now we shall consider other, different types of stability. First, note that $\mathbf{x} = \mathbf{0}$ is a steady state solution (an equilibrium state) to the zero-input linear system $d\mathbf{x}/dt = \mathbf{A}(t)\mathbf{x}$. We shall define a region of state space by $\|\mathbf{x}\| < \epsilon$, and see if there exists a small region of nonzero perturbations surrounding the equilibrium state $\mathbf{x} = \mathbf{0}$ that give rise to a trajectory which remains within $\|\mathbf{x}\| < \epsilon$. If this is true for all $\epsilon > 0$, no matter how small, then we have

Definition 9.2: The equilibrium state $\mathbf{x} = \mathbf{0}$ of $d\mathbf{x}/dt = \mathbf{A}(t)\mathbf{x}$ is *stable in the sense of Liapunov* (for short, stable i.s.L.) if for any t_0 and every real number $\epsilon > 0$, there is some $\delta > 0$, as small as we please, depending on t_0 and ϵ such that if $\|\mathbf{x}_0\| < \delta$ then $\|\mathbf{x}(t)\| < \epsilon$ for all $t > t_0$.

This definition is also valid for nonlinear systems with an equilibrium state $\mathbf{x} = \mathbf{0}$. It is the most common definition of stability, and in the literature "stable i.s.L." is often shortened to "stable". States that are not stable i.s.L. will be called *unstable*. Note stability i.s.L. is a local condition, in that δ can be as small as we please. Finally, since $\mathbf{x} = \mathbf{0}$ is an obvious choice of equilibrium state for a linear system, when speaking about linear systems we shall not be precise but instead will say the system is stable when we mean the zero state is stable.

Example 9.2.

Consider the nonlinear system of Example 9.1. If $x_0 \leq 1$, then

$$|x(t)| = \frac{|x_0|}{|x_0 + e^t(1 - x_0)|} = \frac{|x_0|}{|1 + (e^t - 1)(1 - x_0)|} \leq |x_0|$$

In Definition 9.2 we can set $\delta = \epsilon > 0$ if $\epsilon \leq 1$, and if $\epsilon > 1$ we set $\delta = 1$ to show the zero state of Example 9.1 is stable i.s.L. Hence the zero state is stable i.s.L. even though the response can become unbounded for some x_0. (This situation corresponds to Fig. 9-2(*b*) of Problem 9.1.) Of course if the response became unbounded for all $x_0 \neq 0$, the zero state would be considered unstable. Another point to note is that in the application of Definition 9.2, in the range where ϵ is small there results the choice of a correspondingly small δ.

Example 9.3.

Given the Van der Pol equation

$$\frac{d}{dt}\begin{pmatrix} x_1 \\ x_2 \end{pmatrix} = \begin{pmatrix} x_2 \\ (1 - x_1^2)x_2 - x_1 \end{pmatrix}$$

with initial condition $\mathbf{x}(0) = \mathbf{x}_0$. The trajectories in state space can be plotted as shown in Fig. 9-1. We will call the trajectory in bold the limit cycle. Trajectories originating outside the limit cycle spiral in towards it and trajectories originating inside the limit cycle spiral out towards it. Consider a small circle of any radius, centered at the origin but such that it lies completely within the limit cycle. Call its radius ϵ and note only $\mathbf{x}_0 = \mathbf{0}$ will result in a trajectory that stays within $\|\mathbf{x}\|_2 < \epsilon$. Therefore the zero state of the Van der Pol equation is unstable but any trajectory is bounded. (This situation corresponds to Fig. 9-2(*e*) of Problem 9.1.)

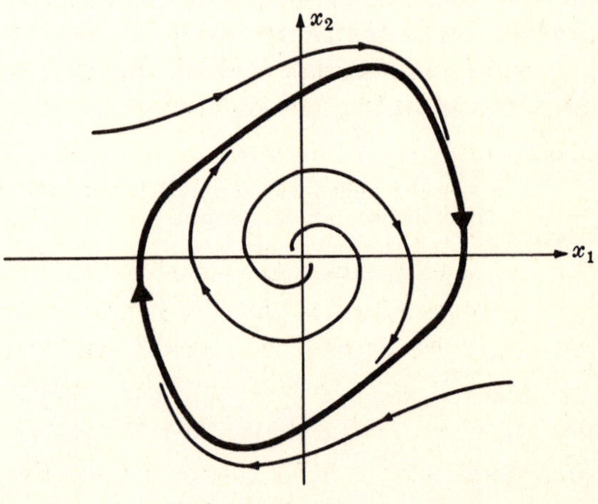

Fig. 9-1

***Theorem 9.2*:** The transition matrix of a linear system is bounded as $\|\boldsymbol{\Phi}(t, t_0)\| < \kappa(t_0)$ for all $t \geq t_0$ if and only if the equilibrium state $\mathbf{x} = \mathbf{0}$ of $d\mathbf{x}/dt = \mathbf{A}(t)\mathbf{x}$ is stable i.s.L.

Note $\|\mathbf{x}(t)\|$ is bounded if $\|\boldsymbol{\Phi}(t, t_0)\|$ is bounded.

Proof: First assume $\|\Phi(t,t_0)\| < \kappa(t_0)$ where κ is a constant depending only on t_0. If we are given any $\epsilon > 0$, then we can always find $\delta = \epsilon/\kappa(t_0)$ such that if $\|\mathbf{x}_0\| < \delta$ then $\epsilon = \kappa(t_0)\delta > \|\Phi(t,t_0)\|\,\|\mathbf{x}_0\| \geq \|\Phi(t,t_0)\mathbf{x}_0\| = \|\mathbf{x}(t)\|$. From Definition 9.2 we conclude stability i.s.L.

Next we assume stability i.s.L. Let us suppose $\Phi(t,t_0)$ is not bounded, so that there is at least one element $\Phi_{ij}(t,t_0)$ that becomes large as t tends to ∞. If $\|\mathbf{x}_0\| < \delta$ for a nonzero δ, then the element x_j of $\mathbf{x}_0$ can be nonzero, which results in a trajectory that eventually leaves any region in state space defined by $\|\mathbf{x}\| < \epsilon$. This results in an unstable system, so that we have reached a contradiction and conclude that $\Phi(t,t_0)$ must be bounded.

Taken together, Theorems 9.1 and 9.2 show that boundedness of $\|\mathbf{x}(t)\|$ is equivalent to stability i.s.L. for linear systems, and is independent of $\mathbf{x}_0$. When any form of stability is independent of the size of the initial perturbation $\mathbf{x}_0$, we say the stability is *global*, or speak of stability *in the large*. Therefore another way of stating Theorem 9.1 is to say (local) stability i.s.L. implies global stability for linear systems. The nonlinear system of Example 9.1 is stable i.s.L. but not globally stable i.s.L.

In practical applications we often desire the response to return eventually to the equilibrium position $\mathbf{x} = \mathbf{0}$ after a small displacement. This is a stronger requirement than stability i.s.L. which only demands that the response stay within a region $\|\mathbf{x}\| < \epsilon$.

Definition 9.3: The equilibrium state $\mathbf{x} = \mathbf{0}$ is *asymptotically stable* if (1) it is stable i.s.L. and (2) for any t_0 and any $\mathbf{x}_0$ sufficiently close to $\mathbf{0}$, $\mathbf{x}(t) \to \mathbf{0}$ as $t \to \infty$.

This definition is also valid for nonlinear systems. It turns out that (1) must be assumed besides (2), because there exist pathological systems where $\mathbf{x}(t) \to \mathbf{0}$ but are not stable i.s.L.

Example 9.4.

Consider the linear harmonic oscillator $\dfrac{d\mathbf{x}}{dt} = \begin{pmatrix} 0 & 1 \\ -1 & 0 \end{pmatrix}\mathbf{x}$ with transition matrix

$$\Phi(t,0) = \begin{pmatrix} \cos t & \sin t \\ -\sin t & \cos t \end{pmatrix}$$

The $\mathbf{A}$ matrix has eigenvalues at $\pm j$. To apply Definition 9.2, $\|\mathbf{x}(t)\|_2 \leq \|\Phi(t,t_0)\|_2\,\|\mathbf{x}_0\|_2 = \|\mathbf{x}_0\|_2 < \epsilon = \delta$ since $\|\Phi(t,t_0)\|_2 = 1$. Therefore the harmonic oscillator is stable i.s.L. However, $\mathbf{x}(t)$ never damps out to $\mathbf{0}$, so the harmonic oscillator is not asymptotically stable.

Example 9.5.

The equilibrium state $x = 0$ is asymptotically stable for the system of Example 9.1, since any small perturbation ($x_0 < 1$) gives rise to a trajectory that eventually returns to 0.

In all cases except one, if the conditions of the type of stability are independent of t_0, then the adjective *uniform* is added to the descriptive phrase.

Example 9.6.

If δ does not depend on t_0 in Definition 9.2, we have uniform stability i.s.L.

Example 9.7.

The stability of any time-invariant system is uniform.

The exception to the usual rule of adding "uniform" to the descriptive phrase results because here we only consider linear systems. In the general framework of stability definitions for time-varying nonlinear systems, there is no inconsistency. To avoid the complexities of general nonlinear systems, here we give

Definition 9.4: If the linear system $d\mathbf{x}/dt = \mathbf{A}(t)\mathbf{x}$ is uniformly stable i.s.L. and if for all t_0 and for any fixed ρ however large $\|\mathbf{x}_0\| < \rho$ gives rise to a response $\mathbf{x}(t) \to \mathbf{0}$ as $t \to \infty$, then the system is *uniformly asymptotically stable.*

The difference between Definitions 9.3 and 9.4 is that the conditions of 9.4 do not depend on t_0, and additionally must hold for all ρ. If ρ could be as small as we please, this would be analogous to Definition 9.3. This complication arises only because of the linearity, which in turn implies that Definition 9.4 is also global.

***Theorem 9.3*:** The linear system $d\mathbf{x}/dt = \mathbf{A}(t)\mathbf{x}$ is uniformly asymptotically stable if and only if there exist two positive constants κ_1 and κ_2 such that $\|\mathbf{\Phi}(t, t_0)\| \leq \kappa_1 e^{-\kappa_2(t-t_0)}$ for all $t \geq t_0$ and all t_0.

The proof is given in Problem 9.2.

Example 9.8.

Given the linear time-varying scalar system $dx/dt = -x/t$. This has a transition matrix $\Phi(t, t_0) = t_0/t$. For initial times $t_0 > 0$ the system is asymptotically stable. However, the response does not tend to 0 as fast as an exponential. This is because for $t_0 < 0$ the system is unstable, and the asymptotic stability is not uniform.

Example 9.9.

Any time-invariant linear system $d\mathbf{x}/dt = \mathbf{A}\mathbf{x}$ is uniformly asymptotically stable if and only if all the eigenvalues of $\mathbf{A}$ have negative real parts.

However for the time-varying system $d\mathbf{x}/dt = \mathbf{A}(t)\mathbf{x}$, if $\mathbf{A}(t)$ has all its eigenvalues with negative real parts for each fixed t, this in general does not mean the system is asymptotically stable or even stable.

Example 9.10.

Given the system

$$\frac{d}{dt}\begin{pmatrix} x_1 \\ x_2 \end{pmatrix} = \begin{pmatrix} 4\kappa & -3\kappa e^{8\kappa t} \\ \kappa e^{-8\kappa t} & 0 \end{pmatrix}\begin{pmatrix} x_1 \\ x_2 \end{pmatrix}$$

with initial conditions $\mathbf{x}(0) = \mathbf{x}_0$. The eigenvalues are $\lambda_1 = \kappa$ and $\lambda_2 = 3\kappa$. Then if $\kappa < 0$, both eigenvalues have real parts less than zero. However, the exact solution is

$$2x_1(t) = 3(x_{10} + x_{20})e^{5\kappa t} - (x_{10} + 3x_{20})e^{7\kappa t} \quad \text{and} \quad 2x_2(t) = (x_{10} + 3x_{20})e^{-\kappa t} - (x_{10} + x_{20})e^{-3\kappa t}$$

For any nonzero real κ the system is unstable.

There are many other types of stability to consider in the general case of nonlinear time-varying systems. For brevity, we shall not discuss any more than Definitions 9.1–9.4 for zero-input systems. Furthermore, these definitions of stability, and those of the next section, carry over in an obvious manner to discrete-time systems. The only difference is that t takes on discrete values.

9.3 DEFINITIONS OF STABILITY FOR NONZERO INPUTS

In some cases we are more interested in the input-output relationships of a system than its zero-input response. Consider the system

$$d\mathbf{x}/dt = \mathbf{A}(t)\mathbf{x} + \mathbf{B}(t)\mathbf{u} \qquad \mathbf{y} = \mathbf{C}(t)\mathbf{x} \tag{9.1}$$

with initial condition $\mathbf{x}(t_0) = \mathbf{x}_0$ and where $\|\mathbf{A}(t)\| < \kappa_{\mathbf{A}}$, a constant, and $\|\mathbf{B}(t)\| < \kappa_{\mathbf{B}}$, another constant, for all $t \geq t_0$, and $\mathbf{A}(t)$ and $\mathbf{B}(t)$ are continuous.

Definition 9.5: The system (*9.1*) is *externally stable* if for any t_0, any $\mathbf{x}_0$, and any $\mathbf{u}$ such that $\|\mathbf{u}(t)\| \leq \delta$ for all $t \geq t_0$, there exists a constant ϵ which depends only on t_0, $\mathbf{x}_0$ and δ such that $\|\mathbf{y}(t)\| \leq \epsilon$ for all $t \geq t_0$.

In other words, if every bounded input produces a bounded output we have external stability.

***Theorem 9.4*:** The system (*9.1*), with $\mathbf{x}_0 = \mathbf{0}$ and single input and output, is uniformly externally stable if and only if there exists a number $\beta < \infty$ such that

$$\int_{t_0}^{t} |h(t, \tau)|\, d\tau \leq \beta$$

for all $t \geq t_0$ where $h(t, \tau)$ is the impulse response.

If β depends on t_0 the system is only externally stable. If $\mathbf{x}_0 \neq \mathbf{0}$, we additionally require the zero-input system to be stable i.s.L. and $\mathbf{C}(t)$ bounded for $t > t_0$ so that the output does not become unbounded. If the system has multiple inputs and/or multiple outputs, the criterion turns out to be $\int_{t_0}^{t} ||\mathbf{H}(t,\tau)||_1 \, d\tau \leq \beta$ where the l_1 norm of a vector $\mathbf{v}$ in $\mathcal{V}_n$ is $||\mathbf{v}||_1 = \sum_{i=1}^{n} |v_i|$ (see Sections 3.10 and 4.6).

Proof: First we show that if $\int_{t_0}^{t} |h(t,\tau)| d\tau \leq \beta$, then we get external stability. Since $y(t) = \int_{t_0}^{t} h(t,\tau)\, u(\tau)\, d\tau$, we can take norms on both sides and use norm properties to obtain

$$|y(t)| \leq \int_{t_0}^{t_1} |h(t,\tau)|\,|u(\tau)|\, d\tau \leq \delta \int_{t_0}^{t_1} |h(t,\tau)|\, d\tau \leq \delta\beta = \epsilon$$

From Definition 9.5 we then have external stability.

Next, if the system is externally stable we shall prove $\int_{t_0}^{t} |h(t,\tau)| d\tau \leq \beta$ by contradiction. We set $u_1(\tau) = \operatorname{sgn} h(t,\tau)$ for $t_0 \leq \tau \leq t$, where sgn is the signum function which has bound 1 for any t. By the hypothesis of external stability,

$$\epsilon \geq |y(t)| = \left| \int_{t_0}^{t} h(t,\tau)\, u_1(\tau)\, d\tau \right| = \int_{t_0}^{t} |h(t,\tau)|\, d\tau$$

Now suppose $\int_{t_0}^{t} |h(t,\tau)| d\tau \leq \beta$ is not true. Then by taking suitable values of t and t_0 we can always make this larger than any preassigned number α. Suppose we choose $t = \theta$ and $t_0 = \theta_0$ in such a way that $\int_{\theta_0}^{\theta} |h(\theta,\tau)|\, d\tau > \alpha$. Again we set $u_2(\tau) = \operatorname{sgn} h(\theta,\tau)$ so that $\alpha < \int_{\theta_0}^{\theta} h(\theta,\tau) u_2(\tau) d\tau = y(\theta)$. Since α is any preassigned number, we can set $\alpha = \epsilon$ and arrive at a contradiction.

Example 9.11.

Consider the system $dy/dt = u$. This has an impulse response $h(t,\tau) = U(t-\tau)$, a unit step starting at time τ. Then $\int_{t_0}^{t} |U(t-\tau)|\, d\tau = t - t_0$ which becomes unbounded as t tends to ∞. Therefore this system is not externally stable although the zero-input system is stable i.s.L.

In general, external stability has no relation whatsoever with zero-input stability concepts because external stability has to do with the time behavior of the output and zero-input stability is concerned with the time behavior of the state.

Example 9.12.

Consider the scalar time-varying system

$$dx/dt = \alpha(t)x + e^{t+\theta(t,t_0)}u \qquad \text{and} \qquad y = e^{-t+\theta(t_0,t)}x$$

where $\theta(t,t_0) = \int_{t_0}^{t} \alpha(\tau)\, d\tau$. Then the transition matrix is $\Phi(t,t_0) = e^{\theta(t,t_0)}$. Therefore

$$h(t,t_0) = e^{-t+\theta(t_0,t)}\, \Phi(t,t_0)\, e^{t_0+\theta(t_0,t_0)} = e^{t_0 - t}$$

so that $\int_{t_0}^{t} |h(t,\tau)|\, d\tau < \int_{-\infty}^{t} |h(t,\tau)|\, d\tau = 1$. Thus the system is externally stable. However, since $\alpha(t)$ can be almost anything, any form of zero-input stability is open to question.

However, if we make $\mathbf{C}(t)$ a constant matrix, then we can identify the time behavior of the output with that of the state. In fact, with a few more restrictions on (*9.1*) we have

***Theorem 9.5*:** For the system (*9.1*) with l_1 norms, $\mathbf{C}(t) = \mathbf{I}$, and nonsingular $\mathbf{B}(t)$ such that $\|\mathbf{B}^{-1}(t)\|_1 < \kappa_{\mathbf{B}^{-1}}$, the system is externally stable if and only if it is uniformly asymptotically stable.

To have $\mathbf{B}(t)$ nonsingular requires $\mathbf{u}$ to be an n-vector. If $\mathbf{B}$ is constant and nonsingular, it satisfies the requirements stated in Theorem 9.5. Theorem 9.5 is proved in Problem 9.2.

9.4 LIAPUNOV TECHNIQUES

The Routh-Hurwitz, Nyquist, root locus, etc., techniques are valid for linear, time-invariant systems. The method of Liapunov has achieved some popularity for dealing with nonlinear and/or time-varying systems. Unfortunately, in most cases its practical utility is severely limited because response characteristics other than stability are desired.

Consider some metric $\rho(\mathbf{x}(t), \mathbf{0})$. This is the "distance" between the state vector and the $\mathbf{0}$ vector. If some metric, any metric at all, can be found such that the metric tends to zero as $t \to \infty$, it can be concluded that the system is asymptotically stable. Actually, Liapunov realized we do not need a metric to show this, because the triangle inequality (Property 4 of Definition 3.45) can be dispensed with.

Definition 9.6: A *time-invariant Liapunov function,* denoted $\nu(\mathbf{x})$, is any scalar function of the state variable $\mathbf{x}$ that satisfies the following conditions for all $t \geq t_0$ and all $\mathbf{x}$ in the neighborhood of the origin:

(1) $\nu(\mathbf{x})$ and its partial derivatives exist and are continuous;

(2) $\nu(\mathbf{0}) = 0$;

(3) $\nu(\mathbf{x}) > 0$ for $\mathbf{x} \neq \mathbf{0}$;

(4) $d\nu/dt = (\mathbf{grad}_\mathbf{x}\, \nu)^T\, d\mathbf{x}/dt < 0$ for $\mathbf{x} \neq \mathbf{0}$.

The problem is to find a Liapunov function for a particular system, and there is no general method to do this.

Example 9.13.

Given the scalar system $dx/dt = -x$. We shall consider the particular function $\eta(x) = x^2$. Applying the tests in Definition 9.6, (1) $\eta(x) = x^2$ and $\partial\eta/\partial x = 2x$ are continuous, (2) $\eta(0) = 0$, (3) $\eta(x) = x^2 > 0$ for all $x \neq 0$, and (4) $d\eta/dt = 2x\, dx/dt = -2x^2 < 0$ for all $x \neq 0$. Therefore $\eta(x)$ is a Liapunov function.

***Theorem 9.6*:** Suppose a time-invariant Liapunov function can be found for the state variable $\mathbf{x}$ of the system $d\mathbf{x}/dt = \mathbf{f}(\mathbf{x}, t)$ where $\mathbf{f}(\mathbf{0}, t) = \mathbf{0}$. Then the state $\mathbf{x} = \mathbf{0}$ is asymptotically stable.

Proof: Here we shall only prove $\mathbf{x} \to \mathbf{0}$ as $t \to \infty$. Definition 9.6 assures existence and continuity of ν and $d\nu/dt$. Now consider $\nu(\boldsymbol{\phi}(t; t_0, \mathbf{x}_0))$, i.e. as a function of t. Since $\nu > 0$ and $d\nu/dt < 0$ for $\boldsymbol{\phi} \neq \mathbf{0}$, integration with respect to t shows $\nu(\boldsymbol{\phi}(t_1; t_0, \mathbf{x}_0)) > \nu(\boldsymbol{\phi}(t_2; t_0, \mathbf{x}_0))$ for $t_0 < t_1 < t_2$. Although ν is thus positive and monotone decreasing, its limit may not be zero. (Consider $1 + e^{-t}$.) Assume ν has a constant limit $\kappa > 0$. Then $d\nu/dt = 0$ when $\nu = \kappa$. But $d\nu/dt < 0$ for $\boldsymbol{\phi} \neq \mathbf{0}$, and when $\boldsymbol{\phi} = \mathbf{0}$ then $d\nu/dt = (\mathbf{grad}_\mathbf{x} \nu)^T \mathbf{f}(\mathbf{0}, t) = 0$. So $d\nu/dt = 0$ implies $\boldsymbol{\phi} = \mathbf{0}$ which implies $\nu = 0$, a contradiction. Thus $\nu \to 0$, assuring $\mathbf{x} \to \mathbf{0}$.

If Definition 9.6 holds for all t_0, then we have uniform asymptotic stability. Additionally if the system is linear or if we substitute "everywhere" for "in a neighborhood of the origin" in Definition 9.6, we have uniform global asymptotic stability. If condition (4) is weakened to $d\nu/dt \leq 0$, we have only stability i.s.L.

Example 9.14.

For the system of Example 9.13, since we have found a Liapunov function $\nu(x) = x^2$, we conclude the system $dx/dt = -x$ is uniformly asymptotically stable. Notice that we did not need to solve the state equation to do this, which is the advantage of the technique of Liapunov.

Definition 9.7: A *time-varying Liapunov function*, denoted $\nu(\mathbf{x}, t)$, is any scalar function of the state variable $\mathbf{x}$ and time t that satisfies for all $t \geq t_0$ and all $\mathbf{x}$ in a neighborhood of the origin:

(1) $\nu(\mathbf{x}, t)$ and its first partial derivatives in $\mathbf{x}$ and t exist and are continuous;

(2) $\nu(\mathbf{0}, t) = 0$;

(3) $\nu(\mathbf{x}, t) \geq \alpha(\|\mathbf{x}\|) > 0$ for $\mathbf{x} \neq 0$ and $t \geq t_0$, where $\alpha(0) = 0$ and $\alpha(\xi)$ is a continuous nondecreasing scalar function of ξ;

(4) $d\nu/dt = (\mathbf{grad}_{\mathbf{x}}\, \nu)^T d\mathbf{x}/dt + \partial\nu/\partial t < 0$ for $\mathbf{x} \neq \mathbf{0}$.

Note for all $t \geq t_0$, $\nu(\mathbf{x}, t)$ must be $\geq$ a continuous nondecreasing, time-invariant function of the norm $\|\mathbf{x}\|$.

***Theorem 9.7*:** Suppose a time-varying Liapunov function can be found for the state variable $\mathbf{x}(t)$ of the system $d\mathbf{x}/dt = \mathbf{f}(\mathbf{x}, t)$. Then the state $\mathbf{x} = \mathbf{0}$ is asymptotically stable.

Proof: Since $d\nu/dt < 0$ and ν is positive, integration with respect to t shows that $\nu(\mathbf{x}(t_0), t_0) > \nu(\mathbf{x}(t), t)$ for $t > t_0$. Now the proof must be altered from that of Theorem 9.6 because the time dependence of the Liapunov function could permit ν to tend to zero even though $\mathbf{x}$ remains nonzero. (Consider $\nu = x^2 e^{-t}$ when $x = t$). Therefore we require $\nu(\mathbf{x}, t) \geq \alpha(\|\mathbf{x}\|)$, and $\alpha = 0$ implies $\|\mathbf{x}\| = 0$. Hence if ν tends to zero with this additional assumption, we have asymptotic stability for some t_0.

If the conditions of Definition 9.7 hold for all t_0 and if $\nu(\mathbf{x}, t) \leq \beta(\|\mathbf{x}\|)$ where $\beta(\xi)$ is a continuous nondecreasing scalar function of ξ with $\beta(0) = 0$, then we have uniform asymptotic stability. Additionally, if the system is linear or if we substitute "everywhere" for "in a neighborhood of the origin" in Definition 9.7 and require $\alpha(\|\mathbf{x}\|) \to \infty$ with $\|\mathbf{x}\| \to \infty$, then we have uniform global asymptotic stability.

Example 9.15.

Given the scalar system $dx/dt = x$. The function $x^2 e^{-4t}$ satisfies all the requirements of Definition 9.6 at any fixed t and yet the system is not stable. This is because there is no $\alpha(\|\mathbf{x}\|)$ meeting the requirement of Definition 9.7.

It is often of use to weaken condition (4) of Definition 9.6 or 9.7 in the following manner. We need $\nu(\mathbf{x}, t)$ to eventually decrease to zero. However, it is permissible for $\nu(\mathbf{x}, t)$ to be constant in a region of state space if we are assured the system trajectories will move to a region of state space in which $d\nu/dt$ is strictly less than zero. In other words, instead of requiring $d\nu/dt < 0$ for all $\mathbf{x} \neq \mathbf{0}$ and all $t \geq t_0$, we could require $d\nu/dt \leq 0$ for all $\mathbf{x} \neq \mathbf{0}$ and $d\nu/dt$ does not vanish identically in $t \geq t_0$ for any t_0 and any trajectory arising from a nonzero initial condition $\mathbf{x}(t_0)$.

Example 9.16.

Consider the Sturm-Liouville equation $d^2y/dt^2 + p(t)\,dy/dt + q(t)y = 0$. We shall impose conditions on the scalars $p(t)$ and $q(t)$ such that uniform asymptotic stability is guaranteed. First, call $y = x_1$ and $dy/dt = x_2$, and consider the function $\nu(\mathbf{x}, t) = x_1^2 + x_2^2/q(t)$. Clearly conditions (1) and (2) of Definition 9.7 hold if $q(t)$ is continuously differentiable, and for (3) to hold suppose $1/q(t) \geq \kappa_1 > 0$ for all t so that $\alpha(\|\mathbf{x}\|) = \|\mathbf{x}\|_2^2 \min\{1, \kappa_1\}$. Here $\min\{1, \kappa_1\} = 1$ if $\kappa_1 \geq 1$, and $\min\{1, \kappa_1\} = \kappa_1$ if $\kappa_1 < 1$. For (4) to hold, we calculate

$$\frac{d\nu}{dt} = 2x_1 \frac{dx_1}{dt} + 2\frac{x_2}{q(t)}\frac{dx_2}{dt} - \frac{x_2^2}{q^2(t)}\frac{dq}{dt}$$

Since $dx_1/dt = x_2$ and $dx_2/dt = -p(t)x_2 - q(t)x_1$, then

$$\frac{d\nu}{dt} = -\frac{2p(t)\,q(t) + dq/dt}{q^2(t)}x_2^2$$

Hence we require $2p(t)\,q(t) + dq/dt > 0$ for all $t \geqq t_0$, since $q^{-2} > \kappa_1^2$. But even if this requirement is satisfied, $d\nu/dt = 0$ for $x_2 = 0$ and any x_1. Condition (4) cannot hold. However, if we can show that when $x_2 = 0$ and $x_1 \neq 0$, the system moves to a region where $x_2 \neq 0$, then $\nu(\mathbf{x}, t)$ will still be a Liapunov function. Hence when $x_2 = 0$, we have $dx_2/dt = -q(t)x_1$. Therefore if $q(t) \neq 0$, then x_2 must become nonzero when $x_1 \neq 0$. Thus $\nu(\mathbf{x}, t)$ is a Liapunov function whenever, for any $t \geqq t_0$, $q(t) \neq 0$ and $1/q(t) \geqq \kappa_1 > 0$, and $2p(t)\,q(t) + dq/dt > 0$. Under these conditions the zero state of the given Sturm-Liouville equation is asymptotically stable.

Furthermore we can show it is uniformly globally asymptotically stable if these conditions are independent of t_0 and if $q(t) \leqq \kappa_2$. The linearity of the system implies global stability, and since the previous calculations did not depend on t_0 we can show uniformity if we can find a $\beta(\|\mathbf{x}\|) \geqq \nu(\mathbf{x}, t)$. If $q(t) \leqq \kappa_2$, then $\|x\|_2^2 \max\{1, \kappa_2^{-1}\} = \beta(\|\mathbf{x}\|) \geqq \nu(\mathbf{x}, t)$.

For the discrete-time case, we have the analog for Definition 9.7 as

Definition 9.8: A *discrete-time Liapunov function*, denoted $\nu(\mathbf{x}, k)$, is any scalar function of the state variable $\mathbf{x}$ and integer k that satisfies, for all $k > k_0$ and all $\mathbf{x}$ in a neighborhood of the origin:

(1) $\nu(\mathbf{x}, k)$ is continuous;

(2) $\nu(\mathbf{0}, k) = 0$;

(3) $\nu(\mathbf{x}, k) \geqq \alpha(\|\mathbf{x}\|) > 0$ for $\mathbf{x} \neq \mathbf{0}$;

(4) $\Delta\nu(\mathbf{x}, k) = \nu(\mathbf{x}(k+1), k+1) - \nu(\mathbf{x}(k), k) < 0$ for $\mathbf{x} \neq \mathbf{0}$.

Theorem 9.8: Suppose a discrete-time Liapunov function can be found for the state variable $\mathbf{x}(k)$ of the system $\mathbf{x}(k+1) = \mathbf{f}(\mathbf{x}, k)$. Then the state $\mathbf{x} = \mathbf{0}$ is asymptotically stable.

The proof is similar to that of Theorem 9.7. If the conditions of Definition 9.7 hold for all k_0 and if $\nu(\mathbf{x}, k) \leqq \beta(\|\mathbf{x}\|)$ where $\beta(\xi)$ is a continuous nondecreasing function of ξ with $\beta(0) = 0$, then we have uniform asymptotic stability. Additionally, if the system is linear or if we substitute "everywhere" for "in a neighborhood of the origin" in Definition 9.8 and require $\alpha(\|\mathbf{x}\|) \to \infty$ with $\|\mathbf{x}\| \to \infty$, then we have uniform global asymptotic stability. Again, we can weaken condition (4) in Definition 9.8 to achieve the same results. Condition (4) can be replaced by $\Delta\nu(\mathbf{x}, k) \leqq 0$ if we have assurance that the system trajectories will move from regions of state space in which $\Delta\nu = 0$ to regions in which $\Delta\nu < 0$.

Example 9.17.

Given the linear system $\mathbf{x}(k+1) = \mathbf{A}(k)\,\mathbf{x}(k)$, where $\|\mathbf{A}(k)\| < 1$ for all k and for some norm. Choose as a discrete-time Liapunov function $\nu(\mathbf{x}) = \|\mathbf{x}\|$. Then

$$\Delta\nu = \nu(\mathbf{x}(k+1)) - \nu(\mathbf{x}(k)) = \|\mathbf{x}(k+1)\| - \|\mathbf{x}(k)\| = \|\mathbf{A}(k)\,\mathbf{x}(k)\| - \|\mathbf{x}(k)\|$$

Since $\|\mathbf{A}(k)\| < 1$, then $\|\mathbf{A}(k)\,\mathbf{x}(k)\| \leqq \|\mathbf{A}(k)\|\,\|\mathbf{x}(k)\| < \|\mathbf{x}(k)\|$ so that $\Delta\nu < 0$. The other pertinent properties of ν can be verified, so this system is uniformly globally asymptotically stable.

9.5 LIAPUNOV FUNCTIONS FOR LINEAR SYSTEMS

The major difficulty with Liapunov techniques is finding a Liapunov function. Clearly, if the system is unstable no Liapunov function exists. There is indeed cause to worry that no Liapunov function exists even if the system is stable, so that the search for such a function would be in vain. However, for nonlinear systems it has been shown under quite general conditions that if the equilibrium state $\mathbf{x} = \mathbf{0}$ is uniformly globally asymptotically stable, then a time-varying Liapunov function exists. The purpose of this section is to manufacture a Liapunov function. We shall consider only linear systems, and to be specific we shall be concerned with the asymptotically stable real linear system

$$d\mathbf{x}/dt = \mathbf{A}(t)\mathbf{x} \qquad (9.2)$$

in which there exists a constant κ_3 depending on t_0 such that $\|\mathbf{A}(t)\|_2 \leq \kappa_3(t_0)$ and where $\int_t^\infty \|\mathbf{\Phi}(\tau, t)\|^2 \, d\tau$ exists for some norm. Using Theorem 9.3 we can show that uniformly asymptotically stable systems satisfy this last requirement because $\|\mathbf{\Phi}(t, \tau)\| \leq \kappa_1 e^{-\kappa_2(t-\tau)}$ so that

$$\int_t^\infty \|\mathbf{\Phi}(\tau, t)\|^2 \, d\tau \quad \leq \quad \kappa_1^2 \int_t^\infty e^{-2\kappa_2(\tau - t)} \, d\tau \quad = \quad \kappa_1^2 / 2\kappa_2 \quad = \quad \kappa_4 \quad < \quad \infty$$

However, there do exist asymptotically stable systems that do not satisfy this requirement.

Example 9.18.

Consider the time-varying scalar system $dx/dt = -x/2t$. This has a transition matrix $\Phi(t, t_0) = (t_0/t)^{1/2}$. The system is asymptotically stable for $t_0 > 0$. However, even for $t \geq t_0 > 0$, we find

$$\int_t^\infty \|\Phi(\tau, t)\|^2 \, d\tau \quad = \quad \int_t^\infty (t/\tau) \, d\tau \quad = \quad \infty$$

Note carefully the position of the arguments of Φ.

***Theorem 9.9*:** For system (*9.2*), $\nu(\mathbf{x}, t) = \mathbf{x}^T \mathbf{P}(t) \mathbf{x}$ is a time-varying Liapunov function, where $\mathbf{P}(t) = \int_t^\infty \mathbf{\Phi}^T(\tau, t) \, \mathbf{Q}(\tau) \, \mathbf{\Phi}(\tau, t) \, d\tau$ in which $\mathbf{Q}(t)$ is any continuous positive definite symmetric matrix such that $\|\mathbf{Q}(t)\| \leq \kappa_5(t_0)$ and $\mathbf{Q}(t) - \epsilon\mathbf{I}$ is positive definite for all $t \geq t_0$ if $\epsilon > 0$ is small enough.

This theorem is of little direct help in the determination of stability because $\mathbf{\Phi}(t, \tau)$ must be known in advance to compute $\nu(\mathbf{x}, t)$. Note $\mathbf{P}(t)$ is positive definite, real, and symmetric.

Proof: Since $\mathbf{\Phi}$ and $\mathbf{Q}$ are continuous, if $\mathbf{P}(t) < \infty$ for all $t \geq t_0$ then condition (1) of Definition 9.7 is satisfied. Use of $\|\mathbf{Q}\| \leq \kappa_5$ gives $\|\mathbf{P}\| \leq \kappa_5 \int_t^\infty \|\mathbf{\Phi}(\tau, t)\|^2 \, d\tau$. Since the integral exists for the systems (*9.2*) under discussion, then condition (1) holds. Condition (2) obviously holds. To show condition (3),

$$\nu(\mathbf{x}, t) \quad = \quad \int_t^\infty (\mathbf{\Phi}(\tau, t) \, \mathbf{x}(t))^T \, \mathbf{Q}(\tau) \, \mathbf{\Phi}(\tau, t) \, \mathbf{x}(t) \, d\tau \quad = \quad \int_t^\infty \mathbf{x}^T(\tau) \, \mathbf{Q}(\tau) \, \mathbf{x}(\tau) \, d\tau$$

Since $\mathbf{Q}(t) - \epsilon\mathbf{I}$ is positive definite, then $\mathbf{x}^T(\mathbf{Q} - \epsilon\mathbf{I})\mathbf{x} \geq 0$ so that $\mathbf{x}^T\mathbf{Q}\mathbf{x} \geq \epsilon\mathbf{x}^T\mathbf{x} > 0$ for any $\mathbf{x}(\tau) \neq \mathbf{0}$. Therefore

$$\nu(\mathbf{x}, t) \quad \geq \quad \epsilon \int_t^\infty \mathbf{x}^T(\tau) \, \mathbf{x}(\tau) \, d\tau \quad = \quad \epsilon \int_t^\infty \|\mathbf{x}\|_2^2 \, d\tau$$

Since $\|\mathbf{A}(t)\|_2 \leq \kappa_3$ for the system (*9.2*), then use of Problem 4.43 gives

$$\nu(\mathbf{x}, t) \quad \geq \quad \epsilon\kappa_3^{-1} \int_t^\infty \|\mathbf{A}(\tau)\|_2 \, \|\mathbf{x}(\tau)\|_2^2 \, d\tau \quad \geq \quad -\epsilon\kappa_3^{-1} \int_t^\infty \mathbf{x}^T\mathbf{A}\mathbf{x} \, d\tau$$

$$= \quad -\epsilon\kappa_3^{-1} \int_t^\infty \mathbf{x}^T(d\mathbf{x}/d\tau) \, d\tau \quad = \quad \epsilon\kappa_3^{-1} \|\mathbf{x}(t)\|_2^2 / 2 \quad = \quad \alpha(\|\mathbf{x}\|) \quad > \quad 0$$

for all $\mathbf{x} \neq \mathbf{0}$. Finally, to satisfy condition (4), since $\nu(\mathbf{x}, t) = \int_t^\infty \mathbf{x}^T(\tau) \, \mathbf{Q}(\tau) \, \mathbf{x}(\tau) \, d\tau$, then $d\nu/dt = -\mathbf{x}^T(t) \, \mathbf{Q}(t) \, \mathbf{x}(t) < 0$ since $\mathbf{Q}$ is positive definite.

Example 9.19.

Consider the scalar system of Example 9.8, $dx/dt = -x/t$. Then $\Phi(t, t_0) = t_0/t$. A Liapunov function can be found for this system for $t_0 > 0$, when it is asymptotically stable, because

$$\|A(t)\|_2 \; = \; t^{-1} \; < \; t_0^{-1} \; = \; \kappa_3(t_0) \qquad \text{and} \qquad \int_t^\infty \|\Phi(\tau, t)\|^2 \, d\tau \; = \; \int_t^\infty (t/\tau)^2 \, d\tau \; = \; t$$

For simplicity, we choose $Q(\tau) = 1$ so that $P(t) = \int_t^\infty (t/\tau)^2 \, d\tau = t$. Then $\nu(x, t) = tx^2$, which is a time-varying Liapunov function for all $t \geq t_0 > 0$.

Consider the system

$$d\mathbf{x}/dt = \mathbf{A}(t)\mathbf{x} + \mathbf{f}(\mathbf{x}, \mathbf{u}(t), t) \tag{9.3}$$

where for $t \geqq t_0$, $\mathbf{f}(\mathbf{x}, \mathbf{u}, t)$ is real, continuous for small $\|\mathbf{x}\|$ and $\|\mathbf{u}\|$, and $\|\mathbf{f}\| \to 0$ as $\|\mathbf{x}\| \to 0$ and as $\|\mathbf{u}\| \to 0$. As discussed in Section 8.9 this equation often results when considering the linearization of a nonlinear system or considering parameter variations. (Let $\delta\mathbf{x}$ from Section 8.9 equal $\mathbf{x}$ in equation (*9.3*).) In fact, most nonswitching control systems satisfy this type of equation, and $\|\mathbf{f}\|$ is usually small because it is the job of the controller to keep deviations from the nominal small.

Theorem 9.10: If the system (*9.3*) reduces to the asymptotically stable system (*9.2*) when $\mathbf{f} = \mathbf{0}$, then the equilibrium state is asymptotically stable for some small $\|\mathbf{x}\|$ and $\|\mathbf{u}\|$.

In other words, if the linearized system is asymptotically stable for $t \geqq t_0$, then the corresponding equilibrium state of the nonlinear system is asymptotically stable to small disturbances.

Proof: Use the Liapunov function $\nu(\mathbf{x}, t) = \mathbf{x}^T\mathbf{P}(t)\mathbf{x}$ given by Theorem 9.9, in which $\mathbf{\Phi}(t, \tau)$ is the transition matrix of the asymptotically stable system (*9.2*). Then conditions (1) and (2) of Definition 9.7 hold for this as a Liapunov function for system (*9.3*) in the same manner as in the proof of Theorem 9.9. For small $\|\mathbf{x}\|$ it makes little difference whether we consider equation (*9.3*) or (*9.2*), and so the lower bound $\alpha(\|\mathbf{x}\|)$ in condition (3) can be fixed up to hold for trajectories of (*9.3*). For condition (4) we investigate $d\nu/dt = d(\mathbf{x}^T\mathbf{P}\mathbf{x})/dt = 2\mathbf{x}^T\mathbf{P}\,d\mathbf{x}/dt + \mathbf{x}^T(d\mathbf{P}/dt)\mathbf{x} = 2\mathbf{x}^T\mathbf{P}\mathbf{A}\mathbf{x} + 2\mathbf{x}^T\mathbf{P}\mathbf{f} + \mathbf{x}^T(d\mathbf{P}/dt)\mathbf{x}$. But from the proof of Theorem 9.9 we know $d\nu/dt$ along motions of the system $d\mathbf{x}/dt = \mathbf{A}(t)\mathbf{x}$ satisfies $d\nu/dt = -\mathbf{x}^T\mathbf{Q}\mathbf{x} = 2\mathbf{x}^T\mathbf{P}\mathbf{A}\mathbf{x} + \mathbf{x}^T(d\mathbf{P}/dt)\mathbf{x}$ so that $-\mathbf{Q} = 2\mathbf{P}\mathbf{A} + d\mathbf{P}/dt$. Hence for the system (*9.3*), $d\nu/dt = -\mathbf{x}^T\mathbf{Q}\mathbf{x} + 2\mathbf{x}^T\mathbf{P}\mathbf{f}$. But $\|\mathbf{f}\| \to 0$ as $\|\mathbf{x}\| \to 0$ and $\|\mathbf{u}\| \to 0$, so that for small enough $\|\mathbf{x}\|$ and $\|\mathbf{u}\|$, the term $-\mathbf{x}^T\mathbf{Q}\mathbf{x}$ dominates and $d\nu/dt$ is negative.

Example 9.20.

Consider the system of Example 9.1. The linearized system $dx/dt = -x$ is uniformly asymptotically stable, so Theorem 9.10 says there exist some small motions in the neighborhood of $x = 0$ which result in the asymptotic stability of the nonlinear system. In fact, we know from the solution to the nonlinear equation that initial conditions $x_0 < 1$ cause trajectories that return to $x = 0$.

Example 9.21.

Consider the system $dx/dt = -x + u(1 + x - u)$, where $u(t) = \epsilon$, a small constant. Then $f(x, \epsilon, t) = \epsilon(1 + x - \epsilon)$ does not tend to zero as $x \to 0$. Consequently Theorem 9.10 cannot be used directly. However, note the steady state value of $x = \epsilon$. Therefore redefine variables as $z = x - \epsilon$. Then $dz/dt = -z + \epsilon z$, which is stable for all $\epsilon \leqq 1$.

Example 9.22.

Consider the system

$$dx_1/dt = x_2 + x_1(x_1^2 + x_2^2)/2 \qquad \text{and} \qquad dx_2/dt = -x_1 + x_2(x_1^2 + x_2^2)/2$$

The linearized system $dx_1/dt = x_2$ and $dx_2/dt = -x_1$ is stable i.s.L. because it is the harmonic oscillator of Example 9.4. But the solution to the nonlinear equation is $x_1^2 + x_2^2 = [(x_{10}^2 + x_{20}^2)^{-1} - t]^{-1}$ which is unbounded for any x_{10} or x_{20}. The trouble is that the linearized system is only stable i.s.L., not asymptotically stable.

9.6 EQUATIONS FOR THE CONSTRUCTION OF LIAPUNOV FUNCTIONS

From Theorem 9.9, if we take the derivative of $\mathbf{P}(t)$ we obtain

$$d\mathbf{P}/dt = -\mathbf{\Phi}^T(t, t)\,\mathbf{Q}(t)\,\mathbf{\Phi}(t, t) + \int_t^\infty [\partial\mathbf{\Phi}(\tau, t)/\partial t]^T\,\mathbf{Q}(\tau)\,\mathbf{\Phi}(\tau, t)\,d\tau + \int_t^\infty \mathbf{\Phi}^T(\tau, t)\,\mathbf{Q}(\tau)\,[\partial\mathbf{\Phi}(\tau, t)/\partial t]\,d\tau$$

Since $\mathbf{\Phi}(t, t) = \mathbf{I}$ and $\partial\mathbf{\Phi}(\tau, t)/\partial t = \partial\mathbf{\Phi}^{-1}(t, \tau)/\partial t = -\mathbf{\Phi}(\tau, t)\,\mathbf{A}(t)$, we obtain

$$d\mathbf{P}/dt + \mathbf{A}^T(t)\,\mathbf{P}(t) + \mathbf{P}(t)\,\mathbf{A}(t) = -\mathbf{Q}(t) \tag{9.4}$$

Theorem 9.11: If and only if the solution $\mathbf{P}(t)$ to equation (*9.4*) is positive definite, where $\mathbf{A}(t)$ and $\mathbf{Q}(t)$ satisfy the conditions of Theorem 9.9, then the system (*9.2*), $d\mathbf{x}/dt = \mathbf{A}(t)\mathbf{x}$, is uniformly asymptotically stable.

Proof: If $\mathbf{P}(t)$ is positive definite, then $\nu = \mathbf{x}^T\mathbf{P}(t)\mathbf{x}$ is a Liapunov function. If the system (*9.2*) is uniformly asymptotically stable, we obtain equation (*9.4*) for $\mathbf{P}$.

Theorem 9.11 represents a rare occurrence in Liapunov theory, namely a necessary and sufficient condition for asymptotic stability. Usually it is a difficult procedure to find a Liapunov function for a nonlinear system, but equation (*9.4*) gives a means to generate a Liapunov function for a linear equation. Unfortunately, finding a solution to equation (*9.4*) means solving $n(n+1)/2$ equations (since $\mathbf{P}$ is symmetric), whereas it is usually easier to solve the n equations associated with $d\mathbf{x}/dt = \mathbf{A}(t)\mathbf{x}$. However, in the constant coefficient case we can take $\mathbf{Q}$ to be a constant positive definite matrix, so that $\mathbf{P}$ is the solution to

$$\mathbf{A}^T\mathbf{P} + \mathbf{P}\mathbf{A} = -\mathbf{Q} \tag{9.5}$$

This equation gives a set of $n(n+1)/2$ linear algebraic equations that can be solved for $\mathbf{P}$ after any positive definite $\mathbf{Q}$ has been selected. It turns out that the solution always exists and is unique if the system is asymptotically stable. (See Problem 9.6.) Then the solution $\mathbf{P}$ can be checked by Sylvester's criterion (Theorem 4.10), and if and only if $\mathbf{P}$ is positive definite the system is asymptotically stable. This procedure is equivalent to determining if all the eigenvalues of $\mathbf{A}$ have negative real parts. (See Problem 9.7.) Experience has shown that it is usually easier to use the Liénard-Chipart (Routh-Hurwitz) test than the Liapunov method, however.

Example 9.23.

Given the system $d^2y/dt^2 + 2dy/dt + y = 0$. To determine the stability by Liapunov's method we set up the state equations

$$\frac{d\mathbf{x}}{dt} = \begin{pmatrix} 0 & 1 \\ -1 & -2 \end{pmatrix}\mathbf{x} \qquad y = (1 \;\; 0)\mathbf{x}$$

We arbitrarily choose $\mathbf{Q} = \mathbf{I}$ and solve

$$\begin{pmatrix} 0 & -1 \\ 1 & -2 \end{pmatrix}\begin{pmatrix} p_{11} & p_{12} \\ p_{12} & p_{22} \end{pmatrix} + \begin{pmatrix} p_{11} & p_{12} \\ p_{12} & p_{22} \end{pmatrix}\begin{pmatrix} 0 & 1 \\ -1 & -2 \end{pmatrix} = \begin{pmatrix} -1 & 0 \\ 0 & -1 \end{pmatrix}$$

The solution is $\mathbf{P} = \frac{1}{2}\begin{pmatrix} 3 & 1 \\ 1 & 1 \end{pmatrix}$. Using Sylvester's criterion we find $p_{11} = 3 > 0$ and $\det \mathbf{P} = 1 > 0$, and so $\mathbf{P}$ is positive definite and the system is stable. But it is much easier to use Routh's test which almost immediately says the system is stable.

We may ask at this point what practical use can be made of Theorem 9.9, other than the somewhat comforting conclusion of Theorem 9.10. The rather unsatisfactory answer is that we might be able to construct a Liapunov function for systems of the form $d\mathbf{x}/dt = \mathbf{A}(t)\mathbf{x} + \mathbf{f}(\mathbf{x}, t)$ where $\mathbf{A}(t)$ has been chosen such that we can use Theorem 9.9 (or equation (*9.5*) when the system is time-invariant). A hint on how to choose $\mathbf{Q}$ is given in Problem 9.7.

Example 9.24.

Given the system $d^2y/dt^2 + (2 + e^{-t})\,dy/dt + y = 0$. We set up the state equation in the form $d\mathbf{x}/dt = \mathbf{A}\mathbf{x} + \mathbf{f}$:

$$\frac{d\mathbf{x}}{dt} = \begin{pmatrix} 0 & 1 \\ -1 & -2 \end{pmatrix}\mathbf{x} + \begin{pmatrix} 0 \\ -e^{-t}x_2 \end{pmatrix} \qquad y = (1 \;\; 0)\mathbf{x}$$

As found in Example 9.23, $2\nu = 3x_1^2 + 2x_1x_2 + x_2^2$. Then

$$d\nu/dt = d(\mathbf{x}^T\mathbf{P}\mathbf{x})/dt = 2\mathbf{x}^T\mathbf{P}\,d\mathbf{x}/dt = 2\mathbf{x}^T\mathbf{P}(\mathbf{A}\mathbf{x} + \mathbf{f})$$

$$= -\mathbf{x}^T\mathbf{Q}\mathbf{x} + 2\mathbf{x}^T\mathbf{P}\mathbf{f} = -x_1^2 - x_2^2 - (x_1 + x_2)e^{-t}x_2$$

$$\frac{dv}{dt} \quad = \quad -\mathbf{x}^T \begin{pmatrix} 1 & e^{-t/2} \\ e^{-t/2} & 1+e^{-t} \end{pmatrix} \mathbf{x}$$

Using Sylvester's criterion, $1 > 0$ and $1 + e^{-t} > e^{-2t}/4$, so the given system is uniformly asymptotically stable.

For discrete-time systems, the analog of the construction procedure given in Theorem 9.9 is

$$\nu(\mathbf{x}, k) \quad = \quad \sum_{m=k}^{\infty} \mathbf{x}^T(k)\, \boldsymbol{\Phi}^T(m, k)\, \mathbf{Q}(m)\, \boldsymbol{\Phi}(m, k)\, \mathbf{x}(k)$$

and the analog of equation (*9.5*) is $\mathbf{A}^T\mathbf{P}\mathbf{A} - \mathbf{P} = -\mathbf{Q}$.

Solved Problems

9.1. A ball bearing with rolling friction rests on a deformable surface in a gravity field. Deform the surface to obtain examples of the various kinds of stability.

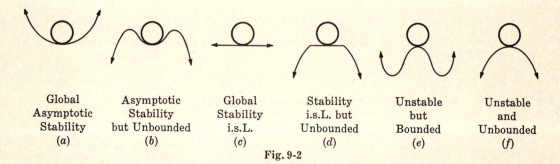

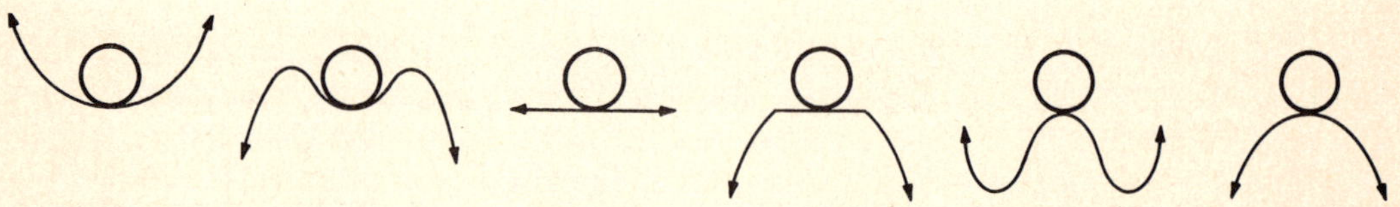

Global Asymptotic Stability (*a*) | Asymptotic Stability but Unbounded (*b*) | Global Stability i.s.L. (*c*) | Stability i.s.L. but Unbounded (*d*) | Unstable but Bounded (*e*) | Unstable and Unbounded (*f*)

Fig. 9-2

We can see that if the equilibrium state is globally asymptotically stable as in Fig. 9-2(*a*), then it is also stable i.s.L., as are (*b*), (*c*) and (*d*), etc. If the shape of the surface does not vary with time, then the adjective "uniform" can be affixed to the description under each diagram. If the shape of the surface varies with time, the adjective "uniform" may or may not be affixed, depending on how the shape of the surface varies with time.

9.2. Prove Theorems 9.3 and 9.5.

We wish to show $d\mathbf{x}/dt = \mathbf{A}(t)\mathbf{x}$ is uniformly asymptotically stable if and only if $\|\boldsymbol{\Phi}(t, t_0)\| \leq \kappa_1 e^{-\kappa_2(t-t_0)}$ for all $t \geq t_0$ and all t_0. Also we wish to show $d\mathbf{x}/dt = \mathbf{A}(t)\mathbf{x} + \mathbf{B}(t)\mathbf{u}$, where $\|\mathbf{B}^{-1}(t)\|_1 \leq \kappa_{\mathbf{B}-1}$, is externally stable if and only if $\|\boldsymbol{\Phi}(t, t_0)\| \leq \kappa_1 e^{-\kappa_2(t-t_0)}$, and hence if and only if the unforced system is uniformly asymptotically stable.

Since $\|\mathbf{x}(t)\| \leq \|\boldsymbol{\Phi}(t, t_0)\|\,\|\mathbf{x}_0\|$, if $\|\boldsymbol{\Phi}(t, t_0)\| \leq \kappa_1 e^{-\kappa_2(t-t_0)}$ and $\|\mathbf{x}_0\| \leq \delta$, then $\|\mathbf{x}(t)\| \leq \kappa_1\delta$ and tends to zero as $t \to \infty$, which proves uniform asymptotic stability. Next, if $d\mathbf{x}/dt = \mathbf{A}(t)\mathbf{x}$ is uniformly asymptotically stable, then for any $\rho > 0$ and any $\eta > 0$ there is a t_1 (independent of t_0 by uniformity) such that if $\|\mathbf{x}_0\| \leq \rho$ then $\|\mathbf{x}(t)\| \leq \eta$ for all $t > t_0 + t_1$. In fact, for $\rho = 1$ and $\eta = e^{-1}$, there is a $t_1 = T$ such that $\eta = e^{-1} \geq \|\mathbf{x}(t_0 + T)\| = \|\boldsymbol{\Phi}(t_0 + T, t_0)\mathbf{x}_0\|$. Since $\|\mathbf{x}_0\| = 1$, from Theorem 4.11, property 2, $\|\boldsymbol{\Phi}(t_0 + T, t_0)\mathbf{x}_0\| = \|\boldsymbol{\Phi}(t_0 + T, t_0)\|$. Therefore for all t_0, $\|\boldsymbol{\Phi}(t_0 + T, t_0)\| \leq e^{-1}$ and for any positive integer k,

$$\|\boldsymbol{\Phi}(t_0 + kT, t_0)\| \;\leq\; \|\boldsymbol{\Phi}(t_0 + kT, t_0 + (k-1)T)\| \cdots \|\boldsymbol{\Phi}(t_0 + 2T, t_0 + T)\|\,\|\boldsymbol{\Phi}(t_0 + T, t_0)\| \;\leq\; e^{-k}$$

Choose k such that $t_0 + kT \leq t < t_0 + (k+1)T$ and define $\kappa_1 = \|\boldsymbol{\Phi}(t, t_0 + kT)\|e < \infty$ from Theorem 9.2. Then

$$\begin{aligned}\|\boldsymbol{\Phi}(t, t_0)\| \;&\leq\; \|\boldsymbol{\Phi}(t, t_0 + kT)\|\,\|\boldsymbol{\Phi}(t_0 + kT, t_0)\| \;=\; \kappa_1 e^{-1}\|\boldsymbol{\Phi}(t_0 + kT, t_0)\| \;\leq\; \kappa_1 e^{-(k+1)} \\ &=\; \kappa_1 e^{-(t_0+(k+1)T-t_0)/T} \;\leq\; \kappa_1 e^{-(t-t_0)/T}\end{aligned}$$

Defining $\kappa_2 = 1/T$ proves Theorem 9.3.

If $||\mathbf{\Phi}(t, t_0)||_1 \leq \kappa_1 e^{-\kappa_2(t-t_0)}$, then

$$\int_{t_0}^{t} ||\mathbf{H}(t,\tau)||_1 \, d\tau \;\leq\; \int_{t_0}^{t} ||\mathbf{\Phi}(t,\tau)||\,||\mathbf{B}(\tau)||\,d\tau \;\leq\; \kappa_{\mathbf{B}} \int_{t_0}^{t} ||\mathbf{\Phi}(t,\tau)||\,d\tau \;\leq\; \kappa_{\mathbf{B}}\kappa_1 \int_{t_0}^{t} e^{-\kappa_2(t-\tau)}\,d\tau$$

$$= \kappa_{\mathbf{B}}\kappa_1(1-e^{-\kappa_2(t-t_0)})/\kappa_2 \;\leq\; \kappa_{\mathbf{B}}\kappa_1/\kappa_2 \;=\; \beta$$

so that use of Theorem 9.4 proves the system is uniformly externally stable.

If the system is uniformly externally stable, then from Theorem 9.4,

$$\beta \;\geq\; \int_{t_0}^{t} ||\mathbf{H}(t,\tau)||_1\,d\tau \;=\; \int_{t_0}^{t} ||\mathbf{\Phi}(t,\tau)\,\mathbf{B}(\tau)||_1\,d\tau$$

Since $||\mathbf{B}^{-1}(\tau)||_1 \leq \kappa_{\mathbf{B}-1}$, then

$$\int_{t_0}^{t} ||\mathbf{\Phi}(t,\tau)||_1\,d\tau \;\leq\; \int_{t_0}^{t} ||\mathbf{\Phi}(t,\tau)\,\mathbf{B}(\tau)||_1\,||\mathbf{B}^{-1}(\tau)||_1\,d\tau \;\leq\; \beta\kappa_{\mathbf{B}-1} \;<\; \infty$$

for all t_0 and all t. Also, for the system (*9.1*) to be uniformly externally stable with arbitrary $\mathbf{x}_0$, it must be uniformly stable i.s.L. so that $||\mathbf{\Phi}(\tau, t_0)||_1 < \kappa$ from Theorem 9.2. Therefore

$$\infty \;>\; \beta\kappa_{\mathbf{B}-1}\kappa \;\geq\; \int_{t_0}^{t} ||\mathbf{\Phi}(t,\tau)||_1\,||\mathbf{\Phi}(\tau,t_0)||_1\,d\tau \;\geq\; \int_{t_0}^{t} ||\mathbf{\Phi}(t,\tau)\,\mathbf{\Phi}(\tau,t_0)||_1\,d\tau$$

$$= \int_{t_0}^{t} ||\mathbf{\Phi}(t,t_0)||_1\,d\tau \;=\; ||\mathbf{\Phi}(t,t_0)||_1\,(t-t_0)$$

Define the time $T = \beta\kappa_{\mathbf{B}-1}\kappa\epsilon$ so that at time $t = t_0 + T$, $\beta\kappa_{\mathbf{B}-1}\kappa \geq ||\mathbf{\Phi}(t_0+T, t_0)||_1\beta\kappa_{\mathbf{B}-1}\kappa\epsilon$. Hence $||\mathbf{\Phi}(t_0+T, t_0)||_1 \leq e^{-1}$, so that using the argument found in the last part of the proof just given for Theorem 9.3 shows that $||\mathbf{\Phi}(t, t_0)||_1 \leq \kappa_1 e^{-(t-t_0)/T}$, which proves Theorem 9.5.

9.3. Is the scalar linear system $dx/dt = 2t(2\sin t - 1)x$ uniformly asymptotically stable?

We should expect something unusual because $|A(t)|$ increases as t increases, and $A(t)$ alternates in sign. We find the transition matrix as $\Phi(t, t_0) = e^{\theta(t)-\theta(t_0)}$ where $\theta(t) = 4\sin t - 4t\cos t - t^2$. Since $\theta(t) < 4 + 4|t| - t^2$, $\lim_{t\to\infty} \theta(t) = -\infty$, so that $\lim_{t\to\infty} \Phi(t, t_0) = 0$ and the system is asymptotically stable. This is true globally, i.e. for any initial condition x_0, because of the linearity of the system (Theorem 9.1). Now we investigate uniformity by plotting $\Phi(t, t_0)$ starting from three different initial times: $t_0 = 0$, $t_0 = 2\pi$ and $t_0 = 4\pi$ (Fig. 9-3). The peaks occur at $\Phi((2n+1)\pi, 2n\pi) = e^{\pi(4-\pi)(4n+1)}$ so that the vertical scale is compressed to give a reasonable plot. We can pick an initial time t_0 large enough that some initial condition in $|x_0| < \delta$ will give rise to a trajectory that will leave $|x| < \epsilon$ for any $\epsilon > 0$ and $\delta > 0$. Therefore although the system is stable i.s.L., it is not uniformly stable i.s.L. and so cannot be uniformly asymptotically stable.

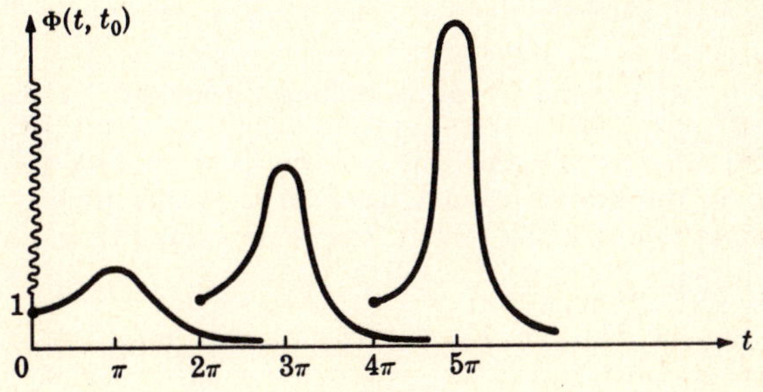

Fig. 9-3

9.4. If $\xi(t) \leq \kappa + \int_{t_0}^{t} [\rho(\tau)\,\xi(\tau) + \mu(\tau)]\,d\tau$ where $\rho(t) \geq 0$ and κ is a constant, show that

$$\xi(t) \;\leq\; \left[\kappa + \int_{t_0}^{t} e^{-\int_{t_0}^{\tau}\rho(\eta)\,d\eta}\,\mu(\tau)\,d\tau\right] e^{\int_{t_0}^{t}\rho(\tau)\,d\tau}$$

This is the Gronwall-Bellman lemma and is often useful in establishing bounds for the response of linear systems.

We shall establish the following chain of assertions to prove the Gronwall-Bellman lemma.

Assertion (i): If $d\omega/dt \leq 0$ and $\omega(t_0) \leq 0$, then $\omega(t) \leq 0$ for $t \geq t_0$.

Proof: If $\omega(t_1) > 0$, then there is a T in $t_0 \leq T \leq t_1$ such that $\omega(T) = 0$ and $\omega(t) > 0$ for $T \leq t \leq t_1$. But using the mean value theorem gives $\omega(t_1) = \omega(t_1) - \omega(T) = (t_1 - T)\,d\omega/dt \leq 0$, which is a contradiction.

Assertion (ii): If $d\omega/dt - \alpha(t)\omega \leq 0$ and $\omega(t_0) \leq 0$, then $\omega(t) \leq 0$ for $t \geq t_0$.

Proof: Multiplying the given inequality by $e^{-\theta(t,t_0)}$, where $\theta(t,t_0) = \int_{t_0}^{t} \alpha(\tau)\,d\tau$, gives

$$0 \geq e^{-\theta(t,t_0)}\,d\omega/dt - \alpha(t)e^{-\theta(t,t_0)} = d(e^{-\theta(t,t_0)}\omega)/dt$$

Applying assertion (i) gives $e^{-\theta(t,t_0)}\omega \leq 0$, and since $e^{-\theta(t,t_0)} > 0$ then $\omega \leq 0$.

Assertion (iii): If $d\phi/dt - \alpha(t)\phi \leq d\gamma/dt - \alpha(t)\gamma$ and $\phi(t_0) \leq \gamma(t_0)$, then $\phi(t) \leq \gamma(t)$.

Proof: Let $\omega = \phi - \gamma$ in assertion (ii).

Assertion (iv): If $d\omega/dt - \alpha(t)\omega \leq \mu(t)$ and $\omega(t_0) \leq \kappa$, then $\omega(t) \leq \left(\kappa + \int_{t_0}^{t} e^{-\theta(\tau,t_0)}\mu(\tau)\,d\tau\right)e^{\theta(t,t_0)}$.

Proof: Set $\omega = \phi$ and $\gamma = \left(\kappa + \int_{t_0}^{t} e^{-\theta(\tau,t_0)}\mu(\tau)\,d\tau\right)e^{\theta(t,t_0)}$ in assertion (iii), because $d\gamma/dt - \alpha(t) = \mu(t)$ and $\gamma(t_0) = \kappa$.

Assertion (v): The Gronwall-Bellman lemma is true.

Proof: Let $\omega(t) = \kappa + \int_{t_0}^{t} [\rho(\tau)\,\xi(\tau) + \mu(\tau)]\,d\tau$, so that the given inequality says $\xi(t) \leq \omega(t)$. Since $\rho(t) \geq 0$ we have $d\omega/dt = \rho(t)\,\xi(t) + \mu(t) \leq \rho(t)\,\omega(t) + \mu(t)$, and $\omega(t_0) = \kappa$. Applying assertion (iv) gives

$$\xi(t) \leq \omega(t) \leq \left[\kappa + \int_{t_0}^{t} e^{-\theta(\tau,t_0)}\mu(\tau)\,d\tau\right]e^{\theta(t,t_0)}$$

9.5. Show that if $\mathbf{A}(t)$ remains bounded for all $t \geq t_0$ ($||\mathbf{A}(t)|| \leq \kappa$), then the system cannot have finite escape time. Also show $||\mathbf{x}(t)|| \leq ||\mathbf{x}_0||e^{\int_{t_0}^{t}||\mathbf{A}(\tau)||\,d\tau}$ for any $\mathbf{A}(t)$.

Integrating $d\mathbf{x}/dt = \mathbf{A}(t)\mathbf{x}$ between t_0 and t gives $\mathbf{x}(t) = \mathbf{x}_0 + \int_{t_0}^{t} \mathbf{A}(\tau)\,\mathbf{x}(\tau)\,d\tau$. Taking the norm of both sides gives

$$||\mathbf{x}(t)|| = ||\mathbf{x}_0 + \int_{t_0}^{t} \mathbf{A}(\tau)\,\mathbf{x}(\tau)\,d\tau|| \leq ||\mathbf{x}_0|| + ||\int_{t_0}^{t} \mathbf{A}(\tau)\,\mathbf{x}(\tau)\,d\tau|| \leq ||\mathbf{x}_0|| + \int_{t_0}^{t} ||\mathbf{A}(\tau)||\,||\mathbf{x}(\tau)||\,d\tau$$

Since $||\mathbf{A}(\tau)|| \geq 0$, we use the Gronwall-Bellman inequality of Problem 9.4 with $\mu(t) = 0$ to obtain $||\mathbf{x}(t)|| \leq ||\mathbf{x}_0||e^{\int_{t_0}^{t}||\mathbf{A}(\tau)||\,d\tau}$. If $||\mathbf{A}(t)|| \leq \kappa$, then $||\mathbf{x}(t)|| \leq ||\mathbf{x}_0||e^{\kappa(t-t_0)}$, which shows $\mathbf{x}(t)$ is bounded for finite t and so the system cannot have finite escape time. However, if $||\mathbf{A}(t)||$ becomes unbounded, we cannot conclude the system has finite escape time, as shown by Problem 9.3.

9.6. Under what circumstances does a unique solution $\mathbf{P}$ exist for the equation $\mathbf{A}^T\mathbf{P} + \mathbf{PA} = -\mathbf{Q}$?

We consider the general equation $\mathbf{BP} + \mathbf{PA} = \mathbf{C}$ where $\mathbf{A}$, $\mathbf{B}$ and $\mathbf{C}$ are arbitrary real $n \times n$ matrices. Define the column vectors of $\mathbf{P}$ and $\mathbf{C}$ as $\mathbf{p}_i$ and $\mathbf{c}_i$ respectively, and the row vectors of $\mathbf{A}$ as $\mathbf{a}_i^T$ for $i = 1, 2, \ldots, n$. Then the equation $\mathbf{BP} + \mathbf{PA} = \mathbf{C}$ can be written as $\mathbf{BP} + \sum_{i=1}^{n} \mathbf{p}_i\mathbf{a}_i^T = \mathbf{C}$, which in turn can be written as

$$\left[\begin{pmatrix} \mathbf{B} & \mathbf{0} & \cdots & \mathbf{0} \\ \mathbf{0} & \mathbf{B} & \cdots & \mathbf{0} \\ \cdots & \cdots & \cdots & \cdots \\ \mathbf{0} & \mathbf{0} & \cdots & \mathbf{B} \end{pmatrix} + \begin{pmatrix} a_{11}\mathbf{I} & a_{21}\mathbf{I} & \cdots & a_{n1}\mathbf{I} \\ a_{12}\mathbf{I} & a_{22}\mathbf{I} & \cdots & a_{n2}\mathbf{I} \\ \cdots & \cdots & \cdots & \cdots \\ a_{1n}\mathbf{I} & a_{2n}\mathbf{I} & \cdots & a_{nn}\mathbf{I} \end{pmatrix}\right]\begin{pmatrix} \mathbf{p}_1 \\ \mathbf{p}_2 \\ \vdots \\ \mathbf{p}_n \end{pmatrix} = \begin{pmatrix} \mathbf{c}_1 \\ \mathbf{c}_2 \\ \vdots \\ \mathbf{c}_n \end{pmatrix}$$

Call the first matrix in brackets $\hat{\mathbf{B}}$ and the second $\hat{\mathbf{A}}$, and the vectors $\mathbf{p}$ and $\mathbf{c}$. Then this equation can be written $(\hat{\mathbf{B}}+\hat{\mathbf{A}})\mathbf{p} = \mathbf{c}$. Note $\hat{\mathbf{B}}$ has eigenvalues equal to the eigenvalues of $\mathbf{B}$, call them β_i, repeated n times. Also, the rows of $\hat{\mathbf{A}} - \lambda\mathbf{I}_n$ are linearly dependent if and only if the rows of $\mathbf{A}^T - \lambda\mathbf{I}_n$ are linearly dependent. This happens if and only if $\det(\mathbf{A}^T - \lambda\mathbf{I}_n) = 0$. So $\hat{\mathbf{A}}$ has eigenvalues equal to the eigenvalues of $\mathbf{A}$, call them α_i, repeated n times. Call $\mathbf{T}$ the matrix that reduces $\mathbf{B}$ to Jordan form $\mathbf{J}$, and $\mathbf{V} = \{v_{ij}\}$ the matrix that reduces $\mathbf{A}$ to Jordan form. Then

$$\begin{pmatrix} v_{11}\mathbf{I} & \cdots & v_{n1}\mathbf{I} \\ \cdots & \cdots & \cdots \\ v_{1n}\mathbf{I} & \cdots & v_{nn}\mathbf{I} \end{pmatrix}^{-1} \begin{pmatrix} \mathbf{T}^{-1} & \cdots & \mathbf{0} \\ \cdots & \cdots & \cdots \\ \mathbf{0} & \cdots & \mathbf{T}^{-1} \end{pmatrix} \left[\begin{pmatrix} \mathbf{B} & \cdots & \mathbf{0} \\ \cdots & \cdots & \cdots \\ \mathbf{0} & \cdots & \mathbf{B} \end{pmatrix} + \begin{pmatrix} a_{11}\mathbf{I} & \cdots & a_{n1}\mathbf{I} \\ \cdots & \cdots & \cdots \\ a_{1n}\mathbf{I} & \cdots & a_{nn}\mathbf{I} \end{pmatrix} \right] \begin{pmatrix} \mathbf{T} & \cdots & \mathbf{0} \\ \cdots & \cdots & \cdots \\ \mathbf{0} & \cdots & \mathbf{T} \end{pmatrix} \begin{pmatrix} v_{11}\mathbf{I} & \cdots & v_{n1}\mathbf{I} \\ \cdots & \cdots & \cdots \\ v_{1n}\mathbf{I} & \cdots & v_{nn}\mathbf{I} \end{pmatrix}$$

$$= \begin{pmatrix} v_{11}\mathbf{I} & \cdots & v_{n1}\mathbf{I} \\ \cdots & \cdots & \cdots \\ v_{1n}\mathbf{I} & \cdots & v_{nn}\mathbf{I} \end{pmatrix}^{-1} \left[\begin{pmatrix} \mathbf{J} & \cdots & \mathbf{0} \\ \cdots & \cdots & \cdots \\ \mathbf{0} & \cdots & \mathbf{J} \end{pmatrix} + \begin{pmatrix} a_{11}\mathbf{I} & \cdots & a_{n1}\mathbf{I} \\ \cdots & \cdots & \cdots \\ a_{1n}\mathbf{I} & \cdots & a_{nn}\mathbf{I} \end{pmatrix} \right] \begin{pmatrix} v_{11}\mathbf{I} & \cdots & v_{n1}\mathbf{I} \\ \cdots & \cdots & \cdots \\ v_{1n}\mathbf{I} & \cdots & v_{nn}\mathbf{I} \end{pmatrix}$$

$$= \begin{pmatrix} \mathbf{J} & \cdots & \mathbf{0} \\ \cdots & \cdots & \cdots \\ \mathbf{0} & \cdots & \mathbf{J} \end{pmatrix} + \begin{pmatrix} \alpha_1\mathbf{I} & & \\ & \ddots & \\ \mathbf{0} & \cdots & \alpha_n\mathbf{I} \end{pmatrix}$$

Hence the eigenvalues of $\hat{\mathbf{B}}+\hat{\mathbf{A}}$ are $\alpha_i + \beta_j$ for $i, j = 1, 2, \ldots, n$.

A unique solution to $(\hat{\mathbf{B}}+\hat{\mathbf{A}})\mathbf{p} = \mathbf{c}$ exists if and only if $\det(\hat{\mathbf{B}}+\hat{\mathbf{A}}) = \prod_{\text{all } i,j} (\alpha_i + \beta_j) \neq 0$. Therefore if and only if $\alpha_i + \beta_j \neq 0$ for all i and j, does a unique solution to $\mathbf{BP}+\mathbf{PA} = \mathbf{C}$ exist. If $\mathbf{B} = \mathbf{A}^T$, then $\beta_j = \alpha_j$, so we require $\alpha_i + \alpha_j \neq 0$. If the real parts of all α_j are in the left half plane (for asymptotic stability), then it is impossible that $\alpha_i + \alpha_j = 0$. Hence a unique solution for $\mathbf{P}$ always exists if the system is asymptotically stable.

9.7. Show that $\mathbf{PA}+\mathbf{A}^T\mathbf{P} = -\mathbf{Q}$, where $\mathbf{Q}$ is a positive definite matrix, has a positive definite solution $\mathbf{P}$ if and only if the eigenvalues of $\mathbf{A}$ are in the left half plane.

This is obvious by Theorem 9.11, but we shall give a direct proof assuming $\mathbf{A}$ can be diagonalized as $\mathbf{M}^{-1}\mathbf{AM} = \mathbf{\Lambda}$. Construct $\mathbf{P}$ as $(\mathbf{M}^{-1})^\dagger\mathbf{M}^{-1}$. Obviously $\mathbf{P}$ is Hermitian, but not necessarily real. Note $\mathbf{P}$ is positive definite because for any nonzero vector $\mathbf{x}$ we have $\mathbf{x}^\dagger\mathbf{Px} = \mathbf{x}^\dagger(\mathbf{M}^{-1})^\dagger\mathbf{M}^{-1}\mathbf{x} = (\mathbf{M}^{-1}\mathbf{x})^\dagger(\mathbf{M}^{-1}\mathbf{x}) > 0$. Also, $\mathbf{x}^\dagger\mathbf{Qx} = -(\mathbf{M}^{-1}\mathbf{x})^\dagger(\mathbf{\Lambda}+\mathbf{\Lambda}^\dagger)(\mathbf{M}^{-1}\mathbf{x}) > 0$ if and only if the real parts of the eigenvalues of $\mathbf{A}$ are in the left half plane. Furthermore, $\nu = \mathbf{x}^\dagger(\mathbf{M}^{-1})^\dagger\mathbf{M}^{-1}\mathbf{x}$ decays as fast as is possible for a quadratic in the state. This is because the state vector decays with a time constant η^{-1} equal to one over the real part of the maximum eigenvalue of $\mathbf{A}$, and hence the square of the norm of the state vector decays in the worst case with a time constant equal to $1/2\eta$. To investigate the time behavior of ν, we find

$$d\nu/dt = \mathbf{x}^\dagger\mathbf{Qx} = -(\mathbf{M}^{-1}\mathbf{x})^\dagger(\mathbf{\Lambda}+\mathbf{\Lambda}^\dagger)(\mathbf{M}^{-1}\mathbf{x}) \leq -2\eta(\mathbf{M}^{-1}\mathbf{x})^\dagger(\mathbf{M}^{-1}\mathbf{x}) = -2\eta\nu$$

For a choice of $\mathbf{M}^{-1}\mathbf{x}$ equal to the unit vector that picks out the eigenvalue with real part η, this becomes an equality and then ν decays with time constant $1/2\eta$.

9.8. In the study of passive circuits and unforced passive vibrating mechanical systems we often encounter the time-invariant real matrix equation $\mathbf{F}\,d^2\mathbf{y}/dt^2 + \mathbf{G}\,d\mathbf{y}/dt + \mathbf{Hy} = \mathbf{0}$ where $\mathbf{F}$ and $\mathbf{H}$ are positive definite and $\mathbf{G}$ is at least nonnegative definite. Prove that the system is stable i.s.L.

Choose as a Liapunov function

$$\nu = \frac{d\mathbf{y}^T}{dt}(\mathbf{F}+\mathbf{F}^T)\frac{d\mathbf{y}}{dt} + \int_0^t \frac{d\mathbf{y}^T}{d\tau}(\mathbf{G}+\mathbf{G}^T)\frac{d\mathbf{y}}{d\tau}\,d\tau + \mathbf{y}^T(\mathbf{H}+\mathbf{H}^T)\mathbf{y}$$

This is positive definite since $\mathbf{F}$ and $\mathbf{H}$ are, and

$$\nu \geq \begin{pmatrix} \mathbf{y}^T & \dfrac{d\mathbf{y}^T}{dt} \end{pmatrix} \begin{pmatrix} \mathbf{H} & \mathbf{0} \\ \mathbf{0} & \mathbf{F} \end{pmatrix} \begin{pmatrix} \mathbf{y} \\ \dfrac{d\mathbf{y}}{dt} \end{pmatrix}$$

which is a positive definite quadratic function of the state. Also,

$$\frac{d\nu}{dt} = \frac{d\mathbf{y}^T}{dt}\left[\mathbf{F}\frac{d^2\mathbf{y}}{dt^2} + \mathbf{G}\frac{d\mathbf{y}}{dt} + \mathbf{H}\mathbf{y}\right] + \left[\mathbf{F}\frac{d^2\mathbf{y}}{dt^2} + \mathbf{G}\frac{d\mathbf{y}}{dt} + \mathbf{H}\mathbf{y}\right]^T\frac{d\mathbf{y}}{dt} = 0$$

By Theorem 9.6, the system is stable i.s.L. The reason this particular Liapunov function was chosen is because it is the energy of the system.

Supplementary Problems

9.9. Given the scalar system $dx/dt = tx + u$ and $y = x$.

(a) Find the impulse response $h(t, t_0)$ and verify $\int_{t_0}^{t} |h(t,\tau)|\,d\tau \le \sqrt{2\pi}\, e^{t^2/2}$.

(b) Show the system is not stable i.s.L.

(c) Explain why Theorem 9.5 does not hold.

9.10. Given the system $dx/dt = -x/t + u$ and $y = x$. (a) Show the response to a unit step function input at time $t_0 > 0$ gives an unbounded output. (b) Explain why Theorem 9.5 does not hold even though the system is asymptotically stable for $t_0 > 0$.

9.11. By altering only one condition in Definition 9.6, show how Liapunov techniques can be used to give sufficient conditions to show the zero state is not asymptotically stable.

9.12. Prove that the real parts of the eigenvalues of a constant matrix $\mathbf{A}$ are $< \sigma$ if and only if given any symmetric, positive definite matrix $\mathbf{Q}$ there exists a symmetric, positive definite matrix $\mathbf{P}$ which is the unique solution of the set of $n(n+1)/2$ linear equations $-2\sigma\mathbf{P} + \mathbf{A}^T\mathbf{P} + \mathbf{P}\mathbf{A} = -\mathbf{Q}$.

9.13. Show that the scalar system $dx/dt = -(1+t)x$ is asymptotically stable for $t \ge 0$ using Liapunov theory.

9.14. Consider the time-varying network of Fig. 9-4. Let $x_1(t) =$ charge on the capacitor and $x_2(t) =$ flux in the inductor, with initial values x_{10} and x_{20} respectively. Then $L(t)\,dx_1/dt = x_2$ and $L(t)\,C(t)\,dx_2/dt + L(t)x_1 + R(t)\,C(t)x_2 = 0$. Starting with

$$\mathbf{P}(t) = \begin{pmatrix} R + (2L/RC) & 1 \\ 1 & 2/R \end{pmatrix}$$

find conditions on R, L, and C that guarantee asymptotic stability.

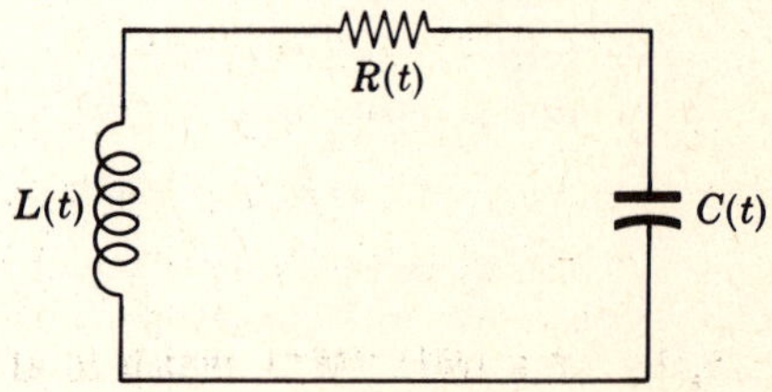

Fig. 9-4

9.15. Using the results of Example 9.16, show that if $\alpha > 0$, $0 < \beta < 1$ and $\alpha^2 > \beta^2(\alpha^2 + \mu^{-2})$, then the Mathieu equation $d^2y/dt^2 + \alpha\,dy/dt + (1 + \beta\cos 2t/\mu)y = 0$ is uniformly asymptotically stable.

9.16. Given the time-varying linear system $d\mathbf{x}/dt = \mathbf{A}(t)\mathbf{x}$ with initial condition $\mathbf{x}_0$ where the elements of $\mathbf{A}(t)$ are continuous in t. Let $\mathbf{H}(t)$ be a symmetric matrix defined by $\mathbf{H}(t) = (\mathbf{A} + \mathbf{A}^T)/2$. Let $\lambda_{\min}(t)$ and $\lambda_{\max}(t)$ be, for each t, the smallest and the largest eigenvalue of $\mathbf{H}(t)$. Using the Liapunov function $\nu(\mathbf{x}) = \mathbf{x}^T\mathbf{x}$, show

$$\|\mathbf{x}_0\|_2\, e^{\int_{t_0}^{t}\lambda_{\min}(\tau)\,d\tau} \le \|\mathbf{x}(t)\|_2 \le \|\mathbf{x}_0\|_2\, e^{\int_{t_0}^{t}\lambda_{\max}(\tau)\,d\tau}$$

9.17. What is the construction similar to that of Problem 9.7 for $\mathbf{P}$ in the discrete-time case $\mathbf{A}^T\mathbf{P}\mathbf{A} - \mathbf{P} = -\mathbf{Q}$?

9.18. Show that if the system of Example 9.22, page 200, is changed slightly to $dx_1/dt = x_2 - \epsilon x_1 + (x_1^2 + x_2^2)/2$ and $dx_2/dt = x_1 - \epsilon x_2 + (x_1^2 + x_2^2)/2$, where $\epsilon > 0$, then the system is uniformly asymptotically stable (but not globally).

9.19. Given the system
$$\frac{d}{dt}\mathbf{x} = \begin{pmatrix} -1 & \theta(t) \\ \alpha & -1 \end{pmatrix} \mathbf{x}$$
Construct a Liapunov function for the system in the case $\theta(t) = 0$. This Liapunov function must give the necessary and sufficient condition for stability, i.e. give the exact stability boundaries on α. Next, use this Liapunov function on the system where $\theta(t)$ is not identically zero to find a condition on $\theta(t)$ under which the system will always be stable.

9.20. Given the system $d\mathbf{x}/dt = (\mathbf{A}(t) + \mathbf{B}(t))\mathbf{x}$ where $d\mathbf{x}/dt = \mathbf{A}(t)\mathbf{x}$ has a transition matrix with norm $\|\mathbf{\Phi}(t,\tau)\| \le e^{-\kappa_2(t-\tau)}$. Using the Gronwall-Bellman inequality with $\mu = 0$, show $\|\mathbf{x}\| \le \|\mathbf{x}_0\| e^{(\kappa_3-\kappa_2)(t-t_0)}$ if $\|\mathbf{B}(t)\| \le \kappa_3 e^{-\kappa_2 t}$.

9.21. Show that $\|\mathbf{x}(t)\| = \|\mathbf{x}_0\| e^{\int_{t_0}^{t} \|\mathbf{A}(\tau)\| d\tau} + \int_{t_0}^{t} e^{\int_{\tau}^{t} \|\mathbf{A}(\eta)\| d\eta} \mathbf{B}(\tau)\mathbf{u}(\tau)\, d\tau$ for the system $d\mathbf{x}/dt = \mathbf{A}(t)\mathbf{x} + \mathbf{B}(t)\mathbf{u}$ with initial condition $\mathbf{x}(t_0) = \mathbf{x}_0$.

Answers to Supplementary Problems

9.9. (*a*) $\int_{-\infty}^{t} |h(t,\tau)|\, d\tau = e^{t^2/2} \int_{-\infty}^{t} e^{-\tau^2/2}\, d\tau \le \sqrt{2\pi}\, e^{t^2/2}$

(*b*) $\phi(t, t_0) = e^{t^2/2} e^{-t_0^2/2}$

(*c*) The system is not externally stable since β depends on t in Theorem 9.4.

9.10. (*a*) $y(t) = (t - t_0^2/t)/2$ for $t \ge t_0 > 0$.

(*b*) The system is not *uniformly* asymptotically stable.

9.11. Change condition (4) to read $d\nu/dt > 0$ in a neighborhood of the origin.

9.12. Replace $\mathbf{A}$ by $\mathbf{A} - \sigma\mathbf{I}$ in equation (*9.5*).

9.13. Use $\nu = x^2$

9.14. $0 < \kappa_1 \le R(t) \le \kappa_2 < \infty$

$0 < \kappa_3 \le L(t) \le \kappa_4 < \infty$

$0 < \kappa_5 \le C(t) \le \kappa_6 < \infty$

$0 < \kappa_7 \le 1 + \dot{R}(L/R^2 - C/2) + \dot{C}L/RC - \dot{L}/R$

$0 < \kappa_8 \le 1 + \dot{R}L/R^2$

9.17. $\mathbf{P} = (\mathbf{M}^\dagger)^{-1}\mathbf{M}^{-1}$ which is always positive definite, and $\mathbf{Q} = \mathbf{M}^{-1}(\mathbf{I} - \mathbf{\Lambda}^\dagger\mathbf{\Lambda})\mathbf{M}^{-1}$ which is positive definite if and only if each eigenvalue of $\mathbf{A}$ has absolute value less than one.

9.18. Use any Liapunov function constructed for the linearized system.

9.19. If $\mathbf{Q} = 2\mathbf{I}$, then $\mathbf{P} = \frac{1}{2}\begin{pmatrix} 2+\alpha^2 & \alpha \\ \alpha & 2 \end{pmatrix}$. The system is stable for all α if $\theta = 0$. If $\theta \ne 0$, we require $4 + 2\alpha\theta - \theta^2(1 + \alpha^2/2)^2 > 0$.

9.20. $\|\mathbf{x}\| \le \|\mathbf{x}_0\| e^{-\kappa_2(t-t_0)} + \int_{t_0}^{t} e^{-\kappa_2(t-\tau)} \|\mathbf{B}(\tau)\|\, \|\mathbf{x}\|\, d\tau$ so apply the Gronwall-Bellman inequality to
$$\|\mathbf{x}\| e^{\kappa_2 t} \le \|\mathbf{x}_0\| e^{\kappa_2 t_0} + \kappa_3 \int_{t_0}^{t} \|\mathbf{x}\|\, d\tau$$

9.21. Use Problem 9.4 with $\|\mathbf{x}(t)\| = \xi(t)$, $\|\mathbf{x}_0\| = \kappa$, $\|\mathbf{A}(t)\| = \rho(t)$, and $\|\mathbf{B}(\tau)\mathbf{u}(\tau)\| = \mu(\tau)$.

Chapter 10

Introduction to Optimal Control

10.1 INTRODUCTION

In this chapter we shall study a particular type of optimal control system as an introduction to the subject. In keeping with the spirit of the book, the problem is stated in such a manner as to lead to a linear closed loop system. A general optimization problem usually does not lead to a linear closed loop system.

Suppose it is a control system designer's task to find a feedback control for an open loop system in which all the states are the real output variables, so that

$$d\mathbf{x}/dt = \mathbf{A}(t)\mathbf{x} + \mathbf{B}(t)\mathbf{u} \qquad \mathbf{y} = \mathbf{I}\mathbf{x} + \mathbf{0}\mathbf{u} \tag{10.1}$$

In other words, we are given the system shown in Fig. 10-1 and wish to design what goes into the box marked *. Later we will consider what happens when $\mathbf{y}(t) = \mathbf{C}(t)\mathbf{x}(t)$ where $\mathbf{C}(t)$ is not restricted to be the unit matrix.

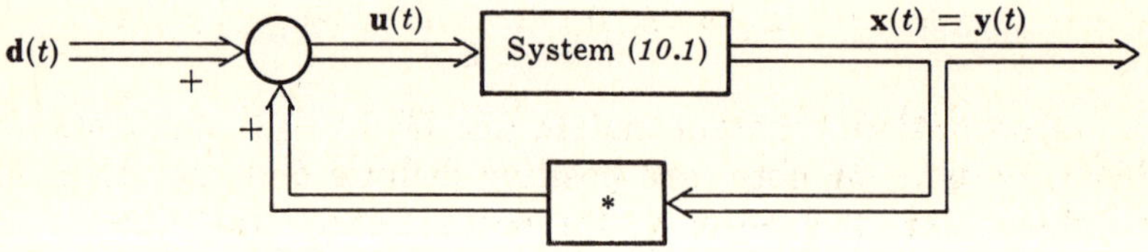

Fig. 10-1. Vector Feedback Control System

A further restriction must be made on the type of system to be controlled.

Definition 10.1: A *regulator* is a feedback control system in which the input $\mathbf{d}(t) = \mathbf{0}$.

For a regulator the only forcing terms are due to the nonzero initial conditions of the state variables, $\mathbf{x}(t_0) = \mathbf{x}_0 \neq \mathbf{0}$ in general. We shall only study regulators first because (1) later the extension to *servomechanisms* (which follow a specified $\mathbf{d}(t)$) will become easier, (2) it turns out that the solution is the same if $\mathbf{d}(t)$ is white noise (the proof of this result is beyond the scope of the book), and (3) many systems can be reduced to regulators.

Example 10.1.

Given a step input of height α, $u(t) = \alpha U(t)$, where the unit step $U(t-t_0) = 0$ for $t < t_0$ and $U(t-t_0) = 1$ for $t \geqq t_0$, into a scalar system with transfer function $1/(s+\beta)$ and initial state x_0. But this is equivalent to a system with a transfer function $1/[s(s+\beta)]$, with initial states α and x_0 and with no input. In other words, we can add an extra integrator with initial condition α at the input to the system flow diagram to generate the step, and the resultant system is a regulator. Note, however, the input becomes a state that must be measured and fed back if the system is to be of the form (*10.1*).

Under the restriction that we are designing a regulator, we require $\mathbf{d}(t) = \mathbf{0}$. Then the input $\mathbf{u}(t)$ is the output of the box marked * in Fig. 10-1 and is the feedback control. We shall assume $\mathbf{u}$ is then some function to be formed of the *present* state $\mathbf{x}(t)$ and time. This is no restriction, because from condition 1 of Section 1.4 upon the representation of the state of this deterministic system, the present state completely summarizes all the past history of the abstract object. Therefore we need not consider controls $\mathbf{u} = \mathbf{u}(\mathbf{x}(\tau), t)$ for $t_0 \leqq \tau \leqq t$, but can consider merely $\mathbf{u} = \mathbf{u}(\mathbf{x}(t), t)$ at the start of our development.

10.2 THE CRITERION FUNCTIONAL

We desire the system to be optimal, but must be very exact in the sense in which the system is optimal. We must find a mathematical expression to measure how the system must be optimal in comparison with other systems. A great many factors influence the engineering utility of a system: cost, reliability, consumer acceptance, etc. The factors mentioned are very difficult to measure and put a single number on for purposes of comparison. Consequently in this introductory chapter we shall simply avoid the question by considering optimality only in terms of system dynamic performance. It is still left to the art of engineering, rather than the science, to incorporate unmeasurable quantities into the criterion of optimality.

In terms of performance, the system response is usually of most interest. Response is the time behavior of the output, as the system tries to follow the input. Since the input to a regulator is zero, we wish to have the time behavior of the output, which in this case is the state, go to zero from the initial condition $\mathbf{x}_0$. In fact, for the purposes of this chapter there exists a convenient means to assign a number to the distance that the response $[\mathbf{x}(\tau)$ for $t_0 \leq \tau \leq t_1]$ is from $\mathbf{0}$. Although the criterion need not be a metric, a metric on the space of all time functions between t_0 and t_1 will accomplish this. Also, if we are interested only in how close we are to zero at time t_1, a metric only on all $\mathbf{x}(t_1)$ is desired. To obtain a linear closed loop system, here we shall only consider the particular quadratic metric

$$\rho^2(\mathbf{x}, \mathbf{0}) \quad = \quad \tfrac{1}{2}\mathbf{x}^T(t_1)\,\mathbf{S}\mathbf{x}(t_1) \;+\; \frac{1}{2}\int_{t_0}^{t_1} \mathbf{x}^T(\tau)\,\mathbf{Q}(\tau)\,\mathbf{x}(\tau)\,d\tau$$

where $\mathbf{S}$ is an $n \times n$ symmetric constant matrix and $\mathbf{Q}(\tau)$ is a $n \times n$ symmetric time-varying matrix. If either $\mathbf{S}$ or $\mathbf{Q}(\tau)$, or both, are positive definite and the other one at least nonnegative definite then $f(\mathbf{x}, \mathbf{0})$ is a norm on the product space $\{\mathbf{x}(t_1), \mathbf{x}(\tau)\}$. It can be shown this requirement can be weakened to $\mathbf{S}$, $\mathbf{Q}$ nonnegative definite if the system $d\mathbf{x}/dt = \mathbf{A}(t)\mathbf{x}$ with $\mathbf{y} = \sqrt{\mathbf{Q}(t)}\,\mathbf{x}$ is observable, but for simplicity we assume one is positive definite. The exact form of $\mathbf{S}$ and $\mathbf{Q}(\tau)$ is to be fixed by the designer at the outset. Thus a number is assigned to the response obtained by each control law $\mathbf{u}(\mathbf{x}(t), t)$, and the optimum system is that whose control law gives the minimum $\rho(\mathbf{x}, \mathbf{0})$.

The choice of $\mathbf{Q}(\tau)$ is dictated by the relative importance of each state over the time interval $t_0 \leq t < t_1$.

Example 10.2.

Consider the system of Problem 2.18, page 34, with the choice of state variables (i). If the angle of attack $\alpha = \phi - K_6\dot{z}$ is to be kept small, we can minimize the integral of $\alpha^2 = \mathbf{x}^T\mathbf{H}^T\mathbf{H}\mathbf{x}$ where $\mathbf{H} = (0 \;\; -K_6 \;\; 1 \;\; 0)$. Then $\mathbf{Q} = \mathbf{H}^T\mathbf{H}$ is nonnegative definite and we must choose $\mathbf{S}$ to be positive definite. Furthermore, if α is of importance only during the first part of the missile's flight, $\mathbf{Q}$ might be chosen as a function of time, such as $\mathbf{Q}(t) = \mathbf{H}^T\mathbf{H}e^{-t}$.

The choice of $\mathbf{S}$ is dictated by the relative importance of each state at the final time, t_1.

Example 10.3.

Consider the missile of Problem 2.18, page 34, with the choice of state variables (i), whose target is stationed at $z = 0$. Then it makes no difference what path the missile flies to arrive near $z = 0$ at $t = t_1$. Therefore choose $\mathbf{Q} = \mathbf{0}$. What matters is how small $z^2(t_1)$ is. Also, we do not want $\dot{z}(t_1)$, $\phi(t_1)$ or $\dot{\phi}(t_1)$ to be too large, so choose

$$\mathbf{S} \quad = \quad \begin{pmatrix} 1 & 0 & 0 & 0 \\ 0 & \epsilon_1 & 0 & 0 \\ 0 & 0 & \epsilon_2 & 0 \\ 0 & 0 & 0 & \epsilon_3 \end{pmatrix}$$

where $\epsilon_1, \epsilon_2, \epsilon_3$ are small fixed positive numbers to be chosen by trial and error after finding the closed loop system for each fixed ϵ_i. If any $\epsilon_i = 0$, an unstable system could result, but might not.

For the system *(10.1)*, the choice of control $\mathbf{u}(\mathbf{x}(t),t)$ to minimize $\rho(\mathbf{x},\mathbf{0})$ is $\mathbf{u} = -\mathbf{B}^{-1}(t)\,\mathbf{x}_0\,\delta(t-t_0)$ in the case where $\mathbf{B}(t)$ has an inverse. Then $\mathbf{x}(t) = \mathbf{0}$ for all $t > t_0$ and $\rho(\mathbf{x},\mathbf{0}) = 0$, its minimum value. If $\mathbf{B}(t)$ does not have an inverse, the optimal control is a sum of delta functions and their derivatives, such as equation *(6.7)*, page 135, and can drive $\mathbf{x}(t)$ to $\mathbf{0}$ almost instantaneously. Because it is very hard to mechanize a delta function, which is essentially achieving infinite closed loop gain, this solution is unacceptable. We must place a bound on the control. Again, to obtain the linear closed loop system, here we shall only consider the particular quadratic

$$\frac{1}{2}\int_{t_0}^{t_1} \mathbf{u}^T(\tau)\,\mathbf{R}(\tau)\,\mathbf{u}(\tau)\,d\tau$$

where $\mathbf{R}(\tau)$ is an $m \times m$ symmetric time-varying positive definite matrix to be fixed at the outset by the designer. $\mathbf{R}$ is positive definite to assure each element of $\mathbf{u}$ is bounded. This is the generalized control "energy", and will be added to $\rho^2(\mathbf{x},\mathbf{0})$. The relative magnitudes of $\|\mathbf{Q}(\tau)\|$ and $\|\mathbf{R}(\tau)\|$ are in proportion to the relative values of the response and control energy. The larger $\|\mathbf{Q}(\tau)\|$ is relative to $\|\mathbf{R}(\tau)\|$, the quicker the response and the higher the gain of the system.

10.3 DERIVATION OF THE OPTIMAL CONTROL LAW

The mathematical problem statement is that we are given $d\mathbf{x}/dt = \mathbf{A}(t)\mathbf{x} + \mathbf{B}(t)\mathbf{u}$ and want to minimize

$$\nu[\mathbf{x},\mathbf{u}] \;=\; \tfrac{1}{2}\mathbf{x}^T(t_1)\,\mathbf{S}\mathbf{x}(t_1) \;+\; \frac{1}{2}\int_{t_0}^{t_1} [\mathbf{x}^T(\tau)\,\mathbf{Q}(\tau)\,\mathbf{x}(\tau) \;+\; \mathbf{u}^T(\tau)\,\mathbf{R}(\tau)\,\mathbf{u}(\tau)]\,d\tau \qquad (10.2)$$

Here we shall give a heuristic derivation of the optimal control and defer the rigorous derivation to Problem 10.1. Consider a real n-vector function of time $\mathbf{p}(t)$, called the costate or Lagrange multiplier, which will obey a differential equation that we shall determine. Note for any $\mathbf{x}$ and $\mathbf{u}$ obeying the state equation, $\mathbf{p}^T(t)[\mathbf{A}(t)\mathbf{x} + \mathbf{B}(t)\mathbf{u} - d\mathbf{x}/dt] = 0$. Adding this null quantity to the criterion changes nothing.

$$\nu[\mathbf{x},\mathbf{u}] \;=\; \tfrac{1}{2}\mathbf{x}^T(t_1)\,\mathbf{S}\mathbf{x}(t_1) \;+\; \int_{t_0}^{t_1} (\tfrac{1}{2}\mathbf{x}^T\mathbf{Q}\mathbf{x} + \tfrac{1}{2}\mathbf{u}^T\mathbf{R}\mathbf{u} + \mathbf{p}^T\mathbf{A}\mathbf{x} + \mathbf{p}^T\mathbf{B}\mathbf{u} - \mathbf{p}^T d\mathbf{x}/d\tau)\,d\tau$$

Integrating the term $\mathbf{p}^T d\mathbf{x}/d\tau$ by parts gives

$$\nu[\mathbf{x},\mathbf{u}] \;=\; \nu_1[\mathbf{x}(t_1)] + \nu_2[\mathbf{x}] + \nu_3[\mathbf{u}]$$

where

$$\nu_1[\mathbf{x}(t_1)] \;=\; \tfrac{1}{2}\mathbf{x}^T(t_1)\mathbf{S}\mathbf{x}(t_1) - \mathbf{x}^T(t_1)\mathbf{p}(t_1) + \mathbf{x}^T(t_0)\mathbf{p}(t_0) \qquad (10.3)$$

$$\nu_2[\mathbf{x}] \;=\; \int_{t_0}^{t_1} (\tfrac{1}{2}\mathbf{x}^T\mathbf{Q}\mathbf{x} + \mathbf{x}^T\mathbf{A}\mathbf{p} + \mathbf{x}^T\,d\mathbf{p}/d\tau)\,d\tau \qquad (10.4)$$

$$\nu_3[\mathbf{u}] \;=\; \int_{t_0}^{t_1} (\tfrac{1}{2}\mathbf{u}^T\mathbf{R}\mathbf{u} + \mathbf{u}^T\mathbf{B}^T\mathbf{p})\,d\tau \qquad (10.5)$$

Introduction of the costate $\mathbf{p}$ has permitted $\nu[\mathbf{x},\mathbf{u}]$ to be broken into ν_1, ν_2 and ν_3, and heuristically we suspect that $\nu[\mathbf{x},\mathbf{u}]$ will be minimum when ν_1, ν_2 and ν_3 are each independently minimized. Recall from calculus that when a smooth function attains a local minimum, its derivative is zero. Analogously, we suspect that if ν_1 is a minimum, then the gradient of *(10.3)* with respect to $\mathbf{x}(t_1)$ is zero:

$$\mathbf{p}(t_1) \;=\; \mathbf{S}\mathbf{x}^{\text{op}}(t_1) \qquad (10.6)$$

and that if ν_2 is a minimum, then the gradient of the integrand of *(10.4)* with respect to $\mathbf{x}$ is zero:

$$d\mathbf{p}/dt \;=\; -\mathbf{A}^T(t)\mathbf{p} - \mathbf{Q}(t)\mathbf{x}^{\text{op}} \qquad (10.7)$$

Here $\mathbf{x}^{op}(t)$ is the response $\mathbf{x}(t)$ using the optimal control law $\mathbf{u}^{op}(\mathbf{x}(t), t)$ as the input to the system (*10.1*). Consequently we *define* $\mathbf{p}(t)$ as the vector which obeys equations (*10.6*) and (*10.7*). It must be realized that these steps are heuristic because the minimum of ν might not occur at the combined minima of ν_1, ν_2 and ν_3; a differentiable function need not result from taking the gradient at each instant of time; and also taking the gradient with respect to $\mathbf{x}$ does not always lead to a minimum. This is why the final step of setting the gradient of the integrand of (*10.5*) with respect to $\mathbf{u}$ equal to zero (equation (*10.8*)) does not give a rigorous proof of the following theorem.

Theorem 10.1: Given the feedback system (*10.1*) with the criterion (*10.2*), having the costate defined by (*10.6*) and (*10.7*). Then if a minimum exists, it is obtained by the optimal control

$$\mathbf{u}^{op} = -\mathbf{R}^{-1}\mathbf{B}^T\mathbf{p} \tag{10.8}$$

The proof is given in Problem 10.1, page 220.

To calculate the optimal control, we must find $\mathbf{p}(t)$. This is done by solving the system equation (*10.1*) using the optimal control (*10.8*) together with the costate equation (*10.7*),

$$\frac{d}{dt}\begin{pmatrix}\mathbf{x}^{op}\\ \mathbf{p}\end{pmatrix} = \begin{pmatrix}\mathbf{A}(t) & -\mathbf{B}(t)\mathbf{R}^{-1}(t)\mathbf{B}^T(t)\\ -\mathbf{Q}(t) & -\mathbf{A}^T(t)\end{pmatrix}\begin{pmatrix}\mathbf{x}^{op}\\ \mathbf{p}\end{pmatrix} \tag{10.9}$$

with $\mathbf{x}^{op}(t_0) = \mathbf{x}_0$ and $\mathbf{p}(t_1) = \mathbf{S}\mathbf{x}^{op}(t_1)$.

Example 10.4.

Consider the scalar time-invariant system $dx/dt = 2x + u$ with criterion

$$\nu = x^2(t_1) + \frac{1}{2}\int_0^1 (3x^2 + u^2/4)\, dt$$

Then $A = 2$, $B = 1$, $R = 1/4$, $Q = 3$, $S = 2$. Hence we solve

$$\frac{d}{dt}\begin{pmatrix}x^{op}\\ p\end{pmatrix} = \begin{pmatrix}2 & -4\\ -3 & -2\end{pmatrix}\begin{pmatrix}x^{op}\\ p\end{pmatrix}$$

with $x^{op}(0) = x_0$, $p(1) = 2x(1)$. Using the methods of Chapter 5,

$$\begin{pmatrix}x^{op}(t)\\ p(t)\end{pmatrix} = \left[\frac{e^{-4t}}{8}\begin{pmatrix}2 & 4\\ 3 & 6\end{pmatrix} + \frac{e^{+4t}}{8}\begin{pmatrix}6 & -4\\ -3 & 2\end{pmatrix}\right]\begin{pmatrix}x^{op}(0)\\ p(0)\end{pmatrix} \tag{10.10}$$

Evaluating this at $t = 1$ and using $p(1) = 2x^{op}(1)$ gives

$$p(0) = \frac{15e^{+8} + 1}{10e^{+8} - 2}x_0 \tag{10.11}$$

Then from Theorem 10.1,

$$u^{op}(x_0, t) = \frac{2e^{+4t} + 30e^{8-4t}}{1 - 5e^{+8}}x_0 \tag{10.12}$$

Note this procedure has generated $\mathbf{u}^{op}$ as a function of t and $\mathbf{x}_0$. If $\mathbf{x}_0$ is known, storage of $\mathbf{u}^{op}(\mathbf{x}_0, t)$ as a function of time only, in the memory of a computer controller, permits open loop control; i.e. starting at t_0 the computer generates an input time-function for $d\mathbf{x}^{op}/dt = \mathbf{A}\mathbf{x}^{op} + \mathbf{B}\mathbf{u}^{op}(\mathbf{x}_0, t)$ and no feedback is needed.

However, in most regulator systems the initial conditions $\mathbf{x}_0$ are not known. By the introduction of feedback we can find a control law such that the system is optimum for arbitrary $\mathbf{x}_0$. For purposes of illustration we will now give one method for finding the feedback, although this is not the most efficient way to proceed for the problem studied in this chapter. We can eliminate $\mathbf{x}_0$ from $d\mathbf{x}^{op}/dt = \mathbf{A}\mathbf{x}^{op} + \mathbf{B}\mathbf{u}^{op}(\mathbf{x}_0, t)$ by solving for $\mathbf{x}^{op}(t)$ in terms of $\mathbf{x}_0$ and t, i.e. $\mathbf{x}^{op}(t) = \boldsymbol{\phi}(t; \mathbf{x}_0, t_0)$ where $\boldsymbol{\phi}$ is the trajectory of the optimal system. Then $\mathbf{x}^{op}(t) = \boldsymbol{\phi}(t; \mathbf{x}_0, t_0)$ can be solved for $\mathbf{x}_0$ in terms of $\mathbf{x}^{op}(t)$ and t, $\mathbf{x}_0 = \mathbf{x}_0(\mathbf{x}^{op}(t), t)$. Substituting for $\mathbf{x}_0$ in the system equation then gives the feedback control system $d\mathbf{x}^{op}/dt = \mathbf{A}\mathbf{x}^{op} + \mathbf{B}\mathbf{u}^{op}(\mathbf{x}^{op}(t), t)$.

Example 10.5.

To find the feedback control for Example 10.4, from *(10.10)* and *(10.11)*

$$x^{\text{op}}(t) \quad = \quad \frac{x_0}{5e^{+8}-1}(5e^{8-4t} - e^{+4t})$$

Solving this for x_0 and substituting into *(10.12)* gives the feedback control

$$u^{\text{op}}(x^{\text{op}}(t), t) \quad = \quad \frac{2 + 30e^{8(1-t)}}{1 - 5e^{8(1-t)}} x^{\text{op}}(t)$$

Hence what goes in the box marked * in Fig. 10-1 is the time-varying gain element $K(t)$ where

$$K(t) \quad = \quad \frac{2 + 30e^{8(1-t)}}{1 - 5e^{8(1-t)}}$$

and the overall closed loop system is

$$\frac{dx^{\text{op}}}{dt} \quad = \quad \frac{4 + 20e^{8(1-t)}}{1 - 5e^{8(1-t)}} x^{\text{op}}$$

10.4 THE MATRIX RICCATI EQUATION

To find the time-varying gain matrix $\mathbf{K}(t)$ directly, let $\mathbf{p}(t) = \mathbf{P}(t)\,\mathbf{x}^{\text{op}}(t)$. Here $\mathbf{P}(t)$ is an $n \times n$ Hermitian matrix to be found. Then $\mathbf{u}^{\text{op}}(\mathbf{x}, t) = -\mathbf{R}^{-1}\mathbf{B}^T\mathbf{P}\mathbf{x}^{\text{op}}$ so that $\mathbf{K} = -\mathbf{R}^{-1}\mathbf{B}^T\mathbf{P}$. The closed loop system then becomes $d\mathbf{x}^{\text{op}}/dt = (\mathbf{A} - \mathbf{B}\mathbf{R}^{-1}\mathbf{B}^T\mathbf{P})\mathbf{x}^{\text{op}}$, and call its transition matrix $\mathbf{\Phi}_{\text{cl}}(t, \tau)$. Substituting $\mathbf{p} = \mathbf{P}\mathbf{x}^{\text{op}}$ into the bottom equation of *(10.9)* gives

$$(d\mathbf{P}/dt)\mathbf{x}^{\text{op}} + \mathbf{P}\,d\mathbf{x}^{\text{op}}/dt \quad = \quad -\mathbf{Q}\mathbf{x}^{\text{op}} - \mathbf{A}^T\mathbf{P}\mathbf{x}^{\text{op}}$$

Using the top equation of *(10.9)* for $d\mathbf{x}^{\text{op}}/dt$ then gives

$$\mathbf{0} \quad = \quad (d\mathbf{P}/dt + \mathbf{Q} + \mathbf{A}^T\mathbf{P} + \mathbf{P}\mathbf{A} - \mathbf{P}\mathbf{B}\mathbf{R}^{-1}\mathbf{B}^T\mathbf{P})\mathbf{x}^{\text{op}}$$

But $\mathbf{x}^{\text{op}}(t) = \mathbf{\Phi}_{\text{cl}}(t, t_0)\mathbf{x}_0$, and since $\mathbf{x}_0$ is arbitrary and $\mathbf{\Phi}_{\text{cl}}$ is nonsingular, we find the $n \times n$ matrix $\mathbf{P}$ must satisfy the matrix Riccati equation

$$-d\mathbf{P}/dt \quad = \quad \mathbf{Q} + \mathbf{A}^T\mathbf{P} + \mathbf{P}\mathbf{A} - \mathbf{P}\mathbf{B}\mathbf{R}^{-1}\mathbf{B}^T\mathbf{P} \qquad (10.13)$$

This has the "final" condition $\mathbf{P}(t_1) = \mathbf{S}$ since $\mathbf{p}(t_1) = \mathbf{P}(t_1)\,\mathbf{x}(t_1) = \mathbf{S}\mathbf{x}(t_1)$. Changing independent variables by $\tau = t_1 - t$, the matrix Riccati equation becomes $d\mathbf{P}/d\tau = \mathbf{Q} + \mathbf{A}^T\mathbf{P} + \mathbf{P}\mathbf{A} - \mathbf{P}\mathbf{B}\mathbf{R}^{-1}\mathbf{B}^T\mathbf{P}$ where the arguments of the matrices are $t_1 - \tau$ instead of t. The equation can then be solved numerically on a computer from $\tau = 0$ to $\tau = t_1 - t_0$, starting at the initial condition $\mathbf{P}(0) = \mathbf{S}$. Occasionally the matrix Riccati equation can also be solved analytically.

Example 10.6.

For the system of Example 10.4, the 1×1 matrix Riccati equation is

$$-dP/dt \quad = \quad 3 + 4P - 4P^2$$

with $P(1) = 2$. Since this is separable for this example,

$$t - t_1 \quad = \quad \int_{t_1}^{t} \frac{dP}{4(P - 3/2)(P + 1/2)} \quad = \quad \frac{1}{8}\ln\frac{P(t) - 3/2}{P(t_1) - 3/2} - \frac{1}{8}\ln\frac{P(t) + 1/2}{P(t_1) + 1/2}$$

Taking antilogarithms and rearranging, after setting $t_1 = 1$, gives

$$P(t) \quad = \quad \frac{15 + e^{8(t-1)}}{10 - 2e^{8(t-1)}}$$

Then $K(t) = -R^{-1}B^TP = -4P(t)$, which checks with the answer obtained in Example 10.5.

Another method of solution of the matrix Riccati equation is to use the transition matrix of (*10.9*) directly. Partition the transition matrix as

$$\frac{\partial}{\partial t}\begin{pmatrix} \mathbf{\Phi}_{11}(t,\tau) & \mathbf{\Phi}_{12}(t,\tau) \\ \mathbf{\Phi}_{21}(t,\tau) & \mathbf{\Phi}_{22}(t,\tau) \end{pmatrix} = \begin{pmatrix} \mathbf{A} & -\mathbf{B}\mathbf{R}^{-1}\mathbf{B}^T \\ -\mathbf{Q} & -\mathbf{A}^T \end{pmatrix}\begin{pmatrix} \mathbf{\Phi}_{11}(t,\tau) & \mathbf{\Phi}_{12}(t,\tau) \\ \mathbf{\Phi}_{21}(t,\tau) & \mathbf{\Phi}_{22}(t,\tau) \end{pmatrix} \tag{10.14}$$

Then

$$\begin{pmatrix} \mathbf{x}(t) \\ \mathbf{p}(t) \end{pmatrix} = \begin{pmatrix} \mathbf{\Phi}_{11}(t,t_1) & \mathbf{\Phi}_{12}(t,t_1) \\ \mathbf{\Phi}_{21}(t,t_1) & \mathbf{\Phi}_{22}(t,t_1) \end{pmatrix}\begin{pmatrix} \mathbf{x}(t_1) \\ \mathbf{S}\mathbf{x}(t_1) \end{pmatrix}$$

Eliminating $\mathbf{x}(t_1)$ from this gives

$$\mathbf{p}(t) = \mathbf{P}(t)\,\mathbf{x}(t) = [\mathbf{\Phi}_{21}(t,t_1) + \mathbf{\Phi}_{22}(t,t_1)\mathbf{S}][\mathbf{\Phi}_{11}(t,t_1) + \mathbf{\Phi}_{12}(t,t_1)\mathbf{S}]^{-1}\mathbf{x}(t) \tag{10.15}$$

so that $\mathbf{P}$ is the product of the two bracketed matrices. A sufficient condition (not necessary) for the existence of the solution $\mathbf{P}(t)$ of the Riccati equation is that the open loop transition matrix does not become unbounded in $t_0 \leq t \leq t_1$. Therefore the inverse in equation (*10.15*) can always be found under this condition.

Example 10.7.

For Example 10.4, use of equation (*10.10*) gives $P(t)$ from (*10.15*) as

$$P(t) = \frac{[3e^{4(t-1)} - 3e^{-4(t-1)} + 2(6e^{4(t-1)} + 2e^{-4(t-1)})]/8}{[2e^{4(t-1)} + 6e^{-4(t-1)} + 2(4e^{4(t-1)} - 4e^{-4(t-1)})]/8}$$

which reduces to the answer obtained in Example 10.6.

We have a useful check on the solution of the matrix Riccati equation.

***Theorem 10.2*:** If $\mathbf{S}$ is positive definite and $\mathbf{Q}(t)$ at least nonnegative definite, or vice versa, and $\mathbf{R}(t)$ is positive definite, then an optimum $\nu[\mathbf{x}, \mathbf{u}^{op}]$ exists if and only if the solution $\mathbf{P}(t)$ to the matrix Riccati equation (*10.13*) exists and is bounded and positive definite for all $t < t_1$. Under these conditions $\nu[\mathbf{x}, \mathbf{u}^{op}] = \frac{1}{2}\mathbf{x}^T(t_0)\,\mathbf{P}(t_0)\,\mathbf{x}(t_0)$.

Proof is given in Problem 10.1, page 220. It is evident from the proof that if both $\mathbf{S}$ and $\mathbf{Q}(t)$ are nonnegative definite, $\mathbf{P}(t)$ can be nonnegative definite if $\nu[\mathbf{x}, \mathbf{u}^{op}]$ exists.

***Theorem 10.3*:** For $\mathbf{S}(t)$, $\mathbf{Q}(t)$ and $\mathbf{R}(t)$ symmetric, the solution $\mathbf{P}(t)$ to the matrix Riccati equation (*10.13*) is symmetric.

Proof: Take the transpose of (*10.13*), note it is identical to (*10.13*), and recall that there is only one solution $\mathbf{P}(t)$ that is equal to $\mathbf{S}$ at time t_1.

Note this means for an $n \times n$ $\mathbf{P}(t)$ that only $n(n+1)/2$ equations need be solved on the computer, because $\mathbf{S}$, $\mathbf{Q}$ and $\mathbf{R}$ can always be taken symmetric. Further aids in obtaining a solution for the time-varying case are given in Problems 10.20 and 10.21.

10.5 TIME-INVARIANT OPTIMAL SYSTEMS

So far we have obtained only time-varying feedback gain elements. The most important engineering application is for time-invariant closed loop systems. Consequently in this section we shall consider $\mathbf{A}$, $\mathbf{B}$, $\mathbf{R}$ and $\mathbf{Q}$ to be constant, $\mathbf{S} = \mathbf{0}$ and $t_1 \to \infty$ to obtain a constant feedback element $\mathbf{K}$.

Because the existence of the solution of the Riccati equation is guaranteed if the open loop transition matrix does not become unbounded in $t_0 \leq t \leq t_1$, in the limit as $t_1 \to \infty$ we should expect no trouble from asymptotically stable open loop systems. However, we wish to incorporate unstable systems in the following development, and need the following existence theorem for the limit as $t_1 \to \infty$ of the solution to the Riccati equation, denoted by $\mathbf{\Pi}$.

Theorem 10.4: If the states of the system *(10.1)* that are not asymptotically stable are controllable, then $\lim_{t_1 \to \infty} \mathbf{P}(t_0; t_1) = \mathbf{\Pi}(t_0)$ exists and $\mathbf{\Pi}(t_0)$ is constant and positive definite.

Proof: Define a control $\mathbf{u}^1(\tau) = -\mathbf{B}^T(\tau)\mathbf{\Phi}^T(t_0, \tau)\mathbf{W}^{-1}(t_0, t_2)\mathbf{x}(t_0)$ for $t_0 \leq \tau \leq t_2$, similar to that used in Problem 6.8, page 142. Note $\mathbf{W}$ drives all the controllable states to zero at the time $t_2 < \infty$. Let $\mathbf{u}^1(\tau) = \mathbf{0}$ for $\tau > t_2$. Defining the response of the asymptotically stable states as $\mathbf{x}_{as}(t)$, then $\int_{t_2}^{\infty} \mathbf{x}_{as}^T\mathbf{Q}\mathbf{x}_{as}\,dt < \infty$. Therefore

$$\nu[\mathbf{x}, \mathbf{u}^1] = \int_{t_0}^{t_2} (\mathbf{x}^T\mathbf{Q}\mathbf{x} + \mathbf{u}^{1T}\mathbf{R}\mathbf{u}^1)\,dt + \int_{t_2}^{\infty} \mathbf{x}_{as}^T\mathbf{Q}_{as}\,dt < \infty$$

Note $\nu[\mathbf{x}, \mathbf{u}^1] \leq \alpha(t_0, t_2)\mathbf{x}^T(t_0)\mathbf{x}(t_0)$ after carrying out the integration, since both $\mathbf{x}(t)$ and $\mathbf{u}^1(t)$ are linear in $\mathbf{x}(t_0)$. Here $\alpha(t_0, t_2)$ is some bounded scalar function of t_0 and t_2. Then from Theorem 10.2,

$$\tfrac{1}{2}\mathbf{x}^T(t_0)\mathbf{\Pi}(t_0)\mathbf{x}(t_0) = \nu[\mathbf{x}, \mathbf{u}^{op}] \leq \alpha(t_0, t_2)\mathbf{x}^T(t_0)\mathbf{x}(t_0) < \infty$$

Therefore $\|\mathbf{\Pi}(t_0)\|_2 \leq 2\alpha(t_0, t_2) < \infty$ so $\mathbf{\Pi}(t_0)$ is bounded. It can be shown that for $\mathbf{S} = \mathbf{0}$, $\mathbf{P}_{\mathbf{S}=\mathbf{0}}(t_0; t) \leq \mathbf{P}_{\mathbf{S}=\mathbf{0}}(t_0; t_1)$ for $t_0 \leq t \leq t_1$, and also that when $\mathbf{S} > \mathbf{0}$, then

$$\lim_{t \to \infty} \|\mathbf{P}_{\mathbf{S}>\mathbf{0}}(t_0; t) - \mathbf{P}_{\mathbf{S}=\mathbf{0}}(t_0; t)\| = 0$$

Therefore, $\lim_{t_1 \to \infty} \mathbf{P}(t_0; t_1)$ must be a constant because for t_1 large any change in $\mathbf{P}(t_0; t_1)$ must be to increase $\mathbf{P}(t_0; t_1)$ until it hits the bound $\mathbf{\Pi}(t_0)$.

Since we are now dealing with a time-invariant closed loop system, by Definition 1.8 the time axis can be translated and an equivalent system results. Hence we can send $t_0 \to -\infty$, start the Riccati equation at $\mathbf{P}(t_1) = \mathbf{S} = \mathbf{0}$ and integrate numerically backwards in time until the steady state constant $\mathbf{\Pi}$ is reached.

Example 10.8.

Consider the system $dx/dt = u$ with criterion $\nu = \frac{1}{2}\int_{t_0}^{t_1} (x^2 + u^2)\,dt$. Then the Riccati equation is $-dP/dt = 1 - P^2$ with $P(t_1) = 0$. This has a solution $P(t_0) = \tanh(t_1 - t_0)$. Therefore

$$\Pi = \lim_{t_1 \to \infty} P(t_0; t_1) = \lim_{t_1 \to \infty} \tanh(t_1 - t_0) = \lim_{t_0 \to -\infty} \tanh(t_1 - t_0) = 1$$

The optimal control is $u^{op} = -R^{-1}B^T\Pi x = -x$.

Since $\mathbf{\Pi}$ is a positive definite constant solution of the matrix Riccati equation, it satisfies the quadratic algebraic equation

$$\mathbf{0} = \mathbf{Q} + \mathbf{A}^T\mathbf{\Pi} + \mathbf{\Pi}\mathbf{A} - \mathbf{\Pi}\mathbf{B}\mathbf{R}^{-1}\mathbf{B}^T\mathbf{\Pi} \qquad (10.16)$$

Example 10.9.

For the system of Example 10.8, the quadratic algebraic equation satisfied by Π is $0 = 1 - \Pi^2$. This has solutions ± 1, so that $\Pi = 1$ which is the only positive definite solution.

A very useful engineering property of the quadratic optimal system is that it is always stable if it is controllable. Since a linear, constant coefficient, stable closed loop system always results, often the quadratic criterion is chosen solely to attain this desirable feature.

Theorem 10.5: If $\mathbf{\Pi}$ exists for the constant coefficient system, the closed loop system is asymptotically stable.

Proof: Choose a Liapunov function $V = \mathbf{x}^T\mathbf{\Pi}\mathbf{x}$. Since $\mathbf{\Pi}$ is positive definite, $V > 0$ for all nonzero $\mathbf{x}$. Then

$$dV/dt = \mathbf{x}^T\mathbf{\Pi}(\mathbf{A}\mathbf{x} - \mathbf{B}\mathbf{R}^{-1}\mathbf{B}^T\mathbf{\Pi}\mathbf{x}) + (\mathbf{A}\mathbf{x} - \mathbf{B}\mathbf{R}^{-1}\mathbf{B}^T\mathbf{\Pi}\mathbf{x})^T\mathbf{\Pi}\mathbf{x} = -\mathbf{x}^T\mathbf{Q}\mathbf{x} - \mathbf{x}^T\mathbf{\Pi}\mathbf{B}\mathbf{R}^{-1}\mathbf{B}^T\mathbf{\Pi}\mathbf{x} < 0$$

for all $\mathbf{x} \neq \mathbf{0}$ because $\mathbf{Q}$ is positive definite and $\mathbf{\Pi}\mathbf{B}\mathbf{R}^{-1}\mathbf{B}^T\mathbf{\Pi}$ is nonnegative definite.

Since there are in general $n(n+1)$ solutions of *(10.16)*, it helps to know that $\mathbf{\Pi}$ is the only positive definite solution.

Theorem 10.6: The unique positive definite solution of *(10.16)* is $\mathbf{\Pi}$.

Proof: Suppose there exist two positive definite solutions $\mathbf{\Pi}_1$ and $\mathbf{\Pi}_2$. Then

$$(\mathbf{\Pi}_1 - \mathbf{\Pi}_2)(\mathbf{A} - \mathbf{B}\mathbf{R}^{-1}\mathbf{B}^T\mathbf{\Pi}_1) + (\mathbf{A} - \mathbf{B}\mathbf{R}^{-1}\mathbf{B}^T\mathbf{\Pi}_2)^T(\mathbf{\Pi}_1 - \mathbf{\Pi}_2) = \mathbf{Q} - \mathbf{Q} = \mathbf{0} \qquad (10.17)$$

Recall from Problem 9.6, that the equation $\mathbf{XF} + \mathbf{GX} = \mathbf{K}$ has a unique solution $\mathbf{X}$ whenever $\lambda_i(\mathbf{F}) + \lambda_j(\mathbf{G}) \neq 0$ for any i and j, where $\lambda_i(\mathbf{F})$ is an eigenvalue of $\mathbf{F}$ and $\lambda_j(\mathbf{G})$ is an eigenvalue of $\mathbf{G}$. But $\mathbf{A} - \mathbf{B}\mathbf{R}^{-1}\mathbf{B}^T\mathbf{\Pi}_i$ for $i = 1$ and 2 are stability matrices, from Theorem 10.5. Therefore the sum of the real parts of any combination of their eigenvalues must be less than zero. Hence *(10.17)* has the unique solution $\mathbf{\Pi}_1 - \mathbf{\Pi}_2 = \mathbf{0}$.

Generally the equation *(10.16)* can be solved for very high-order systems, $n \leq 50$. This gives another advantage over classical frequency domain techniques, which are not so easily adapted to computer solution. Algebraic solution of *(10.16)* for large n is difficult, because there are $n(n+1)$ solutions that must be checked. However, if a good initial guess $\mathbf{P}(t_2)$ is available, i.e. $\mathbf{P}(t_2) = \delta\mathbf{P}(t_2) + \mathbf{\Pi}$ where $\delta\mathbf{P}(t_2)$ is small, numerical solution of *(10.13)* backwards from any t_2 to a steady state gives $\mathbf{\Pi}$. In other words, $\delta\mathbf{P}(t) = \mathbf{P}(t) - \mathbf{\Pi}$ tends to zero as $t \to -\infty$, which is true because:

Theorem 10.7: If $\mathbf{A}$, $\mathbf{B}$, $\mathbf{R}$ and $\mathbf{Q}$ are constant matrices, and $\mathbf{R}$, $\mathbf{Q}$ and the initial state $\mathbf{P}(t_2)$ are positive definite, then equation *(10.13)* is globally asymptotically stable as $t \to -\infty$ relative to the equilibrium state $\mathbf{\Pi}$.

Proof: $\delta\mathbf{P}$ obeys the equation (subtracting *(10.16)* from *(10.13)*)

$$-d\delta\mathbf{P}/dt = \mathbf{F}^T\delta\mathbf{P} + \delta\mathbf{P}\,\mathbf{F} - \delta\mathbf{P}\,\mathbf{B}\mathbf{R}^{-1}\mathbf{B}^T\delta\mathbf{P}$$

where $\mathbf{F} = \mathbf{A} - \mathbf{B}\mathbf{R}^{-1}\mathbf{B}^T\mathbf{\Pi}$. Choose as a Liapunov function $2V = \operatorname{tr}[\delta\mathbf{P}^T(\mathbf{\Pi}\mathbf{\Pi}^T)^{-1}\delta\mathbf{P}]$. For brevity, we investigate here only the real scalar case. Then $2F = -Q\Pi^{-1} - B^2R^{-1}\Pi$ and $2V = \Pi^{-2}\,\delta P^2$, so that $dV/dt = \Pi^{-2}\,\delta P\, d\delta P/dt = \Pi^{-2}\,\delta P^2\,(Q\Pi^{-1} + B^2R^{-1}P) > 0$ for all nonzero δP since Q, R, P and Π are all > 0. Hence $V \to 0$ as $t \to -\infty$, and $\delta P \to 0$. It can be shown in the vector case that $dV/dt > 0$ also.

Example 10.10.

For the system of Example 10.8, consider the deviation due to an incorrect initial condition, i.e. suppose $P(t_1) = \epsilon$ instead of the correct value $P(t_1) = 0$. Then $P(t_0) = \tanh(t_1 - t_0 + \tanh^{-1}\epsilon)$. For any finite ϵ, $\lim_{t_0 \to -\infty} P(t_0) = 1 = \Pi$.

Experience has shown that for very high-order systems, the approximations inherent in numerical solution (truncation and roundoff) often lead to an unstable solution for bad initial guesses, in spite of Theorem 10.7. Another method of computation of $\mathbf{\Pi}$ will now be investigated.

Theorem 10.8: If λ_i is an eigenvalue of $\mathbf{H}$, the constant $2n \times 2n$ matrix corresponding to *(10.9)*, then $-\lambda_i^*$ is also an eigenvalue of $\mathbf{H}$.

Proof: Denote the top n elements of the ith eigenvector of $\mathbf{H}$ as $\mathbf{f}_i$ and the bottom n elements as $\mathbf{g}_i$. Then

$$\lambda_i\begin{pmatrix}\mathbf{f}_i\\ \mathbf{g}_i\end{pmatrix} = \mathbf{H}\begin{pmatrix}\mathbf{f}_i\\ \mathbf{g}_i\end{pmatrix} = \begin{pmatrix}\mathbf{A} & -\mathbf{B}\mathbf{R}^{-1}\mathbf{B}^T\\ -\mathbf{Q} & -\mathbf{A}^T\end{pmatrix}\begin{pmatrix}\mathbf{f}_i\\ \mathbf{g}_i\end{pmatrix} \qquad (10.18)$$

or

$$\lambda_i\mathbf{f}_i = \mathbf{A}\mathbf{f}_i - \mathbf{B}\mathbf{R}^{-1}\mathbf{B}^T\mathbf{g}_i$$

$$\lambda_i\mathbf{g}_i = -\mathbf{Q}\mathbf{f}_i - \mathbf{A}^T\mathbf{g}_i$$

or

$$-\lambda_i\begin{pmatrix}-\mathbf{g}_i\\ \mathbf{f}_i\end{pmatrix} = \begin{pmatrix}\mathbf{A}^T & -\mathbf{Q}\\ -\mathbf{B}\mathbf{R}^{-1}\mathbf{B}^T & -\mathbf{A}\end{pmatrix}\begin{pmatrix}-\mathbf{g}_i\\ \mathbf{f}_i\end{pmatrix} = \mathbf{H}^T\begin{pmatrix}-\mathbf{g}_i\\ \mathbf{f}_i\end{pmatrix}$$

This shows $-\lambda_i$ is an eigenvalue of $\mathbf{H}^T$. Also, since $\det \mathbf{M} = (\det \mathbf{M}^T)^*$ for any matrix $\mathbf{M}$, then for any ξ such that $\det(\mathbf{H} - \xi\mathbf{I}) = 0$ we find that $\det(\mathbf{H}^T - \xi\mathbf{I}) = 0$. This shows that if ξ^* is an eigenvalue of $\mathbf{H}^T$, then ξ is an eigenvalue of $\mathbf{H}$. Since $-\lambda_i$ is an eigenvalue of $\mathbf{H}^T$, it is concluded that $-\lambda_i^*$ is an eigenvalue of $\mathbf{H}$.

This theorem shows that the eigenvalues of $\mathbf{H}$ are placed symmetrically with regard to the $j\omega$ axis in the complex plane. Then $\mathbf{H}$ has at most n eigenvalues with real parts < 0, and will have exactly n unless some are purely imaginary.

Example 10.11.

The $\mathbf{H}$ matrix corresponding to Example 10.8 is $\mathbf{H} = \begin{pmatrix} 0 & -1 \\ -1 & 0 \end{pmatrix}$. This has eigenvalues $+1$ and -1, which are symmetric with respect to the imaginary axis.

Example 10.12.

A fourth-order system might give rise to an 8×8 $\mathbf{H}$ having eigenvalues placed as shown in Fig. 10-2.

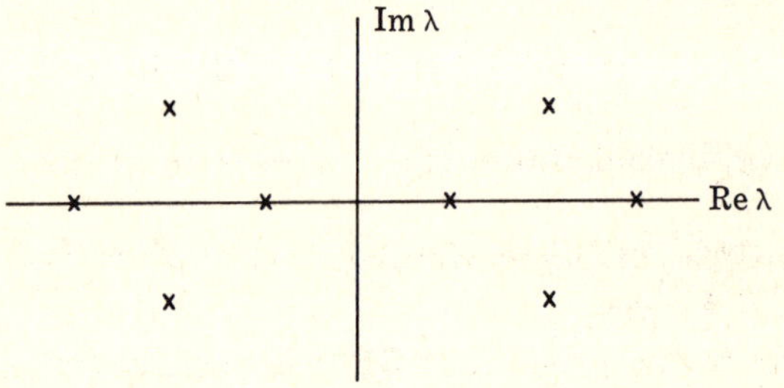

Fig. 10-2

The factorization into eigenvalues with real parts < 0 and real parts > 0 is another way of looking at the Wiener factorization of a polynomial $\rho(\omega^2)$ into $\rho^+(\omega)\,\rho^-(\omega)$.

Theorem 10.9: If $\lambda_1, \lambda_2, \ldots, \lambda_n$ are n distinct eigenvalues of $\mathbf{H}$ having real parts < 0, then $\mathbf{\Pi} = \mathbf{G}\mathbf{F}^{-1}$ where $\mathbf{F} = (\mathbf{f}_1 \mid \mathbf{f}_2 \mid \ldots \mid \mathbf{f}_n)$ and $\mathbf{G} = (\mathbf{g}_1 \mid \mathbf{g}_2 \mid \ldots \mid \mathbf{g}_n)$ as defined in equation (*10.18*).

Proof: Since $\mathbf{\Pi}$ is the unique Hermitian positive definite solution of (*10.16*), we will show first that $\mathbf{G}\mathbf{F}^{-1}$ is a solution to (*10.16*), then that it is Hermitian, and finally that it is positive definite.

Define an $n \times n$ diagonal matrix $\mathbf{\Lambda}$ with $\lambda_1, \lambda_2, \ldots, \lambda_n$ its diagonal elements so that from the eigenvalue equation (*10.18*),

$$\begin{pmatrix} \mathbf{F} \\ \mathbf{G} \end{pmatrix} \mathbf{\Lambda} \;=\; \mathbf{H}\begin{pmatrix} \mathbf{F} \\ \mathbf{G} \end{pmatrix} \;=\; \begin{pmatrix} \mathbf{A} & -\mathbf{B}\mathbf{R}^{-1}\mathbf{B}^T \\ -\mathbf{Q} & -\mathbf{A}^T \end{pmatrix}\begin{pmatrix} \mathbf{F} \\ \mathbf{G} \end{pmatrix}$$

from which

$$\mathbf{F}\mathbf{\Lambda} = \mathbf{A}\mathbf{F} - \mathbf{B}\mathbf{R}^{-1}\mathbf{B}^T\mathbf{G} \tag{10.19}$$

$$\mathbf{G}\mathbf{\Lambda} = -\mathbf{Q}\mathbf{F} - \mathbf{A}^T\mathbf{G} \tag{10.20}$$

Premultiplying (*10.19*) by $\mathbf{F}^{-1}$ and substituting for $\mathbf{\Lambda}$ in (*10.20*) gives

$$\mathbf{G}\mathbf{F}^{-1}\mathbf{A}\mathbf{F} - \mathbf{G}\mathbf{F}^{-1}\mathbf{B}\mathbf{R}^{-1}\mathbf{B}^T\mathbf{G} \;=\; -\mathbf{Q}\mathbf{F} - \mathbf{A}^T\mathbf{G}$$

Postmultiplying by $\mathbf{F}^{-1}$ shows $\mathbf{G}\mathbf{F}^{-1}$ satisfies (*10.16*).

Next, we show $\mathbf{G}\mathbf{F}^{-1}$ is Hermitian. It suffices to show $\mathbf{M} = \mathbf{F}^\dagger\mathbf{G}$ is Hermitian, since $\mathbf{G}\mathbf{F}^{-1} = \mathbf{F}^{\dagger -1}\mathbf{M}\mathbf{F}^{-1}$ is then Hermitian. Let the elements of $\mathbf{M}$ be $m_{jk} = \mathbf{f}_j^\dagger\mathbf{g}_k$; for $j \neq k$,

$$m_{jk} - m_{kj}^* = \begin{pmatrix}\mathbf{f}_j\\ \mathbf{g}_j\end{pmatrix}^\dagger \begin{pmatrix}\mathbf{0} & \mathbf{I}\\ -\mathbf{I} & \mathbf{0}\end{pmatrix}\begin{pmatrix}\mathbf{f}_k\\ \mathbf{g}_k\end{pmatrix}$$

$$= \frac{1}{\lambda_j^* + \lambda_k}\left\{\lambda_j^*\begin{pmatrix}\mathbf{f}_j\\ \mathbf{g}_j\end{pmatrix}^\dagger \begin{pmatrix}\mathbf{0} & \mathbf{I}\\ -\mathbf{I} & \mathbf{0}\end{pmatrix}\begin{pmatrix}\mathbf{f}_k\\ \mathbf{g}_k\end{pmatrix} + \begin{pmatrix}\mathbf{f}_j\\ \mathbf{g}_j\end{pmatrix}^\dagger \begin{pmatrix}\mathbf{0} & \mathbf{I}\\ -\mathbf{I} & \mathbf{0}\end{pmatrix}\lambda_k\begin{pmatrix}\mathbf{f}_k\\ \mathbf{g}_k\end{pmatrix}\right\}$$

$$= \frac{1}{\lambda_j^* + \lambda_k}\begin{pmatrix}\mathbf{f}_j\\ \mathbf{g}_j\end{pmatrix}^\dagger \left\{\mathbf{H}^T\begin{pmatrix}\mathbf{0} & \mathbf{I}\\ -\mathbf{I} & \mathbf{0}\end{pmatrix} + \begin{pmatrix}\mathbf{0} & \mathbf{I}\\ -\mathbf{I} & \mathbf{0}\end{pmatrix}\mathbf{H}\right\}\begin{pmatrix}\mathbf{f}_k\\ \mathbf{g}_k\end{pmatrix}$$

Since the term in braces equals $\mathbf{0}$, we have $m_{jk} = m_{kj}^*$ and thus $\mathbf{GF}^{-1}$ is Hermitian.

Finally, to show $\mathbf{GF}^{-1}$ is positive definite, define two $n \times n$ matrix functions of time as $\boldsymbol{\theta}(t) = \mathbf{F}e^{\mathbf{\Lambda}t}\mathbf{F}^{-1}$ and $\boldsymbol{\psi}(t) = \mathbf{G}e^{\mathbf{\Lambda}t}\mathbf{F}^{-1}$. Then $\boldsymbol{\theta}(\infty) = \mathbf{0} = \boldsymbol{\psi}(\infty)$. Using *(10.19)* and *(10.20)*,

$$d\boldsymbol{\theta}/dt = (\mathbf{AF} - \mathbf{BR}^{-1}\mathbf{B}^T\mathbf{G})e^{\mathbf{\Lambda}t}\mathbf{F}^{-1}$$

$$d\boldsymbol{\psi}/dt = -(\mathbf{QF} + \mathbf{A}^T\mathbf{G})e^{\mathbf{\Lambda}t}\mathbf{F}^{-1}$$

Then

$$\mathbf{GF}^{-1} = \boldsymbol{\theta}^\dagger(0)\,\boldsymbol{\psi}(0) = -\int_0^\infty \frac{d}{dt}(\boldsymbol{\theta}^\dagger\boldsymbol{\psi})\,dt$$

so that

$$\mathbf{GF}^{-1} = \int_0^\infty (e^{\mathbf{\Lambda}t}\mathbf{F}^{-1})^\dagger(\mathbf{F}^\dagger\mathbf{QF} + \mathbf{G}^\dagger\mathbf{BR}^{-1}\mathbf{B}^T\mathbf{G})(e^{\mathbf{\Lambda}t}\mathbf{F}^{-1})\,dt$$

Since the integrand is positive definite, $\mathbf{GF}^{-1}$ is positive definite, and $\mathbf{GF}^{-1} = \mathbf{\Pi}$.

Corollary 10.10: The closed loop response is $\mathbf{x}(t) = \mathbf{F}e^{\mathbf{\Lambda}(t-t_0)}\mathbf{F}^{-1}\mathbf{x}_0$ with costate $\mathbf{p}(t) = \mathbf{G}e^{\mathbf{\Lambda}(t-t_0)}\mathbf{F}^{-1}\mathbf{x}_0$.

The proof is similar to the proof that $\mathbf{GF}^{-1}$ is positive definite.

Corollary 10.11: The eigenvalues of the closed loop matrix $\mathbf{A} - \mathbf{BR}^{-1}\mathbf{B}^T\mathbf{\Pi}$ are $\lambda_1, \lambda_2, \ldots, \lambda_n$ and the eigenvectors of $\mathbf{A} - \mathbf{BR}^{-1}\mathbf{B}^T\mathbf{\Pi}$ are $\mathbf{f}_1, \mathbf{f}_2, \ldots, \mathbf{f}_n$.

The proof follows immediately from equation *(10.19)*. Furthermore, since $\lambda_1, \lambda_2, \ldots, \lambda_n$ are assumed distinct, then from Theorem 4.1 we know $\mathbf{f}_1, \mathbf{f}_2, \ldots, \mathbf{f}_n$ are linearly independent, so that $\mathbf{F}^{-1}$ always exists. Furthermore, Theorem 10.5 assures $\operatorname{Re}\lambda_i < 0$ so no λ_i can be imaginary.

Example 10.13.

The $\mathbf{H}$ matrix corresponding to the system of Example 10.8 has eigenvalues $-1 = \lambda_1$ and $+1 = \lambda_2$. Corresponding to these eigenvalues,

$$\begin{pmatrix}f_1\\ g_1\end{pmatrix} = \alpha\begin{pmatrix}1\\ 1\end{pmatrix} \quad \text{and} \quad \begin{pmatrix}f_2\\ g_2\end{pmatrix} = \beta\begin{pmatrix}1\\ -1\end{pmatrix}$$

where α and β are any nonzero constants. Usually, we would merely set α and β equal to one, but for purposes of instruction we do not here. We discard λ_2 and its eigenvector since it has a real part > 0, and form $F = f_1 = \alpha$ and $G = g_1 = \alpha$. Then $\Pi = GF^{-1} = 1$ because the α's cancel. From Problem 4.41, in the vector case $\mathbf{F} = \mathbf{F}_0\mathbf{K}$ and $\mathbf{G} = \mathbf{G}_0\mathbf{K}$ where $\mathbf{K}$ is the diagonal matrix of arbitrary constants associated with each eigenvector, but still $\mathbf{\Pi} = \mathbf{GF}^{-1} = \mathbf{G}_0\mathbf{K}(\mathbf{F}_0\mathbf{K})^{-1} = \mathbf{G}_0\mathbf{F}_0$.

Use of an eigenvalue and eigenvector routine to calculate $\mathbf{\Pi}$ from Theorem 10.9 has given results for systems of order $n \leq 50$. Perhaps the best procedure is to calculate an approximate $\mathbf{\Pi}_0 = \mathbf{GF}^{-1}$ using eigenvectors, and next use $\operatorname{Re}(\mathbf{\Pi}_0 + \mathbf{\Pi}_0^T)/2$ as an initial guess to the Riccati equation *(10.13)*. Then the Riccati equation stability properties (Theorem 10.7) will reduce any errors in $\mathbf{\Pi}_0$, as well as provide a check on the eigenvector calculation.

10.6 OUTPUT FEEDBACK

Until now we have considered only systems in which the output was the state, $\mathbf{y} = \mathbf{Ix}$. For $\mathbf{y} = \mathbf{Ix}$, all the states are available for measurement. In the general case $\mathbf{y} = \mathbf{C}(t)\mathbf{x}$, the states must be reconstructed from the output. Therefore we must assume the observability of the closed loop system $d\mathbf{x}/dt = \mathbf{F}(t)\mathbf{x}$ where $\mathbf{F}(t) = \mathbf{A}(t) - \mathbf{B}(t)\,\mathbf{R}^{-1}(t)\,\mathbf{B}^T(t)\,\mathbf{P}(t;t_1)$.

To reconstruct the state from the output, the output is differentiated $n-1$ times.

$$\mathbf{y} = \mathbf{C}(t)\mathbf{x} = \mathbf{N}_1(t)\mathbf{x}$$

$$d\mathbf{y}/dt = \mathbf{N}_1(t)\,d\mathbf{x}/dt + d\mathbf{N}_1/dt\,\mathbf{x} = (\mathbf{N}_1\mathbf{F} + d\mathbf{N}_1/dt)\mathbf{x} = \mathbf{N}_2\mathbf{x}$$

$$\cdots\cdots\cdots\cdots\cdots\cdots\cdots\cdots\cdots\cdots\cdots\cdots$$

$$d^{n-1}\mathbf{y}/dt^{n-1} = (\mathbf{N}_{n-1}\mathbf{F} + d\mathbf{N}_{n-1}/dt)\mathbf{x} = \mathbf{N}_n\mathbf{x}$$

where $\mathbf{N}^T = (\mathbf{N}_1^T | \ldots | \mathbf{N}_n^T)$ is the observability matrix defined in Theorem 6.11. Define a $nk \times k$ matrix of differentiation operators $\mathbf{H}(d/dt)$ by $\mathbf{H}(d/dt) = (\mathbf{I} | \mathbf{I}d/dt | \ldots | \mathbf{I}d^{n-1}/dt^{n-1})$. Then $\mathbf{x} = \mathbf{N}^{-I}(t)\,\mathbf{H}(d/dt)\mathbf{y}$. Since the closed loop system is observable, $\mathbf{N}$ has rank n. From Property 16 of Section 4.8, page 87, the generalized inverse $\mathbf{N}^{-I}$ has rank n. Using the results of Problem 3.11, page 63, we conclude the n-vector $\mathbf{x}$ exists and is uniquely determined. The optimal control is then $\mathbf{u} = \mathbf{R}^{-1}\mathbf{B}^T\mathbf{P}\mathbf{N}^{-I}\mathbf{H}\mathbf{y}$.

Example 10.14.

Given the system

$$\frac{d}{dt}\begin{pmatrix} x_1 \\ x_2 \end{pmatrix} = \begin{pmatrix} 0 & 1 \\ -\alpha_2 & -\alpha_1 \end{pmatrix}\begin{pmatrix} x_1 \\ x_2 \end{pmatrix} + \begin{pmatrix} 0 \\ 1 \end{pmatrix} u \qquad y = (1 \;\; 0)\mathbf{x}$$

with optimal control $u = k_1 x_1 + k_2 x_2$. Since $y = x_1$ and $dx_1/dt = x_2$, the optimal control in terms of the output is $u = k_1 y + k_2\,dy/dt$.

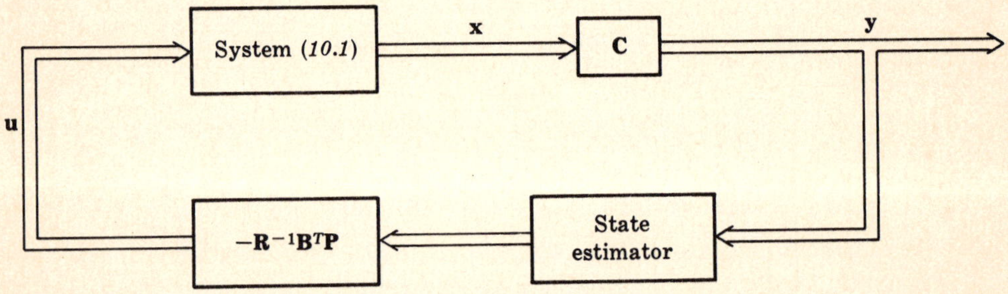

Fig. 10-3

Since $\mathbf{u}$ involves derivatives of $\mathbf{y}$, it may appear that this is not very practical. This mathematical result arises because we are controlling a deterministic system in which there is no noise, i.e. in which differentiation is feasible.

However, in most cases the noise is such that the probabilistic nature of the system must be taken into account. A result of stochastic optimization theory is that under certain circumstances the best estimate of the state can be used in place of the state in the optimal control and still an optimum is obtained (the "separation theorem"). An estimate of each state can be obtained from the output, so that structure of the optimal controller is as shown in Fig. 10-3.

10.7 THE SERVOMECHANISM PROBLEM

Here we shall discuss only servomechanism problems that can be reduced to regulator problems. We wish to find the optimum compensation to go into the box marked ** in Fig. 10-4.

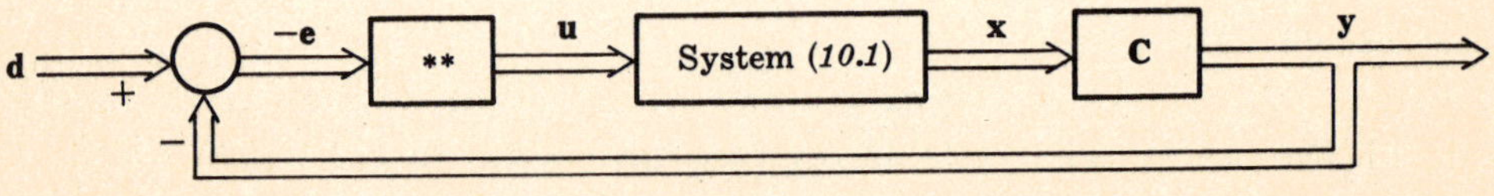

Fig. 10-4. The Servomechanism Problem

The criterion to be minimized is

$$\tfrac{1}{2}\mathbf{e}^T(t_1)\,\mathbf{S}\mathbf{e}(t_1) \;+\; \frac{1}{2}\int_{t_0}^{t_1} [\mathbf{e}^T(\tau)\,\mathbf{Q}(\tau)\,\mathbf{e}(\tau) \;+\; \mathbf{u}^T(\tau)\,\mathbf{R}(\tau)\,\mathbf{u}(\tau)]\,d\tau \qquad (10.21)$$

Note when $\mathbf{e}(t) = \mathbf{y}(t) - \mathbf{d}(t)$ is minimized, $\mathbf{y}(t)$ will follow $\mathbf{d}(t)$ closely.

To reduce this problem to a regulator, we consider only those $\mathbf{d}(t)$ that can be generated by arbitrary $\mathbf{z}(t_0)$ in the equation $d\mathbf{z}/dt = \mathbf{A}(t)\mathbf{z}$, $\mathbf{d} = \mathbf{C}(t)\mathbf{z}$. The coefficients $\mathbf{A}(t)$ and $\mathbf{C}(t)$ are identical to those of the open loop system *(10.1)*.

Example 10.15.

Given the open loop system

$$\frac{d\mathbf{x}}{dt} \;=\; \begin{pmatrix} 0 & 1 & 0 \\ 0 & 0 & 0 \\ 0 & 0 & -2 \end{pmatrix}\mathbf{x} + \begin{pmatrix} b_1 \\ b_2 \\ b_3 \end{pmatrix} u \qquad y \;=\; (1\ \ 1\ \ 1)\mathbf{x}$$

Then we consider only those inputs $\mathbf{d}(t) = z_{10} + z_{20}(t+1) + z_{30}e^{-2t}$. In other words, we can consider as inputs only ramps, steps and e^{-2t} functions and arbitrary linear combinations thereof.

Restricting $\mathbf{d}(t)$ in this manner permits defining new state variables $\mathbf{w} = \mathbf{x} - \mathbf{z}$. Then

$$d\mathbf{w}/dt \;=\; \mathbf{A}(t)\mathbf{w} + \mathbf{B}(t)\mathbf{u} \qquad \mathbf{e} \;=\; \mathbf{C}(t)\mathbf{w} \qquad (10.22)$$

Now we have the regulator problem *(10.22)* subject to the criterion *(10.21)*, and solution of the matrix Riccati equation gives the optimal control as $\mathbf{u} = -\mathbf{R}^{-1}\mathbf{B}^T\mathbf{P}\mathbf{w}$. The states $\mathbf{w}$ can be found from $\mathbf{e}$ and its $n-1$ derivatives as in Section 10.6, so the content of the box marked $**$ in Fig. 10-4 is $\mathbf{R}^{-1}\mathbf{B}^T\mathbf{P}\mathbf{N}^{-1}\mathbf{H}$.

The requirement that the input $\mathbf{d}(t)$ be generated by the zero-input equation $d\mathbf{z}/dt = \mathbf{A}(t)\mathbf{z}$ has significance in relation to the error constants discussed in Section 8.3. Theorem 8.1 states that for unity feedback systems, such as that of Fig. 10-4, if the zero output is asymptotically stable then the class of inputs $\mathbf{d}(t)$ that the system can follow (such that $\lim_{t\to\infty}\mathbf{e}(t) = \mathbf{0}$) is generated by the equations $d\mathbf{w}/dt = \mathbf{A}(t)\mathbf{w} + \mathbf{B}(t)\mathbf{g}$ and $\mathbf{d}(t) = \mathbf{C}(t)\mathbf{w} + \mathbf{g}$ where $\mathbf{g}(t)$ is any function such that $\lim_{t\to\infty}\mathbf{g}(t) = \mathbf{0}$. Unfortunately, such an input is not reducible to the regulator problem in general. However, taking $\mathbf{g} \equiv \mathbf{0}$ assures us the system can follow the inputs that are reducible to the regulator problem if the closed loop zero output is asymptotically stable.

Restricting the discussion to time-invariant systems gives us the assurance that the closed loop zero output is asymptotically stable, from Theorem 10.5. If we further restrict the discussion to inputs of the type $\mathbf{d}_i = (t-t_0)^l U(t-t_0)\mathbf{e}_i$ as in Definition 8.1, then Theorem 8.2 applies. Then we must introduce integral compensation when the open loop transfer function matrix $\mathbf{H}(s)$ is not of the correct type l to follow the desired input.

Example 10.16.

Consider a system with transfer function $G(s)$ in which $G(0) \neq \infty$, i.e. it contains no pure integrations. We must introduce integral compensation, $1/s$. Then the optimal servomechanism to follow a step input is as shown in Fig. 10-5 where the box marked $**$ contains $\mathbf{R}^{-1}\mathbf{b}^T\mathbf{\Pi}\mathbf{N}^{-1}\mathbf{H}(s)$. This is a linear combination of $1, s, \ldots, s^n$ since the overall system $G(s)/s$ is of order $n+1$. Thus we can write the contents of $**$ as $k_1 + k_2 s + \cdots + k_n s^n$. The compensation for $G(s)$ is then $k_1/s + k_2 + k_3 s + \cdots + k_n s^{n-1}$. In a noisy environment the differentiations cannot be realized and are approximated by a filter, so that the compensation takes the form of integral plus proportional plus a filter.

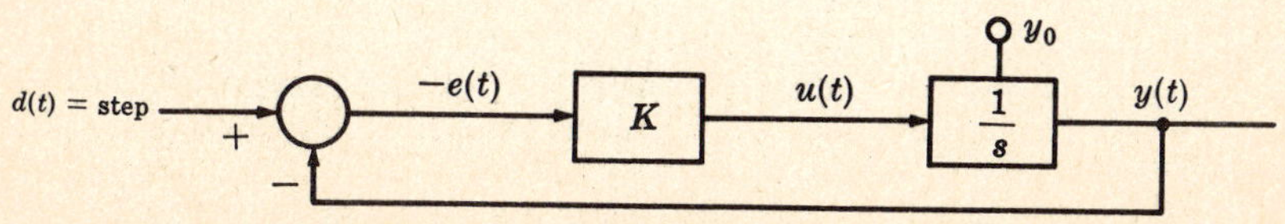

Fig. 10-5

10.8 CONCLUSION

In this chapter we have studied the linear optimal control problem. For time-invariant systems, we note a similarity to the pole placement technique of Section 8.6. We can take our choice as to how to approach the feedback control design problem. Either we select the pole positions or we select the weighting matrices **Q** and **R** in the criterion. The equivalence of the two methods is manifested by Corollary 10.11, although the equivalence is not one-to-one because analysis similar to Problem 10.8 shows that a control $u = \mathbf{k}^T\mathbf{x}$ is optimal if and only if $|\det(j\omega\mathbf{I} - \mathbf{A} - \mathbf{b}\mathbf{k}^T)| \geqq |\det(j\omega\mathbf{I} - \mathbf{A})|$ for all ω. The dual of this equivalence is that Section 8.7 on observers is similar to the Kalman-Bucy filter and the algebraic separation theorem of Section 8.8 is similar to the separation theorem of stochastic optimal control.

Solved Problems

10.1. Prove Theorem 10.1, page 211, and Theorem 10.2, page 213.

The heuristic proof given previously was not rigorous for the following six reasons.

(1) By minimizing under the integral sign at each instant of time, i.e. with t fixed, the resulting minimum is not guaranteed continuous in time. Therefore the resulting optimal functions $\mathbf{x}^{op}(t)$ and $\mathbf{p}(t)$ may not have a derivative and equation *(10.7)* would make no sense.

(2) The open loop time function $\mathbf{u}(t)$ was found, and only later related to the feedback control $\mathbf{u}(\mathbf{x}, t)$.

(3) We wish to take piecewise continuous controls into account.

(4) Taking derivatives of smooth functions gives local maxima, minima, or inflection points. We wish to guarantee a global minimum.

(5) We supposed the minimum of each of the three terms ν_1, ν_2, and ν_3 gave the minimum of ν.

(6) We said in the heuristic proof that if the function were a minimum, then equations *(10.6)*, *(10.7)* and *(10.8)* held, i.e. were necessary conditions for a minimum. We wish to give sufficient conditions for a minimum: we will start with the assumption that a certain quantity $v(\mathbf{x}, t)$ obeys a partial differential equation, and then show that a minimum is attained.

To start the proof, call the trajectory $\mathbf{x}(t) = \boldsymbol{\phi}(t; \mathbf{x}(t_0), t_0)$ corresponding to the optimal system $d\mathbf{x}/dt = \mathbf{A}\mathbf{x} + \mathbf{B}\mathbf{u}^{op}(\mathbf{x}, t)$ starting from t_0 and initial condition $\mathbf{x}(t_0)$. Note $\boldsymbol{\phi}$ does not depend on $\mathbf{u}$ since $\mathbf{u}$ has been chosen as a specified function of $\mathbf{x}$ and t. Then $\nu[\mathbf{x}, \mathbf{u}^{op}(\mathbf{x}, t)]$ can be evaluated if $\boldsymbol{\phi}$ is known, simply by integrating out t and leaving the parameters t_1, t_0 and $\mathbf{x}(t_0)$. Symbolically,

$$\nu[\mathbf{x}, \mathbf{u}^{op}] = \tfrac{1}{2}\boldsymbol{\phi}^T(t_1; \mathbf{x}(t_0), t_0)\mathbf{S}\boldsymbol{\phi}(t_1; \mathbf{x}(t_0), t_0) + \frac{1}{2}\int_{t_0}^{t_1} [\boldsymbol{\phi}^T(\tau; \mathbf{x}(t_0), t_0)\,\mathbf{Q}(\tau)\,\boldsymbol{\phi}(\tau; \mathbf{x}(t_0), t_0) + (\mathbf{u}^{op})^T(\boldsymbol{\phi}, \tau)\mathbf{R}\mathbf{u}^{op}(\boldsymbol{\phi}, \tau)]\,d\tau$$

Since we can start from any initial conditions $\mathbf{x}(t_0)$ and initial time t_0, we can consider $\nu[\mathbf{x}, \mathbf{u}^{op}] = v(\mathbf{x}(t_0), t_0; t_1)$ where v is an explicit function of $n+1$ variables $\mathbf{x}(t_0)$ and t_0, depending on the fixed parameter t_1.

Suppose we can find a solution $v(\mathbf{x}, t)$ to the (Hamilton-Jacobi) partial differential equation

$$\frac{\partial v}{\partial t} + \tfrac{1}{2}\mathbf{x}^T\mathbf{Q}\mathbf{x} - \tfrac{1}{2}(\mathrm{grad}_{\mathbf{x}}\, v)^T\mathbf{B}\mathbf{R}^{-1}\mathbf{B}^T\,\mathrm{grad}_{\mathbf{x}}\, v + (\mathrm{grad}_{\mathbf{x}}\, v)^T\mathbf{A}\mathbf{x} = 0 \qquad (10.23)$$

with the boundary condition $v(\mathbf{x}(t_1), t_1) = \tfrac{1}{2}\mathbf{x}^T(t_1)\,\mathbf{S}\mathbf{x}(t_1)$. Note that for any control $\mathbf{u}(\mathbf{x}, t)$,

$$\tfrac{1}{2}\mathbf{u}^T\mathbf{R}\mathbf{u} + \tfrac{1}{2}(\mathrm{grad}_{\mathbf{x}}\, v)^T\mathbf{B}\mathbf{R}^{-1}\mathbf{B}^T\,\mathrm{grad}_{\mathbf{x}}\, v + (\mathrm{grad}_{\mathbf{x}}\, v)^T\mathbf{B}\mathbf{u}$$
$$= \tfrac{1}{2}(\mathbf{u} + \mathbf{R}^{-1}\mathbf{B}^T\,\mathrm{grad}_{\mathbf{x}}\, v)^T\mathbf{R}(\mathbf{u} + \mathbf{R}^{-1}\mathbf{B}^T\,\mathrm{grad}_{\mathbf{x}}\, v) \geqq 0$$

where the equality is attained only for

$$\mathbf{u}(x, t) = -\mathbf{R}^{-1}\mathbf{B}^T\,\mathrm{grad}_{\mathbf{x}}\, v \qquad (10.24)$$

Rearranging the inequality and adding $\tfrac{1}{2}\mathbf{x}^T\mathbf{Q}\mathbf{x}$ to both sides,

$$\tfrac{1}{2}\mathbf{x}^T\mathbf{Q}\mathbf{x} + \tfrac{1}{2}\mathbf{u}^T\mathbf{R}\mathbf{u} \geqq \tfrac{1}{2}\mathbf{x}^T\mathbf{Q}\mathbf{x} - \tfrac{1}{2}(\mathrm{grad}_{\mathbf{x}}\, v)^T\mathbf{B}\mathbf{R}^{-1}\mathbf{B}^T\,\mathrm{grad}_{\mathbf{x}}\, v - (\mathrm{grad}_{\mathbf{x}}\, v)^T\mathbf{B}\mathbf{u}$$

Using *(10.23)* and $d\mathbf{x}/dt = \mathbf{A}\mathbf{x} + \mathbf{B}\mathbf{u}$, we get

$$\tfrac{1}{2}\mathbf{x}^T\mathbf{Q}\mathbf{x} + \tfrac{1}{2}\mathbf{u}^T\mathbf{R}\mathbf{u} \geqq -\left[\frac{\partial v}{\partial t} + (\mathrm{grad}_{\mathbf{x}}\, v)^T(\mathbf{A}\mathbf{x} + \mathbf{B}\mathbf{u})\right] = -\frac{d}{dt}v(\mathbf{x}(t), t)$$

Integration preserves the inequality, except for a set of measure zero:

$$\frac{1}{2}\int_{t_0}^{t_1} (\mathbf{x}^T\mathbf{Q}\mathbf{x} + \mathbf{u}^T\mathbf{R}\mathbf{u})\,dt \;\geqq\; v(\mathbf{x}(t_0),t_0) - v(\mathbf{x}(t_1),t_1)$$

Given the boundary condition on the Hamilton-Jacobi equation *(10.23)*, the criterion $\nu[\mathbf{x},\mathbf{u}]$ for any control is

$$\nu[\mathbf{x},\mathbf{u}] = \tfrac{1}{2}\mathbf{x}^T(t_1)\,\mathbf{S}\mathbf{x}(t_1) + \frac{1}{2}\int_{t_0}^{t_1} (\mathbf{x}^T\mathbf{Q}\mathbf{x} + \mathbf{u}^T\mathbf{R}\mathbf{u})\,dt \;\geqq\; v(\mathbf{x}(t_0),t_0)$$

so that if $v(\mathbf{x}(t_0),t_0)$ is nonnegative definite, then it is the cost due to the optimal control, *(10.24)*:

$$\mathbf{u}^{\text{op}}(\mathbf{x},t) = -\mathbf{R}^{-1}(t)\,\mathbf{B}^T(t)\,\mathbf{grad}_{\mathbf{x}}\,v(\mathbf{x},t)$$

To solve the Hamilton-Jacobi equation *(10.23)*, set $v(\mathbf{x},t) = \frac{1}{2}\mathbf{x}^T\mathbf{P}(t)\mathbf{x}$ where $\mathbf{P}(t)$ is a time-varying Hermitian matrix to be found. Then $\mathbf{grad}_{\mathbf{x}}\,v = \mathbf{P}(t)\mathbf{x}$ and the Hamilton-Jacobi equation reduces to the matrix Riccati equation *(10.13)*. Therefore the optimal control problem has been reduced to the existence of positive definite solutions for $t < t_1$ of the matrix Riccati equation. Using more advanced techniques, it can be shown that a unique positive definite solution to the matrix Riccati equation always exists if $\|\mathbf{A}(t)\| < \infty$ for $t < t_1$. We say positive definite because

$$\tfrac{1}{2}\mathbf{x}^T(t_0)\,\mathbf{P}(t_0)\,\mathbf{x}(t_0) = v(\mathbf{x}(t_0),t_0) = \nu[\mathbf{x},\mathbf{u}^{\text{op}}] > 0 \qquad \text{for all } t_0 < t_1$$

and all nonzero $\mathbf{x}(t_0)$, which proves Theorem 10.2. Since we know from Section 10.4 that the matrix Riccati equation is equivalent to *(10.9)*, this proves Theorem 10.1.

10.2. Find the optimal feedback control for the scalar time-varying system $dx/dt = -x/2t + u/t$ to minimize $\nu = \frac{1}{2}\int_{t_0}^{t_1} (x^2 + u^2)\,dt$.

Note the open loop system has a transition matrix $\Phi(t,\tau) = (\tau/t)^{1/2}$, which escapes at $t = 0$. The corresponding Riccati equation is

$$-\frac{dP}{dt} = 1 - \frac{P}{t} - \frac{P^2}{t^2}$$

with boundary condition $P(t_1) = 0$. This has a solution

$$P(t) = t\,\frac{t_1^2 - t^2}{t_1^2 + t^2}$$

Then $P(t) > 0$ for $0 < t < t_1$ and for $t < t_1 \leqq 0$. However, for $-t_1 < t < 0 < t_1$, the interval in which the open loop system escapes, $P(t) < 0$. Hence we do not have a nonnegative solution of the Riccati equation, and the control is not the optimal one in this interval. This can only happen when $\|A(t)\|$ is not bounded in the interval considered.

10.3. Given the nonlinear scalar system $dy/dt = y^3 e$. An open loop control $d(t) = -t$ has been found (perhaps by some other optimization scheme) such that the nonlinear system $dy_n/dt = y_n^3 d$ with initial condition $y_n(0) = 1$ will follow the nominal path $y_n(t) = (1+t^2)^{-1/2}$. Unfortunately the initial condition is not exactly one, but $y(0) = 1 + \epsilon$ where ϵ is some very small, unknown number. Find a feedback control that is stable and will minimize the error $y(t_1) - y_n(t_1)$ at some time t_1.

Call the error $y(t) - y_n(t) = x(t)$ at any time. Consider the feedback system of Fig. 10-6 below. The equations corresponding to this system are

$$\frac{dy}{dt} = \frac{dy_n}{dt} + \frac{dx}{dt} = (y_n + x)^3(d + u)$$

$$= y_n^3 d + 3y_n^2 xd + 3y_n x^2 d + x^3 d + y_n^3 u + 3y_n^2 xu + 3y_n x^2 u + x^3 u$$

Assume the error $|x| \ll |y|$ and corrections $|u| \ll |d|$ so that

$$|3y_n^2 xd + y_n^3 u| \gg |3y_n x^2 d + x^3 d + 3y_n^2 xu + 3y_n x^2 u + x^3 u| \qquad (10.25)$$

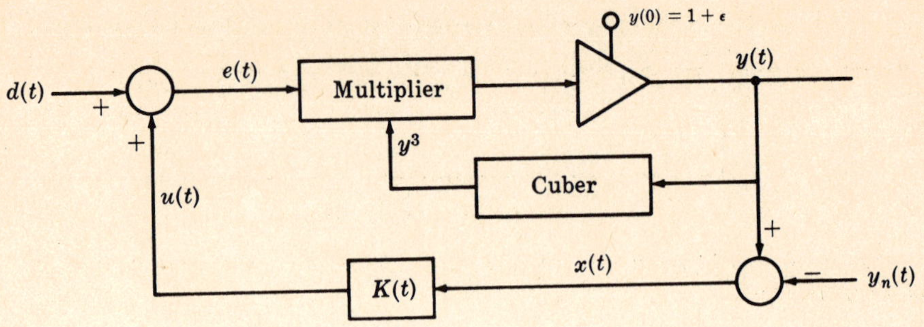

Fig. 10-6

Since $dy_n/dt = y_n^3 d$, we have the approximate relation

$$\frac{dx}{dt} \approx 3y_n^2 xd + y_n^3 u = -\frac{3t}{1+t^2}x + \frac{1}{(1+t^2)^{3/2}}u$$

We choose to minimize

$$\nu = \tfrac{1}{2}x^2(t_1) + \frac{1}{2}\int_0^{t_1} R(t)\, u^2(t)\, dt$$

where $R(t)$ is some appropriately chosen weighting function such that neither $x(t)$ nor $u(t)$ at any time t gets so large that the inequality (*10.25*) is violated. Then $K(t) = -P(t)/[(1+t^2)^{3/2}R(t)]$ where $P(t)$ is the solution to the Riccati equation

$$-\frac{dP}{dt} = -\frac{6tP}{1+t^2} - \frac{P^2}{(1+t^2)^3 R(t)}$$

with $P(t_1) = 1$. This Riccati equation is solved numerically on a computer from t_1 to 0. Then the time-varying gain function $K(t)$ can be found and used in the proposed feedback system.

10.4. Given the open loop system with a transfer function $(s+\alpha)^{-1}$. Find a compensation to minimize

$$\nu = \frac{1}{2}\int_0^\infty [(y-d_0)^2 + (du/dt)^2]\, dt$$

where d_0 is the value of an arbitrary step input.

Since the closed loop system must follow a step, and has no pure integrations, we introduce integral compensation. The open loop system in series with a pure integration is described by $dx_2/dt = -\alpha x_2 + u$, and defining $x_1 = u$ and $du/dt = \mu$ gives $dx_1/dt = du/dt = \mu$ so that

$$\frac{d\mathbf{x}}{dt} = \begin{pmatrix} 0 & 0 \\ 1 & -\alpha \end{pmatrix}\mathbf{x} + \begin{pmatrix} 1 \\ 0 \end{pmatrix}\mu \qquad y = (0\ \ 1)\mathbf{x}$$

Since an arbitrary step can be generated by this, the equations for the error become

$$\frac{d\mathbf{w}}{dt} = \begin{pmatrix} 0 & 0 \\ 1 & -\alpha \end{pmatrix}\mathbf{w} + \begin{pmatrix} 1 \\ 0 \end{pmatrix}\mu \qquad e = (0\ \ 1)\mathbf{w}$$

subject to minimization of $\nu = \frac{1}{2}\int_0^\infty (e^2 + \mu^2)\, dt$. The corresponding matrix Riccati equation is

$$\mathbf{0} = \begin{pmatrix} 0 & 0 \\ 0 & 1 \end{pmatrix} + \begin{pmatrix} 0 & 1 \\ 0 & -\alpha \end{pmatrix}\mathbf{\Pi} + \mathbf{\Pi}\begin{pmatrix} 0 & 0 \\ 1 & -\alpha \end{pmatrix} - \mathbf{\Pi}\begin{pmatrix} 1 & 0 \\ 0 & 0 \end{pmatrix}\mathbf{\Pi}$$

Using the method of Theorem 10.9 gives

$$\mathbf{\Pi} = \Delta\begin{pmatrix} \gamma+\alpha & 1 \\ 1 & \gamma \end{pmatrix}$$

where $\sqrt{2}\,\gamma = \sqrt{\alpha^2 + \sqrt{\alpha^4 - 4}} + \sqrt{\alpha^2 - \sqrt{\alpha^4 - 4}}$ and $\Delta^{-1} = \alpha^2 + \gamma\alpha + 1$. The optimal control is then $\mu = \Delta(w_1 + \gamma w_2)$. Since $e = w_2$ and $w_1 = \alpha e + de/dt$, the closed loop system is as shown in Fig. 10-7 below.

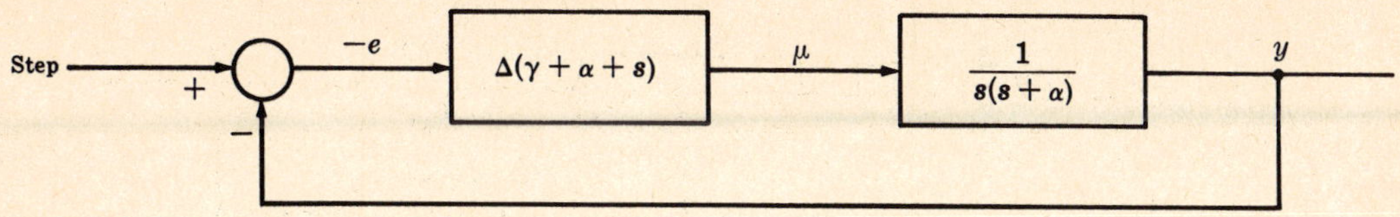

Fig. 10-7

Figure 10-8 is equivalent to that of Fig. 10-7 so that the compensation introduced is integral plus proportional.

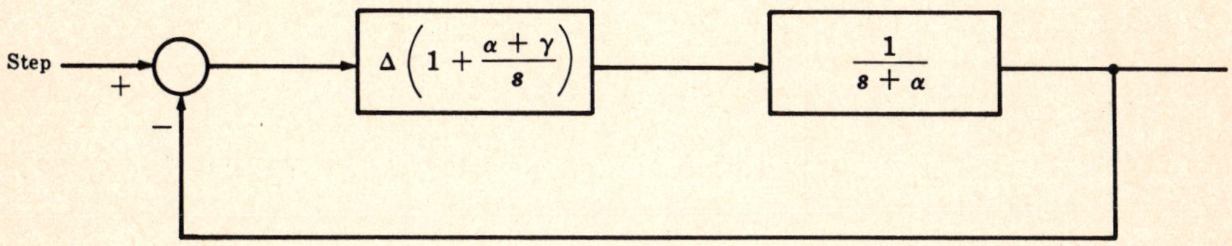

Fig. 10-8

10.5. Given the single-input time-invariant system in partitioned Jordan canonical form

$$\frac{d}{dt}\left(\frac{\mathbf{z}_c}{\mathbf{z}_d}\right) = \left(\begin{array}{c|c} \mathbf{J}_c & \mathbf{0} \\ \hline \mathbf{0} & \mathbf{J}_d \end{array}\right)\left(\frac{\mathbf{z}_c}{\mathbf{z}_d}\right) + \left(\frac{\mathbf{b}}{\mathbf{0}}\right)u$$

where $\mathbf{z}_c$ are the j controllable states, $\mathbf{z}_d$ are the $n-j$ uncontrollable states, $\mathbf{J}_c$ and $\mathbf{J}_d$ are $j \times j$ and $(n-j) \times (n-j)$ matrices corresponding to real Jordan form, Section 7.3, $\mathbf{b}$ is a j-vector that contains no zero elements, u is a scalar control, and the straight lines indicate the partitioning of the vectors and matrices. Suppose it is desired to minimize the quadratic criterion ν where $\nu = \frac{1}{2}\int_{t_0}^{\infty} [\mathbf{z}_c^T\mathbf{Q}_c\mathbf{z}_c + \rho^{-1}u^2]\,d\tau$. Here $\rho > 0$ is a scalar, and $\mathbf{Q}_c$ is a positive definite symmetric real matrix. Show that the optimal feedback control for this regulator problem is of the form $u(t) = -\mathbf{k}^T\mathbf{z}_c(t)$ where $\mathbf{k}$ is a constant j-vector, and no uncontrollable states $\mathbf{z}_d$ are fed back. From the results of this, indicate briefly the conditions under which a general time-invariant single-input system that is in Jordan form will have only controllable states fed back; i.e. the conditions under which the optimal closed loop system can be separated into controllable and uncontrollable parts.

The matrix $\mathbf{\Pi}$ satisfies equation *(10.16)*. For the case given, $\mathbf{A} = \left(\begin{array}{c|c} \mathbf{J}_c & \mathbf{0} \\ \hline \mathbf{0} & \mathbf{J}_d \end{array}\right)$, $\mathbf{B} = \left(\frac{\mathbf{b}}{\mathbf{0}}\right)$, $\mathbf{Q} = \left(\begin{array}{c|c} \mathbf{Q}_c & \mathbf{0} \\ \hline \mathbf{0} & \mathbf{0} \end{array}\right)$, $\mathbf{R}^{-1} = \rho$ and partition $\mathbf{\Pi} = \left(\begin{array}{c|c} \mathbf{\Pi}_c & \mathbf{\Pi}_{cd} \\ \hline \mathbf{\Pi}_{cd}^T & \mathbf{\Pi}_d \end{array}\right)$. Then

$$u = -\rho(\mathbf{b}^T \mid \mathbf{0}^T)\left(\begin{array}{c|c} \mathbf{\Pi}_c & \mathbf{\Pi}_{cd} \\ \hline \mathbf{\Pi}_{cd}^T & \mathbf{\Pi}_d \end{array}\right)\left(\frac{\mathbf{z}_c}{\mathbf{z}_d}\right) = -\rho\mathbf{b}^T\mathbf{\Pi}_c\mathbf{z}_c - \rho\mathbf{b}^T\mathbf{\Pi}_{cd}\mathbf{z}_d$$

Now if $\mathbf{\Pi}_c$ can be shown to be constant and $\mathbf{\Pi}_{cd}$ can be shown $\mathbf{0}$, then $\mathbf{k}^T = \rho\mathbf{b}^T\mathbf{\Pi}_c$. But from *(10.16)*,

$$\left(\begin{array}{c|c} \mathbf{0} & \mathbf{0} \\ \hline \mathbf{0} & \mathbf{0} \end{array}\right) = \left(\begin{array}{c|c} \mathbf{J}_c & \mathbf{0} \\ \hline \mathbf{0} & \mathbf{J}_d \end{array}\right)\left(\begin{array}{c|c} \mathbf{\Pi}_c & \mathbf{\Pi}_{cd} \\ \hline \mathbf{\Pi}_{cd}^T & \mathbf{\Pi}_d \end{array}\right) + \left(\begin{array}{c|c} \mathbf{\Pi}_c & \mathbf{\Pi}_{cd} \\ \hline \mathbf{\Pi}_{cd}^T & \mathbf{\Pi}_d \end{array}\right)\left(\begin{array}{c|c} \mathbf{J}_c & \mathbf{0} \\ \hline \mathbf{0} & \mathbf{J}_d \end{array}\right)$$

$$-\left(\begin{array}{c|c} \mathbf{\Pi}_c & \mathbf{\Pi}_{cd} \\ \hline \mathbf{\Pi}_{cd}^T & \mathbf{\Pi}_d \end{array}\right)\left(\frac{\mathbf{b}}{\mathbf{0}}\right)\rho(\mathbf{b}^T \mid \mathbf{0}^T)\left(\begin{array}{c|c} \mathbf{\Pi}_c & \mathbf{\Pi}_{cd} \\ \hline \mathbf{\Pi}_{cd}^T & \mathbf{\Pi}_d \end{array}\right) + \left(\begin{array}{c|c} \mathbf{Q}_c & \mathbf{0} \\ \hline \mathbf{0} & \mathbf{0} \end{array}\right)$$

or

$$\left(\begin{array}{c|c} \mathbf{0} & \mathbf{0} \\ \hline \mathbf{0} & \mathbf{0} \end{array}\right) = \left(\begin{array}{c|c} \mathbf{J}_c\mathbf{\Pi}_c & \mathbf{J}_c\mathbf{\Pi}_{cd} \\ \hline \mathbf{J}_d\mathbf{\Pi}_{cd}^T & \mathbf{J}_d\mathbf{\Pi}_d \end{array}\right) + \left(\begin{array}{c|c} \mathbf{\Pi}_c\mathbf{J}_c & \mathbf{\Pi}_{cd}\mathbf{J}_d \\ \hline \mathbf{\Pi}_{cd}^T\mathbf{J}_c & \mathbf{\Pi}_d\mathbf{J}_d \end{array}\right)$$
$$-\rho\left(\begin{array}{c|c} \mathbf{\Pi}_c\mathbf{b}\mathbf{b}^T\mathbf{\Pi}_c & \mathbf{\Pi}_c\mathbf{b}\mathbf{b}^T\mathbf{\Pi}_{cd} \\ \hline \mathbf{\Pi}_{cd}^T\mathbf{b}\mathbf{b}^T\mathbf{\Pi}_c & \mathbf{\Pi}_{cd}^T\mathbf{b}\mathbf{b}^T\mathbf{\Pi}_{cd} \end{array}\right) + \left(\begin{array}{c|c} \mathbf{Q}_c & \mathbf{0} \\ \hline \mathbf{0} & \mathbf{0} \end{array}\right)$$

Within the upper left-hand partition,

$$\mathbf{0} = \mathbf{J}_c\mathbf{\Pi}_c + \mathbf{\Pi}_c\mathbf{J}_c - \rho\mathbf{\Pi}_c\mathbf{b}\mathbf{b}^T\mathbf{\Pi}_c + \mathbf{Q}_c$$

the matrix Riccati equation for the controllable system, which has a constant positive definite solution $\mathbf{\Pi}_c$. Within the upper right-hand corner,

$$\mathbf{0} = \mathbf{J}_c\mathbf{\Pi}_{cd} + \mathbf{\Pi}_{cd}\mathbf{J}_d - \rho\mathbf{\Pi}_c\mathbf{b}\mathbf{b}^T\mathbf{\Pi}_{cd} + \mathbf{0}$$

This is a *linear* equation in $\mathbf{\Pi}_{cd}$ and has a solution $\mathbf{\Pi}_{cd} = \mathbf{0}$ by inspection, and thus $u = -\rho\mathbf{b}^T\mathbf{\Pi}_c\mathbf{z}_c$.

Now a general system $\dot{\mathbf{x}} = \mathbf{A}\mathbf{x} + \mathbf{B}\mathbf{u}$ can be transformed to real Jordan form by $\mathbf{x} = \mathbf{T}\mathbf{z}$, where $\mathbf{T}^{-1}\mathbf{A}\mathbf{T} = \mathbf{J}$. Then the criterion

$$\nu = \frac{1}{2}\int_{t_0}^{\infty} (\mathbf{x}^T\mathbf{Q}\mathbf{x} + \rho^{-1}u^2)\,d\tau = \frac{1}{2}\int_{t_0}^{\infty} (\mathbf{z}^T\mathbf{T}^T\mathbf{Q}\mathbf{T}\mathbf{z} + \rho^{-1}u^2)\,d\tau$$

gives the matrix Riccati equation shown for $\mathbf{z}_c$ and $\mathbf{z}_d$ if $\mathbf{T}^T\mathbf{Q}\mathbf{T} = \left(\begin{array}{c|c} \mathbf{Q}_c & \mathbf{0} \\ \hline \mathbf{0} & \mathbf{0} \end{array}\right)$, i.e. only the controllable states are weighted in the criterion. Otherwise uncontrollable states must be fed back. If the uncontrollable states are included, ν diverges if any are unstable, but if they are all stable the action of the controllable states is influenced by them and ν remains finite.

10.6. Given the controllable and observable system

$$d\mathbf{x}/dt = \mathbf{A}(t)\mathbf{x} + \mathbf{B}(t)\mathbf{u}, \quad \mathbf{y} = \mathbf{C}(t)\mathbf{x} \tag{10.26}$$

It is desired that the closed loop response of this system approach that of an ideal or model system $d\mathbf{w}/dt = \mathbf{L}(t)\mathbf{w}$. In other words, we have a hypothetical system $d\mathbf{w}/dt = \mathbf{L}(t)\mathbf{w}$ and wish to adjust $\mathbf{u}$ such that the real system $d\mathbf{x}/dt = \mathbf{A}(t)\mathbf{x} + \mathbf{B}(t)\mathbf{u}$ behaves in a manner similar to the hypothetical system. Find a control $\mathbf{u}$ such that the error between the model and plant output vector derivatives becomes small, i.e. minimize

$$\nu = \frac{1}{2}\int_{t_0}^{t_1}\left[\left(\frac{d\mathbf{y}}{dt} - \mathbf{L}\mathbf{y}\right)^T\mathbf{Q}\left(\frac{d\mathbf{y}}{dt} - \mathbf{L}\mathbf{y}\right) + \mathbf{u}^T\mathbf{R}\mathbf{u}\right]dt \tag{10.27}$$

Note that in the case $\mathbf{A}$, $\mathbf{L}$ = constants and $\mathbf{B} = \mathbf{C} = \mathbf{I}$, $\mathbf{R} = \mathbf{0}$, this is equivalent to asking that the closed loop system $d\mathbf{x}/dt = (\mathbf{A} + \mathbf{B}\mathbf{K})\mathbf{x}$ be identical to $d\mathbf{x}/dt = \mathbf{L}\mathbf{x}$ and hence have the same poles. Therefore in this case it reduces to a pole placement scheme (see Section 8.6).

Substituting the plant equation (*10.26*) into the performance index (*10.27*),

$$\nu = \frac{1}{2}\int_{t_0}^{t_1} \{[(\dot{\mathbf{C}} + \mathbf{C}\mathbf{A} - \mathbf{L}\mathbf{C})\mathbf{x} + \mathbf{C}\mathbf{B}\mathbf{u}]^T\mathbf{Q}[(\dot{\mathbf{C}} + \mathbf{C}\mathbf{A} - \mathbf{L}\mathbf{C})\mathbf{x} + \mathbf{C}\mathbf{B}\mathbf{u}] + \mathbf{u}^T\mathbf{R}\mathbf{u}\}\,dt \tag{10.28}$$

This performance index is not of the same form as criterion (*10.2*) because cross products of $\mathbf{x}$ and $\mathbf{u}$ appear. However, let $\hat{\mathbf{u}} = \mathbf{u} + (\mathbf{R} + \mathbf{B}^T\mathbf{Q}\mathbf{B})^{-1}\mathbf{B}^T\mathbf{C}^T\mathbf{Q}(d\mathbf{C}/dt + \mathbf{C}\mathbf{A} - \mathbf{L}\mathbf{C})\mathbf{x}$. Since $\mathbf{R}$ is positive definite, $\mathbf{z}^T\mathbf{R}\mathbf{z} > 0$ for any nonzero $\mathbf{z}$. Since $\mathbf{B}^T\mathbf{Q}\mathbf{B}$ is nonnegative definite, $0 < \mathbf{z}^T\mathbf{R}\mathbf{z} + \mathbf{z}^T\mathbf{B}^T\mathbf{Q}\mathbf{B}\mathbf{z} = \mathbf{z}^T(\mathbf{R} + \mathbf{B}^T\mathbf{Q}\mathbf{B})\mathbf{z}$ so that $\hat{\mathbf{R}} = \mathbf{R} + \mathbf{B}^T\mathbf{Q}\mathbf{B}$ is positive definite and hence its inverse always exists. Therefore the control $\hat{\mathbf{u}}$ can always be found in terms of $\mathbf{u}$ and $\mathbf{x}$. Then the system (*10.26*) becomes

$$d\mathbf{x}/dt = \hat{\mathbf{A}}\mathbf{x} + \mathbf{B}\hat{\mathbf{u}} \tag{10.29}$$

and the performance index (*10.28*) becomes

$$\nu = \frac{1}{2}\int_{t_0}^{t_1} (\mathbf{x}^T\hat{\mathbf{Q}}\mathbf{x} + \hat{\mathbf{u}}^T\hat{\mathbf{R}}\hat{\mathbf{u}})\,dt \tag{10.30}$$

in which, defining $\mathbf{M}(t) = d\mathbf{C}/dt + \mathbf{C}\mathbf{A} - \mathbf{L}\mathbf{C}$ and $\mathbf{K}(t) = (\mathbf{B}^T\mathbf{Q}\mathbf{B} + \mathbf{R})^{-1}\mathbf{B}^T\mathbf{C}^T\mathbf{Q}\mathbf{M}$,

$$\hat{\mathbf{A}} = \mathbf{A} - \mathbf{B}\mathbf{K}, \quad \hat{\mathbf{Q}} = \mathbf{C}^T[(\mathbf{M} - \mathbf{B}\mathbf{K})^T\mathbf{Q}(\mathbf{M} - \mathbf{B}\mathbf{K}) + \mathbf{K}^T\mathbf{R}\mathbf{K}]\mathbf{C} \quad \text{and} \quad \hat{\mathbf{R}} = \mathbf{R} + \mathbf{B}^T\mathbf{Q}\mathbf{B}$$

Since $\hat{\mathbf{R}}$ has been shown positive definite and $\hat{\mathbf{Q}}$ is nonnegative definite by a similar argument, the regulator problem (*10.29*) and (*10.30*) is in standard form. Then $\hat{\mathbf{u}} = -\hat{\mathbf{R}}^{-1}\mathbf{B}^T\mathbf{P}\mathbf{x}$ is the optimal solution to the regulator problem (*10.29*) and (*10.30*), where $\mathbf{P}$ is the positive definite solution to

$$-d\mathbf{P}/dt \;=\; \hat{\mathbf{Q}} + \hat{\mathbf{A}}^T\mathbf{P} + \mathbf{P}\hat{\mathbf{A}} - \mathbf{P}\mathbf{B}\hat{\mathbf{R}}^{-1}\mathbf{B}^T\mathbf{P}$$

with boundary condition $\mathbf{P}(t_1) = \mathbf{0}$. The control to minimize (*10.27*) is then

$$\mathbf{u} \;=\; -\hat{\mathbf{R}}^{-1}\mathbf{B}^T(\mathbf{P} + \mathbf{B}^T\mathbf{C}^T\mathbf{Q}\mathbf{M})\mathbf{x}$$

Note that cases in which $\hat{\mathbf{Q}}$ is not positive definite or the system (*10.29*) is not controllable may give no solution $\mathbf{P}$ to the matrix Riccati equation. Even though the conditions under which this procedure works have not been clearly defined, the engineering approach would be to try it on a particular problem and see if a satisfactory answer could be obtained.

10.7. In Problem 10.6, feedback was placed around the plant $d\mathbf{x}/dt = \mathbf{A}(t)\mathbf{x} + \mathbf{B}(t)\mathbf{u}$, $\mathbf{y} = \mathbf{C}(t)\mathbf{x}$ so that it would behave similar to a model that might be hypothetical. In this problem we consider actually building a model $d\mathbf{w}/dt = \mathbf{L}(t)\mathbf{w}$ with output $\mathbf{v} = \mathbf{J}(t)\mathbf{w}$ and using it as a prefilter ahead of the plant. Find a control $\mathbf{u}$ to the plant such that the error between the plant and model output becomes small, i.e. minimize

$$\nu \;=\; \frac{1}{2}\int_{t_0}^{t_1} [(\mathbf{y}-\mathbf{v})^T\mathbf{Q}(\mathbf{y}-\mathbf{v}) + \mathbf{u}^T\mathbf{R}\mathbf{u}]\,dt \qquad (10.31)$$

Again, we reduce this to an equivalent regulator problem. The plant and model equations can be written as

$$\frac{d}{dt}\begin{pmatrix}\mathbf{w}\\ \mathbf{x}\end{pmatrix} \;=\; \begin{pmatrix}\mathbf{L} & \mathbf{0}\\ \mathbf{0} & \mathbf{A}\end{pmatrix}\begin{pmatrix}\mathbf{w}\\ \mathbf{x}\end{pmatrix} + \begin{pmatrix}\mathbf{0}\\ \mathbf{B}\end{pmatrix}\mathbf{u} \qquad (10.32)$$

and the criterion is, in terms of the $(\mathbf{w}\ \mathbf{x})$ vector,

$$\nu \;=\; \frac{1}{2}\int_{t_0}^{t_1}\left[\begin{pmatrix}\mathbf{w}\\ \mathbf{x}\end{pmatrix}^T(-\mathbf{J}\ \ \mathbf{C})^T\mathbf{Q}(-\mathbf{J}\ \ \mathbf{C})\begin{pmatrix}\mathbf{w}\\ \mathbf{x}\end{pmatrix} + \mathbf{u}^T\mathbf{R}\mathbf{u}\right]dt$$

Thus the solution proceeds as usual and $\mathbf{u}$ is a linear combination of $\mathbf{w}$ and $\mathbf{x}$.

However, note that the system (*10.32*) is uncontrollable with respect to the respect to the $\mathbf{w}$ variables. In fact, merely by setting $\mathbf{L} = \mathbf{0}$ and using $\mathbf{J}(t)$ to generate the input, a form of general servomechanism problem results. The conditions under which the general solution to the servomechanism problem exists are not known. Hence we should expect the solution to this problem to exist only under a more restrictive set of circumstances than Problem 10.6. But again, if a positive definite solution of the corresponding matrix Riccati equation can be found, this procedure can be useful in the design of a particular engineering system.

The resulting system can be diagrammed as shown in Fig. 10-9.

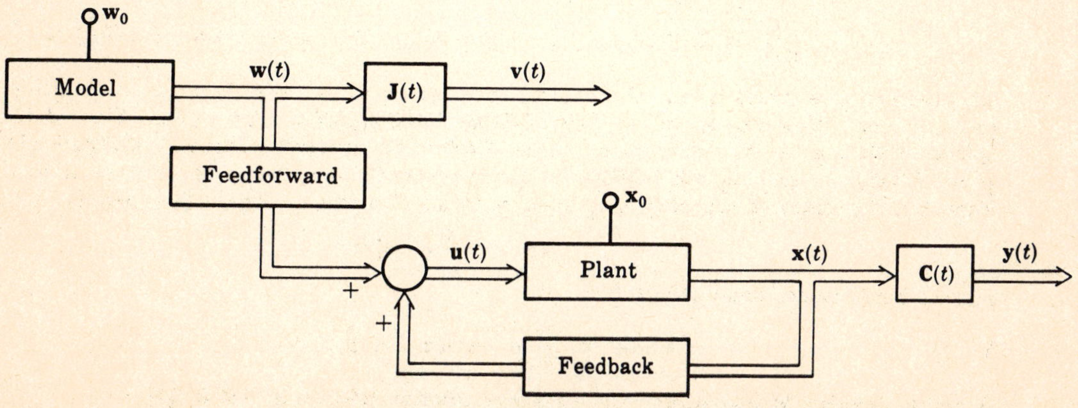

Fig. 10-9

10.8. Given the time-invariant controllable and observable single input–single output system $d\mathbf{x}/dt = \mathbf{A}\mathbf{x} + \mathbf{b}u$, $y = \mathbf{c}^T\mathbf{x}$. Assume that $u = \mathbf{k}^T\mathbf{x}$, where $\mathbf{k}$ has been chosen such that $2\nu = \int_0^\infty (qy^2 + u^2)\,dt$ has been minimized. Find the asymptotic behavior of the closed loop system eigenvalues as $q \to 0$ and $q \to \infty$.

Since the system is optimal, $\mathbf{k} = -\mathbf{b}^T\mathbf{\Pi}$ where $\mathbf{\Pi}$ satisfies (*10.16*), which is written here as

$$\mathbf{0} \;=\; q\mathbf{c}\mathbf{c}^T + \mathbf{A}^T\mathbf{\Pi} + \mathbf{\Pi}\mathbf{A} - \mathbf{\Pi}\mathbf{b}\mathbf{b}^T\mathbf{\Pi} \qquad (10.33)$$

If $q = 0$, then $\mathbf{\Pi} = \mathbf{0}$ is the unique nonnegative definite solution of the matrix Riccati equation. Then $\mathbf{u} = \mathbf{0}$, which corresponds to the intuitive feeling that if the criterion is independent of response, make this control zero. Therefore as $q \to 0$, the closed loop eigenvalues tend to the open loop eigenvalues.

To examine the case $q \to \infty$, add and subtract $s\mathbf{\Pi}$ from (*10.33*), multiply on the right by $(s\mathbf{I} - \mathbf{A})^{-1}$ and on the left by $(-s\mathbf{I} - \mathbf{A}^T)^{-1}$ to obtain

$$\mathbf{0} \;=\; q(-s\mathbf{I} - \mathbf{A}^T)^{-1}\mathbf{c}\mathbf{c}^T(s\mathbf{I} - \mathbf{A})^{-1} - \mathbf{\Pi}(s\mathbf{I} - \mathbf{A})^{-1} - (-s\mathbf{I} - \mathbf{A}^T)^{-1}\mathbf{\Pi} - (-s\mathbf{I} - \mathbf{A}^T)^{-1}\mathbf{\Pi}\mathbf{b}\mathbf{b}^T\mathbf{\Pi}(s\mathbf{I} - \mathbf{A})^{-1} \qquad (10.34)$$

Multiply on the left by $\mathbf{b}^T$ and the right by $\mathbf{b}$, and call the scalar $G(s) = \mathbf{c}^T(s\mathbf{I} - \mathbf{A})^{-1}\mathbf{b}$ and the scalar $H(s) = +\mathbf{b}^T\mathbf{\Pi}(s\mathbf{I} - \mathbf{A})^{-1}\mathbf{b}$. The reason for this notation is that, given an input with Laplace transform $\rho(s)$ to the open loop system, $\mathbf{x}(t) = \mathcal{L}^{-1}\{(s\mathbf{I} - \mathbf{A})^{-1}\mathbf{b}\rho(s)\}$ so that

$$\mathcal{L}\{y(t)\} \;=\; \mathbf{c}^T\mathcal{L}\{\mathbf{x}(t)\} \;=\; \mathbf{c}^T(s\mathbf{I} - \mathbf{A})^{-1}\mathbf{b}\rho(s) \;=\; G(s)\,\rho(s)$$

and

$$-\mathcal{L}\{u(t)\} \;=\; \mathbf{b}^T\mathbf{\Pi}\mathcal{L}\{\mathbf{x}(t)\} \;=\; \mathbf{b}^T\mathbf{\Pi}(s\mathbf{I} - \mathbf{A})^{-1}\mathbf{b}\rho(s) \;=\; H(s)\,\rho(s)$$

In other words, $G(s)$ is the open loop transfer function and $H(s)$ is the transfer function from the input to the control. Then (*10.34*) becomes

$$0 \;=\; qG(-s)\,G(s) - H(s) - H(-s) - H(s)\,H(-s)$$

Adding 1 to each side and rearranging gives

$$|1 + H(s)|^2 \;=\; 1 + q|G(s)|^2 \qquad (10.35)$$

It has been shown that only optimal systems obey this relationship. Denote the numerator of $H(s)$ as $n(H)$ and of $G(s)$ as $n(G)$, and the denominator of $H(s)$ as $d(H)$ and of $G(s)$ as $d(G)$. But $d(G) = \det(s\mathbf{I} - \mathbf{A}) = d(H)$. Multiplying (*10.35*) by $|d(G)|^2$ gives

$$|d(G) + n(H)|^2 \;=\; |d(G)|^2 + q|n(G)|^2 \qquad (10.36)$$

As $q \to \infty$, if there are m zeros of $n(G)$, the open loop system, then $2m$ zeros of (*10.36*) tend to the zeros of $|n(G)|^2$. The remaining $2(n-m)$ zeros tend to ∞ and are asymptotic to the zeros of the equation $s^{2(n-m)} = q$. Since the closed loop eigenvalues are the left half plane zeros of $|d(G) + n(H)|^2$, we conclude that as $q \to \infty$ m closed loop poles tend to the m open loop zeros and the remaining $n - m$ closed loop poles tend to the left half plane zeros of the equation $s^{2(n-m)} = q$.

In other words, they tend to a stable Butterworth configuration of order $n - m$ and radius q^γ where $2(n - m) = \gamma^{-1}$. If the system has 3 open loop poles and one open loop zero, then 2 open loop poles are asymptotic as shown in Fig. 10-10.

This is independent of the open loop pole-zero configuration. Also, note equation (*10.36*) requires the closed loop poles to tend to the open loop poles as $q \to 0$. Furthermore, the criterion $\int_0^\infty (qy^2 + u^2)\,dt$ is quite general for scalar controls since $\mathbf{c}$ can be chosen by the designer (also see Problem 10.19).

Of course we should remember that the results of this analysis are valid only for this particular criterion involving the output and are not valid for a general quadratic in the state.

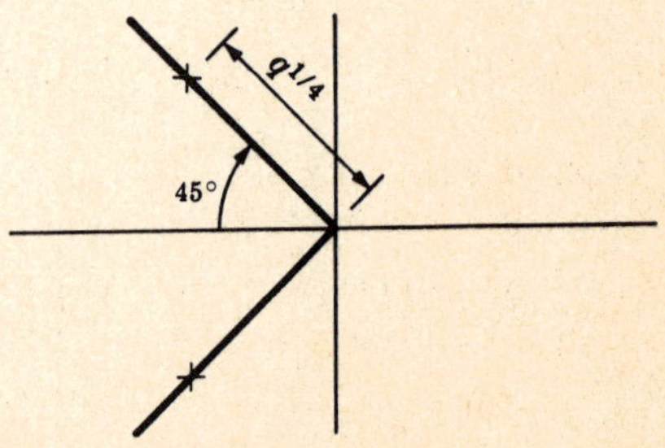

Fig. 10-10

Supplementary Problems

10.9. Given the scalar system $dx/dt = 4x + u$ with initial condition $x(0) = x_0$. Find a feedback control u to minimize $2\nu = \int_0^\infty (9x^2 + u^2)\,dt$.

10.10. Consider the plant

$$\frac{d\mathbf{x}}{dt} = \begin{pmatrix} 0 & 1 \\ 0 & -\sqrt{5} \end{pmatrix}\mathbf{x} + \begin{pmatrix} 0 \\ 1 \end{pmatrix}u$$

We desire to find u to minimize $2\nu = \int_0^\infty (4x_1^2 + u^2)\,dt$.

(*a*) Write the canonical equations (*10.9*) and their boundary conditions.

(*b*) Solve the canonical equations for $\mathbf{x}(t)$ and $\mathbf{p}(t)$.

(*c*) Find the open loop control $u(\mathbf{x}_0, t)$.

10.11. For the system and criterion of Problem 10.10,

(*a*) write three coupled nonlinear scalar equations for the elements ϕ_{ij} of the $\mathbf{P}(t)$ matrix,

(*b*) find $\mathbf{P}(t)$ using the relationship $\mathbf{p}(t) = \mathbf{P}(t)\,\mathbf{x}(t)$ and the results of part (*b*) of Problem 10.10,

(*c*) verify the results of (*b*) satisfy the equations found in (*a*), and that $\mathbf{P}(t)$ is constant and positive definite,

(*d*) draw a flow diagram of the closed loop system.

10.12. Consider the plant

$$\frac{d}{dt}\begin{pmatrix} x_1 \\ x_2 \end{pmatrix} = \begin{pmatrix} 0 & 1 \\ 0 & 0 \end{pmatrix}\begin{pmatrix} x_1 \\ x_2 \end{pmatrix} + \begin{pmatrix} 0 \\ 1 \end{pmatrix}u$$

and performance index

$$\nu = \frac{1}{2}\int_0^T (qx_1^2 + u^2)\,dt$$

Assuming the initial conditions are $x_1(0) = 5$, $x_2(0) = -3$, find the equations (*10.9*) whose solutions will give the control (*10.8*). What are the boundary conditions for these equations (*10.9*)?

10.13. Consider the scalar plant

$$dx/dt = x + u$$

and the performance index

$$\nu = \tfrac{1}{2}Sx^2(T) + \frac{1}{2}\int_0^T (x^2 + u^2)\,dt$$

(*a*) Using the state-costate transition matrix $\Phi(t,\tau)$, find $P(t)$ such that $p(t) = P(t)\,x(t)$.

(*b*) What is $\lim_{t\to-\infty} P(t)$?

10.14. Given the system of Fig. 10-11, find $K(t)$ to minimize

$2\nu = \int_{t_0}^{t_1} (x^2 + u^2/\rho)\,dt$ and find $\lim_{t_1\to\infty} K(t)$.

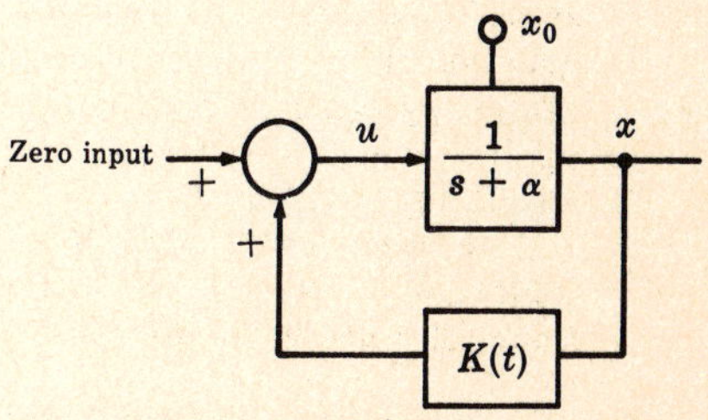

Fig. 10-11

10.15. Given the system

$$\frac{d\mathbf{x}}{dt} = \begin{pmatrix} -2 & 1 \\ 0 & 1 \end{pmatrix}\mathbf{x} + \begin{pmatrix} 1 \\ 0 \end{pmatrix}u$$

Find the control that minimizes

$$2\nu = \int_0^\infty (\mathbf{x}^T\mathbf{x} + u^2)\,dt$$

10.16. Consider the motion of a missile in a straight line. Let $\dot{r} = v$ and $\dot{v} = u$, where r is the position and v is the velocity. Also, u is the acceleration due to the control force. Hence the state equation is $\frac{d}{dt}\begin{pmatrix} r \\ v \end{pmatrix} = \begin{pmatrix} 0 & 1 \\ 0 & 0 \end{pmatrix}\begin{pmatrix} r \\ v \end{pmatrix} + \begin{pmatrix} 0 \\ 1 \end{pmatrix}u$. It is desired to minimize

$$2\nu = q_v v^2(T) + q_r r^2(T) + \int_0^T u^2(\tau)\,d\tau$$

where q_v and q_r are scalars > 0.

Find the feedback control matrix $\mathbf{M}(\theta)$ relating the control $u(t)$ to the states $r(t)$ and $v(t)$ in the form

$$u(t) = \mathbf{M}(\theta)\begin{pmatrix} r(t) \\ v(t) \end{pmatrix}$$

where $\theta = T - t$. Here θ is known as the "time to go".

10.17. Given the system $\dfrac{d\mathbf{x}}{dt} = \begin{pmatrix} 0 & 1 \\ 0 & 0 \end{pmatrix}\mathbf{x} + \begin{pmatrix} 0 \\ 1 \end{pmatrix} u$ with criterion $2\nu = \displaystyle\int_0^\infty (qx_1^2 + u^2)\,dt$. Draw the root locus of the closed loop roots as q varies from zero to ∞.

10.18. Consider the matrix

$$\mathbf{H} = \begin{bmatrix} a & -b^2/r \\ -q & -a \end{bmatrix}$$

associated with the optimization problem

$$dx/dt = ax + bu$$

$$\nu = \frac{1}{2}\int_0^T (qx^2 + ru^2)\,dt$$

$$x(0) = x_0; \qquad p(T) = 0$$

Show that the roots of $\mathbf{H}$ are symmetrically placed with respect to the $j\omega$ axis and show that the location of the roots of $\mathbf{H}$ depends only upon the ratio q/r for fixed a and b.

10.19. Given the system *(10.1)* and let $\mathbf{S} = \mathbf{0}$, $\mathbf{Q}(t) = \eta\mathbf{Q}_0(t)$ and $\mathbf{R}(t) = \rho\mathbf{R}_0(t)$ in the criterion *(10.2)*. Prove that the optimal control law $\mathbf{u}^{\mathrm{op}}$ depends only on the ratio η/ρ if $\mathbf{Q}_0(t)$ and $\mathbf{R}_0(t)$ are fixed.

10.20. Show that the transition matrix *(10.14)* is *symplectic*, i.e.

$$\begin{pmatrix} \mathbf{\Phi}_{11}^T(t,\tau) & \mathbf{\Phi}_{21}^T(t,\tau) \\ \mathbf{\Phi}_{12}^T(t,\tau) & \mathbf{\Phi}_{22}^T(t,\tau) \end{pmatrix}\begin{pmatrix} \mathbf{0} & \mathbf{I} \\ -\mathbf{I} & \mathbf{0} \end{pmatrix}\begin{pmatrix} \mathbf{\Phi}_{11}(t,\tau) & \mathbf{\Phi}_{12}(t,\tau) \\ \mathbf{\Phi}_{21}(t,\tau) & \mathbf{\Phi}_{22}(t,\tau) \end{pmatrix} = \begin{pmatrix} \mathbf{0} & \mathbf{I} \\ -\mathbf{I} & \mathbf{0} \end{pmatrix}$$

and also show $\quad \det\begin{pmatrix} \mathbf{\Phi}_{11}(t,\tau) & \mathbf{\Phi}_{12}(t,\tau) \\ \mathbf{\Phi}_{21}(t,\tau) & \mathbf{\Phi}_{22}(t,\tau) \end{pmatrix} = 1.$

10.21. Using the results of Problem 10.20, show that for the case $\mathbf{S} = \mathbf{0}$, $\mathbf{P}(t) = \mathbf{\Phi}_{11}^{-1}(t,\tau)\,\mathbf{\Phi}_{21}(t,\tau)$. This provides a check on numerical computations.

10.22. For the scalar time-varying system $dx/dt = -x\tan t + u$, find the optimal feedback control $u(x,t)$ to minimize $\nu = \dfrac{1}{2}\displaystyle\int_0^{t_1} (3x^2 + 4u^2)\,dt$.

10.23. Find the solution $P(t)$ to the scalar Riccati equation corresponding to the open loop system with finite escape time $dx/dt = -(x+u)/t$ with criterion $\nu = \dfrac{1}{2}\displaystyle\int_{t_0}^{t_1} (6x^2 + u^2)\,dt$. Consider the behavior for $0 < t_0 < t_1$, for $t_0 < t_1 < 0$ and for $-\gamma t_1 < t_0 < 0 < t_1$ where $\gamma^5 = 3/2$.

10.24. Given the system shown in Fig. 10-12. Find the compensation K to optimize the criterion $2\nu = \displaystyle\int_0^\infty [(y-d)^2 + u^2]\,dt$.

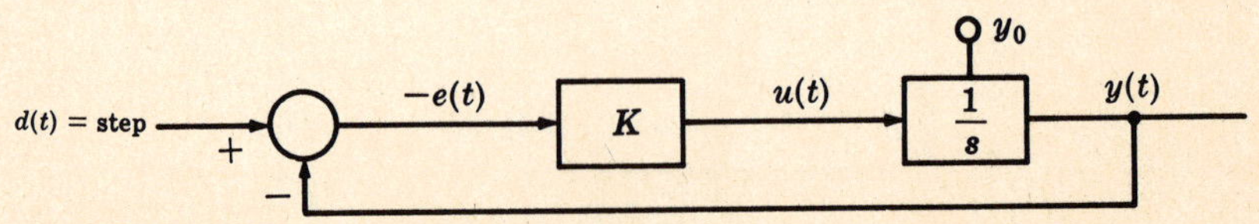

Fig. 10-12

10.25. In Problem 10.6, show that if $\mathbf{C} = \mathbf{B} = \mathbf{I}$ in equation *(10.26)* and if $\mathbf{R} = \mathbf{0}$ and $\mathbf{Q} = \mathbf{I}$ in equation *(10.27)*, then the system becomes identical to the model.

10.26. Use the results of Problem 10.4 to generate a root locus for the closed loop system as α varies. What happens when $\alpha = 0$?

10.27. Given the linear time-invariant system $d\mathbf{x}/dt = \mathbf{Ax} + \mathbf{Bu}$ with criterion $2\nu = \int_{t_0}^{t_1} (\mathbf{x}^T\mathbf{Qx} + \mathbf{u}^T\mathbf{Ru})\,dt$ where $\mathbf{Q}$ and $\mathbf{R}$ are both time-invariant positive definite symmetric matrices, but $\mathbf{Q}$ is $n \times n$ and $\mathbf{R}$ is $m \times m$. Find the feedback control $\mathbf{u}(\mathbf{x}, t; t_1)$ that minimizes ν in terms of the constant $n \times n$ matrices $\mathbf{K}_{11}, \mathbf{K}_{12}, \mathbf{K}_{21}, \mathbf{K}_{22}$ where

$$\left(\begin{array}{c|c} \mathbf{K}_{11} & \mathbf{K}_{12} \\ \hline \mathbf{K}_{21} & \mathbf{K}_{22} \end{array}\right) = \left(\begin{array}{c|c|c|c|c|c} \mathbf{f}_1 & \cdots & \mathbf{f}_n & \mathbf{f}_{n+1} & \cdots & \mathbf{f}_{2n} \\ \mathbf{g}_1 & \cdots & \mathbf{g}_n & \mathbf{g}_{n+1} & \cdots & \mathbf{g}_{2n} \end{array}\right)^{-1}$$

in which the $2n$-vector $\begin{pmatrix} \mathbf{f}_i \\ \mathbf{g}_i \end{pmatrix}$ satisfies the eigenvalue problem given in equation *(10.18)*, where $\lambda_1, \lambda_2, \ldots, \lambda_n$ have negative real parts and $\lambda_{n+1}, \lambda_{n+2}, \ldots, \lambda_{2n}$ have positive real parts and are distinct.

10.28. Given only the matrices $\mathbf{G}$ and $\mathbf{F}$ as defined in Theorem 10.9, note that $\mathbf{G}$ and $\mathbf{F}$ are in general complex since they contain partitions of eigenvectors. Find a method for calculating $\mathbf{\Pi} = \mathbf{GF}^{-1}$ using only the real number field, not complex numbers.

10.29. (*a*) Given the nominal scalar system $d\xi/dt = \xi + \mu$ with $\xi_0 = \xi(0)$, and the performance criterion $J = \frac{1}{2}\int_0^\infty [(\eta^2 - 1)\xi^2 + \mu^2]\,dt$ for constant $\eta > 1$. *Construct* two Liapunov functions for the optimal closed loop system that are valid for all $\eta > 1$. *Hint:* One method of construction is given in Chapter 10 and another in Chapter 9.

(*b*) Now suppose the actual system is $d\xi/dt = \epsilon\xi^3 + \xi + \mu$. Using both Liapunov functions obtained in (*a*), give estimates upon how large $\epsilon > 0$ can be such that the closed loop system remains asymptotically stable. In other words, find a function $f(\eta, \xi)$ such that $\epsilon < f(\eta, \xi)$ implies asymptotic stability of the closed loop system.

10.30. Given the system $d\mathbf{x}/dt = \mathbf{A}(t)\mathbf{x} + \mathbf{B}(t)\mathbf{u}$ with $\mathbf{x}(t_0) = \mathbf{x}_0$ and with criterion function

$$J = \mathbf{x}^T(t_f)\,\mathbf{x}(t_f)/2 + \int_{t_0}^{t_1} \mathbf{u}^T\mathbf{u}\,dt/2$$

(*a*) What are the canonical equations and their boundary conditions?

(*b*) Given the transition matrix $\mathbf{\Phi}(t, \tau)$ where $\partial\mathbf{\Phi}(t, \tau)/\partial t = \mathbf{A}(t)\mathbf{\Phi}(t, \tau)$ and $\mathbf{\Phi}(\tau, \tau) = \mathbf{I}$. Solve the canonical equations in terms of $\mathbf{\Phi}(t, \tau)$ and find a feedback control $\mathbf{u}(t)$.

(*c*) Write and solve the matrix Riccati equation.

Answers to Supplementary Problems

10.9. $u = -9x$

10.10. (a)
$$\begin{aligned} dx_1/dt &= x_2 & x_1(0) &= x_{10} \\ dx_2/dt &= -\sqrt{5}\,x_2 - p_2 & x_2(0) &= x_{20} \\ dp_1/dt &= -4x_1 & p_1(\infty) &= 0 \\ dp_2/dt &= -p_1 + \sqrt{5}\,p_2 & p_2(\infty) &= 0 \end{aligned}$$

(b)
$$\begin{aligned} x_1(t) &= (2x_{10} + x_{20})e^{-t} - (x_{10} + x_{20})e^{-2t} \\ x_2(t) &= -(2x_{10} + x_{20})e^{-t} + 2(x_{10} + x_{20})e^{-2t} \\ p_2(t) &= (\sqrt{5} - 1)(2x_{10} + x_{20})e^{-t} + (4 - 2\sqrt{5})(x_{10} + x_{20})e^{-2t} \\ p_1(t) &= 4(2x_{10} + x_{20})e^{-t} - 2(x_{10} + x_{20})e^{-2t} \end{aligned}$$

(c) $u(t) = (1 - \sqrt{5})(2x_{10} + x_{20})e^{-t} + (2\sqrt{5} - 4)(x_{10} + x_{20})e^{-2t}$

10.11. (a)
$$\begin{aligned} 0 &= 4 - \phi_{12}^2 \\ 0 &= \phi_{11} - \sqrt{5}\,\phi_{12} - \phi_{12}\phi_{22} \\ 0 &= 2\phi_{12} - 2\sqrt{5}\,\phi_{22} - \phi_{22}^2 \end{aligned}$$

(b) $\mathbf{P}(t) = \begin{pmatrix} 6 & 2 \\ 2 & 3 - \sqrt{5} \end{pmatrix}$

(d)

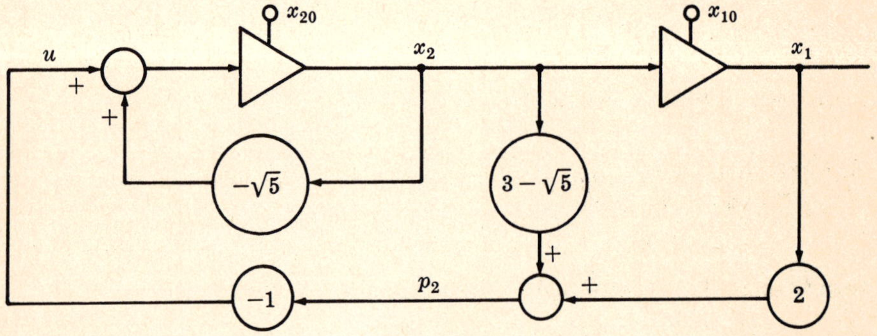

Fig. 10-13

10.12. $$\frac{d}{dt}\begin{pmatrix} x_1 \\ x_2 \\ p_1 \\ p_2 \end{pmatrix} = \begin{pmatrix} 0 & 1 & 0 & 0 \\ 0 & 0 & 0 & -1 \\ -q & 0 & 0 & 0 \\ 0 & 0 & -1 & 0 \end{pmatrix}\begin{pmatrix} x_1 \\ x_2 \\ p_1 \\ p_2 \end{pmatrix}$$ with $x_1(0) = 5$, $x_2(0) = -3$, $p_1(T) = 0$, $p_2(T) = 0$.

10.13. (a) $$P(t) = \frac{[S(1+\sqrt{2}) + 1]e^{2\sqrt{2}(T-t)} - 1 - S(1-\sqrt{2})}{[\sqrt{2} - 1 + S]e^{2\sqrt{2}(T-t)} + 1 + \sqrt{2} - S}$$

(b) $\lim_{t \to -\infty} P(t) = -1 + \sqrt{2}$ regardless of S.

10.14. $$K(t) = \frac{1 - e^{\gamma(t_1 - t)}}{(\alpha + \sqrt{\alpha^2+1})e^{\gamma(t_1-t)} + \sqrt{\alpha^2+1} - \alpha}$$ where $\gamma = (\alpha^2+1)/\rho$. $\lim_{t_1 \to \infty} K(t) = \alpha - \sqrt{\alpha^2+1}$

10.15. Any control results in $\nu = \infty$ since there is an unstable uncontrollable state.

10.16. $\mathbf{M}(\theta) = -\Delta^{-1}(q_r\theta + q_rq_v\theta^2/2 \quad q_vq_r\theta^2 + q_rq_v\theta^3/3)$ where $\Delta = 1 + q_v\theta + q_r\theta^3/3 + q_rq_v\theta^4/12$.

10.17. This is the diagram associated with Problem 10.8.

10.18. $\lambda = \pm\sqrt{a^2 + b^2 q/r}$

10.19. $\mathbf{u}^{\text{op}} = \mathbf{R}_0^{-1}\mathbf{b}^T\mathbf{P}_0\eta/\rho$ where $\mathbf{P}_0$ depends only on the ratio η/ρ as

$$-d\mathbf{P}_0/dt = \mathbf{Q}_0 + \mathbf{A}^T\mathbf{P}_0 + \mathbf{P}_0\mathbf{A} - \mathbf{P}_0\mathbf{b}^T\mathbf{R}_0^{-1}\mathbf{b}\mathbf{P}_0\eta/\rho$$

10.20. Let $\boldsymbol{\psi}(t, t_0) = \boldsymbol{\Phi}^T(t, t_0)\mathbf{E}\boldsymbol{\Phi}(t, t_0) - \mathbf{E}$ where $\mathbf{E} = \begin{pmatrix} \mathbf{0} & \mathbf{I} \\ -\mathbf{I} & \mathbf{0} \end{pmatrix}$. Then $\boldsymbol{\psi}(t_0, t_0) = \mathbf{0}$ and $d\boldsymbol{\psi}/dt = \boldsymbol{\Phi}^T(t, t_0)(\mathbf{H}^T\mathbf{E} + \mathbf{E}\mathbf{H})\boldsymbol{\Phi}(t, t_0) = \mathbf{0}$, so $\boldsymbol{\psi}(t, t_0) = \mathbf{0}$. This shows that for every increasing function of time in $\boldsymbol{\Phi}$ there is a corresponding decreasing function of time.

10.22. $u = \left(\tan t - \dfrac{1}{2}\tan\dfrac{t - t_1}{2}\right)x$

10.23. $P(t) = \dfrac{6t_1^6 t - 6t_1 t^6}{3t_1^5 + 2t^5}$. For $0 < t_0 < t_1$ and $t_0 < t_1 < 0$, $P(t) > 0$. For $-\gamma t_1 < t_0 < 0 < t_1$, $P(t) < 0$ and tends to $-\infty$ as t tends to $-\gamma t_1$ from the right.

10.24. $K = 1$

10.26. At $\alpha = 0$, the closed loop system poles are at $e^{j7\pi/8}$ and $e^{j9\pi/8}$.

10.27. $$\begin{pmatrix} \mathbf{x}(t) \\ \mathbf{p}(t) \end{pmatrix} = \begin{pmatrix} \mathbf{K}_{11} & \mathbf{K}_{12} \\ \mathbf{K}_{21} & \mathbf{K}_{22} \end{pmatrix}^{-1} \begin{pmatrix} e^{\mathbf{A}_s(t-t_1)} & \mathbf{0} \\ \mathbf{0} & e^{\mathbf{A}_u(t-t_1)} \end{pmatrix} \begin{pmatrix} \mathbf{K}_{11} & \mathbf{K}_{12} \\ \mathbf{K}_{21} & \mathbf{K}_{22} \end{pmatrix} \begin{pmatrix} \mathbf{x}(t_1) \\ \mathbf{p}(t_1) \end{pmatrix}$$

Note $\mathbf{p}(t_1) = \mathbf{0}$ and solve for $\mathbf{x}(t_1)$ in terms of $\mathbf{x}(t)$. Then $\mathbf{p}(t) = \mathbf{P}(t)\,\mathbf{x}(t)$.

10.28. Let the first $2m \leq n$ eigenvectors be complex conjugates, so that $\mathbf{f}_{2i-1}^* = \mathbf{f}_{2i}$ and $\mathbf{g}_{2i-1}^* = \mathbf{g}_{2i}$ for $i = 1, 2, \ldots, m$. Define

$$\widetilde{\mathbf{F}} = (\operatorname{Re}\mathbf{f}_1 \mid \operatorname{Im}\mathbf{f}_1 \mid \operatorname{Re}\mathbf{f}_3 \mid \operatorname{Im}\mathbf{f}_3 \mid \ldots \mid \operatorname{Im}\mathbf{f}_{2m-1} \mid \mathbf{f}_{2m} \mid \ldots \mid \mathbf{f}_n)$$

and $\widetilde{\mathbf{G}}$ similarly. Then $\widetilde{\mathbf{F}}$ and $\widetilde{\mathbf{G}}$ are real and $\boldsymbol{\Pi} = \widetilde{\mathbf{G}}\widetilde{\mathbf{F}}^{-1}$.

10.29. (*a*) From Chapter 10, $V_1 = (\eta + 1)\xi^2$ and from Chapter 9, $V_2 = \xi^2/2\eta$.

(*b*) $\epsilon < \eta\xi^{-2}$ for both V_1 and V_2.

10.30. (*a*) $d\mathbf{x}/dt = \mathbf{A}\mathbf{x} - \mathbf{B}\mathbf{B}^T\mathbf{p}$ and $d\mathbf{p}/dt = -\mathbf{A}^T\mathbf{p}$ with $\mathbf{x}(t_0) = \mathbf{x}_0$ and $\mathbf{p}(t_1) = \mathbf{x}(t_1)$.

(*b*) $$\mathbf{u} = -\mathbf{B}^T\boldsymbol{\Phi}^T(t_1, t)\left[\boldsymbol{\Phi}(t, t_1) - \int_{t_1}^{t} \boldsymbol{\Phi}(t, \tau)\mathbf{B}\mathbf{B}^T\boldsymbol{\Phi}^T(t_1, \tau)\,d\tau\right]^{-1}\mathbf{x}(t)$$

(*c*) obvious from (*b*) above.

INDEX

SCHAUM'S OUTLINE SERIES

COLLEGE PHYSICS
including 625 SOLVED PROBLEMS
Edited by CAREL W. van der MERWE, Ph.D.,
Professor of Physics, New York University

COLLEGE CHEMISTRY
including 385 SOLVED PROBLEMS
Edited by JEROME L. ROSENBERG, Ph.D.,
Professor of Chemistry, University of Pittsburgh

GENETICS
including 500 SOLVED PROBLEMS
By WILLIAM D. STANSFIELD, Ph.D.,
Dept. of Biological Sciences, Calif. State Polytech.

MATHEMATICAL HANDBOOK
including 2400 FORMULAS and 60 TABLES
By MURRAY R. SPIEGEL, Ph.D.,
Professor of Math., Rensselaer Polytech. Inst.

First Yr. COLLEGE MATHEMATICS
including 1850 SOLVED PROBLEMS
By FRANK AYRES, Jr., Ph.D.,
Professor of Mathematics, Dickinson College

COLLEGE ALGEBRA
including 1940 SOLVED PROBLEMS
By MURRAY R. SPIEGEL, Ph.D.,
Professor of Math., Rensselaer Polytech. Inst.

TRIGONOMETRY
including 680 SOLVED PROBLEMS
By FRANK AYRES, Jr., Ph.D.,
Professor of Mathematics, Dickinson College

MATHEMATICS OF FINANCE
including 500 SOLVED PROBLEMS
By FRANK AYRES, Jr., Ph.D.,
Professor of Mathematics, Dickinson College

PROBABILITY
including 500 SOLVED PROBLEMS
By SEYMOUR LIPSCHUTZ, Ph.D.,
Assoc. Prof. of Math., Temple University

STATISTICS
including 875 SOLVED PROBLEMS
By MURRAY R. SPIEGEL, Ph.D.,
Professor of Math., Rensselaer Polytech. Inst.

ANALYTIC GEOMETRY
including 345 SOLVED PROBLEMS
By JOSEPH H. KINDLE, Ph.D.,
Professor of Mathematics, University of Cincinnati

DIFFERENTIAL GEOMETRY
including 500 SOLVED PROBLEMS
By MARTIN LIPSCHUTZ, Ph.D.,
Professor of Mathematics, University of Bridgeport

CALCULUS
including 1175 SOLVED PROBLEMS
By FRANK AYRES, Jr., Ph.D.,
Professor of Mathematics, Dickinson College

DIFFERENTIAL EQUATIONS
including 560 SOLVED PROBLEMS
By FRANK AYRES, Jr., Ph.D.,
Professor of Mathematics, Dickinson College

SET THEORY and Related Topics
including 530 SOLVED PROBLEMS
By SEYMOUR LIPSCHUTZ, Ph.D.,
Assoc. Prof. of Math., Temple University

FINITE MATHEMATICS
including 750 SOLVED PROBLEMS
By SEYMOUR LIPSCHUTZ, Ph.D.,
Assoc. Prof. of Math., Temple University

MODERN ALGEBRA
including 425 SOLVED PROBLEMS
By FRANK AYRES, Jr., Ph.D.,
Professor of Mathematics, Dickinson College

LINEAR ALGEBRA
including 600 SOLVED PROBLEMS
By SEYMOUR LIPSCHUTZ, Ph.D.,
Assoc. Prof. of Math., Temple University

MATRICES
including 340 SOLVED PROBLEMS
By FRANK AYRES, Jr., Ph.D.,
Professor of Mathematics, Dickinson College

PROJECTIVE GEOMETRY
including 200 SOLVED PROBLEMS
By FRANK AYRES, Jr., Ph.D.,
Professor of Mathematics, Dickinson College

GENERAL TOPOLOGY
including 650 SOLVED PROBLEMS
By SEYMOUR LIPSCHUTZ, Ph.D.,
Assoc. Prof. of Math., Temple University

GROUP THEORY
including 600 SOLVED PROBLEMS
By B. BAUMSLAG, B. CHANDLER, Ph.D.,
Mathematics Dept., New York University

VECTOR ANALYSIS
including 480 SOLVED PROBLEMS
By MURRAY R. SPIEGEL, Ph.D.,
Professor of Math., Rensselaer Polytech. Inst.

ADVANCED CALCULUS
including 925 SOLVED PROBLEMS
By MURRAY R. SPIEGEL, Ph.D.,
Professor of Math., Rensselaer Polytech. Inst.

COMPLEX VARIABLES
including 640 SOLVED PROBLEMS
By MURRAY R. SPIEGEL, Ph.D.,
Professor of Math., Rensselaer Polytech. Inst.

LAPLACE TRANSFORMS
including 450 SOLVED PROBLEMS
By MURRAY R. SPIEGEL, Ph.D.,
Professor of Math., Rensselaer Polytech. Inst.

NUMERICAL ANALYSIS
including 775 SOLVED PROBLEMS
By FRANCIS SCHEID, Ph.D.,
Professor of Mathematics, Boston University

DESCRIPTIVE GEOMETRY
including 175 SOLVED PROBLEMS
By MINOR C. HAWK, *Head of Engineering Graphics Dept., Carnegie Inst. of Tech.*

ENGINEERING MECHANICS
including 460 SOLVED PROBLEMS
By W. G. McLEAN, B.S. in E.E., M.S.,
Professor of Mechanics, Lafayette College
and E. W. NELSON, B.S. in M.E., M. Adm. E.,
Engineering Supervisor, Western Electric Co.

THEORETICAL MECHANICS
including 720 SOLVED PROBLEMS
By MURRAY R. SPIEGEL, Ph.D.,
Professor of Math., Rensselaer Polytech. Inst.

LAGRANGIAN DYNAMICS
including 275 SOLVED PROBLEMS
By D. A. WELLS, Ph.D.,
Professor of Physics, University of Cincinnati

STRENGTH OF MATERIALS
including 430 SOLVED PROBLEMS
By WILLIAM A. NASH, Ph.D.,
Professor of Eng. Mechanics, University of Florida

FLUID MECHANICS and HYDRAULICS
including 475 SOLVED PROBLEMS
By RANALD V. GILES, B.S., M.S. in C.E.,
Prof. of Civil Engineering, Drexel Inst. of Tech.

FLUID DYNAMICS
including 100 SOLVED PROBLEMS
By WILLIAM F. HUGHES, Ph.D.,
Professor of Mech. Eng., Carnegie Inst. of Tech.
and JOHN A. BRIGHTON, Ph.D.,
Asst. Prof. of Mech. Eng., Pennsylvania State U.

ELECTRIC CIRCUITS
including 350 SOLVED PROBLEMS
By JOSEPH A. EDMINISTER, M.S.E.E.,
Assoc. Prof. of Elec. Eng., University of Akron

ELECTRONIC CIRCUITS
including 160 SOLVED PROBLEMS
By EDWIN C. LOWENBERG, Ph.D.,
Professor of Elec. Eng., University of Nebraska

FEEDBACK & CONTROL SYSTEMS
including 680 SOLVED PROBLEMS
By J. J. DiSTEFANO III, A. R. STUBBERUD,
and I. J. WILLIAMS, Ph.D.,
Engineering Dept., University of Calif., at L.A.

TRANSMISSION LINES
including 165 SOLVED PROBLEMS
By R. A. CHIPMAN, Ph.D.,
Professor of Electrical Eng., University of Toledo

REINFORCED CONCRETE DESIGN
including 200 SOLVED PROBLEMS
By N. J. EVERARD, MSCE, Ph.D.,
Prof. of Eng. Mech. & Struc., Arlington State Col.
and J. L. TANNER III, MSCE,
Technical Consultant, Texas Industries Inc.

MECHANICAL VIBRATIONS
including 225 SOLVED PROBLEMS
By WILLIAM W. SETO, B.S. in M.E., M.S.,
Assoc. Prof. of Mech. Eng., San Jose State College

MACHINE DESIGN
including 320 SOLVED PROBLEMS
By HALL, HOLOWENKO, LAUGHLIN
Professors of Mechanical Eng., Purdue University

BASIC ENGINEERING EQUATIONS
including 1400 BASIC EQUATIONS
By W. F. HUGHES, E. W. GAYLORD, Ph.D.,
Professors of Mech. Eng., Carnegie Inst. of Tech.

ELEMENTARY ALGEBRA
including 2700 SOLVED PROBLEMS
By BARNETT RICH, Ph.D.,
Head of Math. Dept., Brooklyn Tech. H.S.

PLANE GEOMETRY
including 850 SOLVED PROBLEMS
By BARNETT RICH, Ph.D.,
Head of Math. Dept., Brooklyn Tech. H.S.

TEST ITEMS IN EDUCATION
including 3100 TEST ITEMS
By G. J. MOULY, Ph.D., L. E. WALTON, Ph.D.,
Professors of Education, University of Miami